T0180117

Lecture Notes in Computer Science 13374

More information about this series at https://link.springer.com/bookseries/558

Pier Luigi Mazzeo · Emanuele Frontoni ·
Stan Sclaroff · Cosimo Distante (Eds.)

Image Analysis and Processing

ICIAP 2022 Workshops

ICIAP International Workshops
Lecce, Italy, May 23–27, 2022
Revised Selected Papers, Part II

Editors
Pier Luigi Mazzeo
National Research Council
Lecce, Italy

Emanuele Frontoni
Università Politecnica delle Marche
Ancona, Italy

Stan Sclaroff
Boston University
Boston, MA, USA

Cosimo Distante
National Research Council
Lecce, Italy

ISSN 0302-9743 ISSN 1611-3349 (electronic)
Lecture Notes in Computer Science
ISBN 978-3-031-13323-7 ISBN 978-3-031-13324-4 (eBook)
https://doi.org/10.1007/978-3-031-13324-4

This Springer imprint is published by the registered company Springer Nature Switzerland AG
The registered company address is: Gewerbestrasse 11, 6330 Cham, Switzerland

Preface

This volume contains 47 of the papers accepted for presentation at the workshops hosted by the 21st International Conference on Image Analysis and Processing (ICIAP 2022), held in Lecce, Italy, during May 23–27, 2022. ICIAP is organized every two years by CVPL, the group of Italian researchers affiliated with the International Association for Pattern Recognition (IAPR). The aim of the conference is to bring together researchers working on image processing, computer vision, and pattern recognition from around the world. Topics traditionally covered are related to computer vision, pattern recognition, and image processing, addressing both theoretical and applicative aspects.

In total, 16 different workshops were selected to complement ICIAP 2022 in Lecce. All the 16 workshops have received a total of 157 submissions, and after a peer-review selection process, carried out by the individual workshop organizers, ultimately led to the selection of 96 papers, with an overall acceptance rate of 61%.

This volume contains 47 papers (out of 96) from the following workshops:

- Medical Imaging Analysis for Covid-19 (MIA COVID)
- Artificial Intelligence for preterm infants' healthCare (AI-Care)
- Binary is the new Black (and White): Recent Advances on Binary Image Processing
- Towards a Complete Analysis of People: From Face and Body to Clothes (T-CAP)
- Workshop on Small-Drone Surveillance, Detection and Counteraction Techniques (WOSDETC)
- Artificial Intelligence for Digital Humanities (AI4DH)
- Human Behavior Analysis for Smart City Environment Safety (HBAxSCES)
- Learning in Precision Livestock Farming (LPLF)
- Novel Benchmarks and Approaches for Real-World Continual Learning (CL4REAL)
- Medical Transformers (MEDXF)

The papers accepted for the other workshops are included in the companion volume (LNCS 13373).

Medical Imaging Analysis for Covid-19 (MIA COVID), organized by Fares Bougourzi (ISASI-CNR, Italy), Cosimo Distante (ISASI-CNR, Italy), Abdelmalik Taleb-Ahmed (Université Polytechnique Hauts-de-France, France), Fadi Dornaika (University of the Basque Country, Spain), and Abdenour Hadid (Université Polytechnique Hauts-de-France, France), provided an overview of the potential applications of AI in combating this pandemic using medical imaging methods.

Artificial Intelligence for preterm infants' healthCare (AI-Care), organized by Sara Moccia (Scuola Superiore Sant'Anna, Pisa, Italy) along with Emanuele Frontoni and Lucia Migliorelli (Università Politecnica delle Marche, Italy), aimed to group expert AI researchers in the field of preterm infant monitoring to discuss the most recent research work and highlight current challenges and needs.

Binary is the new Black (and White): Recent Advances on Binary Image Processing covered anything using, implementing, or improving binary image analysis, a specific area of image processing, which is less mainstream nowadays, but which still supports most computer vision systems implementations. It was organized by Costantino Grana and Federico Bolelli (Università degli Studi di Modena e Reggio Emilia, Italy).

Towards a Complete Analysis of People: From Face and Body to Clothes (T-CAP), organized by Mohamed Daoudi (IMT Nord Europe, France), Roberto Vezzani (Università degli Studi di Modena e Reggio Emilia, Italy), Marcella Cornia (Università degli Studi di Modena e Reggio Emilia, Italy), Guido Borghi (Università di Bologna, Italy), Claudio Ferrari, (Università di Parma, Italy), Federico Becattini (Università di Firenze, Italy), and Andrea Pilzer (Aalto University, Finland), aimed to improve the communication between researchers and companies and to develop novel ideas that can shape the future of this area, in terms of motivations, methodologies, prospective trends, and potential industrial applications.

The Workshop on Small-Drone Surveillance, Detection and Counteraction Techniques (WOSDETC) aimed at bringing together researchers from both academia and industry, to share recent advances in this field. It was organized by Angelo Coluccia (Università del Salento, Italy), Alessio Fascista (Università del Salento, Italy), Arne Schumann (Fraunhofer IOSB, Germany), Lars Sommer (Fraunhofer IOSB, Germany), Anastasios Dimou (Information Technologies Institute, Greece), Dimitrios Zarpalas (Information Technologies Institute, Greece) Nabin Sharma (University of Technology Sydney, Australia), and Mrunalini Nalamat (University of Technology Sydney, Australia).

Artificial Intelligence for Digital Humanities (AI4DH) aimed to encourage and highlight novel strategies and original research in applying artificial intelligence techniques in digital humanities research, such as data discovery, digital data creation, management, data analytics in literature, linguistics, culture heritage, media, social science, history, music and acoustics, and artificial intelligence for digital humanities in pedagogy and academic curricula. It was organized by Marina Paolanti and Emanuele Frontoni (Università Politecnica delle Marche, Italy), Francesca Matrone (Politecnico di Torino, Italy), and Silvia Cascianelli, Marcella Cornia, and Lorenzo Baraldi (Università degli Studi di Modena e Reggio Emilia, Italy).

Human Behavior Analysis for Smart City Environment Safety (HBAxSCES) focused on smart cities that aim to ensure secure and safe physical and digital environments for the well-being of citizens. Among other things, ICT systems are reliant on evolving artificial intelligence, pattern recognition, computer vision, 3D simulations and digital twin techniques to make environments more resilient. This workshop was organized by Alessandro Bruno, Zoheir Sabeur, Deniz Chetinkaya, Muntadher Sallal, and Banafshe Arbab-Zavar (Bournemouth University, UK).

Learning in Precision Livestock Farming (LPLF) aimed to attract novel and original contributions on the analysis, study, and proposal of innovative machine and deep learning techniques applied to the automatic monitoring of animals in intensive farms, helping to improve the living conditions of the animals. It was organized by Simone Palazzo, Simona Porto, Claudia Arcidiacono, and Giulia Castagnolo (Università di Catania, Italy) together with Marcella Guarino (Università di Milano, Italy).

Novel Benchmarks and Approaches for Real-World Continual Learning (CL4REAL), organized by Simone Palazzo and Giovanni Bellitto (Università di Catania,

Italy), Angelo Porrello and Matteo Boschini (Università degli Studi di Modena e Reggio Emilia, Italy), and Vincenzo Lomonaco (Università di Pisa, Italy), aimed to attract novel and original contributions exploring the intersection of continual learning and real-world applications.

Medical Transformers (MEDXF) focused on the employment of transformer architectures for medical imaging, with the objective of extending the state of the art of this topic, presenting novel solutions to typical problems in medical image analysis and, just as importantly, investigating the limits and pitfalls of these new techniques in specific and socially-critical domains. It was organized by Ulas Bagc and Zheyuan Zhang (Northwestern University, USA) along with Simone Palazzo and Federica Proietto Salanitri (Università di Catania, Italy).

We warmly thank all the workshop organizers who made such an interesting program possible and we hope that ICIAP 2022 has given us a chance to design a future where technologies allow people to live comfortably, healthily, and in peace.

May 2022

Pier Luigi Mazzeo
Emanuele Frontoni

Organization

General Chairs

Cosimo Distante	National Research Council, Italy
Stan Sclaroff	Boston University, USA

Technical Program Chairs

Giovanni Maria Farinella	University of Catania, Italy
Marco Leo	National Research Council, Italy
Federico Tombari	Google and TUM, Germany

Area Chairs

Lamberto Ballan	University of Padua, Italy
Francois Bremond	Inria, France
Simone Calderara	University of Modena and Reggio Emilia, Italy
Modesto Castrillon Santana	University of Las Palmas de Gran Canaria, Spain
Marco Cristani	University of Verona, Italy
Luigi Di Stefano	University of Bologna, Italy
Sergio Escalera	University of Barcelona, Spain
Luiz Marcos Garcia Goncalves	UFRN, Brazil
Javier Ortega Garcia	Universidad Autonoma de Madrid, Spain
Costantino Grana	University of Modena and Reggio Emilia, Italy
Tal Hassner	Facebook AML and Open University of Israel, Israel
Gian Luca Marcialis	University of Cagliari, Italy
Christian Micheloni	University of Udine, Italy
Fausto Milletarì	NVIDIA, USA
Vittorio Murino	Italian Institute of Technology, Italy
Vishal Patel	Johns Hopkins University, USA
Marcello Pelillo	Università Ca' Foscari Venice, Italy
Federico Pernici	University of Florence, Italy
Andrea Prati	University of Parma, Italy
Justus Piater	University of Innsbruck, Austria
Elisa Ricci	University of Trento, Italy
Alessia Saggese	University of Salerno, Italy
Roberto Scopigno	National Research Council, Italy

North Africa Liaison Chair

Dorra Sellami	University of Sfax, Tunisia

Oceania Liaison Chair

Wei Qi Yan	Auckland University of Technology, New Zealand

North America Liaison Chair

Larry S. Davis	University of Maryland, USA

Asia Liaison Chair

Wei Shi Zheng	Sun Yat-sen University, China

Latin America Liaison Chair

Luiz Marcos Garcia Goncalves	UFRN, Brazil

Invited Speakers

Larry S. Davis	University of Maryland and Amazon, USA
Roberto Cipolla	University of Cambridge, UK
Dima Aldamen	University of Bristol, UK
Laura Leal-Taixe	Technische Universität München, Germany

Steering Committee

Virginio Cantoni	University of Pavia, Italy
Luigi Pietro Cordella	University of Napoli Federico II, Italy
Rita Cucchiara	University of Modena and Reggio Emilia, Italy
Alberto Del Bimbo	University of Firenze, Italy
Marco Ferretti	University of Pavia, Italy
Fabio Roli	University of Cagliari, Italy
Gabriella Sanniti di Baja	National Research Council, Italy

Endorsing Institutions

International Association for Pattern Recognition (IAPR)
Italian Association for Computer Vision, Pattern Recognition and Machine Learning (CVPL)
Springer

Institutional Patronage

Institute of Applied Sciences and Intelligent Systems (ISASI)
National Research Council of Italy (CNR)
Provincia di Lecce
Regione Puglia

Contents – Part II

Human Behaviour Analysis for Smart City Environment Safety - HBAxSCES

Binary is the New Black (and White): Recent Advances on Binary Image Processing

Workshop on Small-Drone Surveillance, Detection and Counteraction Techniques - WOSDETC

Medical Imaging Analysis for Covid-19 - MIACOVID 2022

Contents – Part I

Parts Can Worth Like the Whole - PART 2022

Workshop on Fine Art Pattern Extraction and Recognition - FAPER

Workshop on Intelligent Systems in Human and Artificial Perception - ISHAPE 2022

Artificial Intelligence and Radiomics in Computer-Aided Diagnosis - AIRCAD

Human Behaviour Analysis for Smart City Environment Safety - HBAxSCES

A Framework for Forming Middle Distance Routes Based on Spatial Guidelines, Perceived Accessibility and Visual Cues in Smart City

Margarita Zaleshina[1] and Alexander Zaleshin[2](✉)

[1] Moscow Institute of Physics and Technology, Moscow, Russia
[2] Institute of Higher Nervous Activity and Neurophysiology, Moscow, Russia
terbiosorg@gmail.com

Abstract. This study is devoted to ways of forming routes taking into account external natural and artificial factors, and the perception of such factors by the traveler. The route is built based on preliminary knowledge of an area and is updated with information obtained during travel. Orientation along the route is carried out using signs (waymarks, billboards) and other natural and artificial references which clarify preliminary knowledge of the area. The final and intermediate targets along the route are determined by spatial objects—"points of interest" or "points of attraction"—which are either chosen in advance or occur unpredictably during movement along the route. At the same time, the available accuracy, completeness and degree of relevance of local maps do not always provide the information necessary for travelers. The interface of route creation acts as an intermediary between the preliminary idea of the route and the observable external environment. The interface can supplement incomplete or unavailable information; it helps to search for appropriate objects based on given attributes. Currently digital applications are often used as such interfaces. Objects on the route are constantly changing their properties over time—both according to a previously known schedule and as a result of random events. The appearance of unexpected obstacles, and sudden changes in lighting and weather conditions, force travelers to significantly change their routes and choose new route options. The framework can be used both for optimizing navigation and tourism services and for preparing project designs for landscaping and development of suburban terrains.

Keywords: Route formation · Wayfinding · Choice of routes

1 Introduction

A route is a path that specifies movement relative to spatial landmarks or geographical coordinates, indicating the start, end, and intermediate waypoints. When moving along the route, different types of transport infrastructure can be used for different segments. When planning the route, preliminary information can be prepared—both a route plan (map) and a schedule. The scale of the planned route determines the features of its formation While on short-distance route segments, objects of interest can be observed

P. L. Mazzeo et al. (Eds.): ICIAP 2022 Workshops, LNCS 13374, pp. 3–13, 2022.
https://doi.org/10.1007/978-3-031-13324-4_1

immediately, on long-distance routes the points of interest selected and the details of such objects may be changed and adjusted on site. When analyzing route tracks, researchers usually integrate them with other data for an efficient pathway analysis, primarily data on objects of interest and map data. Urban road networks and terrain relief have a huge impact on route formation for people and vehicles [1].

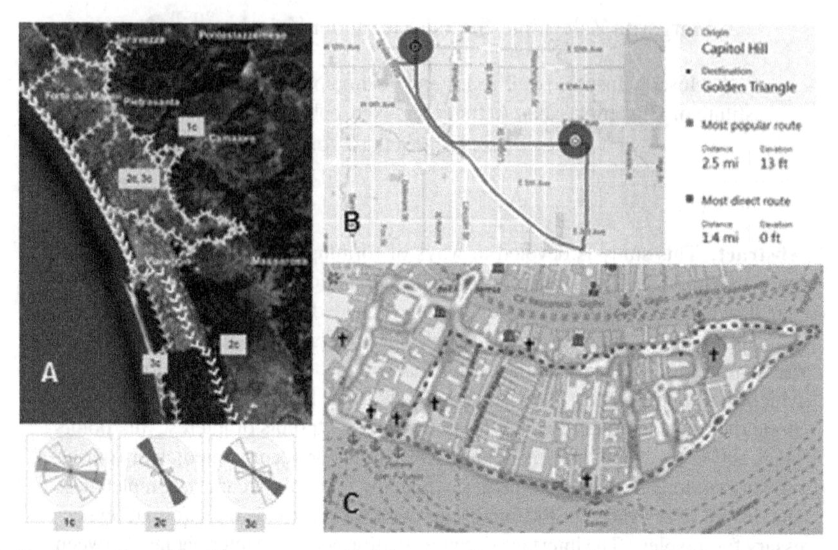

Fig. 1. A chaotic and stable movement B. Most popular route and most direct route. C. Popular routes' heatmap in Venice

Figure 1 shows ways to representation of navigation strategies: A. Difference between chaotic and stable movement (oscillation variability due to POI or hidden object, snap-to-line-object variability, linear segment variability of a track) [2]. B. A sample of difference routes, from Capitol Hill to Golden Triangle. Most popular route: 2.5 miles/Most direct route: 1.4 miles (source: https://metroview.strava.com/map/demo). C. Popular routes' heatmap in Venice (see Sect. 3).

2 Orientation and Perception Along the Route When Using the Navigation Interface

2.1 Mental Maps and Traveler Environment in Route Network Sustainability

A mental map or plan [3] allows a person to imagine a terrain with sets of key elements located therein and to imagine possible ways to interact with the surrounding environment. Most often, such a map is drawn up rather conditionally and unclearly, and for an ordinary survey walk, it may not have an exact route plan or direction selection algorithm.

In the case of medium- or long-distance travel, the endpoint is usually not visible at the start of the route. It is even possible that the target as such is simply absent—if, for instance, a person takes a walk in a large landscaped park that was previously unknown to him or her (Fig. 2).

On the one hand, the person has a general "mental" image of future travel ("mental outline map"). On the other hand, a set of spatial objects can serve as an interactive interface with the outside ambient. Some of these objects serve only as signs (waymarks), helping to choose the route. Other objects can be points of attraction, and intermediate targets on the way.

Transport plans are designed not only by forecasting the processes of freight and passenger transport services but also by taking into account population density, types of development, terrain relief, and climatic conditions.

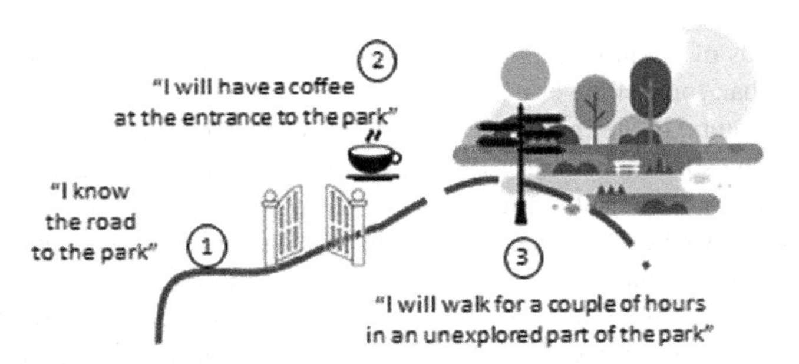

Fig. 2. Standard mental plan of a walk in a little-explored location

Thoughtful plans for the future development of territories have been supplementing the natural growth of urban infrastructure from its inception. The most sustainable solutions always include already existing routes. Thus, pedestrian paths are first created in suburban areas; then, these paths become lanes that are paved with slabs and taken into account in further landscaping. In densely populated areas, the future direction of urban growth is managed by creating new centers of attraction. The targeted attraction of urban residents can promote the investors' interest in purchasing real estate or in the infrastructural development of the territory.

A study by Margara [4] provides an overview of modern algorithms and methods for adapting data structures, which are memory-effective and allow for parallel processing of event flows.

Points of attraction that a person can purposefully move to or accidentally turn to when they see a sign that attracts them can have a significant impact on the route. With similar data being collected every day by mobile providers across the world, the prospect of being able to map contemporary and changing human population distributions over relatively short intervals exists, paving the way for new applications and a near real-time understanding of patterns and processes in human geography [5].

Long-term solutions should be reflected in the real world, not in the information field, by building an infrastructure of tunnels, overpasses, dams that connect difficult and dangerous, but closely located sections of the path.

Areas with a large number of points of attraction can change during the day, seasonally or on weekends (Fig. 3). Peculiarities of the creation of pedestrian and transport infrastructure in many cities are associated with their historical development.

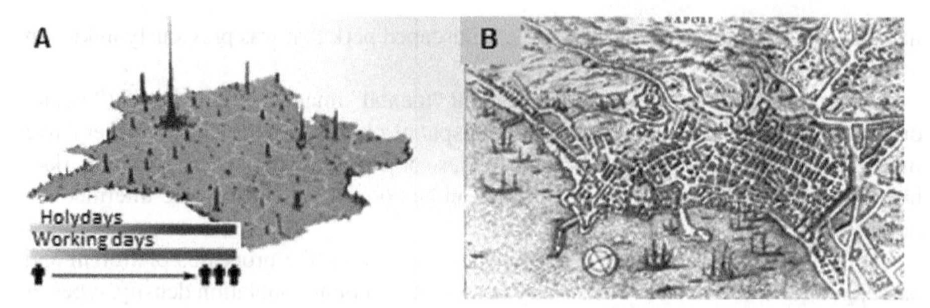

Fig. 3. A. Points of attraction in France (source: https://www.eurekalert.org/pub_releases/2014-10/uos-rpa102814.php). B. Map of Naples, 1761 [6]

The methods of urban studies used in instrumental design in Venice highlight the value of the urban form of the city [7].

Often, the natural of the environment, including elevation differences, boundaries of water bodies, and soil features, served as the basis for the formation of new routes. The most successful and sustainable routes were constantly visited by travelers. Along the roads, tourism service centers were developed, which could later become independent and developed settlements. The formation of routes that differ from each other passes through a series of stable or blank states of the environment or other information. Even if the geometric distance is fixed, due to weather and other unexpected factors, traveling time from one city to another may be variable [8].

The traveler can take a risk and create a short-term transition through the gap. If it is convenient and visible, such a gap transition may start to be used frequently. But the presence of a gap transition does not solve, but rather exacerbates the problem of risk with frequent repetition.

To reduce this risk in practice, if funding is available, an overpass can be built near a place where short-term movements often occur. Such a solution will have greater sustainability and significantly less risk. For example, if an overpass is built along the path of repetitive migration of animals when they cross a highway, then instead of risky trajectories, there is an alternative long-term solution that allows animals to cross the highway above the road. The familiar old solution of running the highway in front of cars does not disappear, but there is a new choice that animals can use.

Many routes constantly encounter gaps in spatiotemporal capabilities, such as rivers, the lack of a path between two neighboring roads, tides, etc. (Fig. 4).

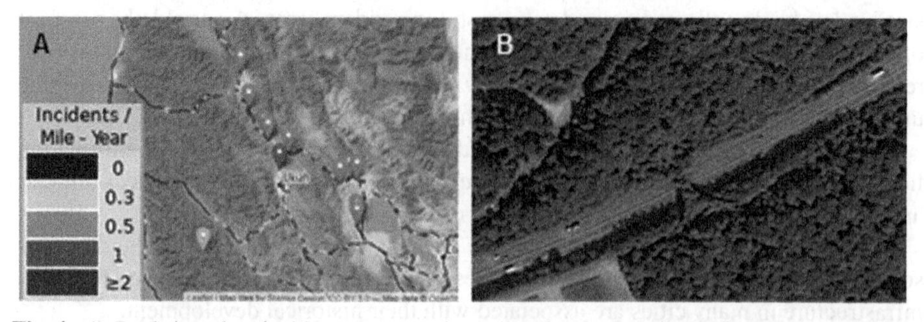

Fig. 4. A. Real-time deer incidents (source: https://roadecology.ucdavis.edu/hotspots/map). B. Overpass: animal crossing bridge in Luxembourg (49°40′01.2"N 6°22′29.0"E)

The interface, consisting of reference points located in certain specific places and having information content, allows people and animals to proceed to the formation of a new trajectory "here and now" based on the waymark signs. This new trajectory may be risky, but it may be the only quick solution to overcome a short section of the path.

2.2 Aggregation of Information for Navigation Tasks

Route data that captures the location of moving objects at certain time intervals has long been an important tool for studying human behavior and solving transport problems. Natural objects that are located close to each other for a long time usually have similar or dependent components that determine their structure and content. To a lesser extent, this applies to artificial objects. Tobler's first law [9] assumes the dependence of some attributes of objects that are close to each other. In an unfamiliar environment, a person searches for previously encountered objects and signs in order to recognize other ones. The uncertainty generated by a little-known situation results in an attempt to orientate and search for fragments of previously encountered elements [10]. The search allows identification of suitable objects and supplements thereto. Missing points can be added based on the available parts of other objects.

Points with known attributes can also serve as reference points. Reference points have stable locations, but their locations can change over time. The complex nature of reference points allows them to be used as a tool for operations with objects and attributes, and as a framework for spatial positioning. The relative positions of points and objects form the structural code of the track points. With small changes, the structural code may remain the same, with significant changes in the data set of the environment; a new structural code is formed from some stable or repetitive components/elements of the environment. Natural perceptual organization can be described in terms of topological invariants; topological perception precedes the perception of other properties of attributes; and primitives of visual form perception are geometric invariants at different levels of structural stability [11].

Going outdoors, a person is guided—when choosing a route—by signs, including road signs, that are known and meaningful to him or her. The sufficiency and redundancy of a set of signs determine the time needed for a person to make a choice and the optimality of the choice itself (Fig. 5). Uncertainty in choosing the route occurs with a lack of known signs and also when the signs are redundant and provide multiple choices.

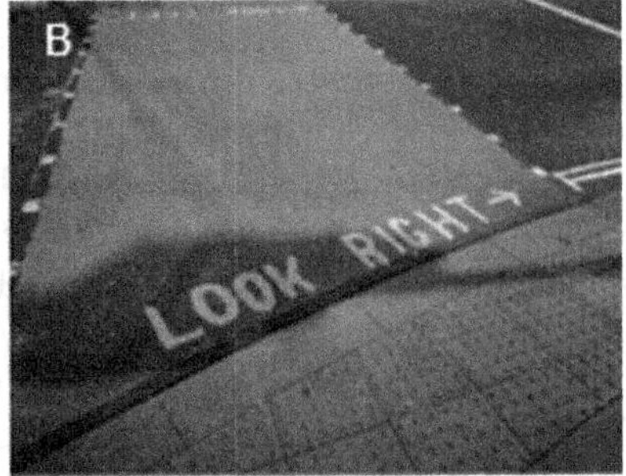

Fig. 5. A. Sign on a hiking trail near Zermatt, Switzerland. B. Crossing signs in London

Signs are not always located in places convenient for data collection. Sometimes signs that are necessary to create a complete picture are located at a considerable distance from others.

In problems of detecting visual changes, researchers have shown that objects embedded in a contextually heterogeneous scene tend to be detected faster than objects embedded in a contextually homogeneous scene. This discovery is very curious, given that a contextually homogeneous scene includes many signals that should aid in perception, e.g. types of objects and their probable location [12].

2.3 Planning, Detailing, and Optimizing a Middle-Distance Route

A person identifies recognizable signs in the environment, both replacing each other in the same place and moving from one place to another. Walking down the street, taking the escalator in the subway, and based on the perceived spatiotemporal series of objects (local attractions and their number), a person determines how much distance he or she has traveled, and how much time has passed. In cities people pay more attention to road signs, and not to other "noisy" information, which allows them to pass any sections of the path quickly.

In addition to the existing "points of interest" on a map of the area, a traveler can find new "points of attraction" that unexpectedly invite attention. They can become intermediate points on the route.

A "wayfinding" interface is a method of interaction between a human, an event on the route and the route itself. Events on a route represented in the form of objects or signs have a real-time impact on choice of the subsequent path. These events create an interface for a person to plan and optimize the route. Furthermore these events are used in the interface when adding new points on the route and when changing the movement along the route. This interface can take various forms, from a mental plan existing only in the mind to a digital map stored in a navigation application. When solving the route optimization problem, a user can change conditions by moving, adding, or fixing route points selected according to given criteria.

Unlike static orientation elements, dynamic elements are not constant in their properties over time. Natural objects can change their visual properties depending on the time of day and season. Artificial objects can change their other attributes without changing visually over time: public facilities (museums, cafes, shops, etc.) have opening hours; public transport runs on a schedule, possibly, with long breaks. Spatial synchronicity of two points on the map, in the absence of other data, makes it possible to determine at least the distance between such points and shows for how long a person will walk on a traversable terrain if the distance is short enough. The issues of a simultaneous search for several objects are discussed in [13].

The presence of a distinguishable choice between two options does not imply the presence of a predisposition to one of them. In this situation, preparation is necessary for the solution: surveying in situation with clarification of route details, external interference to the situation, outside signs or so on. As an alternative to a choice, it is also possible try to avoid a choice. Sometimes, due to an ambiguous situation, draft picks appear (Fig. 6). Besides, there may be unforeseen road incidents requiring a significant change in the route, including heavy traffic or traffic jams, or meeting wild animals (Fig. 4A).

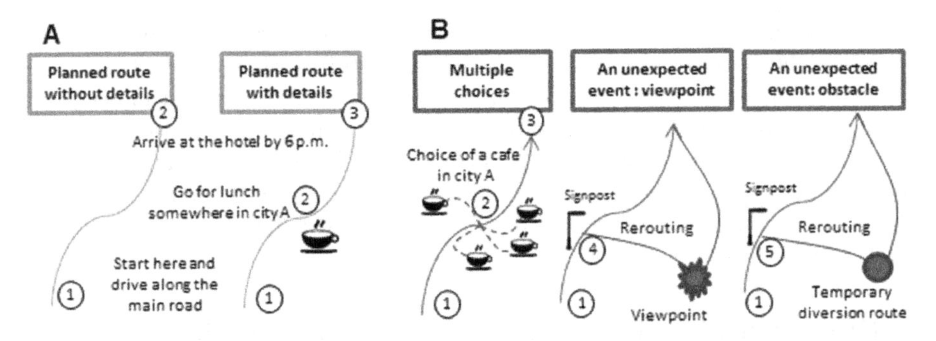

Fig. 6. A. Planned route with and without details. B. Actual route with additional details

An important reason for the change or complete cancellation of a route may be the discrepancy between expected and real events.

Human choice does not necessarily imply selection of the shortest path length or the path that can be traversed in the shortest time. Unexpected signs, such as obstacle signs, can significantly change the route and set of points to be visited. The line of sight of the short route endpoint is obscured by visual obstructions; the route may deviate greatly from the straight line and detour along the visible road section.

Figure 7 shows a comparison of linearity, the shortest route and popular trampling paths near the farm.

Fig. 7. Line-of-sight way (yellow dotted line) and alternative curve way (red dotted line) (site 43°23′10.9"N 10°51′21.0"E) (Color figure online)

2.4 Smart Routes Application Interface

Interface usability is improved by highlighting significant places and the ability to place and remove points in a dynamic way. The interface should allow movement and addition of points, and fixation of points during re-optimization; the main interface functions are selection, placement, and removal of points. The interface is defined by a) data exchange and retrieval, b) the speed of processes and internal exchanges, and c) generalized coordinates. The time for changes and the transport speed should be aligned with the user's response speed and that of the tools and programs used. The interface also needs to align the velocities of perception, movement, and reception of new data, including virtual data along the path, e.g. when using smart applications for a bike ride. (A sample interface for cyclists: https://www.bikemap.net/en/r/3954524/#10.19/48.1851/11.6073).

The user of a smart application interface has a designed set of tools that allows quick responses to events on the road. Additional points in the interface arise from both the human side and external events. It is possible to make adjustments for the user when planning routes, taking into account his or her route history. When optimizing a route, taking into account points of interest, the interface can use a dynamic segmentation of events along the route.

3 GIS Applications in the Study of Navigation Behavior

3.1 Materials and Methods

In this work Strava tracks dataset were studied. Strava is a social fitness service that allows users to share, compare and compete with other users' personal fitness data via mobile and online apps. Focusing on cyclists and runners, Strava lets users track their rides and runs via mobile apps or GPS device to analyze their performance. The Strava API provides several methods through which users can send and receive Strava data from a mobile device. To access Strava dataset the open Strava V3 API was used (API Endpoint: https://www.strava.com/api/v3; Docs Home Page URL: https://developers.strava.com/docs). Spatial data were analyzed using the geographical information system QGIS (https://qgis.org).

The routes on the territory of Venice were selected for a period of 2 years. For this purpose, a list of id pedestrians in historical Venice was obtained. According to this list, a set of tracks in gpx format was requested and received via API. Based on this set of tracks, a heatmap of the distribution density of track points was formed (see Fig. 8). Differences in the length, variability and shape of the tracks were identified for the cases of most popular routes and most direct routes from point A to point B.

In addition, the factors on the environment that influenced the formation of preferences along the routes were revealed (Fig. 8).

The comparative analysis was performed for the following data sets spatially distributed urban structure, including amenities on routes; the main components of tracking, such as location, length, linearity of paths: and heatmap of pedestrian activities.

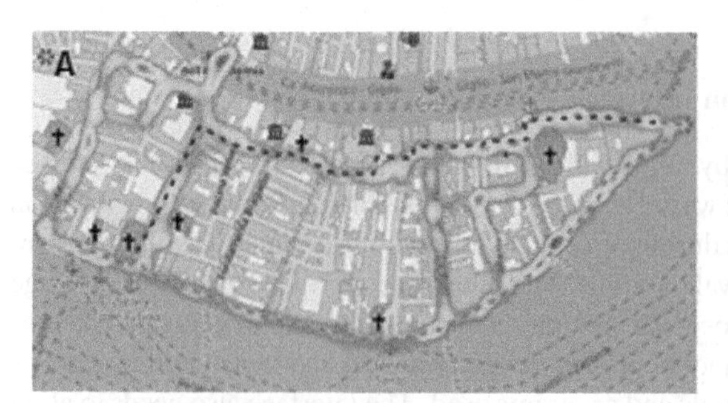

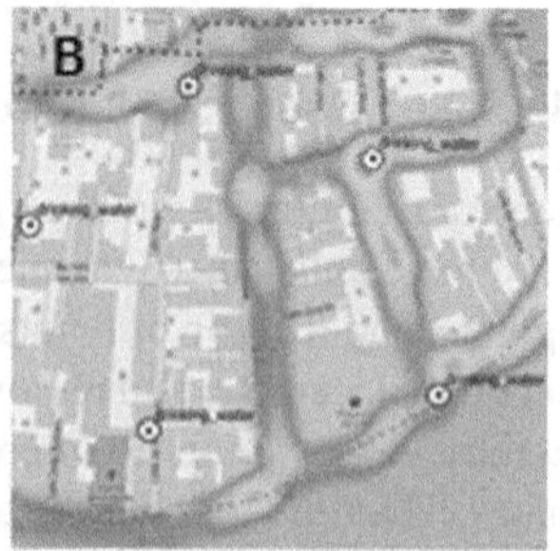

Fig. 8. A. Calculated shortest routes from the Zattere terminal to Punta della Dogana; routes' heatmap in Dorsoduro. B. Influence on the route of drinking water locations

The following operations were performed:

- selection of tracks in Venice from Strava dataset;
- inserting tracks dataset to QGIS project;
- loading a basic map from OpenStreetMap using the OpenLayers Plugin;
- calculation shortest path with Online Routing Mapper Plugin;
- creating a density heatmap for routes;
- selection of drinking water points using Overpass API (https://overpass-turbo.eu);
- comparison shortest paths, heatmap of routes and amenities on routes (drinking water.

The data were processed using QGIS geoalgorithms (http://qgis.org) and Plugin (Table 1).

Table 1. Spatial data processing applications.

Plugin	Description
OpenLayers Plugin https://github.com/sourcepole/qgis-openlayers-plugin	QGIS plugin embeds OpenLayers (http://openlayers.org) functionality
Heatmap https://docs.qgis.org/3.16/en/docs/user_manual/processing_algs/qgis/interpolation.html#qgisheatmapkerneldensityestimation	Creates a density (heatmap) raster of an input point vector layer using kernel density estimation The density is calculated based on the number of points in a location, with larger numbers of clustered points resulting in larger values
Online Routing Mapper https://cbsuygulama.wordpress.com/online-routing-mapper-en	Shortest Path Analysis with Online Routing Mapper Plugin. Generate routes by using online services (Google Directions, Here, MapBox, YourNavigation, OSRM etc.)

3.2 Results

In this study, Strava tracks datasets obtained for popular routes and calculated direct (shortest) routes (about 1–5 km) were analyzed.

Table 2 shows the result of analyzing the parameters of routes calculated using navigation algorithms and the parameters of real tracks from Strava. The comparison between direct and popular routes was tested for different types of territory - the whole of Venice, only Dorsoduro, Venice Lido. More than 1000 real tracks were analyzed for each type of territory. The percent of coincidences of popular routes and direct routes from one to another point is not higher 18% in Venice (if consider all popular routes and determine how many of them are direct). The result is statistically significant ($p < .05$).

Table 2. The parameters of routes and real tracks

	Coincident of popular routes and direct routes, %	Variability of the distance from the real track to the nearest popular route, m[a]	Route variation of real track direction (max-min), angle degree[b]	Route variability of real track direction, angle degree[a]
Set1 (Venice)	18%	98	81	48
Set2 (Dorsoduro)	41%	127	64	46
Set3 (Lido)	57%	193	36	38

[a] Mean value for all 100-m sections in single track (averaged for all tracks in the set).
[b] For the entire track (averaged for all tracks in the set).

4 Conclusion

In this work we demonstrated the usability of spatial data processing methods for comparative analysis of track records. We have shown that detection of typical tracks enriches the possibilities of track analysis processing.

On the basis of an initially indistinct representation of the terrain, it becomes possible to build a route and additionally to identify new significant landmarks and points of attraction in practice. Thus, by varying the positions of signposts and other visual architectural and landscape elements, one can artificially adjust the appeal of certain places, while creating stable points of attraction.

The traveler's choice may be inclined either to routes with a good overview, or to direct routes, or to routes with a large number of attractions.

For the smart spatial development of the urban structure, it is necessary to prepare for travelers the possibilities of alternative choices, with the creation of passages, water crossings, and additional points of interest on the route.

For example, after the construction of bridges, and also after the repair of one of the broken bridges on the Fondamenta Zattere Ai Saloni embankment in Venice, longer straightened routes with increased overview became possible for tourists.

References

1. Kong, X., Li, M., Ma, K., Tian, K., Wang, M., Ning, Z., et al.: Big trajectory data: a survey of applications and services. IEEE Access **6**, 58295–58306 (2018)
2. Zaleshina, M., Zaleshin, A., Galvani, A.: Navigational strategies in transition from initial route. In: Koutsopoulos, K., de Miguel González, R., Donert, K. (eds.) Geospatial Challenges in the 21st Century. Key Challenges in Geography, pp. 335–353. Springer, Cham (2019). https://doi.org/10.1007/978-3-030-04750-4_17
3. Aram, F., Solgi, E., Higueras García, E., Mohammadzadeh, S.D., Mosavi, A., Shamshirband, S.: Design and validation of a computational program for analysing mental maps: aram mental map analyzer. Sustainability **11**, 3790 (2019)
4. Margara, A.: Pattern recognition. In: Sakr, S., Zomaya, A. (eds.) Encyclopedia of Big Data Technologies, p. 1–7. Springer, Cham (2018). https://doi.org/10.1007/978-3-319-77525-8

5. Deville, P, Linard, C., Martin, S., Gilbert, M., Stevens, F.R., Gaughan, A.E., et al.: Dynamic population mapping using mobile phone data. Proc. Natl. Acad. Sci. U. S. A. **111**, 15888–15893 (2014)
6. Scotto, F., Salomoni, G.G., Amidei, F.: Itinerario d'Italia. a spese di Fausto Amidei (1761)
7. Marras, G.: Venice as a Paradigm. Mapp. Urban Spaces, pp. 94–105. Routledge, New York (2021)
8. Neumann, T.: Vessels route planning problem with uncertain data. TransNav Int. J. Mar. Navig. Saf. Sea Transp. **10**, 459–464 (2017)
9. Tobler, A.W.R.: A computer movie simulating urban growth in the detroit region. Science **46**, 234–240 (1970)
10. Tversky, A., Kahneman, D.: Judgment under uncertainty: heuristics and biases. Science **185**, 1124–1131 (1974)
11. Chen, L.: The topological approach to perceptual organization. Vis. cogn. **12**, 553–637 (2005)
12. LaPointe, M.R.P., Lupianez, J., Milliken, B.: Context congruency effects in change detection: opposing effects on detection and identification. Vis. cogn. **21**, 99–122 (2013)
13. Ort, E., Olivers, C.N.L.: The capacity of multiple-target search. Vis. Cogn. **28**, 330–355 (2020)

A Survey on Few-Shot Techniques in the Context of Computer Vision Applications Based on Deep Learning

Miguel G. San-Emeterio(✉)

Atos Research and Innovation, 28037 Madrid, Spain
mglez.sanemeterio@gmail.com

Abstract. This review article about Few-Shot Learning techniques is focused on Computer Vision Applications based on Deep Convolutional Neural Networks. A general discussion about Few-Shot Learning is given, featuring a context-constrained description, a short list of applications, a description of a couple of commonly used techniques and a discussion of the most used benchmarks for FSL computer vision applications. In addition, the paper features a few examples of recent publications in which FSL techniques are used for training models in the context of Human Behaviour Analysis and Smart City Environment Safety. These examples give some insight about the performance of state-of-the-art FSL algorithms, what metrics do they achieve, and how many samples are needed for accomplishing that.

Keywords: Few-Shot Learning · Deep Learning · Computer Vision · Human Behaviour Analysis · Smart City Environment Safety

1 Introduction

In September 2012 the Convolutional Neural Networks (CNNs) were in the spotlight of the Computer Vision community when a model developed by Alex Krizhevsky, Ilya Sutskever and Geoffrey Hinton [18] had an outstanding performance at the ImageNet Large Scale Visual Recognition Challenge (ILSVRC) [33]. This CNN model started the so-called "Deep Learning revolution", which is still booming as of today. During this period, novel computer vision applications have arisen, allowing practitioners and researchers to tackle previously unsolvable problems and to create new out-of-the-box solutions. Although powerful, Deep Learning (DL) techniques have their own set of problems, being data and computer power their most limiting requirements. Since they are Machine Learning (ML) algorithms, they need large amounts of data for training, which in turn require high computational power to process.

Many researchers have focused their work on mitigating these problems, creating a variety of techniques that softens the workload needed to create labelled datasets for supervised training. Some of them consist of using low-complexity labels for training models for high-complexity tasks [49]; others try to augment

P. L. Mazzeo et al. (Eds.): ICIAP 2022 Workshops, LNCS 13374, pp. 14–25, 2022.
https://doi.org/10.1007/978-3-031-13324-4_2

datasets by generating synthetic data [27,28], and others try to reuse previously learnt features [50]. Please note that there are other approaches beside the aforementioned three. Among the last of the three highlighted strategies there is a family of techniques called "Few-Shot Learning" (FSL), which includes "One-Shot Learning" and "Zero-Shot Learning".

For image classification tasks, supervised training of ML models require large amounts of annotated data for each of the classes or categories. Few-Shot Learning comprises those techniques designed for enabling a trained model to learn a new category from few examples. Thus, "One-Shot" would ideally need a single example of the new category to be able to learn how to classify it, whereas "Zero-Shot" would only need its description. Although those techniques are feasible for some tasks and data-types (see [38] for One-Shot and [32] for Zero-Shot), most applications require more samples to perform accurately. For example, a previous work from the author shows how a model trained for classifying different plant species needed 80 samples to classify at 90% accuracy a new-learnt class [3]. The true potential of Few-Shot Learning is its power to provide neural networks with adaptability and agility, which are features where CNNs do not traditionally shine on. This "learning speed" comes at a cost on accuracy and overall performance, although in many cases it is an acceptable trade-off. Also, while under regular conditions data scarcity hinders ML algorithms' performance, potentially down to unacceptable levels, the application of FSL can enable the use of ML algorithms when data is scarce.

This review article discusses the latest developments in Few-Shot Learning techniques applied to Computer Vision models, focusing in the contexts of Human Behaviour Analysis and Smart City Environment Safety applications. In it, there are answers to some questions related to the number of samples needed for training accurate networks using FSL, the trade-off between number of samples and performance in this context, and others. This document is divided in 5 sections. The first one is the introduction. The second, called "Description of Few-Shot Learning", discusses about Few-Shot Learning as a whole, and has 3 subsections titled "Applications of FSL", "FSL Techniques" and "FSL Benchmarks for Computer Vision". The third, "FSL for Human Behaviour Analysis applications", discusses two recent papers featuring FSL in this thematic. The fourth is called "Few-Shot for Smart City Environment Safety applications" and, similarly to the third, includes the discussion of three recent papers on that matter. The fifth and last section is a brief conclusion.

2 Description of Few-Shot Learning

In its broader sense, Few-Shot Learning is a category of Machine Learning techniques where the training dataset contains a limited number of supervised samples.

Since this is a broad description, common ML techniques such as data augmentation would be considered FSL as well, since this method enables accurate training with a limited sampling of annotated data. Nevertheless, in the context of this article the definition of Few-Shot Learning is constrained to the following: a family of Machine Learning methods that reuse previous knowledge for training accurate models using very limited training samples.

It is noteworthy to say that in the context of Computer Vision FSL is usually referred to image classification tasks, although it can be used also for object detection, semantic segmentation and hybrid tasks such as image description with text, among others.

2.1 Applications of FSL

Few-Shot Learning allows to train models with limited training datasets, thus reducing the amount of samples needed for supervised learning processes. There are several scenarios where reduced datasets are needed, and consequently FSL techniques can be applied:

- Extreme data. Data is called "extreme" when it has at least one of these characteristics: great volume, great variations in values, great complexity or sparsity. Few-Shot Learning can be applied in those situations where acquiring sufficient amounts of labelled data is hard or impossible. It is the case of analyzing ancient Chinese documents [22] and hieroglyphs (oracle characters) [14]; or new drug discovery problems as discussed by Altae-Tran *et al.* [1] for example.
- Cut costs. Annotating large amounts of data is a time-consuming process. Also, processing large datasets costs more computational power than smaller ones. Since FSL can be used to reduce the amount of data needed for training, it can be applied to reduce costs in both dataset generation and computer utilization expenses.
- Improve learning processes. In some cases, Few-Shot can be applied to parts of datasets to increase the quality of training rounds. This is the case of imbalanced datasets where some classes are underrepresented in relation to others within the same dataset. It also can be used for *weakly-supervised learning* when the dataset contains mixtures of annotated and non-annotated data. In addition, FSL is capable of boosting some *transfer learning* methods such as *domain adaptation.*

Also, beside the general applications mentioned above, there are many possible fields of application. This document focuses on Computer Vision, but it is noteworthy to mention that FSL is widely applied in all sorts of Machine Learning related methods: Natural Language Processing (NLP), language translation, regression, reinforcement learning, data classification, sound recognition [8], sensor calibration [43], smart city insight transfer [39], etc.

2.2 FSL Techniques

Some techniques use the knowledge for reducing the complexity of the Neural Networks graphs themselves, while others use it for optimizing the weights and parameters of the artificial neurons and the gradients. In this section the two most used FSL techniques are presented. Both intend to reuse prior knowledge to reduce the amount of data needed for successful learning.

Different taxonomies have been proposed for FSL techniques in several surveys and review articles, offering formal descriptions for transfer learning, meta learning, distance learning, embedding learning, etc. [2,16,25,41]. It is noteworthy to say that many of these techniques share common traits and, depending on the context of application, some of them do not present sufficient unique characteristics to be distinguished from one another. For that reason, at this point in time it cannot be assumed that the categories themselves are standardized and fixed. In this regard, what it is called "multi-task learning" in one paper can be part of the "meta-learning" family in another. In this document there are short discussions about a couple of techniques but the category names are not intended to be taken as fixed, immutable references.

Metric Learning. In some papers it is also referred as "embedding learning" [41]. This technique consists on transforming the input data (the samples) into a lower-dimensional embedding located in a space where distances represent similarity between the different encoded samples. In the embedding space, two similar samples would lay at small distances whereas dissimilar samples will be far from each other. This sample discrimination can be applied to effectively reuse prior knowledge about the samples themselves, and thus drastically reduce the training samples needed for learning. This is possible mainly because the embedding space allows for fast an simple comparisons of samples, whereas in other approaches complex operations are needed to effectively process the embeddings into predictions. In this case the encoding functions are already pre-trained, and can be fine-tuned with few examples by applying a specific loss function that forces the neural network's parameters to encode the samples in a given space in where the distances are related to the samples' similarity. In this sense, metric learning can be used for other applications besides Few-Shot Learning [15]. It has proven to be effective for learning tasks that involve comparison like signature verification [6], product design [4] or face recognition [26], among others.

In distance metric learning, the main feature is the training loss function, as it is responsible for setting the characteristics of the embedding space. The most used loss functions in metric learning are contrastive loss [7], triplet loss [34,42] and multi-class N-pair loss [36]. Recent research works explore novel loss functions for specific tasks, like Additive Angular Margin Loss (ArcLoss) for face recognition [10]. Other approaches use ensemble loss functions for better generalization of metric learning models [45]. The encoding space is usually Euclidean for simplicity reasons, but there are other research lines were both known and novel geometries are explored.

Multi-task Learning. In some research articles it is also called "meta-learning" [2], because "multi-task learning" is part of the that family of ML algorithms. This approach uses learning procedures where different learning tasks are trained simultaneously, thus achieving to share parameters between them which gradually increases the generalization capabilities of the model. In order to learn different tasks, the model needs to make abstractions and learn more generic information. In multi-task learning there are two types of knowledge: task-agnostic and task-specific information. The first of them refers to all the preprocessing steps needed to aggregate and interpret features into embeddings. The second is about how to interpret such embeddings into actual predictions, which is specific for each given task.

The most common approach of multi-task learning is to directly share common layers (the encoder) and put different ones at the end in order to process each task's output. In many cases, constrains would be applied to the artificial neurons so the Few-Shot task can only update parameters from the task-specific layers, while the encoder's parameters (task-generic or task-agnostic) can be changed by all of the tasks. Another approach would be to have separate CNN encoders for each task and them concatenate the resulting feature vectors at the end of the feature extractor. In this case all encoders would contribute to the same loss function, and regularization terms would be applied to force the parameters of the different Neural Network layers to be similar, thus aligning the training process of each individual encoder.

2.3 FSL Benchmarks for Computer Vision

As of today, the most important benchmark for testing Few-Shot Learning methods applied to Computer Vision is Google Research's Meta-Dataset [37], "a dataset of datasets for learning to learn from few examples". More specifically, the most complete benchmark is its second version, named VTAB+MD [11], which is a combination of the Visual Task Adaptation Benchmark (VTAB) [47] and the Meta-Dataset. The Meta-Dataset is composed of a selection of images from many other datasets, namely ImageNet, Omniglot, FGVC-Aircraft, Birds-CUB-200-2011, DTD, Quick Draw, FGVCx Fungi, VGG Flower, Traffic Signs (GTSRB) and MSCOCO. It has a total of 4934 different classes. In addition to the collection of images, it has an unique protocol for evaluating Few-Shot algorithms, which improves the two previous FSL benchmark dataset, Omniglot [20] and Mini-ImageNet [38], in several ways:

1. The combination of several datasets of different domains results in a more realistic data heterogeneity, which allows for testing the model's capacity to generalize unseen datasets. It also tests their ability for training with unbalanced datasets in terms of number of samples per class.
2. Its large scale and dataset mixture allows the evaluation protocol to take into account the ability of the different few-shot techniques to form relationships between classes. For example, if the model can do fine-grain classification for detecting different species of birds while being able to distinguish birds from common objects like dinner tables.
3. The data has been selected and structured in a such a way that it mimics realistic class unbalance, allowing the evaluation of the different models in different numbers of samples per class.

4. This benchmark has a clear evaluation guideline that combines different ML tasks, which in turn allows to evaluate the models' capacity of learning from different sources. It also has a set of baselines aimed to measure the benefit of meta-learning, i.e. whether the training process benefits from using more data, learning from different sources, reusing knowledge from pre-trained weights and meta-training parameters.

The Meta-Dataset protocol computes the model's rank by decreasing order of accuracy. As of today, the best performing model on Meta-Dataset benchmark 2021's model code-name "TSA", from Task-Specific Attention [23,24], achieved number 1 position on the Meta-Dataset with 1.65 mean rank across all datasets. The paper, titled "Cross-domain Few-shot Learning with Task-specific Adapters", proposes a novel multi-task learning methodology that combines task-agnostic with task-specific approaches. This is done by directly attaching task-specific adapters to pre-trained task-agnostic models. The article also features a new architecture for said adapters.

3 FSL for Human Behaviour Analysis Applications

In the context of Computer Vision, there are several tasks than can help Human Behaviour Analysis (HBA). All of them can be predicted using Deep Neural Networks, and for most of them there are recent research articles featuring FSL. Here are some examples of such tasks along with a Few-Show citation: face detection, face recognition [29,40,48], facial expression recognition [9], person detection, person recognition, pose estimation, action recognition [12], person re-identification [13], person tracking [21], hand gesture recognition [30], motion prediction [46], etc. In this section two recent research articles on FSL are shortly discussed, each featuring a different HBA application.

Face Recognition. Few-Shot Learning is a natural match for face recognition. As explained in Sect. 2.2, metric learning is a technique that is not only used for FSL, but also other applications such as face recognition. In this regard, the use of the same loss functions such as triplet loss [34,42] or ArcFace loss [10] make face recognition to naturally be a Few-Shot Learning application as well.

A. Putra and S. Setumin published in 2021 an article in which they made a performance comparison of different activation functions on Siamese Networks for Face Recognition [29]. Their Few-Shot technique was to combine multi-task learning (parameter tying) with metric learning. In the article, the researchers trained a Siamese network using different activation functions for the final embedding prediction, which would in turn be used to measure distance (similarity) between the outputs of the two branches of the network. The tested activation functions were sigmoid, softmax, tanh, softplus and softsign. They tested each activation function by learning 1 to 19 classes with only 1 supervised sample (One-Shot Learning). In average, sigmoid activation function achieved 92% accuracy to recognize 19 new faces with only one sample for training using this FSL approach.

Action Recognition. In a recent publication (2021) by Mark Haddad [12], he explored a method for One-Shot and Few-Shot learning for action recognition tasks. The models had to learn classification of 10 different actions: bend, jack, jump, pump-jump, run, side, skip, walk, wave 1 and wave 2. For that matter, he encoded the movement of a few frames into an optical flow vectorized with KMeans method. Therefore, movements were parametrized as KMeans clusters, that can be treated as probability distrbutions for different sets of points that represent each action. Using Kullback-Leibler divergence loss, a model can be trained to perform metric learning with the aforementioned KMeans clusters. In his thesis, Haddad performed tests for both One-Shot Learning and for K-shot, being K a number between 1 and 8 samples. On the Weizmann Human Action dataset, he achieved an average of 89.4% classification accuracy for One-Shot Learning, increased to 98% with 8-Shot Learning (see Fig. 1).

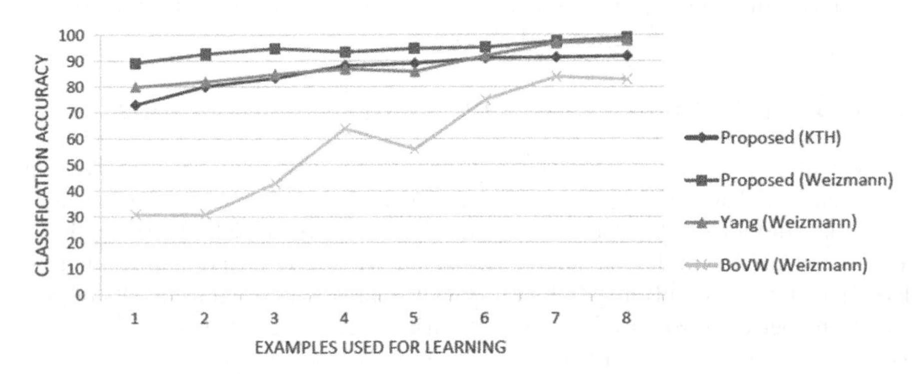

Fig. 1. Extracted Fig. 3.5 from Haddad's thesis [12]. Quote: "Classification accuracy comparison between proposed method and others." It shows results for two different datasets: KTH and Weizmann.

4 Few-Shot for Smart City Environment Safety Applications

In the context of Smart City Environment Security, there are some Computer Vision applications that are extensively applied, specially related to traffic and pedestrian surveillance. Yet another application is assessment for disasters and emergencies that threaten safety of cities, both for its population or its infrastructures. Emergency response is a good fit for the application of Few-Shot Learning methods since they are events where on-site data is usually scarce; and whose changing nature make the models require flexibility and adaptability. There are recent examples where FSL was effectively applied in emergencies such as the ongoing pandemic caused by Covid-19. As one of many examples, Lai *et al.* [19] used Deep Learning algorithms trained using Few-Shot techniques to classify lung lesions caused by COVID-19 using the small datasets available at the time. In this section three research articles were FSL was applied are discussed.

The first of them is related to Smart Cities Environment Safety, while the other two are focused on disaster damage assessment as part of emergency response applications.

Face Anti-spoofing. Recent developments in Deep Learning technology for face reenactment have made possible very realistic impersonating. This is a risk for governments and administrations, because it denies one of the traditional identity proofs, which is face identification in live video. For that reason, many researchers are working towards face anti-spoofing.

In 2021 Yang *et al.* published a paper on face anti-spoofing where they used FSL techniques for achieving One-Shot Learning of new unseen face reenactments [44]. They proposed a FSL technique called Few-Shot Domain Adaptation, which combines data augmentation using photorealistic style transfer and a complex compound loss function $\mathcal{L}_{total}$.

$$\mathcal{L}_{total} = \mathcal{L}_{ClS} + \lambda_1 \mathcal{L}_{Cont} + \lambda_2 \mathcal{L}_{Adv} + \lambda_3 \mathcal{L}_{Lfc} \tag{1}$$

$\mathcal{L}_{ClS}$ is cross-entropy loss. $\mathcal{L}_{Cont}$ is contrastive semantic alignment, which is a term used for reducing distances in the encoding space for positive pairs and increase them for the negatives (a form of metric learning). $\mathcal{L}_{Adv}$ is the progressive adversarial learning, a term used for constraining the parameters of the different domain discriminators. $\mathcal{L}_{Lfc}$ is the less-forgetting constrain, which is a mean square error penalization to prevent the parameters from shifting excessively during training. The λs are trade-off parameters.

With this method they were able to detect face spoofing by training with a single sample and they managed to beat the previous best working model by 5% points in one of the tests (CMOS-ST benchmark when targeting HTER using protocol $C \rightarrow O$).

Disaster Damage Assessment. In this case there are two articles of interest that are noteworthy to discuss.

E. Koukouraki, L. Vanneschi and M. Painho published in 2021 a research article about urban damage detection after earthquake incidents from satellite images, trained using FSL techniques. In it, they tested four different methodologies: cost-sensitive learning, oversampling, undersampling and Prototypical Networks [35], each of them employing different data balancing methods. The best working model was Prototypical Networks (ProtoNets), which combine multi-task learning with metric learning by using distance loss functions to train Siamese neural networks (see Fig. 2). It achieved precision and recall superior to 50% in all four classes (undamaged, minor damage, major damage and destroyed) with average F-score of 64% over all classess.

In a similar work, J. Bowman and L. Yang studied further Few-Shot Learning methods for post-disaster damage detection from satellite images [5]. In this case, they use a FSL technique called "feature re-weighting" that is used for parameter tying (a type of multi-task learning approach) between classification and

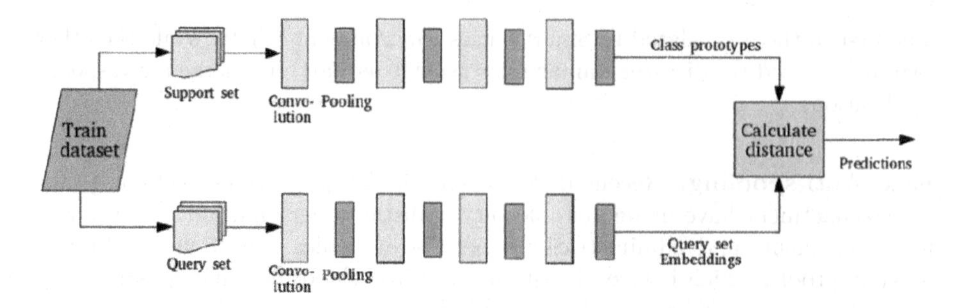

Fig. 2. Extracted Fig. 5 from Koukouraki *et al.* article [17]. It shows the network architecture used for the ProtoNets approach

object detection tasks. They used the encoder (feature extractor) from YOLO [31], and then added an additional CNN model, the re-weighting module, to generate re-weighting vectors for each class. These vectors are concatenated to the embeddings from the feature extractor in order to extend them before the final prediction takes place. In this manner, the re-weighting effectively fuses task-agnostic and task-specific functionalities thus providing FSL capabilities. With the addition of this module, they achieved to improve the mean average precision (mAP) of the baseline YOLO from 0.270 to 0.289 when training with 30 samples. This means that when applying feature re-weighing, the model gave a mAP almost 2% points better, when training with only 30 samples per class.

5 Conclusion

This review article about Few-Shot Learning techniques is different from others because it is heavily focused on Computer Vision Applications based on Deep Convolutional Neural Networks. Moreover, Sects. 3 and 4 feature commented examples of articles were FSL techniques were used for face recognition, action recognition, face anti-spoofing and post-disaster damage assessment from satellite. These examples show how FSL algorithms can be effectively used to learn from few supervised samples with competitive metrics in comparison to baseline models. They show how metric learning and multi-task learning are FSL techniques that combined offer current State of the Art results when training with few samples. As final though, it is noteworthy to mention that this paper emphasizes the importance of the loss function when it comes to *meta-learning* and *transfer learning*.

References

1. Altae-Tran, H., Ramsundar, B., Pappu, A.S., Pande, V.: Low data drug discovery with one-shot learning. ACS Cent. Sci. **3**(4), 283–293 (2017)
2. Antonelli, S., et al.: Few-shot object detection: a survey. ACM Comput. Surv. (CSUR), 6–7 (2021)

3. Argüeso, D., et al.: Few-shot learning approach for plant disease classification using images taken in the field. Comput. Electron. Agric. **175**, 105542 (2020)
4. Bell, S., Bala, K.: Learning visual similarity for product design with convolutional neural networks. ACM Trans. Graph. (TOG) **34**(4), 1–10 (2015)
5. Bowman, J., Yang, L.: Few-shot learning for post-disaster structure damage assessment. In: Proceedings of the 4th ACM SIGSPATIAL International Workshop on AI for Geographic Knowledge Discovery, pp. 27–32 (2021)
6. Bromley, J., Guyon, I., LeCun, Y., Säckinger, E., Shah, R.: Signature verification using a "Siamese" time delay neural network. In: Advances in Neural Information Processing Systems, vol. 6 (1993)
7. Chopra, S., Hadsell, R., LeCun, Y.: Learning a similarity metric discriminatively, with application to face verification. In: 2005 IEEE Computer Society Conference on Computer Vision and Pattern Recognition (CVPR 2005), vol. 1, pp. 539–546. IEEE (2005)
8. Chou, S.Y., Cheng, K.H., Jang, J.S.R., Yang, Y.H.: Learning to match transient sound events using attentional similarity for few-shot sound recognition. In: ICASSP 2019–2019 IEEE International Conference on Acoustics, Speech and Signal Processing (ICASSP), pp. 26–30. IEEE (2019)
9. Ciubotaru, A.N., Devos, A., Bozorgtabar, B., Thiran, J.P., Gabrani, M.: Revisiting few-shot learning for facial expression recognition. arXiv preprint arXiv:1912.02751 (2019)
10. Deng, J., Guo, J., Xue, N., Zafeiriou, S.: ArcFace: additive angular margin loss for deep face recognition. In: Proceedings of the IEEE/CVF Conference on Computer Vision and Pattern Recognition, pp. 4690–4699 (2019)
11. Dumoulin, V., et al.: Comparing transfer and meta learning approaches on a unified few-shot classification benchmark. arXiv preprint arXiv:2104.02638 (2021)
12. Haddad, M.: An instance-based learning statistical framework for one-shot and few-shot human action recognition. Ph.D. thesis, Concordia University (2021)
13. Han, P., et al.: HMMN: online metric learning for human re-identification via hard sample mining memory network. Eng. Appl. Artif. Intell. **106**, 104489 (2021)
14. Han, W., Ren, X., Lin, H., Fu, Y., Xue, X.: Self-supervised learning of ORC-BERT augmentator for recognizing few-shot oracle characters. In: Proceedings of the Asian Conference on Computer Vision (2020)
15. Koch, G., Zemel, R., Salakhutdinov, R., et al.: Siamese neural networks for one-shot image recognition. In: ICML Deep Learning Workshop, vol. 2, Lille (2015)
16. Köhler, M., Eisenbach, M., Gross, H.M.: Few-shot object detection: a survey. arXiv preprint arXiv:2112.11699 (2021)
17. Koukouraki, E., Vanneschi, L., Painho, M.: Few-shot learning for post-earthquake urban damage detection. Remote Sens. **14**(1), 40 (2021)
18. Krizhevsky, A., Sutskever, I., Hinton, G.E.: ImageNet classification with deep convolutional neural networks. In: Advances in Neural Information Processing Systems, vol. 25 (2012)
19. Lai, Y., et al.: 2019 novel coronavirus-infected pneumonia on CT: a feasibility study of few-shot learning for computerized diagnosis of emergency diseases. IEEE Access **8**, 194158–194165 (2020)
20. Lake, B.M., Salakhutdinov, R., Tenenbaum, J.B.: Human-level concept learning through probabilistic program induction. Science **350**(6266), 1332–1338 (2015)
21. Lee, J., Ramanan, D., Girdhar, R.: MetaPix: few-shot video retargeting. arXiv preprint arXiv:1910.04742 (2019)
22. Li, B., Wei, J., Liu, Y., Chen, Y., Fang, X., Jiang, B.: Few-shot relation extraction on ancient Chinese documents. Appl. Sci. **11**(24), 12060 (2021)

23. Li, W.H., Liu, X., Bilen, H.: Cross-domain few-shot learning with task-specific adapters. arXiv preprint arXiv:2107.00358 (2021)
24. Li, W.H., Liu, X., Bilen, H.: Improving task adaptation for cross-domain few-shot learning. arXiv preprint arXiv:2107.00358 (2021)
25. Li, X., Yang, X., Ma, Z., Xue, J.H.: Deep metric learning for few-shot image classification: a selective review. arXiv preprint arXiv:2105.08149 (2021)
26. Liu, W., Wen, Y., Yu, Z., Li, M., Raj, B., Song, L.: SphereFace: deep hypersphere embedding for face recognition. In: Proceedings of the IEEE Conference on Computer Vision and Pattern Recognition, pp. 212–220 (2017)
27. Nikolenko, S.I.: Synthetic Data for Deep Learning. SOIA, vol. 174. Springer, Cham (2021). https://doi.org/10.1007/978-3-030-75178-4
28. Picon, A., San-Emeterio, M.G., Bereciartua-Perez, A., Klukas, C., Eggers, T., Navarra-Mestre, R.: Deep learning-based segmentation of multiple species of weeds and corn crop using synthetic and real image datasets. Comput. Electron. Agric. **194**, 106719 (2022)
29. Putra, A.A.R., Setumin, S.: The performance of Siamese neural network for face recognition using different activation functions. In: 2021 International Conference of Technology, Science and Administration (ICTSA), pp. 1–5. IEEE (2021)
30. Rahimian, E., Zabihi, S., Asif, A., Atashzar, S.F., Mohammadi, A.: Trustworthy adaptation with few-shot learning for hand gesture recognition. In: 2021 IEEE International Conference on Autonomous Systems (ICAS), pp. 1–5. IEEE (2021)
31. Redmon, J., Divvala, S., Girshick, R., Farhadi, A.: You only look once: unified, real-time object detection. In: Proceedings of the IEEE Conference on Computer Vision and Pattern Recognition, pp. 779–788 (2016)
32. Romera-Paredes, B., Torr, P.: An embarrassingly simple approach to zero-shot learning. In: International Conference on Machine Learning, pp. 2152–2161. PMLR (2015)
33. Russakovsky, O., et al.: ImageNet large scale visual recognition challenge. Int. J. Comput. Vision **115**(3), 211–252 (2015). https://doi.org/10.1007/s11263-015-0816-y
34. Schroff, F., Kalenichenko, D., Philbin, J.: FaceNet: a unified embedding for face recognition and clustering. In: Proceedings of the IEEE Conference on Computer Vision and Pattern Recognition, pp. 815–823 (2015)
35. Snell, J., Swersky, K., Zemel, R.: Prototypical networks for few-shot learning. In: Advances in Neural Information Processing Systems, vol. 30 (2017)
36. Sohn, K.: Improved deep metric learning with multi-class n-pair loss objective. In: Advances in Neural Information Processing Systems, vol. 29 (2016)
37. Triantafillou, E., et al.: Meta-dataset: a dataset of datasets for learning to learn from few examples. arXiv preprint arXiv:1903.03096 (2019)
38. Vinyals, O., Blundell, C., Lillicrap, T., Wierstra, D., et al.: Matching networks for one shot learning. In: Advances in Neural Information Processing Systems, vol. 29 (2016)
39. Wang, J., Li, W., Qi, X., Ren, Y.: Transfer knowledge between cities by incremental few-shot learning. In: Gao, H., Wang, X. (eds.) CollaborateCom 2021. LNICST, vol. 407, pp. 241–257. Springer, Cham (2021). https://doi.org/10.1007/978-3-030-92638-0_15
40. Wang, L., Li, Y., Wang, S.: Feature learning for one-shot face recognition. In: 2018 25th IEEE International Conference on Image Processing (ICIP), pp. 2386–2390. IEEE (2018)
41. Wang, Y., Yao, Q., Kwok, J.T., Ni, L.M.: Generalizing from a few examples: a survey on few-shot learning. ACM Comput. Surv. (CSUR) **53**(3), 1–34 (2020)

42. Weinberger, K.Q., Saul, L.K.: Distance metric learning for large margin nearest neighbor classification. J. Mach. Learn. Res. **10**(2), 207–244 (2009)
43. Yadav, K., Arora, V., Jha, S.K., Kumar, M., Tripathi, S.N.: Few-shot calibration of low-cost air pollution (PM2. 5) sensors using meta-learning. arXiv preprint arXiv:2108.00640 (2021)
44. Yang, B., Zhang, J., Yin, Z., Shao, J.: Few-shot domain expansion for face anti-spoofing. arXiv preprint arXiv:2106.14162 (2021)
45. Zabihzadeh, D.: Ensemble of loss functions to improve generalizability of deep metric learning methods. arXiv preprint arXiv:2107.01130 (2021)
46. Zang, C., Pei, M., Kong, Y.: Few-shot human motion prediction via learning novel motion dynamics. In: Proceedings of the Twenty-Ninth International Conference on International Joint Conferences on Artificial Intelligence, pp. 846–852 (2021)
47. Zhai, X., et al.: A large-scale study of representation learning with the visual task adaptation benchmark. arXiv preprint arXiv:1910.04867 (2019)
48. Zheng, W., Gou, C., Wang, F.Y.: A novel approach inspired by optic nerve characteristics for few-shot occluded face recognition. Neurocomputing **376**, 25–41 (2020)
49. Zhou, X., Girdhar, R., Joulin, A., Krähenbühl, P., Misra, I.: Detecting twenty-thousand classes using image-level supervision. arXiv preprint arXiv:2201.02605 (2021)
50. Zhuang, F., et al.: A comprehensive survey on transfer learning. Proc. IEEE **109**(1), 43–76 (2020)

Decision-Support System for Safety and Security Assessment and Management in Smart Cities

Javier González-Villa[1](✉), Arturo Cuesta[1], Marco Spagnolo[2], Marisa Zanotti[2], Luke Summers[3], Alexander Elms[3], Anay Dhaya[3], Karel Jedlička[4], Jan Martolos[5], and Deniz Cetinkaya[6]

[1] GIDAI Group, University of Cantabria, Santander, Spain
javier.gonzalezvilla@unican.es
[2] EnginSoft, s.p.a., Trento, Italy
[3] Crowd Dynamics International Limited, Oxted, UK
[4] University of West Bohemia, Plzeň, Czech Republic
[5] Plan4All z.s., Horní Bříza, Czech Republic
[6] Department of Computing and Informatics, Bournemouth University, Poole, UK

Abstract. Counter-terrorism and its preventive and response actions are crucial factors in security planning and protection of mass events, soft targets and critical infrastructures in urban environments. This paper presents a comprehensive Decision Support System developed under the umbrella of the S4AllCitites project, that can be integrated with legacy systems deployed in the Smart Cities. The system includes urban pedestrian and vehicular evacuation, considering ad-hoc predictive models of the evolution of incendiary and mass shooting attacks in conjunction with a probabilistic model for threat assessment in case of improvised explosive devices. The main objective of the system is to provide decision support to public or private security operators in the planning and real time phases in the prevention or intervention against a possible attack, providing information on evacuation strategies, the probability or expected impact of terrorist threats and the state of the traffic network in normal or unusual conditions allowing the emergency to be managed throughout its evolution.

Keywords: Security and safety · Evacuation · Terrorism · Threats · Fire and smoke · Traffic · Simulation · Decision support system

1 Introduction

International terrorism has many dimensions and characteristics depending on factors such as the historical and geographical context, political links or factors related to different terrorist groups and organizations (Tuman 2009). Today security and terrorism are one of the most widespread problems that requires the attention of law enforcement agencies, policy makers and political institutions due to the social and economic impact it generates. Despite this dependence on factors, terrorist attacks have the purpose of creating great harm and consternation in the population. Cities are a spotlight as they

P. L. Mazzeo et al. (Eds.): ICIAP 2022 Workshops, LNCS 13374, pp. 26–36, 2022.
https://doi.org/10.1007/978-3-031-13324-4_3

cluster large populations in small areas susceptible to terrorist attacks. Therefore, in the field of anti-terrorist urban security, and more broadly in the context of mass events, critical infrastructures and soft targets, it is mandatory to have adequate planning and response strategies to deal with such emergencies. Hence, this paper proposes a Decision Support System (DSS) that can be used during planning and response phases anticipating terrorist threats and while helping to address emergency management issues within the context of smart cities.

According to the (Global Terrorism Database™ (GTD) 2021) more than half of the attacks worldwide are Improvised Explosive Devices (IED), mass shooting, arsons or incendiary/smoke devices attacks. The expected evolution of this kind of attacks (Martin 2016) (EUROPOL 2021) are the necessary foundation for the development of models that can help to minimize their consequences.

More specifically, between 2010 and 2019 29.4% of terrorist attacks were targeted against the population (Global Terrorism Database™ (GTD) 2021), with cities being a major attraction for the terrorists. In this context there is an increment of smart cities that use Information and Communication Technologies (ICT) to increase operational efficiency, share information with the public and improve both the quality of government services and citizen welfare. An important point is that smart cities also need to ensure a secure and safe physical and digital ecosystem for the well-being of citizens. Therefore, it is mandatory to utilize the capabilities already available in smart cities to improve security and safety. These include, for example, anomaly detection, authentication and identification of individuals, threat localisation, behavioural profiling, suspect tracking, traffic monitoring, emergency management and many other capabilities related with awareness, prevention and response (Laufs et al. 2020).

These capabilities have been studied from different perspectives leading to a wide range of results including threats and individuals detection (Chackravarthy et al. 2018) (Bellini et al. 2017), screening and tracking (Brust et al. 2017) (Anees and Kumar 2017), recognition-based authentication (Balla and Jadhao 2018; Boukerche et al. 2017) or the improvement of legacy systems deployed throughout the city endowing them with intelligence (Zingoni et al. 2017; Zhou et al. 2015). However, due to our particular approach, we must emphasise that there are hardly any studies (Dbouk et al. 2014; Bonatsos et al. 2013) that propose a comprehensive DSS involving at the same time emergency management, real-time decision support and forecasting of threats evolution and impact of most common terrorist attacks. The closest in these terms to the existing literature is focused on the management of common crimes such as vandalism and violence, both in terms of management (Fernández et al. 2013), information systems (Truntsevsky et al. 2018), unusual traffic management (Hartama et al. 2017), evacuation (Zhang et al. 2018) and to a lesser extent on the prediction of events such as robbery or homicide (Noor et al. 2013; Araujo et al. 2017).

Thus, this study jointly addresses the facets of predicting and assess the impact of terrorist attacks (IED, mass shooting and arson), together with the management of emergency situations in terms of pedestrian and vehicular intervention, evacuation and monitoring by proposing a comprehensive conceptual and computational model that implements a DSS. This system involves different data sources and computer simulations providing support to decision makers/operators to make appropriate planning, management or response decision (Turban 1995).

2 Material and Method

2.1 Conceptual Model

On the basis of the initial definition of a smart city, a three-layers structure can be used to formalise the mathematical model of the proposed DSS (Decision Support System), see Fig. 1.

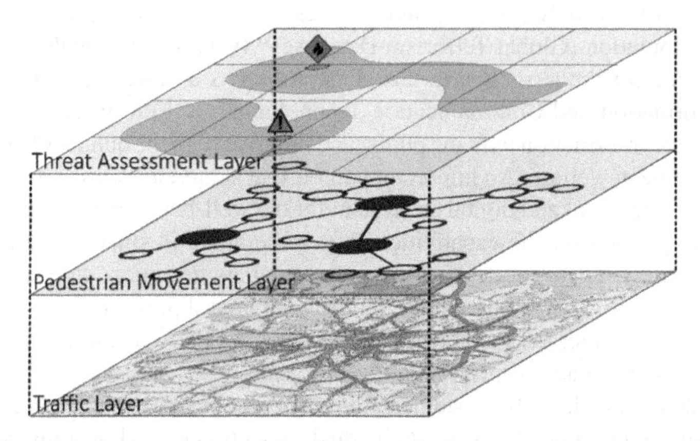

Fig. 1. Conceptual model – mathematical modelled schema of layer in smart cities.

Threat Assessment Layer: Comprises a set of soft targets, crowded areas and infrastructures $S = \{s_0, s_1, \ldots, s_n\}$ where security monitoring is desired. A soft-target can be defined as $s_k = \{B, P, A, D, O\}$, where $B = \{(\phi_0, \theta_0), (\phi_1, \theta_1), \ldots, (\phi_k, \theta_k)\}$ represents an enclosed and geographically defined area (longitude, latitude), P is the spatial distribution of people, A is the security assets deployed (e.g. controls, cameras or patrols) and $D$and O is the set of safe areas and obstacles inside the scenario that are defined by geographical coordinates. Threats being monitored in these areas are therefore defined as $T_i = \{L, C\}$, where $L = (\phi, \theta)$ is the location and $C \in \{Arson, Smoke, IED, Weapon\}$ is the category.

Pedestrian Movement Layer: Topological definition of pedestrian transitable areas is replicated through the graph $G_p = \{N, E\}$, which is arranged by $N = \{n_0, n_1, \ldots, n_n\}$ set of nodes and $E = \{e_0, e_1, \ldots, e_m\}$ set of edges. Each node $n_i = \{L, d, s\}$ is defined by its geographic location and occupant density as well as its current status $s \in \{Passable, Impassable, Evacuate, Safe\}$. Likewise, each edge $e_i = \{d, n_o, n_d, f\}$ represents transitable zones and it is defined by people density, origin and destination nodes and available flow.

Traffic Layer: Traffic network is represented through the graph $G_t = \{V, E\}$, where $V = \{v_0, v_1, \ldots, v_n\}$ are the vehicular transit reference points associated with physical locations and $E = \{e_0, e_1, \ldots, e_m\}$ represents the reachability associations similar to the pedestrian layer but within a traffic environment where the density and flow measurements represent vehicles instead of people. For the generation of traffic profiles, this layer considers the different usual zones Z of origin and destination of trips, which in turn are related by proximity to a node of the traffic network, generating a set of paths P between them and an origin-destination weighted matrix $W = Z \times Z$ (Fig. 2).

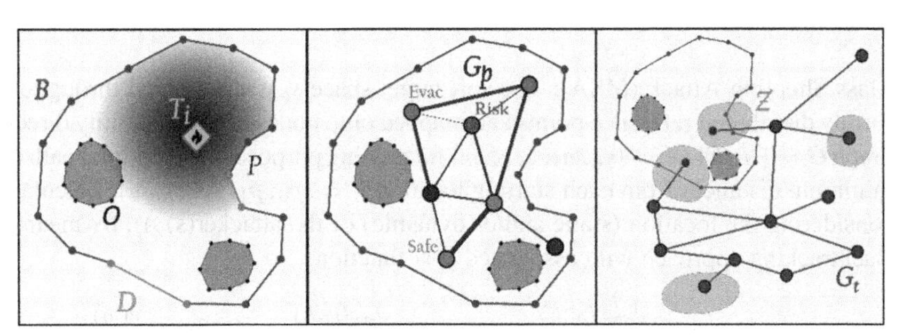

Fig. 2. Conceptual model – graphical representation of mathematical layer-based model. From left to right: threat assessment layer, pedestrian movement layer and traffic layer.

Threat Assessment Layer

This layer assesses the threats and possible impacts/consequences of three type of attacks:

1. Arson and Smoke Bomb: Fire Dynamics Simulator (McGrattan et al. 2017) is used for the most likely locations of this type of attacks by simulating several scenarios changing the actual combustion parameters, different wind and fire loads. The generated results providing artificial measurements $M_f(s_i) = \{m_0, m_1, \ldots, m_k\}$, (e.g. visibility, Fractional Effective Dose (FED)) are classified and stored in a structured way for further use.
2. Improvise Explosive Device (IED): This approach is based on (Cuesta et al. 2019). The boundary box of each soft-target s_k is calculated and subdivided into small regions shaping a fine grid of squared cells. For each cell c_{ij} within the grid, the risk function is calculated as follows:

$$R(s_k, c_{ij}) = w_{dt} \cdot d_t(c_{ij}, E) + w_{da} \cdot d_a(c_{ij}, A) + w_{dr} \cdot d_r(c_{ij}, B) + w_p \cdot p(c_{ij}, P),$$

where the functions studied have associated weights $\{w_{dt}, w_{da}, w_{dr}, w_p\}$ that can be modified (e.g. to give more weight to one or another parameter) but, as a general rule, balance the risk function. The rest of the functions that are measured in the equation are:

- $d_t(c_{ij}, E) \rightarrow$ Inverse (1-p) normalized distance from cell c_{ij} to the nearest exit (negative correlation).
- $d_a(c_{ij}, A) \rightarrow$ Inverse normalized distance from cell c_{ij} to the nearest asset.
- $d_r(c_{ij}, B) \rightarrow$ Normalized radial distance from cell c_{ij} to the boundary box (positive correlation).
- $p(c_{ij}, P) \rightarrow$ Normalized population density inside cell c_{ij}.

After processing all the cells, a matrix $M_r(s_i)$ of risk values is provided which is associated with the threat level, resulting in a probability map with critical locations of IEDs for each soft-target.

3. Mass Shooting Attack (MSA): The soft-target space s_i is discretized through uniformly distributed reference points and mapped onto nodes of a reachability directed graph $G = \{N, E\}, N = \{n_0, n_1, \ldots, n_k\}$ for pathing purposes. The optimal path (i.e. minimum distance) from each starting location $P = \{p_0, p_1, \ldots, p_m\}$ is calculated, considering the location (static and/or dynamic) of the attacker(s) A_l, by means of Backtracking approach with associated cost function:

$$c_f(n_i, n_j) = \frac{\frac{d_{mean}(n_j,E)}{\max_n^{nghbs(n_i)}(d_{mean}(n,E))} + \frac{d_{min}(n_j,E)}{\max_n^{nghbs(n_i)}(d_{min}(n,E))} + \frac{u(n_j)}{\max_n^{nghbs(n_i)}(u(n))}}{\frac{d_{mean}(n_j,A_l)}{\max_n^{nghbs(n_i)}(d_{mean}(n_j,A_l))}}$$

where function $nghbs(n_i)$ represents the neighbours of a particular node, function $d_{mean/max}(n, S)$ is the mean/max distance from node n to a set S of locations and $u(n)$ is the density of population in the surrounding of n. In conclusion, this function represents three important factors: 1) the proximity of a node to an exit/safe area, 2) the spatial availability of that node, and 3) the risk associated with the location of the attacker(s). Following these paths, a microsimulation approach is used to represent the movement and behaviour of people involved considering interactions between agents and repulsion forces between terrorists, people, scenario boundaries and obstacles through a Social Force model (Helbing and Molnár 1995). A physical shooting dynamics approach is followed (Abreu et al. 2019) to represent persons hit by gunfire, where the probability of being hit is estimated and the number of casualties $M_v(s_i)$ are calculated through a stochastic approach.

Results generated by these methodologies can be summarised as a set of geographic locations linked to counter-terrorism security-related information enabling the lower layers to increase their level of intelligence to enable more accurate modelling results.

Pedestrian Movement Layer

This layer uses threat assessment layer inputs $\{M_f, M_r, M_v\}$ and the pedestrian movement layer the status s for each node n_i in the associated graph G_p to update nodes to be evacuated, safe nodes and affected nodes that are impassable. Also, the occupant densities of the different nodes and edges of the network are updated through one of the following approaches depending on the capabilities of the smart city: 1) historical-based estimates of the expected occupancy, 2) real-time monitoring of occupancy through cameras, Wi-Fi location devices, access controls or similar, and 3) random assumptions of occupancy following expected distributions. This graph is considered as an active graph and it is used to produce a preliminary calculation of shortest paths using Dijkstra's algorithm. Its subsequent optimization is carried out considering nodes availability and through a weighted Multiple Criteria Decision Analysis (MCDA) for the assessment of conflicting nodes through its score function:

$$S(n_i) = w_f \cdot \sum_{n}^{nghbs(n_i)} (F(n)) + w_c \cdot C(n_i) + w_{fn} \cdot \sum_{n}^{nghbs(n_i)} (C(n)) + w_{dt} \cdot d_t(n_i)$$

where $\{w_f, w_c, w_{fn}, w_{dt}\}$ are the associated weights with the MCDA and although they are generally assigned the same weight for each variable, they can be modified in each iteration of the optimisation to obtain the required results. Functions $F(n)$ and $C(n)$ represent the available flow for a particular node considering all the related edges and congestions per node. After this, a set of candidate graphs $S_g = \{G_0, G_1, \ldots, G_k\}$ solving these conflicts is generated following an iterative process and another score function is applied to choose the optimal graph considering the total estimation of evacuation time $t_e(G_i)$ for each graph and the sum of individual node congestions with associated weights.

$$S(G_i) = w_t \cdot t_e(G_i) + w_{sc} \cdot \sum_{n}^{n \in N_i} (C(n))$$

Once the optimal graph has been found, it becomes the active graph again, which can be iteratively re-optimised when the model inputs change. This model provides evacuation routing, estimated egress times and mobility profiles, forecasting the number of people who will go to specific locations in a precise time period by determining and modelling the initial impact on the traffic network.

Traffic Layer

This layer provides a real time expected traffic evolution on different road sections according to date and time after a calibration of the network based on traffic historical data or data obtained through the traffic monitoring sensors deployed in the smart city. This calibration process starts from an uncalibrated network represented by the graph G_t which solves for the shortest paths considering availability constraints of the road sections and updating the W origin-destination matrix via path-based (Jayakrishnan et al. 1994) and bush-bashed B algorithms (Dial 2006). Accordingly, following an iterative process for origin-destination matrix adjustment based on gradient approach (Spiess 1990) with some adjustments for large traffic models (Kolovský et al. 2018), the model optimises the set of paths P and the matrix W based on real traffic data, paying attention to discrepancies between model and reality.

2.2 System Architecture

All these methodologies have been integrated together in a comprehensive DSS that, following the architecture presented in Fig. 3, assists security decisionmakers in the planning and response phases by leveraging some of the resources and devices already deployed in the smart cities. Examples of resources and devices include cameras, monitoring Wi-Fi devices, access control sensors, etc. These devices could help to estimate the number of people in specific locations or for example traffic monitoring systems make real-time simulation of unusual traffic flow more reliable.

The architecture follows a producer-consumer approach with a centralized distributed data stream platform (Apache Kafka) for the exchange of information between layers. Each layer in turn is implemented as an independent module that has a Graphical User Interface (GUI) for configuration purposes and Application Programming Interface (API) that provides on-demand service to the rest of the layers, except for fire and smoke simulations that, due to the computational cost, must be pre-simulated and stored locally for further use in specific scenarios, if needed.

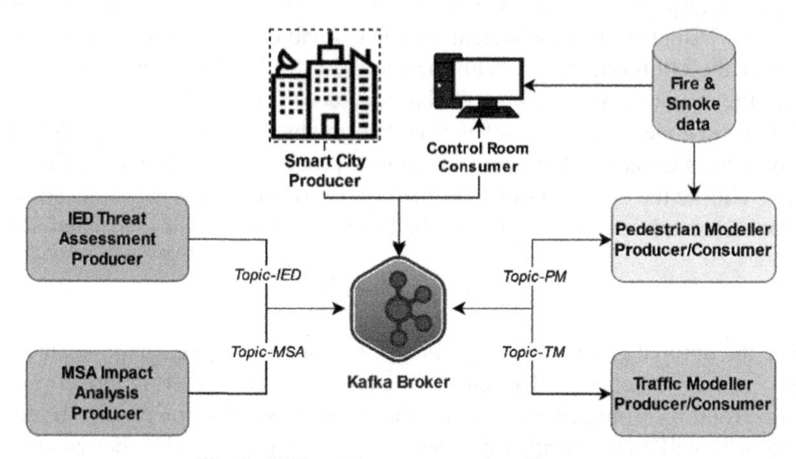

Fig. 3. DSS architecture overview diagram.

3 Case Study

After the development of the system, a comprehensive case study was performed based on data provided by Správa Informačnich Technologii Města Plzně, p.o. as partner of the S4AllCities project. The soft-target scenario was the Doosan Arena stadium in the city of Pilsen (Czech Republic). A detailed description of the stadium and its surrounding areas as well as the city of Pilsen itself was available, including information such as:

- Doosan Arena 3D model obtained via Lidar and RGB scanning using DJI Zenmuse L1 and DJI Zenmuse P1 cameras.
- Initial locations of a possible smoke bomb as well as its device-like specifications (Antari Z 3000 II fog machine).
- One-year traffic data providing a dataset of 250 million observations
- from 627 road built-in sensors, with a 90 s granularity in time, traffic model calibrated by the traffic data (Jedlicka et al. 2020).
- 2D map of the areas surrounding the stadium with expected attendance (11700 spectators + 3300 people), transit locations, security assets usually deployed, car parks and other minor details (Fig. 4).

Fig. 4. City of Pilsen case study schema with initial "smoke bomb" explosive device location. 1) Doosan Arena stadium (green), 2) Surrounding areas (red) and 3) Parking spaces (blue). (Color figure online)

The next step after the simulation, training and calibration of the models with the provided data, was the definition of four use cases to validate all the capabilities of the system listed in Table 1.

Table. 1. Use cases considered in the case study of the city of Pilsen.

#	Scenario	Feature	Details
1	Stadium interior	Smoke bomb	"Smoke bomb" type device triggering evacuation of the stadium
2	Adjacent area in front of the stadium	IED + MSA	Risks of IED and MSA attacks, due to a possible combined attack by two perpetrators. 3300 people uniform distributed in transitable areas are considered
3	Stadium interior and adjacent areas	Pedestrian Evacuation	Evacuation of the stadium and neighbouring areas to the car parks considering risks
4	Pilsen city	Vehicular Evacuation	Simulation of the impact of unusual vehicle flows on the city's traffic network

For each of these use cases, the DSS provided results, starting by retrieving the output of the smoke propagation FDS analysis (virtual smoke machine tuned on Antari Z 3000 II fog machine, North-West wind direction) followed by the simulation of threats and likely impact of attacks (Fig. 5). It is important to note that the information provided to the system operator includes the visualization of both the evolution of the different incidents and the data associated with the artificial scenario measurements (IED probability, FED, visibility and casualties). On the other hand, a simulation of pedestrian evacuation and traffic network unusual behaviour impact is also generated as shown in Fig. 5, where the operator is informed about the predicted evacuation times, routes, recognition of traffic/pedestrian congestion and the possibility of dynamically recalculating these results according to the risks deemed appropriate by, for example, cutting roads or blocking pedestrian evacuation nodes.

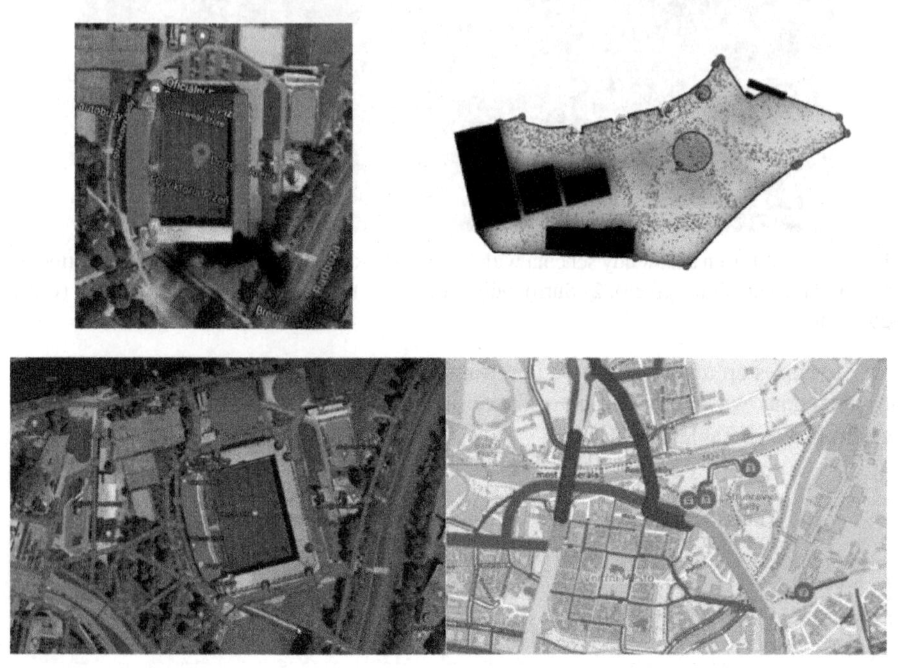

Fig. 5. DSS Graphical user interfaces. Top-Left: Smoke propagation from FDS simulation, Top-Right: IED and MSA threat assessment and impact analysis, Bot-Left: Pedestrian evacuation management and Bot-Right: Traffic network unusual status simulation.

The simulation of our use cases showed that for the case study of the Doosan Arena stadium and the city of Pilsen the most likely dispersion of the smoke from the device would be initially south-east without affecting additional exits from the stadium (after about 5 min, smoke whirls start sticking to the outer South tribune, Fig. 5, Top-Left). In the case of explosive device threat levels and potential casualties, this would result in the blocking of the front exits of the stadium, as a two-shooter attack in this area would result in an estimated of 139 casualties (dead + wounded) considering two minutes intervention time, in addition to considering the potential locations of explosive devices, as shown in the lighter areas of the heat map in Fig. 5. As shown in Fig. 5, the pedestrian evacuation would be directed mainly to the nearest car parks by using the remaining exits available in the stadium. The effect of these pedestrian evacuation profiles would increase in approx. 700 vehicles per the first hour in the northern traffic section and 900 vehicles per the first hour in the southern section, leaving a high density of vehicles in both directions, as shown in Fig. 5.

4 Conclusions and Discussion

The emerging technologies implemented in smart cities as well as new tools and methodologies for computer simulation applied to threat analysis and citizen security are a breakthrough in the fight against terrorism. In this paper we present the methodological design based on three layers (threat, pedestrian and traffic layers) and implementation of a DSS that allows private operators, law enforcement agencies and local authorities to efficiently protect city soft-targets. Within this system, support is provided for both threat analysis and emergency management of pedestrian evacuation and its impact on the metropolitan traffic network. In addition, this paper presents a case study based on real data in the city of Pilsen where the correct functioning of the different layers that make up the system was evaluated and the benefits and characteristics of the system were presented in a more illustrative way, among which is the study of the analysis of the main terrorist threats, the complete management of an evacuation and the monitoring for decision-making of the state of the traffic network.

It must also be considered that this system has certain limitations that can be corrected in later developments. The first of these is that it does not cover all types of threats within the city. Actual reports suggest that future trends (EUROPOL 2021) in terrorism will evolve to simpler and less expensive (knife attacks) or combined attacks (cascading attacks or sabotage of critical infrastructures). Conversely, there would be an exploratory branch of the possible direct interaction of terrorist threats with the traffic network, being able to carry out developments in the field of anti-ramming measures in urban planning. From our point of view, all these limitations are not an obstacle but rather open up future branches of research and lead to the development of increasingly complete security and safety systems.

Acknowledgements. The project (S4AllCities) has received funding from the European Union's H2020 research and innovation programme under grant agreement No. 883522.

References

Abreu, O., Cuesta, A., Balboa, A., Alvear, D.: On the use of stochastic simulations to explore the impact of human parameters on mass public shooting attacks. Saf. Sci. **120**, 941–949 (2019)

Anees, V., Kumar, G.: Direction estimation of crowd flow in surveillance videos. In: 2017 IEEE Region 10 Symposium (TENSYMP), pp. 1–5 (2017)

Araujo, A., Cacho, N., Thome, A., Medeiros, A., Borges, J.: A predictive policing application to support patrol planning in smart cities. In: 2017 International Smart Cities Conference (ISC2), pp. 1–6 (2017)

Balla, P.B., Jadhao, K.: IoT based facial recognition security system. In: 2018 International Conference on Smart City and Emerging Technology (ICSCET), pp. 1–4 (2018)

Bellini, P., Cenni, D., Nesi, P., Paoli, I.: Wi-Fi based city users' behaviour analysis for smart city. J. Vis. Lang. Comput. **42**, 31–45 (2017)

Bonatsos, A., Middleton, L., Melas, P., Sabeur, Z.: Crime open data aggregation and management for the design of safer spaces in urban environments. In: International Symposium on Environmental Software Systems, pp. 311–320 (2013)

Boukerche, A., Siddiqui, A., Mammeri, A.: Automated vehicle detection and classification: models, methods, and techniques. ACM Comput. Surv. (CSUR) **50**(5), 1–39 (2017)
Brust, M.R., Danoy, G., Bouvry, P., Gashi, D., Pathak, H., Gonçalves, M.P.: Defending against intrusion of malicious UAVs with networked Uav defense swarms. In: 2017 IEEE 42nd Conference on Local Computer Networks Workshops (LCN Workshops), pp. 103–111 (2017)
Chackravarthy, S., Schmitt, S., Yang, L.: Intelligent crime anomaly detection in smart cities using deep learning. In: 2018 IEEE 4th International Conference on Collaboration and Internet Computing (CIC), p. 399–404 (2018)
Cuesta, A., Abreu, O., Balboa, A., Alvear, D.: A new approach to protect soft-targets from terrorist attacks. Saf. Sci. **120**, 877–885 (2019)
Dbouk, M., Mcheick, H., Sbeity, I.: CityPro; an integrated city-protection collaborative platform. Procedia Comput. Sci. **37**, 72–79 (2014)
Dial, R.B.: A path-based user-equilibrium traffic assignment algorithm that obviates path storage and enumeration. Transp. Res. Part B Methodol. **40**(10), 917–936 (2006)
EUROPOL: European union terrorism situation and trend report (2021)
Fernández, J., et al.: An intelligent surveillance platform for large metropolitan areas with dense sensor deployment. Sensors **13**(6), 7414–7442 (2013)
Global Terrorism Database™ (GTD): Obtenido de (2021). https://www.start.umd.edu/gtd/
Hartama, D., et al.: A research framework of disaster traffic management to Smart City. In: 2017 Second International Conference on Informatics and Computing (ICIC), pp. 1–5 (2017)
Helbing, D., Molnár, P.: Social force model for pedestrian dynamics. Phys. Rev. E **51**(5), 4282 (1995)
Jayakrishnan, R., Tsai, W.T., Prashker, J.N., Rajadhyaksha, S.: A faster path-based algorithm for traffic assignment (1994)
Jedlicka, K., et al.: Traffic modelling for the smart city of Pilsen (2020)
Kolovský, F., Ježek, J., Kolingerová, I.: The origin-destination matrix estimation for large transportation models in an uncongested network. In: International Conference on Mathematical Applications, pp. 17–22 (2018)
Laufs, J., Borrion, H., Bradford, B.: Security and the smart city: a systematic review. Sustain. Cities Soc. **55**, 102023 (2020)
Martin, R.H.: Soft targets are easy terror targets: increased frequency of attacks, practical preparation, and prevention. Forensic Res. Criminol. Int. J. **3**(2), 1–7 (2016)
McGrattan, K., et al.: Fire Dynamics Simulator User's Guide. National Institute of Standards and Technology (2017)
Noor, M., Nawawi, W., Ghazali, A.: Supporting decision making in situational crime prevention using fuzzy association rule. In: 2013 International Conference on Computer, Control, Informatics and Its Applications (IC3INA), pp. 225–229 (2013)
Spiess, H.: A gradient approach for the OD matrix adjustment problem. a $\in \hat{A}$, 1 (1990)
Truntsevsky, Y.V., Lukiny, I., Sumachev, A., Kopytova, A.: A smart city is a safe city: the current status of street crime and its victim prevention using a digital application. In: MATEC Web of Conferences, vol. 170, p. 01067 (2018)
Tuman, J.S.: Communicating Terror: The Rhetorical Dimensions of Terrorism. Sage Publications (2009)
Turban, E.: Decision Support and Expert Systems Management Support Systems. Prentice-Hall, Inc., Hoboken (1995)
Zhang, W., et al.: Agent-based modeling of a stadium evacuation in a smart city. In: 2018 Winter Simulation Conference (WSC), pp. 2803–2814 (2018)
Zhou, W., Saha, D., Rangarajan, S.: A system architecture to aggregate video surveillance data in Smart Cities. In: 2015 IEEE Global Communications Conference (GLOBECOM), pp. 1–7 (2015)
Zingoni, A., Diani, M., Corsini, G.: A flexible algorithm for detecting challenging moving objects in real-time within IR video sequences. Remote Sens. **9**(11), 1128 (2017)

Embedded Intelligence for Safety and Security Machine Vision Applications

Panagiotis Lioupis[1], Aris Dadoukis[1], Evangelos Maltezos[1], Lazaros Karagiannidis[1], Angelos Amditis[1], Maite Gonzalez[2], Jon Martin[2], David Cantero[2], and Mikel Larrañaga[2(✉)]

[1] Institute of Communication and Computer Systems (ICCS), 15773 Zografou, Greece
{panagiotis.lioupis,aristeidis.dadoukis,evangelos.maltezos, lkaragiannidis,a.amditis}@iccs.gr

[2] Fundación Tekniker, Iñaki Goenaga 5, 20600 Eibar, Spain
{maite.gonzalez,jon.martin,david.cantero, mikel.larranaga}@tekniker.es

Abstract. Artificial intelligence (AI) has experienced a recent increase in use across a wide variety of domains, such as image processing for security applications. Deep learning, a subset of AI, is particularly useful for those image processing applications. Deep learning methods can achieve state-of-the-art results on computer vision for image classification, object detection, and face recognition applications. This allows to automate video surveillance reducing human intervention.

At the same time, although deep learning is a very intensive task in terms of computing resources, hardware and software improvements have emerged, allowing embedded systems to implement sophisticated machine learning algorithms at the edge. Hardware manufacturers have developed powerful co-processors specifically designed to execute deep learning algorithms. But also, new lightweight open-source middleware for constrained resources devices such as EdgeX foundry have emerged to facilitate the collection and processing of data at sensor level, with communication capabilities to cloud enterprise applications.

The aim of this work is to show and describe the development of Smart Camera Systems within S4AllCities H2020 project, following the edge approach.

Keywords: Horizon2020 · Edge · EdgeX foundry · Machine vision · Artificial intelligence · Deep learning

1 Introduction

The aim of S4AllCities EU-funded project is to make cities' infrastructures, services, ICT systems and Internet of Things more resilient while promoting intelligence and information sharing amongst security stakeholders, to foster good safety and security practices in European cities. The project will integrate advanced technological and organizational solutions, being edge-oriented resource-constrained image-based processing systems for surveillance one of them.

P. L. Mazzeo et al. (Eds.): ICIAP 2022 Workshops, LNCS 13374, pp. 37–46, 2022.
https://doi.org/10.1007/978-3-031-13324-4_4

Traditional video surveillance systems demand human intervention to some extent. However, as the number of IP or other types of cameras increases explosively, a fully automatic video recognition framework becomes essential, replacing the manual monitoring. The video data captured by the camera are transmitted to the cloud server to do the entire recognition processes, which may hamper real-time video recognition due to transmission delays through the communication channel.

In this context, recent trends in IoT applications adopt edge computing that appears to decrease latency and computational processing. The edge computing technology allows computation to be performed at the network edge so that computing happens near data sources or directly in the real-world application as an end device. This is possible due to advances in the manufacturing of new processors [1]. Such devices request services and information from the cloud as well as perform several real-time computing tasks (e.g., storage, caching, filtering, processing, etc.) of the data sent to and from the cloud. Hence, edge computing is able to fully contribute to situational awareness applications [2, 3] that continuously generate enormous amounts of data of several types providing also a homogeneous approach for data processing and generation of associated alerts or events or raw information. In summary, edge computing can be used for real-time smart city environments under the public safety aspect enabling: i) context-awareness, ii) geo-distributed capabilities, iii) low latency, and iv) migration of computing resources from the remote cloud to the network edge.

In this background, [4], provided in their study a modular architecture with deep neural networks as a solution for real-time video analytics in an edge-computing environment. In their modular architecture two networks of Front-CNN (Convolutional Neural Network) and Back-CNN were exploited. Experimental results on the public datasets of UCF-Crime [5] and UR-Fall Detection [6] highlighted the potential of their approach. In [7] a video streaming optimization method in the IIoT environment was proposed under the edge computing concept. In the same framework, the author of [8] designed an edge enhanced deep learning system for large-scale video stream analytics system. In their proposed methodology, they performed an initial processing of the data close to the data source at edge and fog nodes, resulting in significant reduction in the data that is transferred and stored in the cloud. The results on the adopted object recognition scenario showed high efficiency gain in the throughput of the system by employing a combination of edge, in-transit and cloud resources when compared to a cloud-only approach. The authors of [9] focused on leveraging edge intelligence for video analytics in smart city applications. Their approach encompasses architecture, methods, and algorithms for: i) dividing the burdensome processing of large-scale video streams into various machine learning tasks; and ii) deploying these tasks as a workflow of data processing in edge devices equipped with hardware accelerators for neural networks. In [8], the authors investigated an architectural approach for supporting scalable real-time video stream processing using edge and in-transit computing. Concerning the privacy aspect, the authors of [11] proposed how to consider to a privacy-oriented framework when video feeds are exploited for surveillance applications.

The aim of this work is to provide an overview of two different edge computing machine vision systems developed withing S4AllCities EU funded project [12].

2 Edge Computing Machine Vision Systems in S4AllCities Project

Within S4AllCities project, an edge computing platform namely, Distribute Edge Computing Internet of Things (DECIoT) is being developed. The DECIoT is a scalable, secure, flexible, fully controlled, potentially interoperable, and modular open-source framework that ensures information sharing with other platforms or systems [13]. Through the DECIoT, computation, data storage, and information sharing are performed together, directly in the edge device, in a real-time manner.

To complement the DECIoT, two different edge computing based machine vision systems are being developed:

- Dedicated ad-hoc hardware platform based on the I.MX8M PLUS [14] integrated with the DECIoT which focuses on machine learning and vision algorithms. The focus was on the development and manufacture of the device more than in the application itself. Anyway, to test the suitability of the platform, a people detection algorithm was embedded (Fig. 1).

Fig. 1. S4AllCities Hardware platform based on the I.MX8M PLUS.

- Conducting additional experiments in terms of real-world applications associated with the video analytics system for person detection combined with edge computing, namely Video Analytics Edge Computing (VAEC) system, that documented in [17]. The VAEC system is also integrated with the DECIoT in order to provide enhanced situation awareness for person detection through a video streaming feed on an embedded edge device with GPU processing and a lightweight object detection deep learning scheme.

3 Integration with the Distributed Edge Computing Platform (DECIoT)

In general, the DECIoT platform is able to address among others the problem of gathering, filtering and aggregating data, interacts with the IoT devices, provides security and system management, provides alerts and notifications, executes commands, stores data temporarily for local persistence, transforms/process data and in the end exports the data in formats and structures that meet the needs of other platforms. This whole process is being done by using open source microservices that are state-of-the-art in the area of distributed edge IoT solutions.

The DECIoT is based on the open, highly flexible and scalable edge computing framework namely EdgeX foundry [15]. The DECIoT platform consist of multiple layers and each layer contains multiple microservices. The communications between the microservice within the same or different layers can be done either directly with the use of REST APIs or with the use of a message bus that follows a pub/sub mechanism. Both of them are being exploited in this study. DECIoT consists of a collection of reference implementation services and SDK tools. The micro services and SDKs are written in Go [15] or C programming languages. A detailed documentation and implementation of the DECIoT has been provided in [17]. In the following we present the different layers of DECIoT adopted in this study (see also Fig. 2).

- The Device Service Layer: The MQTT Device Service was used to receive information from the object detection process. Between the object detection process and the MQTT Device Service, there is a MQTT broker (Mosquito) [18].
- The Core Services Layer: For storing data, as well as commanding and registering devices. This microservices are implemented with the use of Consul, Redis, and adapters developed in Go for integration with all other microservices.
- The Support Services Layer: Here, the relevant microservices were not exploited as no logging, scheduling, and data filtering was needed.
- The Application Services Layer: A new Application Service has been implemented to send data to a smart city's platform using Go language [16]. In the Smart Spaces Safety and Security for All Cities project [19] the Apache Kafka [20] was considered that is the middleware of a smart city platform.

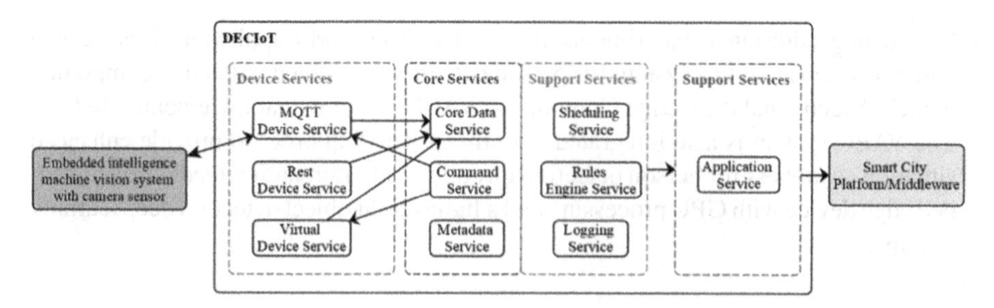

Fig. 2. DECIoT architecture.

4 Hardware Platform for Machine Vision Based on the I.MX8M PLUS

The embedded platform for machine vision based on the i.MX8M integrates a Neural Co-Processing Unit (NPU) which is able to process 2.3 Tera-operations per second (TOPS), enabling the NPU to run deep learning model inference in real time. In order to develop and deploy the Deep Learning application in the embedded processors, the eIQ Machine Learning Framework provided by NXP was used. The framework integrates a Tensorflow Lite library with a python (also C++) API that allows the integration of deep learning models into embedded application. This API is also responsible for converting and deploying the highly parallel tensor calculus into NPU to optimize both performance and processing time.

In order to demonstrate the embedded platform capabilities, an object detection application was implemented, specifically oriented to people detection. This application uses state-of-the-art SSD object detection model with mobilenet feature extractor backbone, whose architecture is shown in Fig. 3.

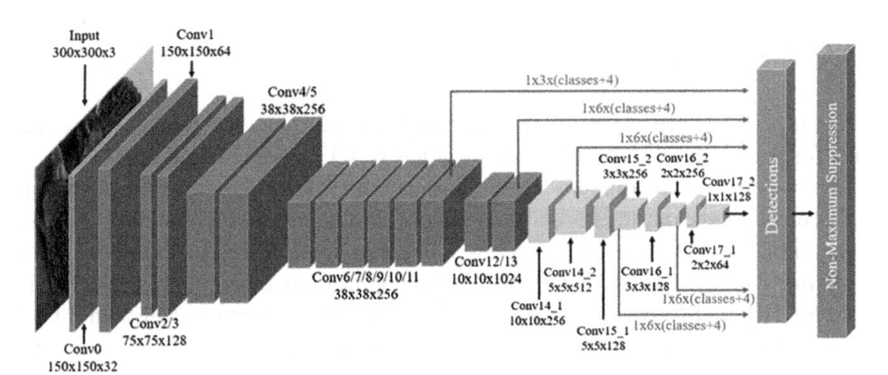

Fig. 3. Mobilenet SSD model architecture.

The application processes video frames searching the presence of the people in the image using a second CMOS camera that provides a second video stream. The performance test gives a 30 ms execution time average resulting in approximately 30 fps video processing. This video processing is modified adding the detection boxes and scores in real time and displayed (only for demonstration purposes) in the evaluation kit display as shown in Fig. 4.

The application computes the number of people in the image, a square with object position in the image and a reliability score for each of the object in the image. This information is also transmitted via MQTT to the DECIoT. However, it is not part of this work to provide evaluation setup for this model.

Fig. 4. People detected.

5 VAEC System

In [16], the performance of VAEC system was evaluated through several real-time experiments for person detection in the following terms: (i) in several light conditions, (ii) using several types of camera sensors, and (iii) in several viewing perspectives. However, the aforementioned experiments considered only in outdoor cases. Here, we expand the evaluation of VAEC system in indoor cases utilizing some representative videos from UCF-Crime dataset associated with real-world applications (arson and burglary/space violation). The VAEC system adopts a lightweight deep learning model with a CNN architecture for object detection, that is the pre-trained YOLOv5s [21], with high inference speed. In the literature, YOLOv5 has a great potential for object detection tasks in several applications with various challenges such as complexity of the scene, shadows, light conditions, viewing perspective of the objects, etc. [22–24]. Table 1 depicts representative consecutive video frames for five selected videos (V1 to V5) of the UCF-Crime dataset associated with the object detection results (colored bounding boxes and the relevant detection probability percentages superimposed to the video frames). For the quantitative assessment of the object detection process, four objective criteria were adopted according to the ISPRS guidelines [24], namely, completeness (CM), correctness (CR), quality (Q), and F1 score measures per object (person) based on the True Positive (TP), False Positive (FP), and False Negative (FN) entries,), given as:

$$CM = \frac{\|TP\|}{\|TP\| + \|FN\|} \tag{1}$$

and

$$CR = \frac{\|TP\|}{\|TP\| + \|FP\|} \tag{2}$$

Table 1. Results of person detection per video of the UCF-Crime dataset [5]. Quantitative person detection results per video for representative videos

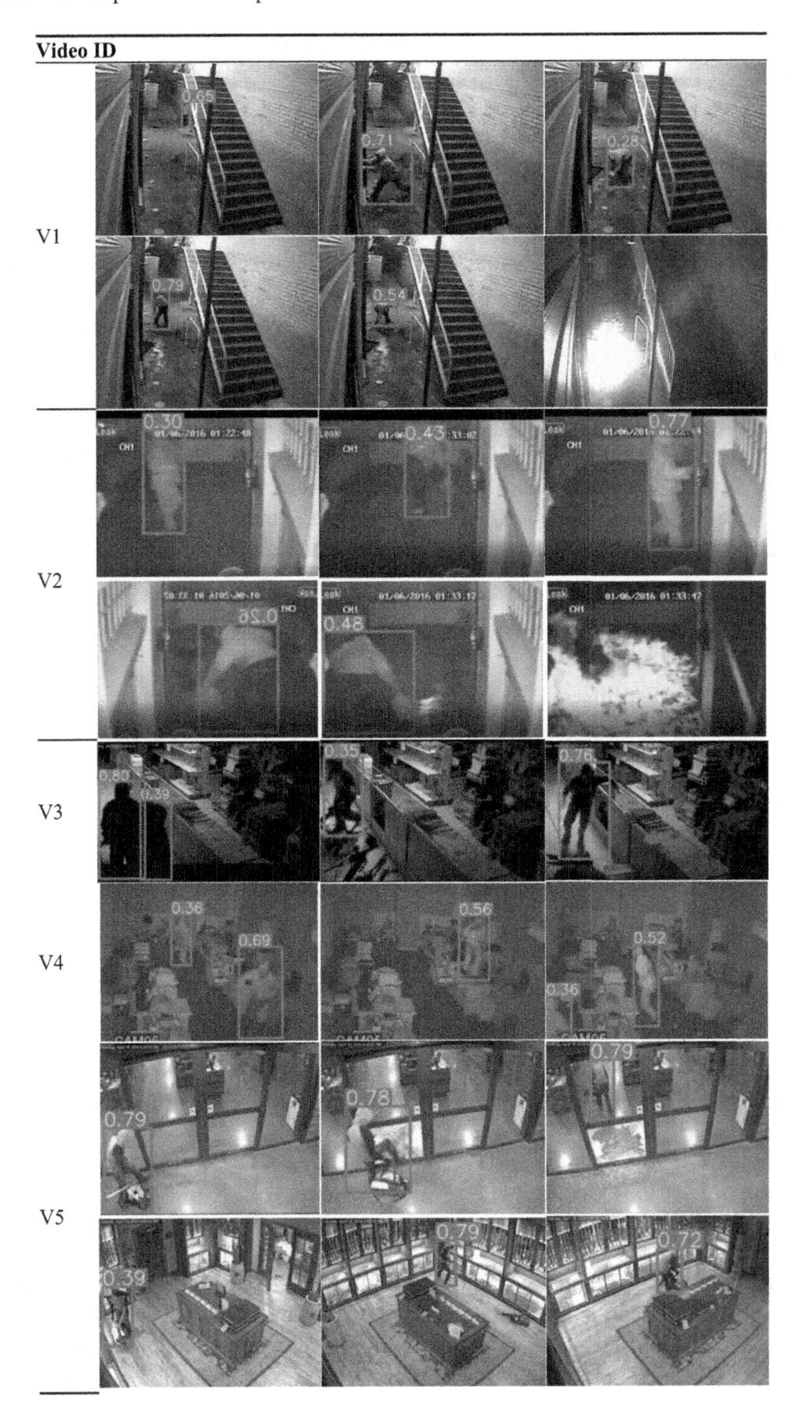

and

$$Q = \frac{\|TP\|}{\|TP\| + \|FP\| + \|FN\|} \quad (3)$$

and

$$F1 = 2 \times \frac{C_R \times C_M}{C_R + C_M} \quad (4)$$

where TP, FP, and FN denote true positives, false positives, and false negatives, respectively. The TP entries are the persons that exist in the scene and thus, were correctly detected. The FP entries are the persons that do not exist in scene and thus, were incorrectly detected. The FN entries are the persons that exist in the scene but were not detected.

The quantitative assessment results for the person detection process for videos V1 to V5 are provided in Table 2. The achieved results of the YOLOv5s (average values of CM = 81.2%, CR = 90.2%, Q = 74.9%, and F1 = 85.3%) are considered to be satisfactory, proving its suitability and efficiency for such applications with several challenges. Also, the achieved results are quite similar with those of [17] indicating a homogeneous and stable performance of the YOLOv5s both in indoor and outdoor environments.

Table 2. Quantitative person detection results per video for representative videos frames.

Experiment ID	Object detection process			F1 score
	C_M (%)	C_R (%)	Q (%)	
VID1	83.3	100.0	83.3	90.9
VID2	80.0	94.1	76.2	86.5
VID3	81.8	81.8	69.2	81.8
VID4	75.0	75.0	60.0	75.0
VID5	85.7	100.0	85.7	92.3
Average	81.2	90.2	74.9	85.3

The achieved results of the YOLOv5s adopted from the VAEC system, are considered to be satisfactory, proving its suitability and efficiency for real-word applications with several challenges.

6 Conclusions

The current approach demonstrates the suitability of edge computing systems for machine vision applications for safety and security. This approach was demonstrated through two different edge computing systems developed within the framework of S4AllCities EU funded project. The evaluation of the object detection model was only provided by the VAEC system through ISPRS guidelines, as it was not pretended to perform this evaluation for the I.MX8M PLUS hardware platform at this stage. However, both systems integrate the open-source framework, EdgeX foundry, a framework that allows the interoperability between devices and applications in the edge.

Acknowledgments. This work is a part of the S4AllCities project. This project has received funding from the European Union's Horizon 2020 research and innovation programme under grant agreement No. 883522. Content reflects only the authors' view and the Research Executive Agency (REA)/European Commission is not responsible for any use that may be made of the information it contains.

References

1. NXP. https://www.nxp.com/applications/enabling-technologies/edge-computing:EDGE-COMPUTING. Accessed 9 Mar 2022
2. Patrikar, D.R., Parate, M.R.: Anomaly detection using edge computing in video surveillance system: review. arXiv 2107, arXiv:2107.02778. Accessed 13 Jan 2022
3. Geraldes, R., et al.: UAV-based situational awareness system using deep learning. IEEE Access **7**, 122583–122594 (2019). https://doi.org/10.1109/ACCESS.2019.2938249
4. Kim, J.-H., Kim, N., Won, C.S.: Deep edge computing for videos. IEEE Access **9**, 123348–123357 (2021). https://doi.org/10.1109/ACCESS.2021.3109904
5. Sultani, W., Chen, C., Shah, M.: Real-world anomaly detection in surveillance videos. In: Proceedings of the IEEE/CVF Conference on Computer Vision and Pattern Recognition, pp. 6479–6488, June 2018
6. Kwolek, B., Kepski, M.: Human fall detection on embedded platform using depth maps and wireless accelerometer. Comput. Methods Programs Biomed. **117**(3), 489–501 (2014)
7. Dou, W., Zhao, X., Yin, X., Wang, H., Luo, Y., Qi, L.: Edge computing-enabled deep learning for real-time video optimization in IIoT. IEEE Trans. Ind. Inf. **17**(4), 2842–2851 (2021). https://doi.org/10.1109/TII.2020.3020386
8. Ali, M., et al.: Edge enhanced deep learning system for large-scale video stream analytics. In: 2018 IEEE 2nd International Conference on Fog and Edge Computing (ICFEC), pp. 1–10, May 2018. https://doi.org/10.1109/CFEC.2018.8358733
9. Rocha Neto, A., Silva, T.P., Batista, T., Delicato, F.C., Pires, P.F., Lopes, F.: Leveraging edge intelligence for video analytics in smart city applications. Information **12**(1), Article no. 1 (2021). https://doi.org/10.3390/info12010014
10. Ali, M., Anjum, A., Rana, O., Zamani, A.R., Balouek-Thomert, D., Parashar, M.: RES: real-time video stream analytics using edge enhanced clouds. IEEE Trans. Cloud Comput., 1 (2020). https://doi.org/10.1109/TCC.2020.2991748
11. Chen, A.T.-Y., Biglari-Abhari, M., Wang, K.I-K.: Trusting the computer in computer vision: a privacy-affirming framework. In: Proceedings of the 2017 IEEE Conference on Computer Vision and Pattern Recognition Workshops (CVPRW), Honolulu, HI, USA, 21–26 July 2017, pp. 1360–1367 (2017). https://doi.org/10.1109/CVPRW.2017.178
12. S4ALLCities project, Smart Spaces Safety and Security, Greece. www.s4allcities.eu. Accessed 19 Apr 2022
13. Maltezos, E., et al.: Public safety in smart cities under the edge computing concept. In: 2021 IEEE International Mediterranean Conference on Communications and Networking (MeditCom), pp. 88–93, September 2021. https://doi.org/10.1109/MeditCom49071.2021.9647550

14. NXP. https://www.nxp.com/products/processors-and-microcontrollers/arm-processors/i-mx-applications-processors/i-mx-8-processors/i-mx-8m-plus-arm-cortex-a53-machine-learning-vision-multimedia-and-industrial-iot:IMX8MPLUS. Accessed 9 Mar 2022
15. T. L. Foundation: Welcome. https://www.edgexfoundry.org. Accessed 13 Jan 2022
16. The Go Programming Language. https://go.dev/. Accessed 10 Feb 2022
17. Maltezos, E., et al.: A video analytics system for person detection combined with edge computing. Computation **10**, 35 (2022). https://doi.org/10.3390/computation10030035
18. Eclipse Mosquitto: Eclipse Mosquitto, 08 January 2018. https://mosquitto.org/. Accessed 26 Jan 2022
19. S4AllCities: Smart Spaces Safety and Security. https://www.s4allcities.eu. Accessed 03 Mar 2022
20. Apache Kafka: Apache Kafka. https://kafka.apache.org/. Accessed 03 Mar 2022
21. Github: Ultralytics. ultralytics/yolov5 (2022). https://github.com/ultralytics/yolov5. Accessed 13 Jan 2022
22. Zhu, X., Lyu, S., Wang, X., Zhao, Q.: TPH-YOLOv5: improved YOLOv5 based on transformer prediction head for object detection on drone-captured scenarios. In: Proceedings of the 2021 IEEE/CVF International Conference on Computer Vision Workshops (ICCVW), Montreal, QC, Canada, 11–17 October 2021, pp. 2778–2788 (2021). https://doi.org/10.1109/ICCVW54120.2021.00312
23. Nepal, U., Eslamiat, H.: Comparing YOLOv3, YOLOv4 and YOLOv5 for autonomous landing spot detection in faulty UAVs. Sensors **22**, 464 (2022). https://doi.org/10.3390/s22020464
24. Phadtare, M., Choudhari, V., Pedram, R., Vartak, S.: Comparison between YOLO and SSD mobile net for object detection in a surveillance drone. Int. J. Sci. Res. Eng. Man. **5** (2021). https://doi.org/10.13140/RG.2.2.34029.51688
25. Rottensteiner, F., Sohn, G., Gerke, M., Wegner, J.D.: ISPRS test project on urban classification and 3D building reconstruction. In: ISPRS-Commission III-Photogrammetric Computer Vision and Image Analysis, Working Group III/4–3D Scene Analysis (2013)

Supporting Energy Digital Twins with Cloud Data Spaces: An Architectural Proposal

Chiara Rucco(✉), Antonella Longo(✉), and Marco Zappatore(✉)

Department of Innovation Engineering, University of Salento, via Monteroni sn, 73100 Lecce, Italy
{chiara.rucco,antonella.longo,marcosalvatore.zappatore}@unisalento.it

Abstract. The concept of Digital Twins offers the possibility of moving work from a physical environment to a virtual or digital environment and the ability to predict asset conditions in the future, or when it is physically undesirable, by exploiting the digital model. This in turn leads to significant reductions in the resources required to design, produce and maintain assets and resources. In the field of energy management, DTs are also starting to be considered as valuable analysis tools, as a digital twin facilitates real-time synchronisation between a real-world model (physical model) and its virtual copy for improved energy monitoring, prediction, and efficiency enhancement; thus, it can significantly reduce the overall energy consumption. A typical problem of DTs is the management of the data to be fed from the physical twin to the DT (and possibly the other way around), as one has to decide whether to store them within the DT or not, and one also has to decide whether to use different (depending on the data sources) or unified data governance models. To this end, an energy data space is proposed to allow the management of the necessary data in a way that is more functional to the DT concept.

Keywords: Energy · Datalake · Digital Twin · Open Data · Fairness

1 Introduction

Global energy requirements are continuously increasing. Conventional methods of producing more energy to meet this growth pose a great threat to the environment. CO_2 emissions and other bi-products of energy production have direct consequences on everyday life. Therefore, we need to understand and improve the energy efficiency at both producer and consumer sides. ICT-enabled smart energy grids and sensors are being installed globally to measure energy consumption and limit the environmental impact: these smart objects produce large volumes of data, generated by different devices and in different formats, so that they embody the concept of Gartner's 'Big Data 3Vs' [15] - volume, velocity and variety. For the purpose of knowledge discovery, this data needs to be collected and

P. L. Mazzeo et al. (Eds.): ICIAP 2022 Workshops, LNCS 13374, pp. 47–58, 2022.
https://doi.org/10.1007/978-3-031-13324-4_5

analyzed, and the extracted insights from the analysis need to be visualized for easy and effective understanding. More specifically, the integration of AI models (of physical objects) and Big Data analytics for IoT data processing [4] is driving one of the most recent and probably one of the most important advances in the technology, namely Digital Twin (DT). DT models are gaining more and more interest because of their potential and impact in application areas such as manufacturing aerospace, healthcare and medicine. DTs can be defined as (physical and/or virtual) machines or computer models that simulate, emulate, mirror or "twin" the life of a physical entity, which may be an object, a process, a human or a human-related property. Each DT is connected to its physical twin through a unique key, which identifies the physical twin and thus establishes a bijective relationship between the DT and its twin. To face these challenges, a highly scalable and flexible data analysis platform for automating the whole process is required. A first model that can meet these requirements is an architecture that draws elements from classic data warehouse systems on the one hand and from pure data lake systems on the other hand. This model, defined as Data Lakehouse [2], together with other paradigms like polystore databases, can be implemented and tested with real life data from smart energy devices in order to contribute to the realization of a society that follows the innovative Circular Economy paradigm, a system where resource input and waste, emission, and energy leakage are minimized by slowing, closing, and narrowing material and energy loops [9]. The new schema will contain heterogeneous data sources and will be processed in order to be compliant to FAIR (Findable, Interoperable, Accessible, Reusable) principles: a well-documented and highly re-usable data set enables the ultimate aim to trusted, effective and sustained reuse of research resources [18]. This paper aims at identifying featuring architectural aspects and modelling challenges for an Energy Data Space to be adopted nationwide in Italy. The work is structured as in the following: the second section presents an overview of the background and the state of art, with reference to the rise of the new 'Energy of Things' characterised by heterogeneous large data sets, and to three cutting-edge projects on this topic. The main goal of the this architectural proposal is described in the third section. The last part is devoted to discussing the impact and results of the solution's development.

2 Background and State of Art

Nowadays, the concept of "Data Lake" is popular for accumulating data from heterogeneous sources. Data lakes are used for storing large scale raw data as a single big data repository, providing ingestion, exploration, and monitoring functionality [17]. Data lakes, in contrast to data warehouses, are databases containing data from different sources in structured, unstructured and semi-structured formats, along with capabilities of handling batch and real-time streams. Moreover, data lakes exhibit different implementation forms (e.g., on premises, cloud or multi-cloud, and hybrid) [22]. Currently, data lakes have been exploited in several application domains, ranging from digital humanities [7] to power grid management [16]. In order to get a full insight on the scenario analyzed, a general overview has been built: at the beginning, the Circular Energy paradigm is discussed, with a view on its heterogeneous energy-related data and on data key principles for achieving *FAIRness* [8]. Then, as a starting point for this research, some existent projects for the creation of a National Energy data repository are explored: the first is an initiative for creating a digital twin of Earth, the second is a Danish proposal for renewable energies, the third is a novel US initiative to make energy data usable and discoverable by researchers.

2.1 Energy of Things

According to the United Nations Sustainable Development Goals agenda [19], energy efficiency is one of the key factors for sustainable development: "ensuring access to affordable, reliable, sustainable and modern energy for all by 2030 will open a new world of opportunities for billions of people through new economic opportunities and jobs"[1]. Furthermore, energy efficiency brings long-term economic benefits by reducing the cost of fuel import/supply, energy production and energy sector emissions. Effective analysis of real-time data in the energy supply chain plays a key role in improving energy efficiency and more optimal energy management [13]. Modern technologies, such as the Internet of Things (IoT), offer a wide range of applications in the energy sector, i.e. in the areas of energy supply, transmission, distribution and demand. With the new surging of portable smart devices, consistently equipped with sensors, supported by more and more performing cloud computing solutions, and densely used for mobile social networking, *human as sensor* has become a promising sensing paradigm. For this reason, the term ENERNET (Energy of Things) was recently introduced by Steve Collier in an IEEE webinar on the future of energy: it is defined as a convergence and a marriage between Smart Grids and IoT [6]. Emerging ENERNET opens up other possibilities in order to have an affordable, reliable, secure and sustainable supply of electrical power and energy. The novel sensing technologies promote the data source into a new information space paradigm, which seamlessly integrates cyber-space (CS), physical space (PS) and social space (SS), namely Cyber-Physical-Social Systems (CPSS) [21]. CPSS has a crucial role in improving energy efficiency, increasing the share of renewable energy and reducing the environmental impact of energy use; this can be compliant to a context of Circular Economy, a concept that has been framed by the Ellen MacArthur Foundation as *an industrial economy that is restorative or regenerative by intention and design* [9]: it represents a system in which resource input and waste, emission, and energy leakage are minimized by slowing, closing, and narrowing material and energy loops. This can be achieved through long-lasting design, maintenance, repair, reuse, re-manufacturing, refurbishing, and recycling. The amount of available data for energy analysis is growing rapidly due to a large number of data sources, such as smart cities installing sensors, IoT and personal devices capturing regular behaviour, human curated datasets (e.g. Open

[1] https://sdgs.un.org/topics/energy.

Maps), large-scale collaborative data-driven research, satellite imagery, multi-agent computer systems, and open government initiatives. This abundance of data is diverse both in format (e.g. structured, images, graph-based, matrix, time-series, geo-spatial, and textual) and in types of analysis performed (e.g. linear algebra, classification, graph algorithms, and relational algebra). To help different types of data and analysis activities, scientists and analysts often rely on ad-hoc procedures to integrate various data sources. This typically means manually curating how to clean, convert and integrate data. Such approaches are delicate and time consuming. In addition, to perform the analysis, they require bringing both data and computation into a single architecture, which is typically a (distributed) system not suitable for all necessary computation. Most analysts and programmers, however, are not well prepared to handle a multitude of systems, handle transitions between systems robustly, or define the correct framework for the assignment.

2.2 Open Initiatives in Energy Computing Field

DestinE Data Lake. As part of the European Commission's Green Deal and Digital Strategy, Destination Earth (DestinE) [20] is a project focused on contribution to achieving the goals of the double transition, green and digital. DestinE is designed to unlock the potential of digital Earth system modelling. It will focus on the impacts of climate change, aquatic and marine ecosystems, polar regions, the cryosphere, biodiversity or extreme weather events, as well as possible adaptation and mitigation strategies. It will help predict major environmental disasters and environmental degradation with unprecedented accuracy and reliability. The heart of Destination Earth will be a unified cloud-based modelling and simulation platform that will provide access to data, advanced computing infrastructure, software, artificial intelligence applications and analytics. As seen in Fig. 1, the project will integrate digital twins (DTs) - digital replicas of different aspects of the Earth system, such as weather and climate change projections, food and water security, global ocean circulation and ocean biogeochemistry, among others - and provide users with access to thematic information, services, models, scenarios, simulations, forecasts and visualisations. The platform will also allow the development of applications and the integration of user data.

The project, which is currently only submitted as a proposal to the European Commission in line with the European Data Strategy, will be implemented gradually over the next 7–10 years starting in 2021. The basic operational platform, digital twins and services will be developed as part of the Commission's digital programme, while Horizon Europe will provide research and innovation opportunities that will support the further development of DestinE.

Flexible Energy Denmark. Flexible Energy Denmark (FED) [12] is a digitisation project that aims to make Danish electricity consumption flexible, so that it becomes possible to use excess electricity from wind turbines and solar cells. The project brings together leading researchers, organisations, utilities, software

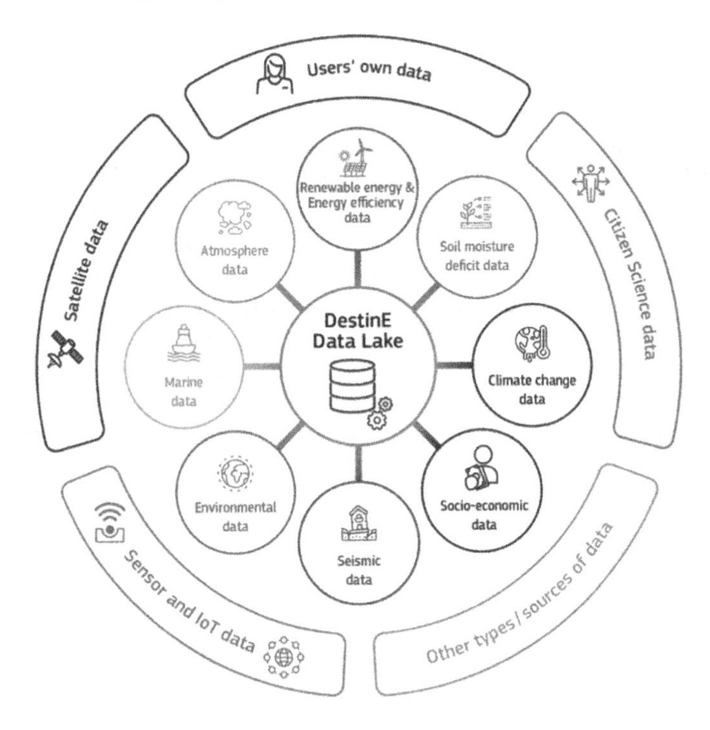

Fig. 1. DestinE Data Lake as proposed in [20]

companies and numerous living laboratories in the country that provide real data for the project. Specifically, FED collects data from a series of Living Labs (LLs) in physical environments representative of real life. Raw data on electricity, water and district heating consumption of many thousands of households, as well as indoor climate data of two primary schools and 155 households in Aalborg end up 1–4 times a day in a Data lake, called FED Data Lake (FEDDL), which is operated by the independent, non-profit national research centre *Center Denmark in Fredericia*[2] and enables efficient and advanced analysis. The FED ecosystem includes:

- A data ecosystem (the Datalake containing a variety of energy-related data that are mainly collected from the living labs in the project, but also from other sources such as BBR (the Danish Registry of Buildings and Houses) and DMI (Danish Meteorological Institute)
- An ecosystem for digital tools (tools based on artificial intelligence, are enabled by Big Data from the data ecosystem)
- An ecosystem for digitisation solutions combining some of the tools developed, with the aim of managing energy flexibility in Denmark.

FEDDL is built using only open source tools that can be run either on-premise or in cloud environments.

[2] https://www.centerdenmark.com/.

Open Energy Data Initiative. The Open Energy Data Initiative (OEDI) [1] is a centralized repository of valuable energy research datasets collected by U.S. Department of Energy programs, offices, and national laboratories. Designed to enable data discovery, OEDI facilitates access to a wide network of results, including data available in technology-specific catalogs such as the Geothermal Data Repository and the Marine Hydrokinetic Data Repository. The initiative aims to improve and automate access to high-value energy datasets across U.S. Department of Energy (DOE) programs, offices, and national laboratories. This platform is being deployed by the National Renewable Energy Laboratory (NREL) to make data usable and discoverable by researchers and industry to accelerate analysis and innovation development. Not only does the data lake provide tools to create actionable insights for analysts and to provide high-value open data, but it can also be used to conduct interesting data mashups or calculations to develop new and expanded data sets (Fig. 2).

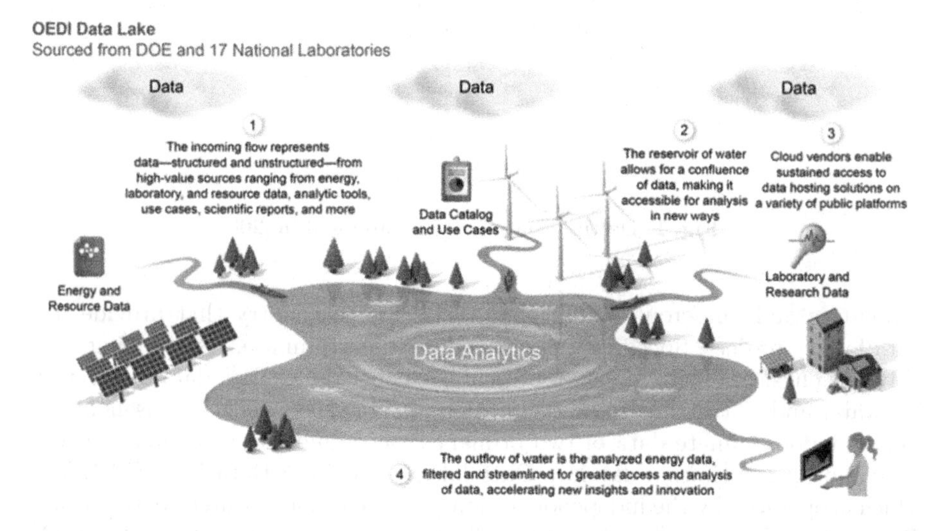

Fig. 2. Open Energy Data Initiative as depicted in [1]

OEDI leverages on Amazon Web Services (AWS) to enable analytics capabilities, innovative dataset access and to trigger new relationships among cloud partners. The data lake is based on consolidated AWS storage solutions for datasets (i.e., AWS S3 buckets) with elastic load balancing, and AWS cloud-optimized analytics tools (e.g., AWS Glue, AWS Athena) that to help users consolidate data into non-standard formats, speed up analytics, and allow users to pull or move small parts of analytics into their AWS accounts.[3]

[3] https://openei.org/wiki/.

3 Design of the Italian Energy Data Space

3.1 Logical Architecture

The ability to equip smart city services and infrastructures with sensors and monitor them with IoT devices is extremely valuable for any kind of future preparedness. It can help in the design and development of existing smart cities and in the ongoing development of other smart cities. In addition to planning, there are also benefits in the area of energy saving. This data provides an excellent insight into how our utilities are distributed and how they are used [10]. The future of the smart city offers the opportunity to use digital twin technologies. It can drive growth by creating a living testbed within a virtual twin that can do two things: first, test scenarios and second, allow the digital twin to learn about the environment by analysing changes in the data it collects. The collected data can be used for data analysis and monitoring. The potential of digital twins will increase as the development of smart cities increases connectivity and the amount of usable data [5]. In this context, the aim of the work is to design a resource for the Internet of Energy, capable of collecting energy data from Italian agencies, consortia and research centres, in order to develop a "Google of Energy", a system capable of indexing and searching energy Big Data. It can be used to facilitate future studies in the energy sector and all reliable infrastructures. Developing and consolidating a new approach to energy management, throughout the analysis of data from institutional databases, sensors, IoT devices, Industry 4.0 infrastructures, in the field of energy and its eco-system, will help to use the Big Data potential to support the Green Deal's priority actions on issues such as climate change, circular economy, pollution, biodiversity and deforestation.

Based on the model developed in the Danish National Energy Data Lake [12], where a national repository for energy data is created, Fig. 3 gives an abstract overview of the proposed Data Lake logical architecture: it is composed of five separated layers, i.e., Data Sources, Data Collection/Ingestion, Data Storage, Data Exploration, and Data Consumers, and four cross-cutting layers, i.e., Privacy and Data Protection layer, Access Management, Meta Data Governance and Resources Management. Layers Data Sources and Data Consumers represent systems which are external to the Data LakeHouse structure.

Data Sources: Data sources considered for this purpose are Mobile Sensors and IoT sensors capable of collecting energy and environmental data, Open Data made available by Public Administrations or research centres and Living Labs. Some of the sources for collecting source data are as follows:

- Open data: Energy Production and consumption
  - GSE: it provides data at national and regional data about renewable sources, transportation, energy counts;
  - ARERA: monitoring of novel generation plants at national level, Data about market, clients, production, consumption;
  - ISTAT: energy production from renewable sources at national level and consumption from families;

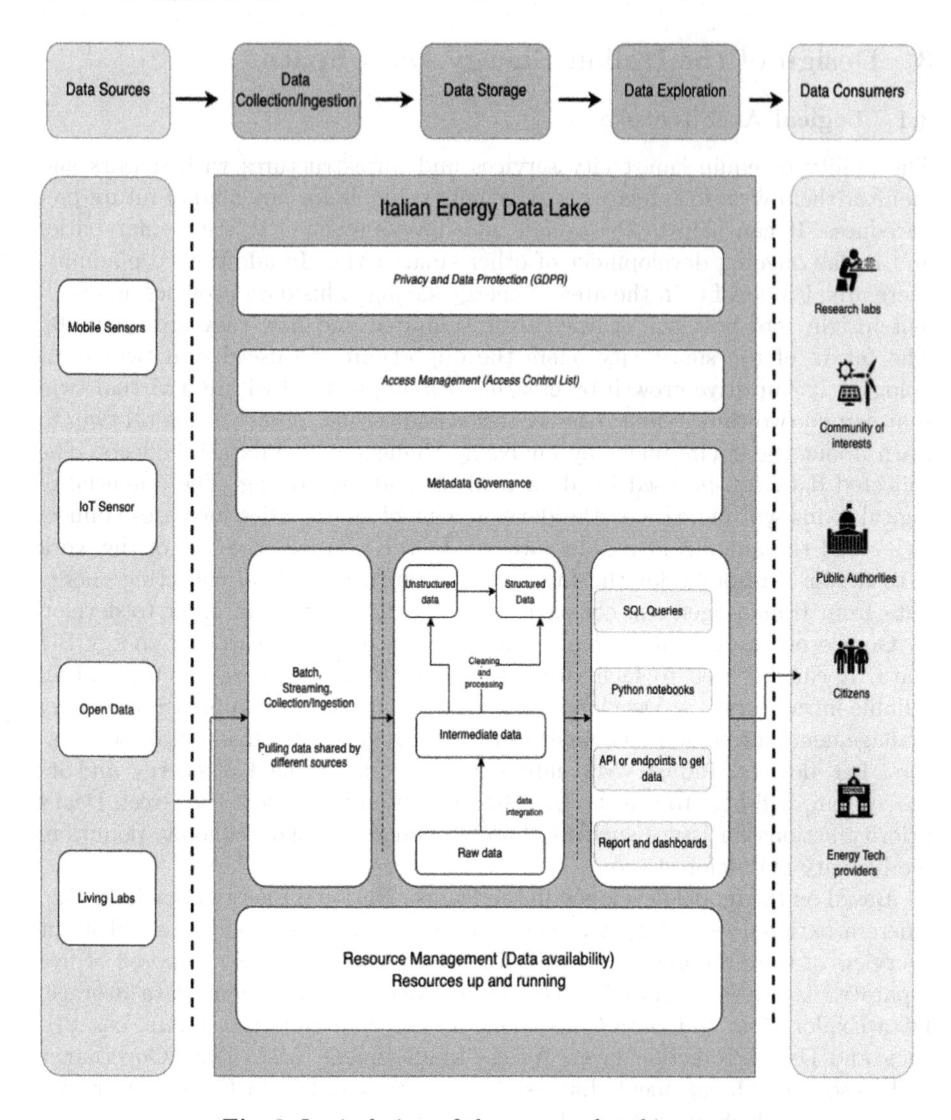

Fig. 3. Logical view of the proposed architecture

- Terna: national data related to production, generation plants, international benchmarks, peaks, consumption;
- Eurostat
  - ∗ Energy statistics section: share of renewable energies, energy productivity, energy supply bu product, energy consumption by product;
  - ∗ Sustainable development: primary energy consumption, population lacking energy due to poverty;

– Industry 4.0: smart meters, data coming from power generation plants.

Data Collection/Ingestion: The custom data collection enables data retrieval from data sources requiring custom scripts, e.g. if the data is embedded into HTML pages, APIs, or when files containing data are provided manually in CSV, TSV and PDF formats. Data ingestion is the transportation of data from assorted sources to a storage medium where it can be accessed, used, and analyzed by an organization. There are different ways of ingesting data, and the design of a particular data ingestion layer can be based on various models or architectures. The two considered kind of data ingestion are batch processing and streaming processing. In the first one, the ingestion layer periodically collects and groups source data and sends it to the destination system. Groups may be processed based on any logical ordering, the activation of certain conditions, or a simple schedule. Real-time processing (also called stream processing or streaming) involves no grouping at all. Data is sourced, manipulated, and loaded as soon as it's created or recognized by the data ingestion layer.

Data Storage: The data lake storage problem asks for selecting appropriate data stores to preserve ingested datasets. There are many solutions in the literature and they apply various relational and NoSQL databases [14], and present different manners of data storage organization. There are solutions considering heterogeneous data sources while others target at a particular type, e.g., relational tables. In order to host different types of data, the solutions could be a universal format or allowing multiple formats. Some approaches rely on the common relational or NoSQL stores while others have developed new storage systems. The data storage systems could be on-premise or cloud-based [22].

This layer can be divided in three different zones:

- *Raw data zone*: all types of data are ingested without processing and stored in their native format. This zone allows users to find the original version of data for their analytics to facilitate subsequent treatments. The stored raw data format can be different from the source format.
- *Intermediate zone*: after ingestion, the data lake is a vast collection of raw datasets with certain metadata. To make the data usable for querying, a number of solutions are proposed for further processing of the raw data, e.g., find more metadata, discover hidden relationships, and perform data integration, transformation or cleaning if necessary.
- *Structured and unstructured zones*: they stores all the available data for data analytics and provides the access of data. This zone allows self-service data consumption for different analytics (reporting, statistical analysis, business intelligence analysis, machine learning algorithms) according to their format.

Data Exploration: The top layer focuses at the interaction of users with the DataLake. It is important that useful information can be retrieved out of data lakes. However, this is challenging due to a large number of ingested sources, and the heterogeneity of data. Given data lake systems with a large number of datasets, users may have knowledge for one or a few data sources, but rarely all the datasets. The query formulation component should support users in creating formal queries that express his information requirement. The data interaction should cover all the functionalities which are required to work with the data, including visualization, annotation, selection and filtering of data, and basic analytical methods. Users can first browse the existing data sources, including their description, statistics, and schema; then she can write a query (SQL or JSONiq[4]) for a single dataset, or use the user interface to make a keyword search over the schema or the data. Alternatively, with certain knowledge of the datasets, which could be learned through the previous exploration processes, they can choose to integrate a subset of relevant datasets, and query them using formal queries or keyword search [11].

An important kind of output that the architecture could provide, is the Fair Data API: according to the FAIR data principles, research outputs are shared in a way that enables and enhances reuse by humans and machines. The characteristics of these resources can be oriented to achieve compliance with FAIR guidelines. For example, output generated uses globally unique identifiers and can assign other identifiers. The data elements described in FAIR correspond to concepts and (meta)data objects modeled, as our DataLake resources and described with rich metadata and context information. In the output of the Data Lake, resources are retrievable via open APIs, that is, absolute URIs and standard Representational State Transfer (REST) protocols.

Data Consumers: Human users play an essential role in the management of data lakes. The users of a data lake are also data providers; the insight provided by the human helps the data in the data lake to mature over time. Data consumers range from communities of interest (e.g., citizens groups and associations interested in performing pollution measurement, and factories and industries interested in their level of environmental pollution) to public authorities, and from citizens (both single individuals and associations) to other end-user categories such as schools or research-labs.

3.2 The Proposed Development Process

The steps for the development of the platform are based on the following phases.

1. The first part will focus on an in-depth study of the state of the art and analysis of existing architectures proposed in the second section. A first phase of heterogeneous data collection from the various energy sources will be carried out.
2. The second phase, aimed at scenario definition, will focus on interviews with SMEs and stakeholders that can help in the design of a use case to prototype the research results. This is aimed at an understanding to elicit the needs, the current state of the art in energy generation, distribution and use.
3. Design of pilot projects and use case, also creating living labs for involving prosumers and providers
4. Development of the digital platform for collecting data and providing data services and tools
5. Incremental extension of use cases and further involvement of new providers, consumers, stakeholders.

[4] https://www.jsoniq.org/.

4 Conclusion

Knowledge of consumers' energy consumption and indoor climate is worth its weight in gold to utilities, industry and researchers. They use the data to plan production and develop services and algorithms that control energy consumption so that it becomes more flexible and renewable energy is not wasted. Our ambition with the national platform is that it can form the basis for the release of data from electricity, water, heat and potentially also gas, so that the data can be used by commercial suppliers to develop new business models that support data-driven models for the green transition. Creating a repository of national energy data that respects the Fairness' key principles, is the starting point to provide an open and extensible platform to enable secure, resilient acquisition and sharing of information with the aim to improve the well-being and inclusion of citizens, produce a more effective response to pollution or other environmental emergencies, and make Smart Cities and extended urban areas feel more secure and safe to the citizens living in them. Further, endeavors from citizens and joint academic-community science can assist with distinguishing environmental health problems related with air quality in metropolitan regions. Unfortunately, there remains a gap between the development and the effective utilization of these cutting-edge technologies within communities of proactive decision-making [3]. The importance of this topic will help to raise public awareness of energy problems, to highlight the importance of citizens' engagement and to inspire citizens to adopt sustainable consumption habits and behavior patterns. These habits will promote new sustainable services, e.g. lengthening product life cycles through reuse, repair and refurbishment and encourage waste reduction, energy savings and circular thinking: the so-called 'citizen science' is emerging.

References

1. re3data.org: OEDI (2020)
2. Alonso, P.J.G.: SETA, a suite-independent agile analytical framework. Master's thesis, Universitat Politecnica de Catalunya (2016)
3. Bales, E., Nikzad, N., Quick, N., et al.: Citisense: mobile air quality sensing for individuals and communities design and deployment of the citisense mobile air-quality system. In: 2012 6th International Conference on Pervasive Computing Technologies for Healthcare (PervasiveHealth) and Workshops, pp. 155–158 (2012)
4. Barricelli, B.R., Casiraghi, E., Fogli, D.: A survey on digital twin: definitions, characteristics, applications, and design implications. IEEE Access **7**, 167653–167671 (2019)
5. Brosinsky, C., Westermann, D., Krebs, R.: Recent and prospective developments in power system control centers: adapting the digital twin technology for application in power system control centers. In: 2018 IEEE International Energy Conference (ENERGYCON), pp. 1–6 (2018)
6. Collier, S.E.: The emerging Enernet: convergence of the smart grid with the internet of things. IEEE Ind. Appl. Mag. **23**(2), 12–16 (2017)

7. Darmont, J., Favre, C., Loudcher, S., Noûs, C.: Data lakes for digital humanities. In: Proceedings of the 2nd International Conference on Digital Tools & Uses Congress, DTUC 2020, New York, NY, USA. Association for Computing Machinery (2020)
8. Wilkinson, M.D., et al.: The fair guiding principles for scientific data management and stewardship. Sci. Data **3**, 160018 (2016)
9. Ellen MacArthur Foundation: Towards the circular economy. Technical report, EMF, McKinsey Company (2013)
10. Fuller, A., Fan, Z., Day, C., Barlow, C.: Digital twin: enabling technologies, challenges and open research. IEEE Access **8**, 108952–108971 (2020)
11. Hai, R., Quix, C., Zhou, C.: Query rewriting for heterogeneous data lakes. In: 22nd European Conference on Advances in Databases and Information Systems, pp. 35–49 (2018)
12. Ben Hamadou, H., Pedersen, T., Thomsen, C.: The Danish National Energy Data Lake: requirements, technical architecture, and tool selection. In: 2020 IEEE International Conference on Big Data (Big Data), pp. 1523–1532 (2020)
13. Motlagh, N.H., Mohammadrezaei, M., Hunt, J., Zakeri, B.: Internet of things (IoT) and the energy sector. Energies **13**(2), 1–27 (2020)
14. Khan, S., Liu, X., Ali, S., Alam, M.: Storage solutions for big data systems: a qualitative study and comparison. arXiv (April 2019)
15. Laney, D.: 3D data management: controlling data volume, velocity and variety. META Group Research Note, vol. 6 (2001)
16. Li, Y., Zhang, A.M., Zhang, X., Wu, Z.: A data lake architecture for monitoring and diagnosis system of power grid. In: Proceedings of the 2018 Artificial Intelligence and Cloud Computing Conference, AICCC 2018, pp. 192–198. Association for Computing Machinery, New York (2018)
17. Madera, C., Laurent, A.: The next information architecture evolution: the data lake wave. In: Proceedings of the 8th International Conference on Management of Digital EcoSystems (MEDES), pp. 174–180 (2016)
18. Mardiansjah, F.H.: Extended urbanization in smaller-sized cities and small town development in Java: the case of the Tegal region. IOP Conf. Ser. Earth Environ. Sci. **447**, 012030 (2020)
19. United Nations. Progress towards the sustainable development goals (2017)
20. Nativi, S., Mazzetti, P., Craglia, M.: Destination earth (destine) architecture validation workshop. Technical report, European Commission (2021)
21. Wang, P., Yang, L.T., Li, J., Chen, J., Hu, S.: Data fusion in cyber-physical-social systems: state-of-the-art and perspectives. Inf. Fus. **51**, 42–57 (2019)
22. Zagan, E., Danubianu, M.: Cloud data lake: the new trend of data storage. In: 2021 3rd International Congress on Human-Computer Interaction, Optimization and Robotic Applications (HORA), pp. 1–4 (2021)

High-Level Feature Extraction for Crowd Behaviour Analysis: A Computer Vision Approach

Alessandro Bruno(✉), Marouane Ferjani, Zoheir Sabeur, Banafshe Arbab-Zavar, Deniz Cetinkaya, Liam Johnstone, Muntadher Sallal, and Djamel Benaouda

Department of Computing and Informatics, Bournemouth University, Poole, UK
{abruno,mferjani,zsabeur,barbabzavar,dchetinkaya,ljohnstone, msallal,dbenaouda}@bournemouth.ac.uk
https://www.bournemouth.ac.uk/

Abstract. The advent of deep learning has brought in disruptive techniques with unprecedented accuracy rates in so many fields and scenarios. Tasks such as the detection of regions of interest and semantic features out of images and video sequences are quite effectively tackled because of the availability of publicly available and adequately annotated datasets. This paper describes a use case scenario with a deep learning models' stack being used for crowd behaviour analysis. It consists of two main modules preceded by a pre-processing step. The first deep learning module relies on the integration of YOLOv5 and DeepSORT to detect and track down pedestrians from CCTV cameras' video sequences. The second module ingests each pedestrian's spatial coordinates, velocity, and trajectories to cluster groups of people using the Coherent Neighbor Invariance technique. The method envisages the acquisition of video sequences from cameras overlooking pedestrian areas, such as public parks or squares, in order to check out any possible unusualness in crowd behaviour. Due to its design, the system first checks whether some anomalies are underway at the microscale level. Secondly, It returns clusters of people at the mesoscale level depending on velocity and trajectories. This work is part of the physical behaviour detection module developed for the S4AllCities H2020 project.

Keywords: Crowd behaviour · Computer Vision · Artificial Intelligence · Deep Learning

1 Introduction

Over the last decade, the scientific community observed a lot of progress in Artificial Intelligence and Computer Vision. Consequently, several application domains spanning object modelling, detection, segmentation, healthcare, crowd dynamics are addressed using computer vision approaches [5,6,20,22].

P. L. Mazzeo et al. (Eds.): ICIAP 2022 Workshops, LNCS 13374, pp. 59–70, 2022.
https://doi.org/10.1007/978-3-031-13324-4_6

The advent of Deep Learning [16] prompted both academics and industry to push the bar on the proposed solutions for several scenarios and use-cases. Since the introduction of AlexNet in 2012 [15], much attention has been focused on Deep Neural Networks to achieve increasingly higher accuracy rates on the topics above and tasks. Some architectures represent milestones in the deep learning literature, namely GoogleNet [24], Inception-V4 and ResNet [23], GANs [10], YOLO [18]. As the literature review shows, AI allowed achieving unprecedented accuracy rates in so many research fields, albeit some paradigms exhibit drawbacks [29]. For instance, supervised learning relies on the availability of a great deal of manually annotated data. Big-sized datasets such as ImageNet [8] come along with millions of images and the corresponding annotations, making supervised learning a suitable paradigm to perform different tasks. Generally speaking, the hand-labelling of images and video sequences is labour intensive and time-consuming. That especially applies to all those domains such as biomedical imaging, behaviour understanding, visual perception, where in-depth knowledge and expertise are required. Some object detection and segmentation tasks are easily extended to video sequences by optimising the image-related version.

Research interest in crowd behaviour analysis has grown remarkably over the last decades. As a result, crowd behaviour analysis has become a multidisciplinary topic involving psychology, computer science, physics. A crowd can be thought of as a collection of individuals showing movements that might be temporarily coordinated upon a common goal or focus of attention [2]. That could apply to both spectators and moving people. Consequently, there are three main levels at which crowds can be described: microscale, mesoscale, macroscale. At the microscale level, pedestrians are identified individually. The state of each of such individuals is delivered by position and velocity. At the mesoscale level, the description of pedestrians is still identified by position and velocity, but it is represented statistically through a distribution function. At the macroscale level, The crowd is considered as a continuum body. Furthermore, it is described with average and observable quantities such as spatial density, momentum, kinetic energy and collectiveness. This paper describes a use case scenario for crowd behaviour analysis and provides an integrated solution. The proposed solution relies on both supervised and unsupervised learning paradigms depending on the task to work out. The proposed solution has been developed within the research activities for the European Research Project S4AllCities [1]. The experiments have been carried out on the publicly available UCSD Anomaly Detection Dataset [27].

2 Related Work

One of the main goals of crowd behaviour analysis is to predict whether some unusual phenomenon takes place to ensure peaceful event organizations and minimize the number of casualties in public areas. This section summarises the scientific literature on the topic by looking into approaches relying on different principles and methodologies. The more traditional methods of crowd behaviour analysis build on the extraction of handcrafted features either to set up expert systems or to feed neural networks and classification systems. For instance, texture analysis tackles the detection of regular and near regular patterns in images [3]. Saqib et al. [21] carried out crowd density estimation using texture descriptors while conversely, some methods address crowd analytics using physics concepts and fluid dynamics as in [9]. However, images and videos in real scenarios contain nonlinearities that have to be faced efficiently for gaining accuracies in the results. [25] Some computer vision-based methods face the challenging topic by checking groups of people exhibiting coherent movements [27]. Other techniques focus on path analysis using mathematical approaches while psychologists highlighted some aspects regarding emergency and situational awareness [19]. A shared line in the methods above is the increase in demand for security measures and monitoring of crowded environments. Therefore, by zooming in on the topic, one can unearth several applications that are closely related to crowd analysis: person tracking [19], anomaly detection [28], behaviour pattern analysis [7], and context-aware crowd counting [17]. As briefly mentioned in the previous section, despite the introduction of deep learning solutions being with high accuracy rates, some open issues related to density variation, irregular distribution of objects, occlusions, pose estimation remain open in the topic of crowd analysis [14]. The following section introduces the integrated solution developed for the S4AllCities project [1].

3 Proposed Method

In this section, the proposed method is thoroughly described by highlighting the role played by each module. The overall architecture for the integrated solution is depicted in Fig. 1 with three main blocks: homographic projection, supervised deep learning models, unsupervised learning module. The following subsections focus on each of the steps mentioned above.

3.1 Pre-processing

The first step of the proposed integrated solution consists of planar homography to project head-plane points onto the ground-plane. As widely described by Hartley and Zisserman [11], planar homography relates the transformation between two planes (up to a scale factor). The homography matrix H has 8 degrees of freedom. That means that four matches are enough to calculate the transformation. The main goal here is to remove or correct the perspective of the given view of the pedestrian-area-overlooking camera. In the use-case scenario, at least four coordinates of pedestrians are needed. They can be easily fetched by enacting YOLOv5 until the four pedestrians are detected. Then, the approach will generate an approximation on the plane-to-plane projection depending on the average height of pedestrians in the given camera's field of view.

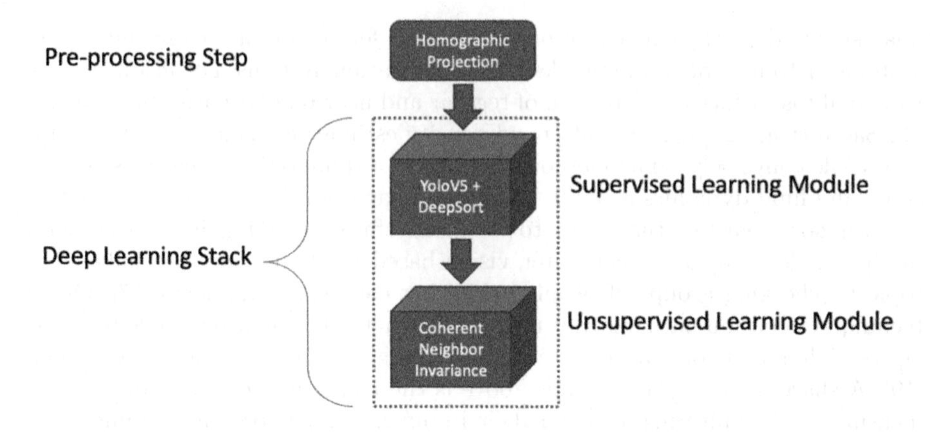

Fig. 1. Deep Learning Stack is depicted in the figure.

3.2 Supervised Deep Learning Module

Inspired by Hou et al.'s method [12] on vehicle tracking, the first of two deep learning modules sees the integration of two popular models such as YOLOv5 [13] and DeepSORT [26]. The former is one of the most accurate models for object detection. At the same time, the latter tracks down human crowd movements over video sequences, which is the extension of the popular YOLOv4 by Bochkovskiy et al. [4]. For a given frame having N pedestrians, $P(x, y)_{i=1,,N}$ represents the ith pedestrian' spatial coordinates. YOLOv5 is quite accurate in detecting pedestrians (see Fig. 2; it does not perform re-identification though. That is why it has been necessary to integrate DeepSORT, which is responsible for tracking down the pedestrians in a video sequence by assigning them a specific reference number. DeepSORT keeps trace of $P(x, y)_{i=1,,N}$ across different times (t0, t1, $\cdots$, tn). In Figs. 3 and 4 an example referring to ID 1 pedestrian is shown. YOLOv5 returns all spatial coordinates of the pedestrians detected as a sequence of bounding boxes. They will be then ingested by DeepSORT, which runs measurement-to-track associations using nearest neighbour queries in visual appearance space (see Fig. 5). On top of both modules, the system is capable of retrieving the spatial coordinates, and the reference number of the pedestrians tracked across the area overlooked by a CCTV camera. The extraction of the details mentioned above is taken every second. Having timestamps, spatial coordinates and reference number allows extracting velocity and storing trajectories. A time frame Δt is taken as a reference to work out the detection of anomalies in the crowd behaviour at the microscale level. Being t_0 the initialisation time of the system, $\mathrm{t}_0 + \Delta t$ is the earliest time where it is possible to detect any anomalies in crowds. Gaussian distributions are considered to analyse pedestrian velocity within the Δt time range. An example of trajectories out of video sequences is given in Fig. 6. The system evaluates anomalies as the samples that deviate from the normal distribution. The more a sample is distant from the distribution, the more likely an anomaly is within the crowd behaviour.

Fig. 2. An example of pedestrian detection from video frames is given above.

Fig. 3. Pedestrian detection at time t_0

Fig. 4. Pedestrian detection at time t_1

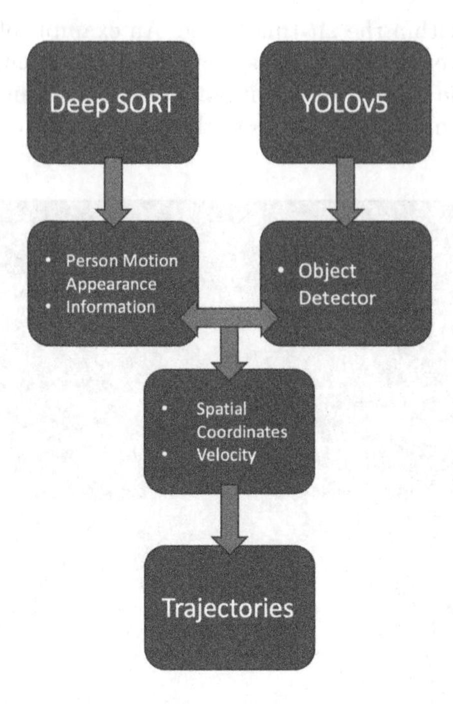

Fig. 5. The first deep learning module consists of the integration of DeepSORT and YOLOv5

Fig. 6. The first deep learning module consists of the integration of DeepSORT and YOLOv5

3.3 Unsupervised Learning for Trajectory Clustering

Due to the advances in detection and tracking techniques, the ability to extract high-quality features of moving objects such as trajectories and velocities is now possible. These features can be critical in understanding and detecting coherent motions in various physical and biological systems. Furthermore, the extraction of these motions enables a deeper understanding of self-organized biological systems. For instance, in surveillance videos, capturing coherent movements exhibited by moving pedestrians permits acquiring a high-level representation of crowd dynamics. These representations can be utilized for a plethora of applications such as object counting, crowd segmentation, action recognition and scene understanding, etc. (Fig. 8).

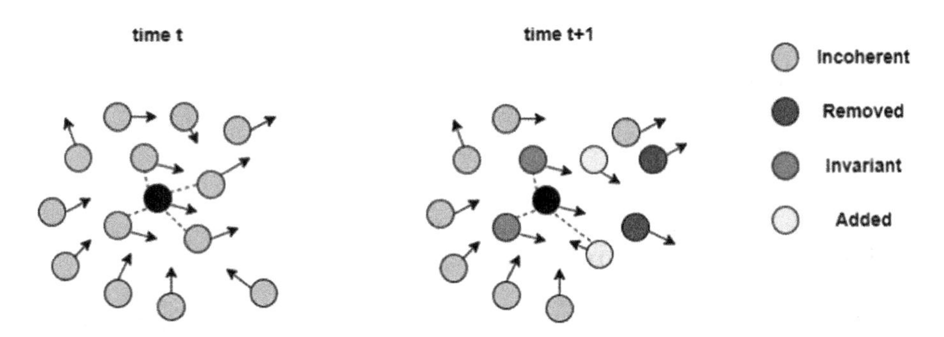

Fig. 7. An exhibition of coherent neighbour invariance. The green dots are viewed as invariant neighbors of the centered black dot (for K = 4). (Color figure online)

Fig. 8. Coherent motion detection in action

Whilst coherent motions are regarded as macroscopic observations of pedestrians' congregational activities, these motions can be distinguished through the interaction among individuals in local neighbourhoods. Inspired from Zhou [30], the Coherent Neighbor Invariance technique is deployed to capture the coherent motion of crowd clutters. The key characteristics that establish the difference between cohesive and arbitrary movements are listed below:

- **Neighborship Invariance**: the spatial-temporal relationship among individuals is inclined to prevail overtime.

- **Velocity Correlations Invariance**: neighboring individuals exhibiting coherent movement showcase high velocity correlations.

Conversely, incoherent individuals that showcase relative independence tend to lack the mentioned properties. To illustrate the Neighborship Invariance property, Fig. 7 displays the use of K nearest neighbour to highlight the emergence of global coherence in local neighborships. The equation below quantifies the velocity correlations between neighbouring individuals, which allows discerning coherent motions.

$$g = \frac{1}{d+1} \sum_{\lambda=t}^{t+d} \frac{v_\lambda^i \cdot v_\lambda^{i_k}}{\|v_\lambda^i\|^2 \cdot \|v_\lambda^{i_k}\|^2} \tag{1}$$

where:

- g : velocity correlation between i and i_k
- v_λ^i : velocity of individual i at time λ
- $v_\lambda^{i_k}$: velocity of individual i_k at time λ
- d : duration of the experiment

4 Experimental Results

An experimental campaign has been carried out over the publicly available UCSD Anomaly Detection Dataset [27]. The dataset consists of video sequences acquired with a stationary camera overlooking pedestrian areas. The dataset offers videos with variable conditions of crowd density, and cameras' field of view. Most of videos contains only pedestrians, still anomalies are represented by bikers, skaters, small carts, pedestrian entities crossing a walkway or walking in the grass that surrounds it.

$$Precision = \frac{TP}{TP + FP} \tag{2}$$

$$Recall = \frac{TP}{TP + FN} \tag{3}$$

The experiments were run on five video sequences from UCSD. Two of which do not contain any anomalies, while the remaining three do. A quantitative analysis of results is conducted over the first deep learning module, which is responsible for the microscale analysis. In Tables 1 and 2 precision and recall (see Eqs. 2 and 3) for YOLOv5 and DeepSORT are reported. The second deep learning module is still currently being developed. Only qualitative results can be shown 7 to give the big picture of the consistency of clusters of people. As it can be noticed in Table 1, YOLOv5 reaches high precision rates on all tests up to 0.98 while recall is penalised by some false negatives. Occlusion and overlapping cause a drop of performances on pedestrian detection. DeepSORT also achieves good precision rates even though sometimes the tracking shows some mismatch.

Recall values drop by 10% on average if compared to precision. Nevertheless, the combination of the two supervised learning modules gains decent performances. As described in Sect. 3.2, the supervised deep learning module allows the extraction of high-level features such as spatial coordinates, velocity and trajectories. On top of that, some parameters are to be fine-tuned, respectively, Δt and the distance from the normal distribution. The latter has a sample evaluated as anomaly, trigger a sort of alert to the crowd behaviour analysis system. Some fine-tuning has been necessary in order to find the right trade-off performances and computational load. δt has been set to 5 s, while 5 pixel/second has been selected as the distance threshold from the normal distribution of velocities.

The experiments on the automatic optimisation of the given advertisement layouts and images have been carried out on a 13-in. Mac-book Pro with 16 GB of RAM, 2.4 GHz Quad-Core Intel Core i5, Intel Iris Plus Graphics 655 1536 MB.

Table 1. YOLOv5 Precision and Recall in 5 tests over UCSD

No. of test	Precision	Recall
Test 1	0.98	0.75
Test 2	0.93	0.72
Test 3	0.95	0.71
Test 4	0.94	0.78
Test 5	0.92	0.70

Table 2. DeepSORT Precision and Recall in 5 tests over UCSD

No. of test	Precision	Recall
Test 1	0.85	0.74
Test 2	0.89	0.72
Test 3	0.83	0.69
Test 4	0.86	0.68
Test 5	0.87	0.72

5 Conclusions

This paper showcases the effectiveness of an integrated solution consisting of three main modules: pre-processing, supervised learning, unsupervised learning. The main goal is to perform crowd behaviour analysis by considering several variables such as velocity, spatial coordinates and trajectories. The first two have been used to detect anomalies in the test set at the microscale level. Successively, the unsupervised learning module ingests velocities and trajectories to initialise clusters of people according to cohesive movements. The microscale analysis task has been entirely carried out with supervised deep learning models such as YOLOv5 and DeepSORT. Cohesive movement-based clustering has been tackled by the Coherent Neighbour Invariance technique. Further experiments are underway to improve precision and recall rates, especially on the pedestrian tracking task. Furthermore, some other alternatives are in consideration to detect anomalies by combining physical properties like velocity and trajectories and semantic features such as objects whose only presence might represent a danger within a given environment. Furthermore, some work is to be done to adapt the method to different datasets and environments.

Acknowledgements. This work is a part of the S4AllCities project. This project has received funding from the European Union's Horizon 2020 research and innovation programme under grant agreement No 883522. Content reflects only the authors' view and the Research Executive Agency (REA)/European Commission is not responsible for any use that may be made of the information it contains.

References

1. Smart spaces safety and security: Greece. https://www.s4allcities.eu/
2. Arbab-Zavar, B., Sabeur, Z.A.: Multi-scale crowd feature detection using vision sensing and statistical mechanics principles. Mach. Vis. Appl. **31**(4), 1–16 (2020). https://doi.org/10.1007/s00138-020-01075-4
3. Ardizzone, E., Bruno, A., Mazzola, G.: Scale detection via keypoint density maps in regular or near-regular textures. Pattern Recogn. Lett. **34**(16), 2071–2078 (2013)
4. Bochkovskiy, A., Wang, C.Y., Liao, H.Y.M.: YOLOV4: optimal speed and accuracy of object detection. arXiv preprint arXiv:2004.10934 (2020)
5. Bruno, A., Ardizzone, E., Vitabile, S., Midiri, M.: A novel solution based on scale invariant feature transform descriptors and deep learning for the detection of suspicious regions in mammogram images. J. Med. Signals Sens. **10**(3), 158 (2020)
6. Bruno, A., Greco, L., La Cascia, M.: Video object recognition and modeling by sift matching optimization. In: Proceedings of the 3rd International Conference on Pattern Recognition Applications and Methods, pp. 662–670 (2014)
7. Cheng, Z., Qin, L., Huang, Q., Yan, S., Tian, Q.: Recognizing human group action by layered model with multiple cues. Neurocomputing **136**, 124–135 (2014)
8. Deng, J., Dong, W., Socher, R., Li, L.J., Li, K., Fei-Fei, L.: ImageNet: a large-scale hierarchical image database. In: 2009 IEEE Conference on Computer Vision and Pattern Recognition, pp. 248–255. IEEE (2009)
9. Dogbe, C.: On the modelling of crowd dynamics by generalized kinetic models. J. Math. Anal. Appl. **387**(2), 512–532 (2012)
10. Goodfellow, I., et al.: Generative adversarial nets. In: Advances in Neural Information Processing Systems, vol. 27 (2014)
11. Hartley, R., Zisserman, A.: Multiple View Geometry in Computer Vision. Cambridge University Press, Cambridge (2003)
12. Hou, X., Wang, Y., Chau, L.P.: Vehicle tracking using deep sort with low confidence track filtering. In: 2019 16th IEEE International Conference on Advanced Video and Signal Based Surveillance (AVSS), pp. 1–6. IEEE (2019)

13. Jocher, G., et al.: ultralytics/YOLOV5: v6.1 - TensorRT, TensorFlow Edge TPU and OpenVINO Export and Inference, February 2022. https://doi.org/10.5281/zenodo.6222936
14. Khan, A., Ali Shah, J., Kadir, K., Albattah, W., Khan, F.: Crowd monitoring and localization using deep convolutional neural network: a review. Appl. Sci. **10**(14), 4781 (2020)
15. Krizhevsky, A., Sutskever, I., Hinton, G.E.: ImageNet classification with deep convolutional neural networks. In: Pereira, F., Burges, C.J.C., Bottou, L., Weinberger, K.Q. (eds.) Advances in Neural Information Processing Systems, vol. 25. Curran Associates, Inc. (2012). https://proceedings.neurips.cc/paper/2012/file/c399862d3b9d6b76c8436e924a68c45b-Paper.pdf
16. LeCun, Y., Bengio, Y., Hinton, G.: Deep learning. Nature **521**(7553), 436–444 (2015)
17. Liu, W., Salzmann, M., Fua, P.: Context-aware crowd counting. In: Proceedings of the IEEE/CVF Conference on Computer Vision and Pattern Recognition, pp. 5099–5108 (2019)
18. Redmon, J., Divvala, S., Girshick, R., Farhadi, A.: You only look once: unified, real-time object detection. In: Proceedings of the IEEE Conference on Computer Vision and Pattern Recognition, pp. 779–788 (2016)
19. Rodriguez, M., Laptev, I., Sivic, J., Audibert, J.Y.: Density-aware person detection and tracking in crowds. In: 2011 International Conference on Computer Vision, pp. 2423–2430. IEEE (2011)
20. Sabeur, Z., Arbab-Zavar, B.: Crowd behaviour understanding using computer vision and statistical mechanics principles. In: Bellomo, N., Gibelli, L. (eds.) Crowd Dynamics, Modeling and Simulation in Science, Engineering and Technology, vol. 3, pp. 49–71. Springer, Cham (2021). https://doi.org/10.1007/978-3-030-91646-6_3
21. Saqib, M., Khan, S.D., Blumenstein, M.: Texture-based feature mining for crowd density estimation: a study. In: 2016 International Conference on Image and Vision Computing New Zealand (IVCNZ), pp. 1–6. IEEE (2016)
22. Singh, U., Determe, J.F., Horlin, F., De Doncker, P.: Crowd monitoring: state-of-the-art and future directions. IETE Tech. Rev. **38**(6), 578–594 (2021)
23. Szegedy, C., Ioffe, S., Vanhoucke, V., Alemi, A.A.: Inception-v4, inception-ResNet and the impact of residual connections on learning. In: Thirty-First AAAI Conference on Artificial Intelligence (2017)
24. Szegedy, C., et al.: Going deeper with convolutions. In: Proceedings of the IEEE Conference on Computer Vision and Pattern Recognition, pp. 1–9 (2015)
25. Tripathi, G., Singh, K., Vishwakarma, D.K.: Convolutional neural networks for crowd behaviour analysis: a survey. Vis. Comput. **35**(5), 753–776 (2018). https://doi.org/10.1007/s00371-018-1499-5
26. Wojke, N., Bewley, A., Paulus, D.: Simple online and realtime tracking with a deep association metric. In: 2017 IEEE International Conference on Image Processing (ICIP), pp. 3645–3649. IEEE (2017). https://doi.org/10.1109/ICIP.2017.8296962
27. Wu, S., Moore, B.E., Shah, M.: Chaotic invariants of Lagrangian particle trajectories for anomaly detection in crowded scenes. In: 2010 IEEE Computer Society Conference on Computer Vision and Pattern Recognition, pp. 2054–2060. IEEE (2010)
28. Xu, D., Song, R., Wu, X., Li, N., Feng, W., Qian, H.: Video anomaly detection based on a hierarchical activity discovery within spatio-temporal contexts. Neurocomputing **143**, 144–152 (2014)

29. Zhang, C., Vinyals, O., Munos, R., Bengio, S.: A study on overfitting in deep reinforcement learning. arXiv preprint arXiv:1804.06893 (2018)
30. Zhou, B., Tang, X., Wang, X.: Coherent filtering: detecting coherent motions from crowd clutters. In: Fitzgibbon, A., Lazebnik, S., Perona, P., Sato, Y., Schmid, C. (eds.) ECCV 2012. LNCS, pp. 857–871. Springer, Heidelberg (2012). https://doi.org/10.1007/978-3-642-33709-3_61

Binary is the New Black (and White): Recent Advances on Binary Image Processing

A Simple yet Effective Image Repairing Algorithm

Lidija Čomić[1] and Paola Magillo[2(✉)]

[1] Faculty of Technical Sciences, University of Novi Sad, Novi Sad, Serbia
comic@uns.ac.rs

[2] Department of Computer Science, Bioengineering, Robotics, and Systems Engineering, University of Genova, Genova, Italy
magillo@dibris.inige.it

Abstract. A 2D binary image is well-composed if it does not contain 2×2 blocks of two diagonal black and two diagonal white pixels, called critical configurations. Some image processing algorithms are simpler on well-composed images. The process of transforming an image into a well-composed one is called repairing.

We propose a new topology-preserving approach, which produces two well-composed images starting from an image I depending on the chosen adjacency (vertex or edge adjacency), in the same original square grid space as I. The size of the repaired images depends on the number and distribution of the critical configurations. A well-composed image I is not changed, while in the worst case the size increases at most two times (or four times if we want to preserve the aspect ratio). The advantage of our approach is in the small size of the repaired images, with a positive impact on the execution time of processing tasks. We demonstrate this experimentally by considering two classical image processing tasks: contour extraction and shrinking.

Keywords: Well-composed images · Repairing 2D digital images · Contour tracing · Shrinking

1 Introduction

We consider two dimensional black-and-white (binary) images, where black is foreground and white is background. A critical configuration is a block of 2×2 pixels within an image, where two pixels are white and two are black, in a chessboard configuration. An image with no critical configurations is called well-composed.

The presence of critical configurations introduces ambiguity in the topology of the image, as the topological (homological) properties of the foreground and of the background of the image (the number of connected components and the number of holes) depend on the used adjacency type (edge- or vertex-adjacency, a.k.a. 4- or 8-adjacency). Opposite adjacency types must be used for foreground

P. L. Mazzeo et al. (Eds.): ICIAP 2022 Workshops, LNCS 13374, pp. 73–83, 2022.
https://doi.org/10.1007/978-3-031-13324-4_7

and background pixels in order to maintain some similarity between continuous and digital topology.

Furthermore, many topological image analysis and processing algorithms are simpler and easier to implement if their input image is known to be well-composed. For such reasons, the research community has been working for years on the topic of image repairing, i.e., the process of transforming an arbitrary image into a well-composed one. All the proposed approaches which preserve the image topology, increase the image size.

We propose a simple approach for topology-preserving image repairing, which is based on inserting new rows or new columns inside the image, just where critical configurations are present. In this way, the growing rate of image size depends on the number and distribution of critical configurations present in the image. We introduce two algorithms based on such an approach. Algorithm A guarantees less than 200% of size growth on all images, but modifies the aspect ratio. Algorithm B preserves the aspect ratio, but cannot guarantee the same bound. The theoretical worst case growing rate is 400%, while it is much less in practical cases.

2 Background Notions

A 2D (square) grid [1,2,4–6] is a tessellation of the plane into closed unit squares (pixels) centered at points in $\mathbb{Z}^2$, with edges parallel to the coordinate axes. Two types of adjacency relation are defined in the grid. Two pixels that share an edge or a vertex are called 4- or 8-adjacent, respectively.

A 2D digital object O is a finite set of pixels in the square grid. The pixels in O are called black (foreground). The pixels in the complement of O are called white (background). The carrier (or continuous analogue) of O is the union (as point sets) of the pixels in O. We will denote it also as O.

A vertex v is critical for a 2D digital object O if v is incident to two white and two black pixels, where black and white pixels alternate cyclically around v. The 2×2 pixels incident with a critical vertex are called a critical configuration, a.k.a. a gap.

3 Related Work

Several image repairing algorithms have been proposed. Here, we restrict our attention to the ones which preserve the topology, and whose output is still in the square grid.

The method by Rosenfeld et al. [9] scales the image by factor 3 in both x and y directions. In the rescaled grid, all black (white) pixels involved in a critical configuration are changed to white (black), for repairing the image according to 8-adjacency (4-adjacency). An example is shown in Fig. 1 (b).

The algorithm by Stelldinger et al. [10] increases the grid resolution twice in both coordinate directions, by creating an additional square for each edge and each vertex in the grid. Therefore, the image size increases four times. If

4-adjacency (8-adjacency) is considered for the black pixels, the squares corresponding to the edges and vertices are black only if all the incident squares are black (at least one incident square is black). An example is shown in Fig. 1 (c).

The algorithm by Čomić and Magillo [3] produces the output in a new square grid, rotated 45 °C with respect to the original one and therefore called the (2D) diamond grid. The pixels of the diamond grid correspond to the pixels and vertices of the original square grid. Thus, the image size is increased two times. Depending on the choice of the color of the pixels associated with the vertices, the repaired image is homotopy equivalent with the original one with either 8- or 4-adjacency. An example is shown in Fig. 1 (d).

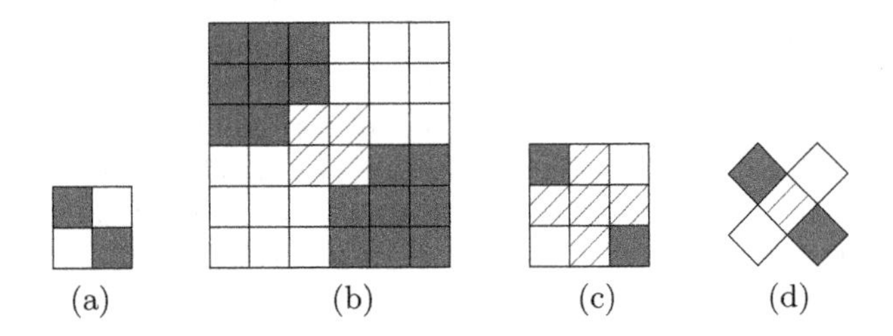

Fig. 1. (A) A critical configuration and the way it is repaired by (b) Rosenfeld et al. [9], (c) Stelldinger et al. [10], and (d) Čomić and Magillo [3]. The dashed pixels are black (white) for preserving 8-adjacency (4-adjacency).

4 Our Approach to Image Repairing

The idea is in a sense similar to [10], but, instead of adding a new row and a new column between each original row and column, we add a new row (or column) only where some critical configuration exists. For each critical configuration, one of the involved pixels changes color in the stretched image.

Our Algorithm A adds the minimum number of rows necessary to eliminate the critical configurations. Algorithm B adds both rows and columns, with the aim of preserving the aspect ratio of the image.

4.1 Algorithm A

Our basic idea is shown in Fig. 2. For each pair of consecutive rows i and $i+1$ in the input image, such that some critical configuration exists across them, we add a new row between i and $i+1$. Such new row is a copy of the row $i+1$ in all pixels, with the exception of the pairs of consecutive pixels involved in a critical configuration. The color of such pairs is set to black (white) to obtain an equivalent image according to 8-adjacency (4-adjacency).

Equivalently, we can add new columns instead of new rows. We first compute the number of necessary new rows and the number of necessary new columns to be added, and then choose the option which gives the smaller image size. Algorithm A, obtained from this simple idea, has the following good properties:

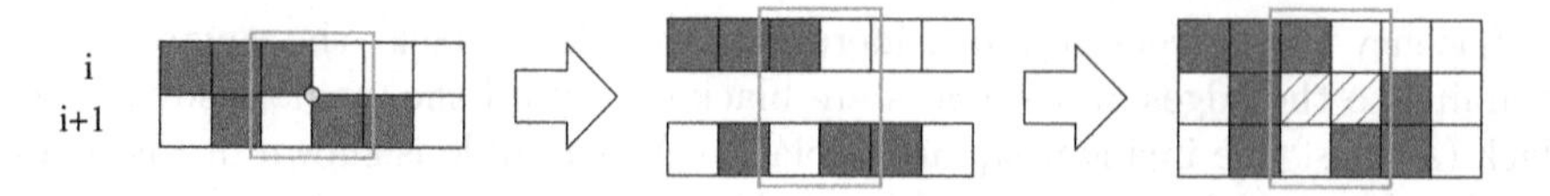

Fig. 2. When a critical configuration (marked with the red box) exists across the rows i and $i+1$, a new row is inserted in between. The color of each pixel in the new row is copied from the row $i+1$, but the two pixels adjacent to those involved in the critical configuration (dashed) are both set to black or to white, for repairing the image according to 8-adjacency or 4-adjacency, respectively. (Color figure online)

1. It preserves the image topology with 4- or 8-adjacency.
2. The size of the repaired images is less than twice the size of the original one. In the worst case, for an image of size $N \times M$, a new row will be added between any two rows, and the size of the image will increase to $N \times (2M - 1)$.
3. The size increment depends on the number of critical configurations present in the input image. At most one row is added for each critical configuration. Intuitively, this is the first image repairing algorithm sensitive to the amount of repairing needed by the image.

The main drawback of Algorithm A is that the aspect ratio of the image is dramatically changed. So, the output of Algorithm A is feasible for processing the image to compute other information, but not for displaying it.

4.2 Algorithm B

In order to preserve the aspect ratio of the input image, Algorithms B adds both rows and columns in a balanced way across the image. Some critical configurations will be repaired by inserting a row, and other critical configurations by inserting a column.

We scan the image diagonally, as shown in Fig. 3. Each time we find a critical configuration involving the four pixels across the rows $i, i+1$ and columns $j, j+1$, we compare the aspect ratio of the original image with that of the new image obtained with the already planned additions of rows and columns. We decide to add a new row or a new column, based on the choice that keeps the new aspect ratio more similar to the original one.

Figure 4 shows the same image repaired by adding just rows or just columns, or by adding both rows and columns.

Compared with Algorithm A, Algorithm B gives an aspect ratio which is very similar to the original one, but it does not guarantee the same bound on size growth. In the worst case, a chessboard pattern, Algorithm B would add a new row between any two rows, and a new column between any two columns. The image size would increase from $N \times M$ to $(2N - 1) \times (2M - 1)$, i.e., four times, as in [10].

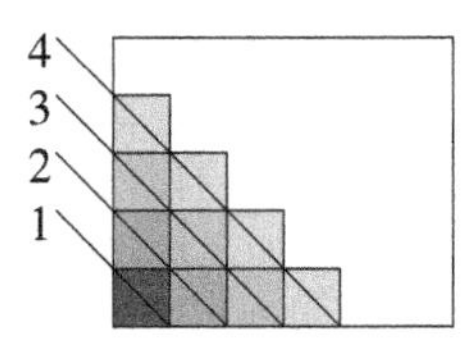

Fig. 3. The image is examined diagonally. While scanning a diagonal of pixels, we check if the upper-right vertex of each pixel is critical.

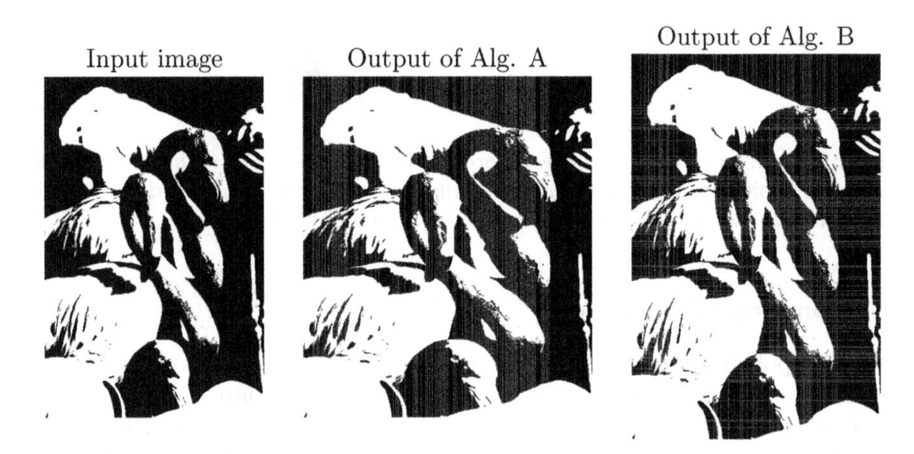

Fig. 4. Test image *flamingo* at low resolution, and its repaired versions with Algorithms A and B. Critical configurations in the input image, and added rows and columns in the repaired ones, are rendered in red. In order to fit in the page, the images have been scaled 20% of their actual size. (Color figure online)

5 Proof of Correctness

We show that the output of our repairing algorithms with respect to 8-adjacency is well-composed and homotopy equivalent to O. The claim for 4-adjacency follows by duality [6]. We focus on the insertion of a new row. The insertion of a column is symmetric. For the insertion of more rows (columns), it is sufficient to repeat the reasoning. For simplicity, we assume the critical configuration is as in Fig. 5.

5.1 Well-Composedness

We show that after processing the rows i and $i+1$, all critical configurations between them are removed and no new critical configurations are created. Let X be the pixel in the duplicated row whose color is changed from white to black (the cyan pixel in Fig. 5), and let us consider the 3×3 neighborhood of X. This color change will remove a critical configuration at the upper-right vertex of X (marked with the yellow dot), and will not create a new critical configuration at any of the other three vertices of X, whatever is the color of A and of B (the latter copied to C), as shown in Fig. 5. The lower-right vertex of X is incident with exactly three black pixels (X and the two black pixels involved in the critical configuration), its lower-left vertex is incident with two 4-adjacent pixels of the same color (the pixels B and C) and its upper-left vertex is incident with two 4-adjacent black pixels (X and the one above it).

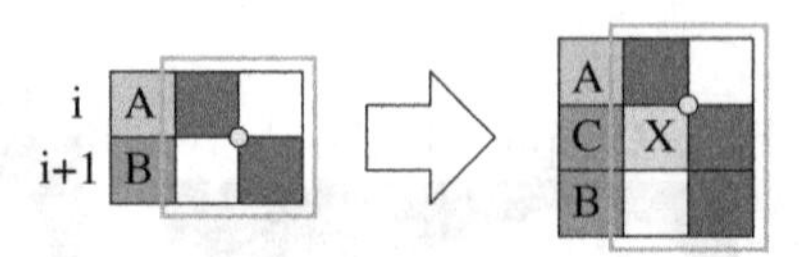

Fig. 5. The configuration at a critical vertex (yellow). The color of the pixels of the new row is copied from the bottom row. Only the cyan pixel is changed to black in order to repair the critical configuration. (Color figure online)

5.2 Homotopy Equivalence

Given a topological space X and its subspace A, a continuous function $F : X \times [0,1] \rightarrow X$ is a (strong) deformation retraction of X onto A if $F(x,0) = x$, $F(x,1) \in A$, $F(a,t) = a$ for all $x \in X$, $a \in A$ and $t \in [0,1]$. If a deformation retraction exists, then X and A are homotopy equivalent.

We show the homotopy equivalence by constructing a deformation retraction from the three rows processed at each step of the algorithms to the two rows of the original image.

Let $f(P,t) = (1-t)P + tP'$ for each point P in each added black pixel X, where P' is its radial projection on the border of that pixel from the center of the pixel below X if the pixel to the left of X is black, or from the center of the lower left neighbor of X if that pixel is white (see Fig. 6 (a)). Let $f(P,t) = P$ for all other points in the black pixels. Let $h(P,t) = (1-t)P + tP'$ for each point P in each black pixel in the third row, where P' is its vertical projection on the upper edge of that pixel, and let $h(P,t) = P$ for all other points in the black pixels in the first two rows (see Fig. 6 (b)). The required deformation retraction is the composition of the maps f and h.

6 Experimental Results and Discussion

We used ten images from the Pixabay repository [8], with two different resolutions for each image. The images were gray-scale with gray values ranging from 0 to 255, and they have been converted to binary images by applying a threshold equal to 128 or 64 (depending on the darkness of the image). The low resolution versions of the images are shown in Fig. 7.

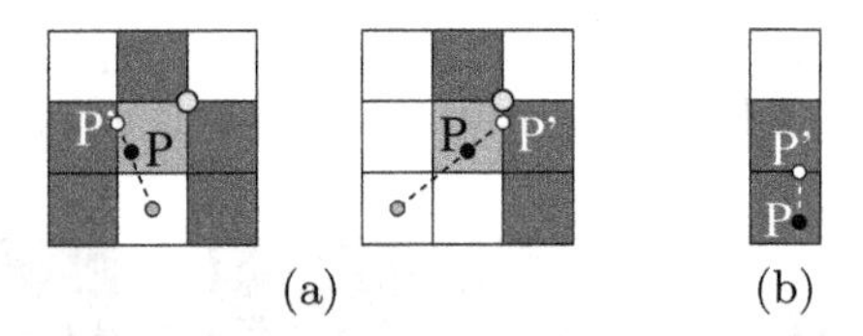

Fig. 6. The cyan pixel has been set to black in order to repair a critical configuration. The black dot is a point P inside a black pixel and the white dot is its image P' through f in (a) and through h in (b). The red dot is the center of the white pixel. (Color figure online)

In Sect. 6.1 we show and comment the results of Algorithms A and B. In Sect. 6.2 we consider the impact of the size of the repaired images in further processing algorithms, including a comparison with images repaired by other algorithms at the state of the art. All presented results refer to image repairing with 8-adjacency. The numbers would be the same with 4-adjacency, the only difference being the color of some pixels.

6.1 Results of Algorithms A and B

We developed a prototype implementation of Algorithms A and B, in Python. The results of the two algorithms on the test images are shown in Table 1. The suffix L or H refers to the same image at low and high resolution, respectively.

As expected, with Algorithm A, the size of all repaired images is less than twice the original one, and it depends heavily on the number of critical configurations in the input image. Also, the aspect ratio is relevantly changed. The size increment seems to be connected more with the number of critical configurations and less to resolution.

With Algorithm B, the aspect ratio is almost preserved, with changes occurring from the second decimal digit. On half of the images, the increment of image size is comparable to Algorithm A. On the other half, the size increases up to three times or more, especially on images with fine-grained patterns and many critical configurations (cfr. *birch*, *bird* and *train* in Fig. 7).

6.2 Processing The Repaired Images

Image repairing is often used as a preprocessing stage, as many image analysis algorithms are simpler on well-composed images. In the following, we study the impact of the size of the repaired images produced by our Algorithms A and B and by the algorithms in [10] and [3] on the performance of image processing algorithms. We have chosen these competitors because they use the smallest additional memory among those which preserve image homotopy and produce an image in the square grid.

The algorithm by Stelldinger et al. [10] produces a repaired image whose size is four times the original one. The algorithm by Čomić and Magillo [3] doubles the size of the image, but the repaired image is in a grid rotated by 45°C. The size of the images repaired by Algorithms A and B is as in Table 1.

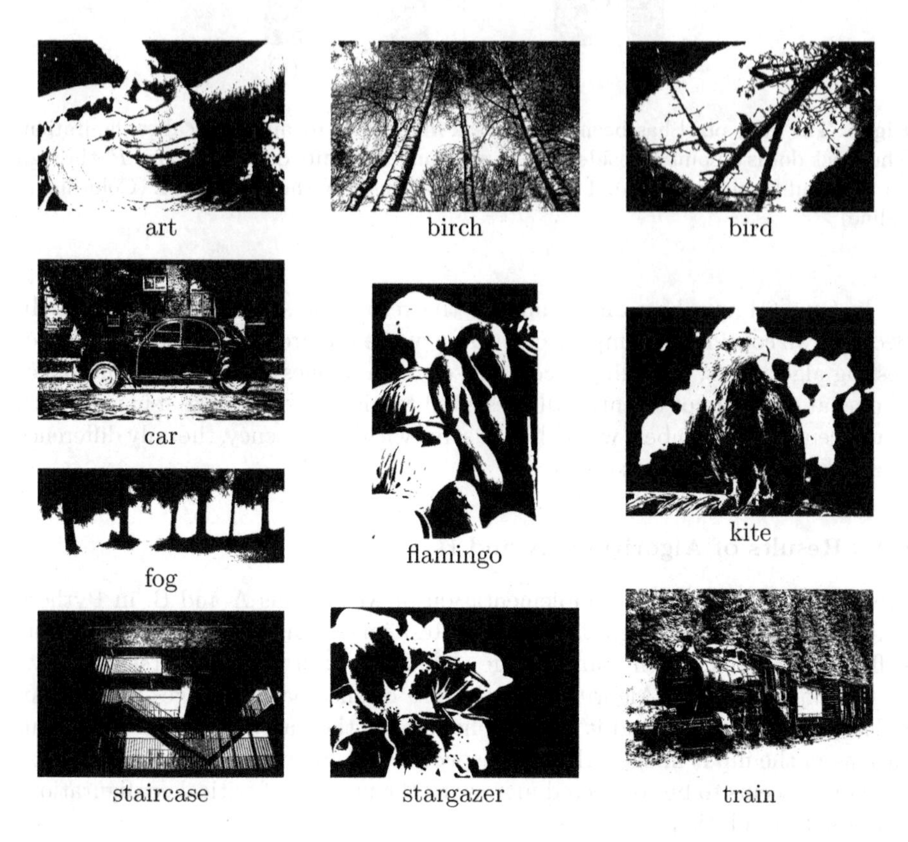

Fig. 7. Our binary versions of the original images. The shown images correspond to the low resolution and are scaled 15% to fit the page width.

As meaningful examples of image processing tasks, we consider contour extraction and shrinking, i.e., a very simple and a rather complex task. A contour is a circular list of black pixels 8-adjacent to at least one white pixel, and 4-adjacent to the previous black pixel in the list (see [7], Chap. 7.5). Shrinking iteratively changes the color of simple (removable) black pixels into white (see [4], Chap. 16.2). For these programs, we used the C implementation from [3]. The results, obtained on a PC equipped with an Intel CPU i7-2600K CPU at 3.4 GHz with 32 GB RAM, are in Tables 2 and 3.

Both processing tasks are faster on the images repaired by Algorithms A and B, than on the ones repaired by [10]. On the output images of Algorithm A, they are faster than on the output images of [3] in all cases with the exception of some images, where the times are comparable. On the images repaired by Algorithm B, contour extraction and shrinking are faster than on the images repaired by [3] in half of the cases.

Table 1. Results with Algorithms A and B. Aspect ratios are rounded to the fourth decimal digit. The percentage of size growth is rounded to integer.

Image	Input			Version A			Version B		
	Size	crit. conf.	Aspect ratio	Size	Aspect ratio	% incr. size	Size	Aspect ratio	% size
artL	640 × 426	378	1.5023	640 × 644	0.9938	151	785 × 523	1.5010	151
artH	1920 × 1280	5390	1.5	1920 × 2403	0.7990	188	2957 × 1972	1.4995	237
birchL	640 × 426	8253	1.5023	640 × 851	0.7521	200	1209 × 805	1.5019	257
birchH	1920 × 1279	74917	1.5012	1920 × 2557	0.7509	200	3702 × 2466	1.5012	372
birdL	640 × 426	1283	1.5023	640 × 801	0.7990	188	968 × 644	1.5031	229
birdH	1920 × 1279	8465	1.5012	1920 × 2531	0.7586	198	3217 × 2143	1.5012	280
carL	640 × 401	1003	1.5960	640 × 714	0.8964	178	928 × 581	1.5972	210
carH	1920 × 1205	5617	1.5934	1920 × 2314	0.8297	191	3037 × 1906	1.5934	250
flamingoL	425 × 640	175	0.6641	553 × 640	0.8641	130	480 × 722	0.6648	127
flamingoH	1276 × 1920	565	0.6646	1663 × 1920	0.8661	130	1441 × 2169	0.6644	128
fogL	640 × 235	320	2.7234	640 × 378	1.6931	161	786 × 289	2.7197	151
fogH	1920 × 706	2170	2.7195	1920 × 1298	1.4792	184	2655 × 977	2.7175	191
kiteL	640 × 524	717	1.2214	640 × 826	0.7748	158	824 × 674	1.2226	166
kiteH	1920 × 1571	3147	1.2222	2928 × 1571	1.8638	153	2549 × 2085	1.2225	176
staircaseL	640 × 417	1012	1.5348	640 × 646	0.9907	155	830 × 541	1.5342	168
staircaseH	1920 × 1253	6946	1.5323	1920 × 1970	0.9746	157	2598 × 1695	1.5327	183
stargazerL	640 × 426	108	1.5023	716 × 426	1.6808	112	690 × 459	1.5033	116
stargazerH	1920 × 1280	343	1.5	1920 × 1507	1.2741	118	2064 × 1377	1.4989	116
trainL	640 × 471	3438	1.3588	640 × 885	0.7232	188	1036 × 762	1.3596	262
trainH	1920 × 1421	15814	1.3512	1920 × 2752	0.6977	194	3235 × 2394	1.3513	284

Repaired images by Algorithm B give a better performance in cases where the original image had a small number of critical configurations (e.g., *flamingo, fog, stargazer*), but also in some other cases (e.g., *kite*). In these latter cases, probably many critical configurations were aligned, and therefore repaired by adding a single new row or column.

In presence of many critical configurations (e.g., *birch, bird, car, train*), the time for processing the output images of Algorithm B can be up to twice w.r.t. [3] (e.g., see shrinking on *birch* and *bird*). In the presence of few critical configurations, the opposite may happen (e.g., *flamingo* and *stargazer*). We remember that, unlike Algorithms A and B, the algorithm in [3] rotates the grid.

Table 2. Execution times (in milliseconds) of contour extraction on the images repaired by the four algorithms.

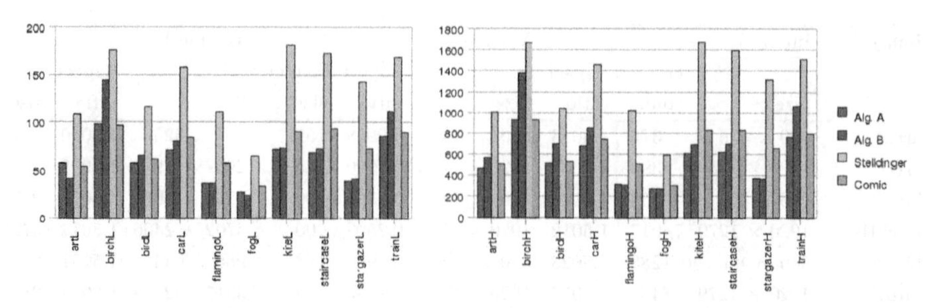

Table 3. Execution times (in milliseconds) of shrinking on the images repaired by the four algorithms.

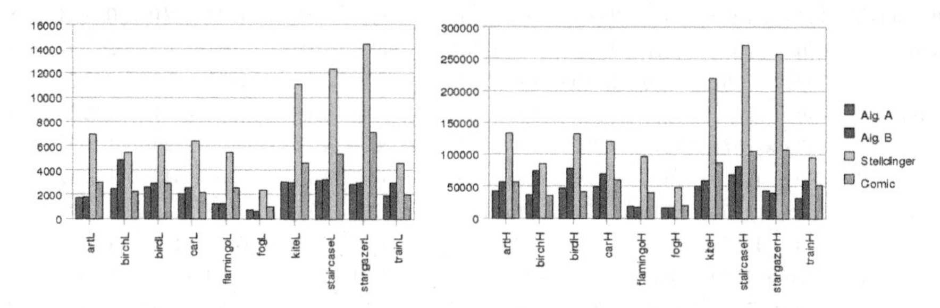

7 Conclusion

We proposed two repairing algorithms having the advantages that the obtained well-composed images lie in the square grid (thus, all existing image processing tools can be applied to the repaired images) and that the size of the output is sensitive to the number of critical configurations in the input.

Algorithm A has the further advantage of a reduced size of the repaired images, compared with all previous approaches. Its main drawback is a directional asymmetry: the repaired images are non-uniformly stretched in either vertical or horizontal direction. This makes Algorithm A less suitable for tasks that require the computation of numerical image properties such as area or perimeter, or of the Boolean operations on two distinct images, while algorithms for the computation of the topological properties can benefit from it.

Algorithm B solves this drawback at the expense of a larger size of the repaired images, if many critical configurations were present. It can be a good compromise for those cases where critical configurations are known to be few (e.g., coming from rasterization of vector formats and conversion errors).

In the future, we plan to introduce a mechanism to balance the number of added black and white pixels, to preserve the darkness/shininess of the image as well. This will improve the similarity of the repaired image with he original one. We also plan to port our prototype implementation into an efficient programming language, such as C.

Acknowledgements. This research has been partially supported by the Ministry of Education, Science and Technological Development through project no. 451-03-68/2022-14/200156

References

1. Brimkov, V.E., Maimone, A., Nordo, G., Barneva, R.P., Klette, R.: The number of gaps in binary pictures. In: Bebis, G., Boyle, R., Koracin, D., Parvin, B. (eds.) ISVC 2005. LNCS, vol. 3804, pp. 35–42. Springer, Heidelberg (2005). https://doi.org/10.1007/11595755_5
2. Brimkov, V.E., Moroni, D., Barneva, R.: Combinatorial relations for digital pictures. In: Kuba, A., Nyúl, L.G., Palágyi, K. (eds.) DGCI 2006. LNCS, vol. 4245, pp. 189–198. Springer, Heidelberg (2006). https://doi.org/10.1007/11907350_16
3. Čomić, L., Magillo, P.: Repairing binary images through the 2D diamond grid. In: Lukić, T., Barneva, R.P., Brimkov, V.E., Čomić, L., Sladoje, N. (eds.) IWCIA 2020. LNCS, vol. 12148, pp. 183–198. Springer, Cham (2020). https://doi.org/10.1007/978-3-030-51002-2_13
4. Klette, R., Rosenfeld, A.: Digital Geometry. Geometric Methods for Digital Picture Analysis, Morgan Kaufmann Publishers, San Francisco (2004)
5. Kong, T.Y., Rosenfeld, A.: Digital topology: introduction and survey. Comput. Vis. Graph. Image Process. **48**(3), 357–393 (1989)
6. Latecki, L.J., Eckhardt, U., Rosenfeld, A.: Well-composed sets. Comput. Vis. Image Underst. **61**(1), 70–83 (1995)
7. Pavlidis, T.: Algorithms for Graphics and Image Processing, 1st edn., p. 448. Springer, Heidelberg (1982). https://doi.org/10.1007/978-3-642-93208-3
8. PIXABAY: Image repository. https://pixabay.com/en/photos/grayscale/
9. Rosenfeld, A., Kong, T.Y., Nakamura, A.: Topology-preserving deformations of two-valued digital pictures. Graph. Models Image Process. **60**(1), 24–34 (1998)
10. Stelldinger, P., Latecki, L.J., Siqueira, M.: Topological equivalence between a 3D object and the reconstruction of its digital image. IEEE Trans. Pattern Anal. Mach. Intell. **29**(1), 126–140 (2007)

A Novel Method for Improving the Voxel-Pattern-Based Euler Number Computing Algorithm of 3D Binary Images

Bin Yao[1], Dianzhi Han[1], Shiying Kang[2], Yuyan Chao[3], and Lifeng He[1,4](✉)

[1] Artificial Intelligence Institute, School of Electronic Information and Artificial Intelligence, Shaanxi University of Science and Technology, Xi'an 710021, Shaanxi, China
helifeng@ist.aichi-pu.ac.jp

[2] School of Computer Science, Xianyang Normal University, Xianyang, Shaanxi 712000, China

[3] Faculty of Environment, Information and Business, Nagoya Sangyo University, Aichi 4888711, Japan

[4] Faculty of Information Science and Technology, Aichi Prefectural University, Aichi 4801198, Japan

Abstract. As an important topological property of a 3D binary image, the Euler number can be calculated by counting certain $2 \times 2 \times 2$ voxel patterns in the image. This paper presents a novel method for improving the voxel-pattern-based Euler number computing algorithm of 3D binary images. In the proposed method, by changing the accessing order of voxels in $2 \times 2 \times 2$ voxel patterns and combining the voxel patterns which provide the same Euler number increments for the given image, the average numbers of voxels to be accessed for processing a $2 \times 2 \times 2$ voxel pattern can be decreased from 8 to 4.25, which will lead to an efficient processing. Experimental results demonstrated that the proposed method is much more efficient than the conventional voxel-pattern-based Euler number computing algorithm.

Keywords: Euler number · Topological property · 3D image · Computer vision · Pattern recognition

1 Introduction

A 3D binary image can be represented by a three-dimensional array of voxels. For every pair of voxels $X = (x_1, x_2, x_3)$ and $Y = (y_1, y_2, y_3)$, X and Y are said to be 6-adjacent if $|x_1 - y_1| + |x_2 - y_2| + |x_3 - y_3| = 1$, while X and Y are said to be 26-adjacent if $max\ (|x_1 - y_1|, |x_2 - y_2|, |x_3 - y_3|) = 1$, i.e., as shown in Fig. 1, $p_1, p_3, p_5, p_7, p_{17}$ and p_{26} are 6-adjacent voxels of voxel p and $p_1, p_2, p_3, \ldots, p_{26}$ are 26-adjacent voxels of voxel p.

In recent years, there have arisen many requirements for 3D image processing with advances in various image recognizing and analyzing communities. As an important topological property of a binary image, the Euler number will not change when the image is stretched, flexed or rotated. Therefore, it has been used in many applications in images processing: processing cell images in medical diagnosis [1], shadow detection

P. L. Mazzeo et al. (Eds.): ICIAP 2022 Workshops, LNCS 13374, pp. 84–94, 2022.
https://doi.org/10.1007/978-3-031-13324-4_8

[2], reflectance-based object recognition [3] and crack detection [4]. For convenience, hereafter, whenever an "image" is mentioned, it refers to the "binary image".

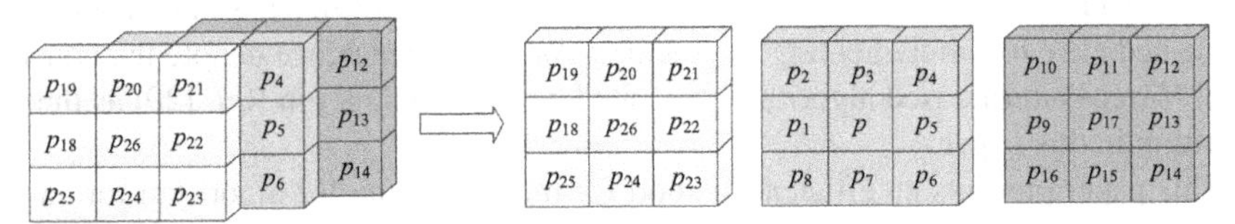

Fig. 1. Adjacent voxels in a 3D image.

The Euler number E of a 3D image is defined as follows.

$$E = O - H + C \tag{1}$$

where O is the total number of objects, H is the total number of holes (or tunnels), and C is the total number of cavities (or bubbles) in the image, respectively [5, 6]. For example, Akira and Aizawa [7] proposed a one-pass algorithm for calculating the numbers of objects, holes and cavities by utilizing an $n \times n$ array of finite-state automata, and then calculating the Euler number by formula (1).

Obviously, for calculating the Euler number according to the above definition, we need to label the objects, the holes and the cavities in the 3D image. In recent years, many novel and efficient labeling algorithms are proposed for solving the problem, such as in [8–10]. However, the labeling work need to have a comprehensive understanding of the images and it cannot be completed by the local information of the images. Experimental results verified that only for obtaining the Euler number of an image, the labeling algorithm is less efficient than other types of Euler number computing algorithms in most cases [11]. To avoid the complicated labeling work, several algorithms have been proposed for calculating the Euler number of a 3D image by alternative methods. In Ref. [12], a perimeter-based Euler number computing algorithm was proposed which calculates the Euler number of a unit-voxel-width 3D image by using of the perimeters and contact perimeters of the objects in the image. In Ref. [13], Lee and Poston proposed an algorithm based on smoothing the 3D image being processed to a differentiable object and applying theorems of differential geometry and algebraic topology. In practice, this method is suitable for both of 2D images and 3D images. In Ref. [14, 15], Lin et al. proposed a method for calculating the Euler number of a 3D image by the number of runs and neighboring runs found in the image and it is more efficient for images with high densities of object voxels. In Ref. [16], Sánchez-Cruz et al. presented a new method to calculate the Euler number of a 3D image by considering a voxelized object with tunnels and/or cavities and the relationship between contact voxel faces with enclosing surfaces. In this method, $2 \times 2 \times 2$ voxel patterns, $2 \times 2 \times 1$ voxel patterns, $1 \times 2 \times 2$ voxel patterns and $2 \times 1 \times 2$ voxel patterns need to be counted in the image.

Recently, two novel methods for computing the Euler number of a 3D image were presented. In Ref. [17], Sossa proposed a novel method in terms of a codification of the vertices of the object voxels, which can be treated as the extension of Ref. [18]. Čomića proposed a surface-based formula for computing the Euler characteristic of an arbitrary object (well-composed or not) in the cubical grid, with either vertex- or face-adjacency in Ref. [19]. The method is based on counting only the boundary vertices and faces in the object, with the vertex count adjusted for the two adjacency relations.

Especially, Park and Rosenfeld [5] presented a method for calculating the Euler number of a 3D image by counting certain 2 × 2 × 2 voxel patterns for 6-adjacent. Morgenthaler extended this method to 26-adjacent in Ref. [20]. These algorithms are simple and easy for implementing in practice. For convenience, we denote the conventional Voxel-Pattern-based algorithm presented by Morgenthaler in Ref. [20] as the *VP* algorithm in this paper.

This paper presents an efficient method for improving the *VP* algorithm for calculating the Euler number of a 3D image. By changing the accessing order of voxels in 2 × 2 × 2 voxel patterns and combining the voxel patterns which provide the same Euler number increments for the given image, the average number of voxels to be accessed for processing a 2 × 2 × 2 voxel pattern can be decreased from 8 to 4.25, which will lead to an efficient processing. Experimental results demonstrated that the proposed method is much more efficient than the conventional voxel-pattern-based Euler number computing algorithm.

The rest of this paper is organized as follows: in the next section, we review the *VP* algorithm, and propose our method in Sect. 3. In Sect. 4, we use experimental results on various resolutions of noise images to evaluate the performance of our method and compare it with the *VP* algorithm. Lastly, we give the conclusion in Sect. 5.

2 Review of Conventional Voxel-Pattern-Based Euler Number Computing Algorithm of a 3D Image

For an $X \times Y \times Z$-size 3D image, we assume that the foreground (object) voxels and background (non-object) voxels are represented by 1 and 0, respectively. As in most image processing algorithms, we assume that all voxels on the border of an image are background voxels, i.e., $f(i, j, k) = 0$ if at least one of these (i, j or k) is equal to 1, X, Y or Z. Moreover, we only consider 26-adjacent for foreground voxels in this paper.

For calculating the Euler number of a 3D image, the *VP* algorithm needs to count certain 2 × 2 × 2 voxel patterns shown in Fig. 2 in the image. Let $\#[i]$ $(1 \leq i \leq 22)$ be the number of times that the voxel pattern i occurs in a 3D image, then the Euler number of the image can be calculated by use of formula (2).

$$E = \Psi_1 - \Psi_2 + \Psi_3 - \Psi_4 + \Psi_5 - \Psi_6 + \Psi_7 - \Psi_8 \tag{2}$$

where

$\Psi_1 = \#[1];$
$\Psi_2 = \#[2] + \#[3] + \#[4];$
$\Psi_3 = \#[5] + \#[6] + \#[7];$
$\Psi_4 = \#[8] + \#[9] + \#[10] + \#[11] + \#[12] + \#[13] + \#[14];$
$\Psi_5 = \#[15] + \#[16] + \#[17];$
$\Psi_6 = \#[18] + \#[19] + \#[20];$
$\Psi_7 = \#[21];$
$\Psi_8 = \#[22].$

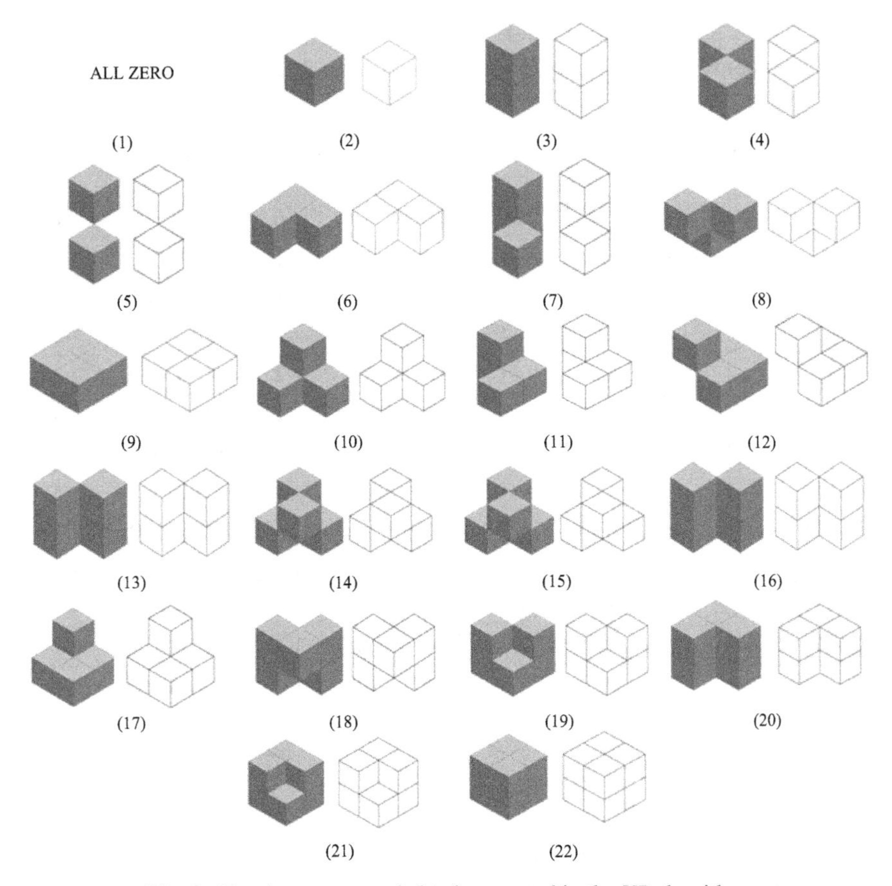

Fig. 2. Voxel patterns needed to be counted in the *VP* algorithm.

As is known, there are eight voxels in a $2 \times 2 \times 2$ voxel pattern and each voxel is either a background voxel or a foreground voxel. Therefore, there are 256 different types of configurations for a $2 \times 2 \times 2$ voxel pattern theoretically. For confirming the configuration of a voxel pattern, we need to check it in all orientations. In the *VP* algorithm, Morgenthaler summarized the Euler number increments of all types of voxel patterns and presented the increment table, as shown in Table 1, where the Euler number increment $\triangle E$ of a voxel pattern which is not equal to zero in a 3D image is tabulated by a binary index. The index is derived from the values of voxels as a binary string *abcdefgh*. For example, for a voxel pattern shown in Fig. 3, if the values of voxels *a*, *b*, *c*, *d*, e, f, *g* and *h* are 0, 0, 0, 0, 1 0, 0 and 1, respectively, the binary string will be 00001001 and the index of the voxel pattern will be a decimal number 9. According to Table 1, the Euler number increment $\triangle E$ of this voxel pattern is -1. The indexes and the Euler number increments $\triangle E$ of rest voxel patterns can be determined in the same manner.

From Table 1, we can find that although there are 256 possible types of configurations for a $2 \times 2 \times 2$ voxel pattern, only 49 voxel patterns' Euler number increments are not equal to zero. Among those voxel patterns which will affect the Euler number of a 3D image, 30 voxel patterns' Euler number increments are −1, 17 voxel patterns' Euler number increments are 1 and 2 voxel patterns' Euler number increments are −2.

Table 1. Indexes of the Euler number increments of voxel patterns.

Index	ΔE	Index	ΔE	Index	ΔE	Index	ΔE
00000010	1	00001001	−1	00001011	−1	00011000	−1
00011001	−1	00011010	−1	00011011	−1	00100001	−1
00100011	−1	00100100	−1	00100101	−1	00100110	−1
00100111	−1	00101000	−1	00101001	−2	00101010	−1
00101011	−2	00101100	−1	00101101	−1	00101110	−1
00101111	−1	00111000	−1	00111001	−1	00111010	−1
00111011	−1	10000001	−1	10000011	−1	10001001	−1
10001011	−1	10010100	1	10010101	1	10010110	1
10010111	1	10011100	1	10011101	1	10011110	1
10011111	1	10100001	−1	10100011	−1	10101001	−1
10101011	−1	10110100	1	10110101	1	10110110	1
10110111	1	10111100	1	10111101	1	10111110	1
10111111	1	Others	0				

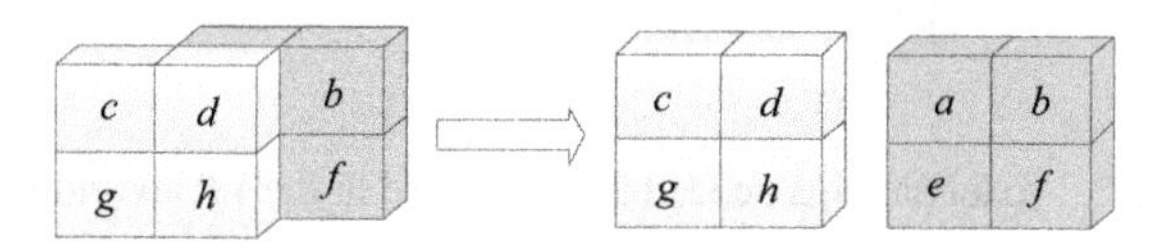

Fig. 3. Accessing order of voxel in a voxel pattern.

According to the *VP* algorithm, for calculating the Euler number of a 3D image, we only need to scan the image, access voxels one by one in the image, confirm the indexes of the corresponding voxel patterns and look for the corresponding ΔE in Table 1. When all voxels are processed and all the indexes of voxel patterns contained in the 3D image are confirmed, we can obtain the increments of all voxel patterns and the sum of all increments will be the Euler number of the image. Obviously, for confirming the voxel patterns which will affect the Euler number of the image, we have to process all voxel patterns and access all the eight voxels in each voxel pattern. Thus, for an $X \times Y \times Z$-size 3D image, the *VP* algorithm needs to access voxels $8 \times X \times Y \times Z$ times for calculating the Euler number.

3 Our Proposed Method

In the *VP* algorithm mentioned above, although only 49 types of voxel patterns need to be counted, it is necessary to confirm all voxel patterns' indexes to determine whether a voxel pattern should be counted or not. For confirming the index of a voxel pattern, there are eight voxels need to be accessed.

As can be seen from Table 1, when the indexes of voxel patterns are presented by a binary string *abcdefgh*, it can be found that some indexes of the voxel patterns are consecutive, i.e., 00011000–00011011, 00100100–00100111, 00101100–00101111, 00111000–00111011, 10010100–10010111, 10011100–10011111, 10110100–10110111 and 10111100–10111111. Noticed that these consecutive indexes of voxel patterns have one thing in common, that is, each group of indexes has the same Euler number increment. Obviously, the difference among these indexes only lays in the last two bits, which vary from 00 to 11.

Based on the above observation, when calculating the Euler number of a 3D image, we can access fewer voxels for confirming the indexes of the voxel patterns. For example, if the values of voxels *a, b, c, d, e* and *f* in the current processing voxel pattern are 0, 0, 0, 1, 1 and 0, respectively, then, we can conclude immediately that the Euler number should be decreased by 1 without accessing the remain two voxels. Thus, in such a case, for processing a voxel pattern, we only need to access six voxels. The other groups of voxel patterns with consecutive indexes can be processed in the same manner. As shown in Table 2, following this consideration, 32 types of voxel patterns listed in the table can be processed by accessing six voxels, which will lead to an efficient processing.

Table 2. The Euler number increment of groups of voxel patterns whose indexes are consecutive.

Index	ΔE	Index	ΔE	Index	ΔE	Index	ΔE
00011000–00011011	−1	10010100–10010111	1	00100100–00100111	−1	10011100–10011111	1
00101100–00101111	−1	10110100–10110111	1	00111000–00111011	−1	10111100–10111111	1

In addition, when further analyzing the indexes of the voxel patterns which will affect the Euler number of an image listed in Table 1, we found an important fact that for each of such voxel patterns, the voxel at the second position of the binary index *abcdefgh*, i.e., "*b*" is certainly a background voxel. According to this fact, when processing a voxel pattern to confirm its index, we should access the voxel at the position "*b*" firstly. If the voxel at the position "*b*" in a voxel pattern is a foreground voxel, we can conclude that the processing voxel pattern should not be counted immediately, i.e., we do not need to access any of other voxels in the voxel pattern. Otherwise, we go on to access other voxels for confirming the index of the voxel pattern. In this way, for any voxel pattern with a foreground voxel at the position "*b*" in its index, we only need to access one voxel to confirm that the voxel pattern does not need to be counted.

In implementation, we can take advantages of the above two strategies simultaneously for improving the *VP* algorithm. The procedure for processing a voxel pattern can be given as follows.

For the voxel pattern being processed, we access the voxel at the position "*b*" in the binary index *abcdefgh* of the voxel pattern firstly.

(1) If the voxel at the position "*b*" is a foreground voxel, we can conclude the processing voxel pattern should not be concerned because its Euler number increment must be 0. Accordingly, we do not need to access other voxels in the current voxel pattern any more. Then, we go on to process the next voxel pattern.
(2) If the voxel at the position "*b*" is a background voxel, we proceed to access the other voxels *a*, *c*, *d*, *e* and *f* in sequence in the voxel pattern. If the value set {*a*, *c*, *d*, *e*, *f*} is one of {0, 0, 1, 1, 0}, {0, 1, 0, 0, 1}, {0, 1, 0, 1, 1}, {0, 1, 1, 1, 0}, {1, 0, 1, 0, 1}, {1, 0, 1, 1, 1}, {1, 1, 1, 0, 1} or {1, 1, 1, 1, 1}, according to Table 2, we only need to access six voxels for confirming the Euler number increments in these cases. Then, we can go to process the next voxel pattern.
(3) Otherwise, we need to access the two remaining voxels to confirm the index of the current voxel pattern. In these cases, we need to access eight voxels for confirming the Euler number increment of the voxel pattern.

The pseudo codes of our method for processing each voxel pattern can be given as follows:

```
For the processing voxel pattern {a, b, c, d, e, f, g, h}
if voxel b is a foreground voxel
|  go to process the next voxel pattern;
else
|  access the voxels a, c, d, e and f in sequence;
|  if the value set of voxels {a, c, d, e, f} is listed in Table 2
|  |  E = E + ΔE;
|  else
|  |  access the voxels g and h;
|  |  confirm ΔE according to Table 1;
|  |  E = E + ΔE;
|  end of if
end of if
```

When all voxel patterns are processed, we can obtain the increments of all voxel patterns and the sum of the increments is the Euler number of the image. For convenience, we denote this Changing-Order-based algorithm as the *CO* algorithm in this paper.

In our proposed method, if the voxel "*b*" in the index *abcdefgh* of a voxel pattern is a foreground voxel, for confirming its index, we do not need to access any of other voxels in the voxel pattern. In this case, we only need to access one voxel for processing a voxel pattern and obtaining its Euler number increment. Furthermore, if the values of voxels *a*, *b*, *c*, *d*, *e* and *f* in the current processing voxel pattern are listed in Table 2, in these cases, we only need to access six voxels for processing a voxel pattern. Otherwise, we need to access all eight voxels for processing the current voxel pattern. If there is the same probability of background voxels and foreground voxels occurring in a voxel pattern, for half of the voxel patterns contained in the image, we only need to access one voxel for confirming their indexes. Moreover, 32 types of voxel patterns listed in Table 2 can be processed by accessing six voxels and the rest 96 types of voxel patterns need to access eight voxels for confirming their indexes. Therefore, for an $X \times Y \times Z$-size 3D image, the total accesses of voxels for calculating the Euler number will be $1 \times 128/256 \times X \times Y \times Z + 6 \times 32/256 \times X \times Y \times Z + 8 \times 96/256 \times X \times Y \times Z = 4.25 \times X \times Y \times Z$, which is much less than that in the *VP* algorithm. In the next section, we will evaluate the performance of the proposed method by experiments.

4 Experimental Results

In this section, we compare the performance of our proposed *CO* algorithm with the conventional voxel-pattern-based *VP* algorithm. All algorithms used for comparison were implemented in the C language on a PC-based workstation (Intel Core i7-6770 CPU@ 3.20 GHz, 8 GB Memory, Ubuntu Linux OS), and compiled by the GNU C compiler (version 4.6.1) with the option −O3.

Because noise images have complicated geometric shapes and complex connectivity, severe evaluations of algorithms will be performed with these images. Five sizes ($32 \times 32 \times 32$, $64 \times 64 \times 64$, $128 \times 128 \times 128$, $256 \times 256 \times 256$, $512 \times 512 \times 512$ voxels) of noise images were used for the experiments. For each size, the 41 noise images were generated by thresholding of the images containing uniform random noise with random values from 0 to 1000 in steps of 25. All experimental results presented in this section were obtained by averaging of the execution time for 1000 runs.

4.1 Execution Time Versus Image Sizes

All noise images were used for this experiment. The experimental results of execution time versus image sizes are shown in Fig. 4. From Fig. 4, we can find that for both the maximum execution time and the average execution time, two compared algorithms have the ideal linear characteristics versus image sizes. Moreover, we can find that our proposed algorithm is more efficient than the *VP* algorithm.

4.2 Execution Time Versus Image Densities

We used forty-one noise images with a size of $64 \times 64 \times 64$ voxels for evaluating the execution time versus the densities of the foreground voxels in an image. The experimental results are shown in Fig. 5 and Table 3. From Fig. 5 and Table 3, we can find that for almost all images, our proposed algorithm is more efficient than the *VP* algorithm, especially for the images whose densities are from 0.4 to 0.8.

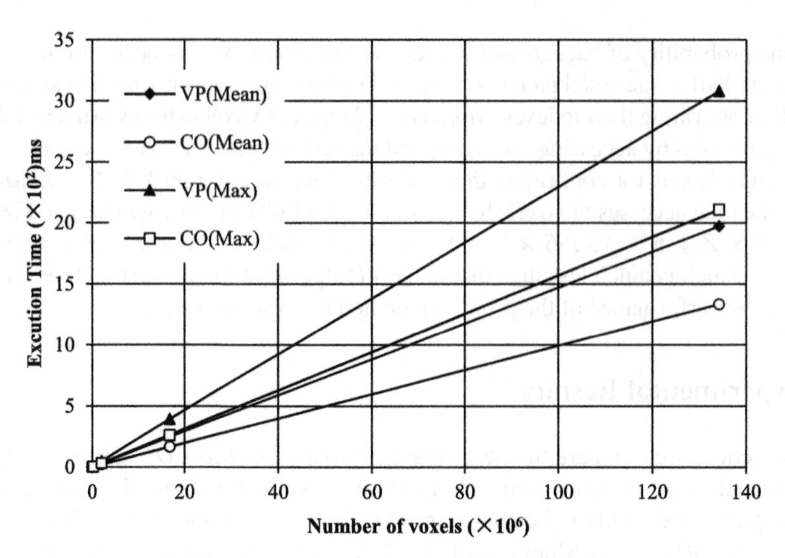

Fig. 4. Execution time versus image sizes.

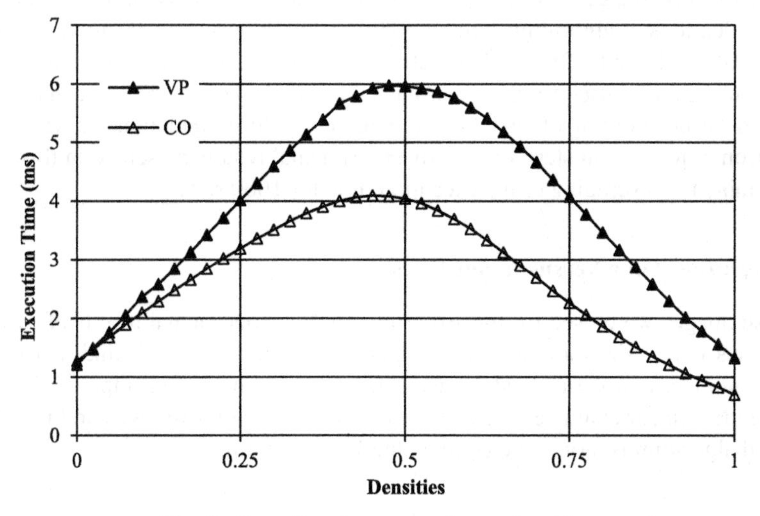

Fig. 5. Execution time versus image densities.

Table 3. Execution time (*ms*) versus the density of 64 × 64 × 64-sized noise images.

Density	VP	CO	Density	VP	CO	Density	VP	CO
0.000	1.156	1.173	0.025	1.404	1.361	0.050	1.638	1.557
0.075	1.880	1.745	0.100	2.122	1.927	0.125	2.372	2.105
0.150	2.624	2.278	0.175	2.904	2.463	0.200	3.174	2.633
0.225	3.442	2.789	0.250	3.701	2.952	0.275	3.965	3.099
0.300	4.226	3.248	0.325	4.478	3.387	0.350	4.721	3.508
0.375	4.929	3.610	0.400	5.121	3.697	0.425	5.270	3.762
0.450	5.388	3.794	0.475	5.468	3.784	0.500	5.479	3.737
0.525	5.425	3.658	0.550	5.359	3.542	0.575	5.25	3.405
0.600	5.110	3.243	0.625	4.928	3.064	0.650	4.731	2.879
0.675	4.505	2.660	0.700	4.262	2.473	0.725	4.001	2.276
0.750	3.730	2.070	0.775	3.466	1.885	0.800	3.186	1.705
0.825	2.915	1.539	0.850	2.642	1.379	0.875	2.372	1.231
0.900	2.123	1.096	0.925	1.865	0.968	0.950	1.642	0.859
0.975	1.432	0.759	1.000	1.234	0.654			

5 Conclusion

In this paper, we presented a novel method for improving the conventional voxel-pattern-based algorithm in order to calculate the Euler number of a 3D binary image. By changing the accessing order of voxels in 2 × 2 × 2 voxel patterns and combining the voxel patterns which provide the same Euler number increments for the given image, the average numbers of voxels to be accessed for processing a 2 × 2 × 2 voxel pattern can be decreased from 8 to 4.25, which will lead to an efficient processing. Experimental results demonstrated that the proposed method is much more efficient than the conventional voxel-pattern-based Euler number computing algorithm.

In fact, there are many voxels accessed repeatedly by use of the raster scan of the given image in the *VP* algorithm. While in our proposed method, we pay more attention to the features existing in some voxel patterns. On the other hand, the essence of the proposed method is checking voxel patterns in the image in sequence, and it is very suitable for parallel implementation on GPUs as done in [21] to boost the performance. In our future work, we will find an alternative method of reducing the number of voxels accessed repeatedly and consider parallel implement on SIMD CPU, GPU or FPGA for improving the voxel-pattern-based Euler number computing algorithm further.

Acknowledgments. This work was supported in part by the National Natural Science Foundation of China under Grant No. 61971272, No. 61603234 and the Scientific Research Foundation of Shaanxi University of Science & Technology under Grant No. 2020BJ-18.

References

1. Hashizume, A., Suzuki, R., Yokouchi, H., et al.: An algorithm of automated RBC classification and its evaluation. Biomed. Eng. **28**(1), 25–32 (1990)
2. Rosin, P., Ellis, T.: Image difference threshold strategies and shadow detection. In: British Machine Vision Conference, Birmingham, UK, pp. 10–13 (1995)
3. Nayar, S., Bolle, R.: Reflectance-based object recognition. Int. J. Comput. Vis. **17**(3), 219–240 (1996)
4. Liu, Y., Cho, S., Spencer, B., et al.: Concrete crack assessment using digital image processing and 3D scene reconstruction. J. Comput. Civ. Eng. **30**(1), 04014124 (2014)
5. Park, C., Rosenfeld, A.: Connectivity and genus in three dimensions. Computer Science Center, Technical report TR-156. Univ. Maryland, College Park (1971)
6. Toriwaki, J., Yonekura, T.: Euler number and connectivity indexes of a three dimensional digital picture. Forma **17**, 183–209 (2002)
7. Akira, N., Aizawa, K.: On the recognition of properties of three-dimensional pictures. IEEE Trans. Pattern Anal. Mach. Intell. **7**(6), 708–713 (1985)
8. Lemaitre, F., Lacassagne, L.: A new run-based connected component labeling for efficiently analyzing and processing holes, p. 09299 (2020). https://doi.org/10.48550/arXiv.2006
9. Bolelli, F., Allegretti, S., Grana, C.: One DAG to rule them all. IEEE Trans. Pattern Anal. Mach. Intell. (2021). https://doi.org/10.1109/TPAMI.2021.3055337
10. He, L., Chao, Y.: A very fast algorithm for simultaneously performing connected-component labeling and Euler number computing. IEEE Trans. Image Process. **24**(9), 2725–2735 (2015)
11. Yao, B., He, L., Kang, S., Zhao, X., Chao, Y.: A new run-based algorithm for Euler number computing. Pattern Anal. Appl. **20**(1), 49–58 (2017). https://doi.org/10.1007/s10044-015-0464-4
12. Bribiesca, E.: Computation of the Euler number using the contact perimeter. Comput. Math. Appl. **60**(5), 1364–1373 (2010)
13. Lee, C., Poston, T.: Winding and Euler numbers for 2D and 3D digital images. Graph. Models Image Process. **53**(6), 522–537 (1991)
14. Lin, X., Xiang, S., Gu, Y.: A new approach to compute the Euler number of 3D image. In: IEEE Conference on Industrial Electronics and Applications, Singapore, 3–5 June 2008
15. Lin, X., Ji, J., Huang, S., et al.: A proof of new formula for 3D images Euler number. Pattern Recognit. Artif. Intell. **23**(1), 52–58 (2010)
16. Sánchez, H., Sossa, H., Braumann, U., et al.: The Euler-Poincaré formula through contact surfaces of voxelized objects. J. Appl. Res. Technol. **11**(1), 65–78 (2013)
17. Sossa, H., Rubío, E., Ponce, V., Sánchez, H.: Vertex codification applied to 3-D binary image Euler number computation. In: Martínez-Villaseñor, L., Batyrshin, I., Marín-Hernández, A. (eds.) MICAI 2019. LNCS, vol. 11835, pp. 701–713. Springer, Cham (2019). https://doi.org/10.1007/978-3-030-33749-0_56
18. Sossa, J., Santiago, R., Pérez, M., et al.: Computing the Euler number of a binary image based on a vertex codification. J. Appl. Res. Technol. **11**(3), 360–370 (2013)
19. Čomića, L., Magillo, P.: Surface-based computation of the Euler characteristic in the cubical grid. Graph. Models **112**, 101093 (2020)
20. Morgenthaler, D.: Three-Dimensional Digital Image Processing. Univ. Maryland, College Park (1981)
21. Lemaitre, F., Hennequin, A., Lacassagne, L.: Taming voting algorithms on Gpus for an efficient connected component analysis algorithm. In: 2021 IEEE International Conference on Acoustics, Speech and Signal Processing, Toronto, Canada, April, pp. 7903–7907 (2021)

Event-Based Object Detection and Tracking - A Traffic Monitoring Use Case -

Simone Mentasti(✉), Abednego Wamuhindo Kambale, and Matteo Matteucci

Department of Electronics Information and Bioengineering of Politecnico di Milano, p.zza Leonardo da Vinci 32, Milan, Italy
{simone.mentasti,abednego.kambale,matteo.matteucci}@polimi.it

Abstract. Traffic monitoring is an important task in many scenarios, in urban roads to identify dangerous behavior and on-highway to check for vehicles moving in the wrong direction. This task is usually performed using conventional cameras but these sensors suffer from fast illumination changes, particularly at night, and extreme weather conditions. This paper proposes a solution for object detection and tracking using event-based cameras. This new technology presents many advantages to address traditional cameras limitations; the most evident are the high dynamic range and temporal resolution. However, due to the different nature of the provided data, solutions need to be implemented to process them in an efficient way. In this work, we propose two solutions for object detection, one based on standard geometrical approaches and one using a deep learning framework. We also release a novel dataset for this task, and present a complete application for road monitoring using event cameras (Dataset available at: https://airlab.deib.polimi.it/datasets-and-tools/).

Keywords: Event-cameras · Object detection and tracking · Road monitoring

1 Introduction

Event cameras, also known as neuromorphic cameras, are a new type of sensor inspired by the working process of the human retina. Unlike traditional cameras, they do not acquire a complete frame at a constant speed. Instead, each sensor pixel works independently and returns a value representing the brightness change only when it occurs for that specific pixel. This new type of sensor possess several advantages over traditional frame cameras; they have a high temporal resolution in the µs range, high pixel bandwidth, low power consumption, and high dynamic range [1]. These properties enable them to give an high throughput in different applications such as pose estimation [2], Simultaneous Localisation And Mapping (SLAM) [3], autonomous robots perception, and object tracking [4]. In particular, event-based object tracking algorithms can be divided into

P. L. Mazzeo et al. (Eds.): ICIAP 2022 Workshops, LNCS 13374, pp. 95–106, 2022.
https://doi.org/10.1007/978-3-031-13324-4_9

two classes: the event-driven tracking method and the pseudo-frame-based tracking method. The first one involves processing each event without accumulating it in a frame and requires ad-hoc solutions designed especially for these types of sensors. The second accumulates events into a binary pseudo-frames that can be used as input to more traditional computer vision algorithms.

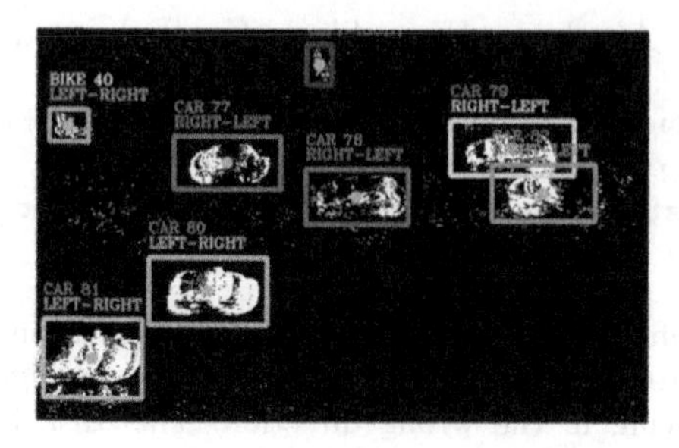

Fig. 1. Example image of the working system in a two lane scenario, for each object the system computes a 2D bounding box, assigns an ID, and performs tracking displaying the moving direction

In this work, we propose a pseudo-frame-based detection and tracking approach using a Prophesee third-generation event-based camera [5]. The goal of our project is to develop a robust system for object detection and tracking in outdoor scenarios, with a particular interest in road monitoring, as shown in Fig. 1. Compared to a traditional camera, event-based sensors provide different advantages in this specific task. In particular, the high dynamic range allowed us to develop a solution that works equally well during night and day, without the risks of having the sensor dazzled by traffic lights. Similarly, due to the high frame rate, rainy conditions are not an issue since the drop of rain will be just some tiny dots on the image, which can be easily filtered.

This work investigates two different approaches to the problem: a model-free algorithm that uses only geometrical features to detect objects and a YOLOv4-based [6] approach which uses a deep learning model to detect cars, bikes, and pedestrians. The final component of the pipeline is the tracking module [6], which uses a Kalman filter [7] and the Hungarian algorithm [8] for detection, association, and ID assignment. The proposed solution is a robust pipeline that can track moving objects in the camera's field of view. In particular, the system has been designed for road monitoring. It can track cars, bikes, and pedestrians and monitors a set of useful parameters like the object's direction and the number of objects by category. Moreover, it can raise alarms in case of vehicles moving in the wrong direction. When the system is mounted on the highway, it is a handy feature to identify vehicles entering ramps in the opposite direction. This work aims to demonstrate the effectiveness of event cameras in the detection and tracking context, their suitability for deep learning, and provide the community with an annotated dataset to train such architectures.

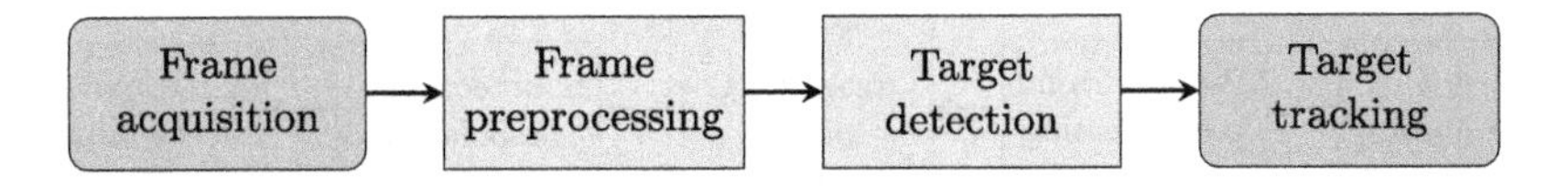

Fig. 2. Frame-based object tracking pipeline.

This paper is structured as follows. In Sect. 2 we present the state of the art on object detection and event-based sensors, although, due to the novelty of the sensor, the literature is fairly limited. Then in Sect. 3 we propose our pipeline for object detection and tracking from event data. For the detection component, two different solutions are proposed. Finally, in Sect. 4 we present some numerical results to highlight the advantages of the best performing solution, and Sect. 5 draws some conclusions and highlights future development.

2 State of the Art

Traditionally, frame-based object detection and tracking consist of four steps, as shown in Fig. 2. First, the *Frame acquisition process* outputs a frame that feeds the *preprocessing* stage, in charge of removing noise and keeping relevant information for further steps.

The *target detection* phase consists of separating the object's pixels from the unwanted signal. The detection can be performed either without a machine learning based object detector or with the help of one. In the first case, a common approach is to use nearest neighbor algorithm [9,10] to perform Connected Component Labeling (CCL) [11], clustering groups of pixels to form objects. Then performing Connected Component Analysis (CCA) [12,13] to determine the area, the centroid, and the bounding box of every detection. Finally it is possible to use these features to categorize different objects in the scene [14,15]. This can be done using geometrical solutions, but also classical machine learning models like Support Vector Machine [16] and AdaBoost [17]. A second approach is to use deep learning (DL) methods to perform object detection [18].

Most DL-based object detection models treat object detection as a regression/classification problem and use end-to-end architectures built on regression/classification to map pixels to bounding box locations together with class probabilities. This operation has the advantage of reducing computational time. One significant and representative framework is You Only Look Once (YOLO) [19]. The YOLO framework predicts both bounding boxes and confidences for multiple categories. The original architecture works with 416×416 images at 45 fps in real-time using consumer GPUs. A simpler variant called Fast YOLO [20] can process images at 155 fps with a slight loss in accuracy. YOLOv2 is an enhanced version that was later proposed in [21], which presents some improvements to the original architecture, like batch normalization, multi-scale training, and anchor boxes.

The final stage of the *association and target tracking* aim to connect single frame detections to a common target and create a track that captures target's

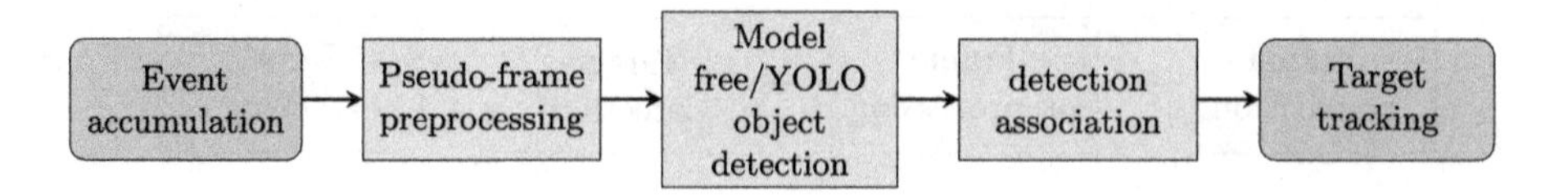

Fig. 3. Schema of the object detection pipeline.

current position and prior movements. The goal is to find an optimal matching between the tracked objects and these predictions. A widely used algorithm for this task is the Hungarian algorithm [22].

While all the presented techniques were designed to work on traditional images, proper adaptation is required to process event-based streams. The literature identifies two distinct ways to achieve target detection and tracking for event-based cameras [23]. One approach uses pseudo-frames, whereas the other performs direct detection on filtered events. The corresponding techniques for frame-based sensors previously highlighted differ slightly for event-based detection when using pseudo-frames. The process starts naturally by capturing the raw event data; undesirable events are filtered to extract meaningful object motion, and frames are built. Then traditional techniques can be employed to process the reconstructed image. The solutions based directly on events, process instead each event asynchronously and independently. Two different methods have been used in the literature for this second approach, one using adaptive time surfaces [24] and another using events association in 3D Point [25,26]. The main advantage of these approaches is that the event-based sensor's high temporal resolution is fully utilized, still, a custom solution needs to be developed, and it is impossible to adapt from standard computer-vision approaches.

3 Object Detection and Tracking from Event Data

This paper proposes a pipeline for object detection and tracking from event data acquired by a Porphesee event-based camera. In particular, we present two alternative approaches; the first one, which we call "Model-free", is based on geometrical operations on the pseudo-frame to retrieve the objects. The second instead leverages a convolutional neural network to extract a list of targets. Nevertheless, the initial processing phase and the clusters tracking are common between the two solutions, as shown in Fig. 3.

3.1 Preprocessing

Pseudo-frames are event-based versions of conventional video frames created by dividing continuous event data into evenly spaced time bins, or frames, to create an image. They are similar to standard binary frames except that they are sparse, with areas with no events set as black. This similarity with binary images enables traditional detection and tracking algorithms to use pseudo-frames as their input with minimal changes.

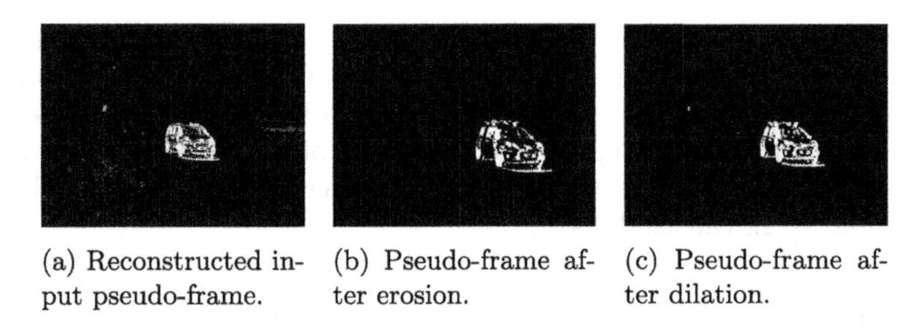

(a) Reconstructed input pseudo-frame. (b) Pseudo-frame after erosion. (c) Pseudo-frame after dilation.

Fig. 4. Evolution of the pseudo-frame through the preprocessing pipeline.

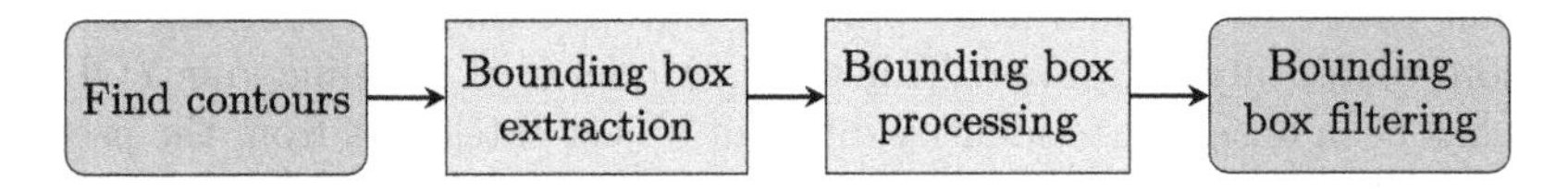

Fig. 5. Schema of the "model-free" pipeline.

The accumulation of events into pseudo-frames starts by receiving the stream of events from the event-based sensors. Each event from the received stream is added into a queue used to generate frames depending on the user-defined event accumulation time (for our test, we employed a window of 20 ms, $dt = 20$ ms, which guarantees a frame rate of 50 Hz). The generation of a frame at a specific time T takes all the events whose timestamp t lies between $T - dt$ and T.

Due to the sensor's high sensitivity, not all received events are valid data; some are just noise. Therefore, the preprocessing step helps filtering the data and discards noisy points that may negatively affect detection. Moreover, an object represented by the events is not always continuous; some object regions may appear without events affecting the object detection phase. For this reason, the last step of the preprocessing pipeline is a sequence of morphological operations, as shown in Fig. 4. In particular, the best results were achieved by applying first an erosion filter to remove single noisy points from the sensor, which are fairly frequent in this type of camera, (Fig. 4a). Then we apply a dilation filter to combine clusters close together, and therefore belonging to the same obstacle. This preprocessing phase is fundamental, particularly for the "Model-free" obstacle detector, since this is based only on geometrical constraints and separate components can negatively affect system performance.

3.2 Model-Free Detection

The "Model-free" detector relies on the frames provided by the pseudo-frame filtering process. Figure 5 shows the pipeline used for this approach. Starting from the filtered pseudo-frame, the first operation finds the objects in the scene by detecting the contours using the algorithm presented in [27]. From the obtained contours, we extract the associated rectangular bounding boxes by retrieving the rectangle's location, width, and height that best fit each contour area. A final step of bounding box processing is required to group bounding boxes near each other which are more likely to represent the same object. Also, in this last step, we discard bounding boxes that lie inside others. The latter is a fairly common issue since event cameras generally return obstacles' contours and color discontinuity on an obstacle can generate a smaller target inside the obstacle bounding box (e.g., car windows).

3.3 YOLO-Based Detector

A YOLOv4 model has been applied to provide a more reliable and accurate detection than the "Model-free" pipeline, but it requires a training dataset and more computational power. Unlike the "Model-free" pipeline, our YOLOv4 model has been designed to identify different object categories from the pseudo-frame. However, accurate detection results require a well-trained model, which depends on the availability of a good dataset. Therefore, we built our dataset from multiple recordings of real-world scenarios to train YOLOv4 and YOLOv4-tiny models. We used a dataset consisting of 3.1k images, which has been obtained manually annotating 1255 images and then performing data augmentation using the Roboflow tool [28]. Data augmentation techniques included flipping, rotation, translation and cropping. We were interested only in detecting cars, bikes, and pedestrians for this task, so these are the only label used in the dataset. We used transfer learning to speed up the training phase, starting from the YOLOv4 pre-trained weights provided online. We used Recall, Precision, and F1-score to evaluate the performance of our YOLOv4 model. In addition, the mean of Average Precision (mAP) metric is used to compute the average of interpolated Precision to Recall for each detected object (see Sect. 4).

3.4 Object Tracking

The object detection phase computes the bounding boxes locations associated with the targets of interest in the visual scene. Then the object tracking phase associates these detections to the trackers in each video frame. A good tracking algorithm should assign one and only one ID to each object in the visual scene. For each new object entering the scene, a new tracker should be assigned independent from the other objects already in the scene.

For each new detection, we assign a predictor whose role is to model the object's movement in the scene. Then, each predictor is associated with a tracker. In this work, we used Kalman filters as predictors. A Kalman filter tracks a single object by modeling its movement. The filter holds the target state x with information on its center coordinate (x, y), its width-height ratio, its area, the variation of its area over time, and its velocity. The observation z holds only observable information from a single frame, i.e., the center coordinates, the aspect ration and the area. Furthermore, the filter keeps some additional data related to the object, such as the time the object has been visible, and for how long the position has been predicted due to absence of measurement.

The detection association step aims to determine the best possible fit between predictors and detections. First, we generate a cost matrix whose element c_{ij} denotes the similarity (or dissimilarity) between a predictor i and a detection j. We used the Intersection over Union (IoU) between two bounding boxes to compute similarity. We also combined other similarity measures to compute the cost, such as shape and distance measures. The shape and distance features consider the geometry and the position of the bounding box. We adopted the approach proposed in [29] to compute the similarity measures between two bounding boxes, and these similarity measures are multiplied to give advantages to the bounding boxes that have similarities both in position and shape instead of considering just one feature. These terms are the Euclidian distance of the shape and the position as from Eqs. (1), (2) and (3), where A represents the detection and B the tracked object. In particular Eq. (2) computes the similarity value on the size, therefore H and W indicates the two bounding box dimensions. Equation (3) computes the position similarity, with x and y as the coordinates of the bounding box centroid.

$$C(A,B) = c_{shape}(A,B) \times c_{dist}(A,B) \tag{1}$$

$$c_{shape}(A,B) = e^{-w_2 \times (\frac{|H_A - H_B|}{H_A + H_B} + \frac{|W_A - W_B|}{W_A + W_B})} \tag{2}$$

$$c_{dist}(A,B) = e^{-w_1 \times ((\frac{x_A - x_B}{W_A})^2 + (\frac{y_A - y_B}{H_A})^2)} \tag{3}$$

In our case, we set $w_1 = 0.5$ and $w_2 = 1.5$. Obtained the cost matrix, we then use the Hungarian algorithm to assign a detection to a predictor and update it. We choose the Hungarian algorithm [22] because it gives a solution to the assignment problem in polynomial time.

We use a different tracking engine for each object category to achieve precise object tracking when objects categories are mixed in the same environment. Thus, when an object is detected as a car, bike, or pedestrian, it is passed directly to the corresponding tracking engine. The framework developed provides the moving object's direction and counts the different objects considering their categories. To extract the object direction, we use its associated predictor as it models the object's movement on the image plane. Indeed, the state x_k contains the vertical component of the velocity v_y and the horizontal component v_x.

To retrieve the object's direction, we compute the norm V of the velocity from v_x and v_y and the angle between v_x and v. First we perform a check on the norm, if it is outside a reasonable threshold it is discarded, and we assume a wrong data association was performed. Otherwise in the case the *angle between* v_x *and* $V <= 60°$, we return $LEFT - RIGHT$ if $v_x > 0$ or $RIGHT - LEFT$ if $v_x < 0$. If the *angle* $> 60°$ we return $UP - DOWN$ if $v_y > 0$ or $DOWN - UP$ if $v_y < 0$.

We also allowed the user to select areas to specify the correct moving direction between two points. An alarm is raised when an object is detected moving in the opposite direction. For example, Fig. 6 presents the interface for two opposite direction lanes. The top one in green moves from left to right, the bottom one in purple from right to left.

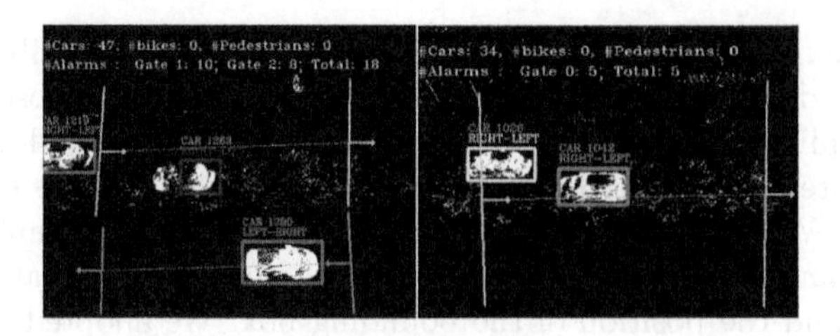

Fig. 6. Example image of the working system in a two lane scenario (left) and on a single lane scenario (right). (Color figure online)

4 Experimental Results

The two proposed approaches have been validated using a dataset acquired with the event-based camera (third-generation Prophesee event-based camera [5]). The experimental setup did not employ a GPU to have a lightweight system, but both the "Model-free" and the Yolov4 have been run on an i7-9750H, 2.60 GHz CPU.

The model-free approach presented above has as its main advantage a tailorable frame rate that can reach 300 fps, on the testing hardware, including also the processing time (i.e., pseudo-frame filtering, bounding box extraction, bounding box filtering). This is a considerably high value for tasks like road monitoring. However, this approach presents some significant limitations. Since the only features used to detect and classify an object are the bounding box's width, height, and area, it is challenging to differentiate multiple object categories accurately. Also, this approach is not particularly robust in very noisy scenarios, and it has many parameters to fine-tune to obtain promising results. Nevertheless, it can run at high frequency on any low-power device, thanks to its simple rule-based approach.

In the case of the YOLOv4 approach, the model training has been performed using a TK80 GPU on Google Colab. For the training phase, we employed the following hyper-parameters: a batch size of 64, with 12 subdivisions that denote the number of pieces the batch is broken into for GPU memory, and a maximum number of batches of 6000. The network was trained to detect only three classes of interest, pedestrians, bicycles, and cars. The best achieved results for the model during the validation with mAP0.5 (mean Average Precision, with Intersection over Union threshold for the bounding boxes of 0.5) was 0.88. For this model, on the testing set, we had a value of 0.83. Table 1 provides the AP (Average Precision) per class for the YOLOV4 model. As it can be noticed, we achieved a good AP for each class.

Our YOLOv4 based detection model is also robust when facing a noisy environment since only objects are detected, and all the noise is discarded. The only disadvantage of this solution is that the YoloV4 model is relatively complex and drastically limits the system framerate. In particular, it achieved only 4 fps on the CPU i7-9750H, 2.60 GHz, while reaching 16 fps on the Google Colab GPU

Table 1. YOLOv4 validation and testing AP value per class.

Class	Validation (%)	Testing (%)
Car	97.3	98.55
Bike	88.69	87.58
Pedestrian	66.98	62.51

Table 2. YOLOv4-tiny validation and testing AP value per class.

Class	Validation (%)	Testing (%)
Car	98.46	94.82
Bike	90.82	44.59
Pedestrian	75.70	48.7

Table 3. Detection metrics for confidence threshold of 0.25.

	P	R	F1-score	IOU (%)
Validation	0.85	0.85	0.85	66.63
Testing	0.68	0.72	0.7	47.94

TK80. To address this issue, we trained a new model on a smaller network, which would lead to less accurate predictions but shows an increased maximum framerate. In particular, we tested the YOLOv4-tiny model. The main difference between YOLOv4-tiny and YOLOv4 is a significant reduction in the network size, making this model considerably faster and suited even for CPU-only tasks. The training process performed is similar to the one of YOLOv4. The global performances are shown in the Table 2. The achieved results are worse than the Yolov4 model, particularly for small objects, but are still acceptable. In particular, if we consider that a Kalman Filter-based tracking is performed after the detection phase. Moreover, this solution is faster, even when working only on the CPU.

Considering our application we decided a confidence threshold of 0.25 to be enough to achieve good results in tracking. Table 3 provides the model Precision, Recall, and F1-score and the average IoU considering the confidence threshold of 0.25. The values are always above 0.5, which indicates the potential of this solution. In particular, not as a standalone one, since the detector might miss some object, but as an accurate source for the Kalman Filter to track the moving target in the scene.

It was impossible to evaluate our model against other available models since the literature related to object detection from event-based cameras is extremely poor, and no available model could be found while performing this test. Instead, a significant comparison can be performed between the two trained solutions, YOLOv4-tiny and Yolov4. While the latter was slow and therefore discarded for

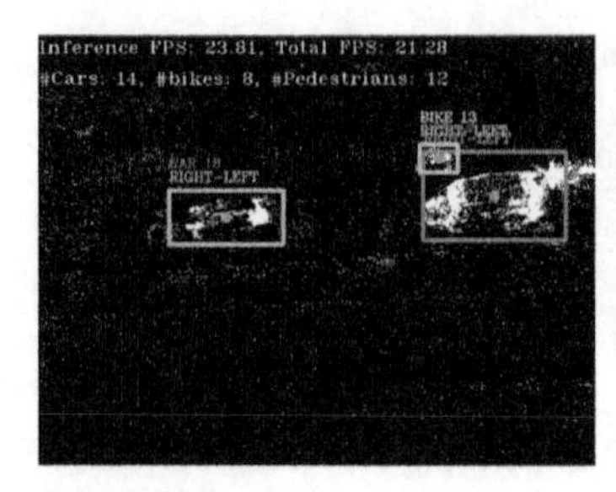

(a) Detection in a night scenario.

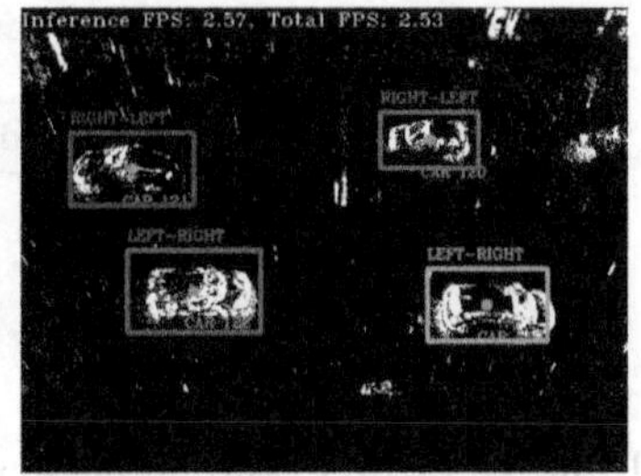

(b) Detection in a rainy scenario.

Fig. 7. Example images of the YoloV4-tiny detections in some challenging scenarios.

this task, Yolov4-tiny could achieve an inference frame rate of 37 fps on CPU i7-9750H, 2.60 GHz, and 76 fps on the TK80 Google Colab GPU, making it suitable for our object detection task. Moreover, the drop in accuracy between the two approaches was not significant.

To conclude, we compared this solution against the original "model-free" approach. The deep-learning approach shows its robustness in noisy scenarios like rain, night (with the light from the vehicle), and camera movement, as shown in Fig. 7. Moreover, compared to the model-free scenario, the detection confidence threshold is the only hyper-parameter to consider while using this model. The limitation of this approach is the frame rate since the model inference time constitutes the bottleneck, but moving the inference to a GPU-based system can be considered a viable solution if needed. Finally, it is particularly interesting to notice how the proposed solution, based on an event-based camera, can perform extremely well in this scenario where traditional cameras would struggle due to weather and lighting conditions.

5 Conclusions

To conclude, in this paper, we presented two solutions for object detection and tracking from event-based data. Events acquired by the camera are accumulated using a binary pseudo-frame that can be processed using two different algorithms. The model-free approach fits well scenarios without a lot of noise and can run at high speed. Therefore it offers the possibility to adjust the frame rate depending on application constraints. The model-free approach is also suitable for low-power systems; nevertheless, it performs poorly with excessive noise and challenging weather conditions. Contrary, the two YOLOv4 models offer better performance and can be used in very noisy scenarios too. But, they are resource-eager, and that leads to a power-consuming system. Moreover, they work at best on systems that use GPUs since, on CPU, they offer a relatively low frame rate. However, the modified version, the YOLOv4-tiny model, can run at an acceptable framerate on a system equipped with a good CPU but no GPU, like the testing hardware. The tracking module has proven its ability to perform the intra-frame data association correctly. In such a way, we could compute each object moving direction and identify vehicles moving in the wrong direction. Moreover, using a Kalman Filter to predict the obstacle's position is extremely useful for consistently providing an object position, even when the detector misses the object for a single frame.

References

1. Gallego. G., et al: Event-based vision: a survey. IEEE Trans. Patt. Anal. Mach. Intell. **44** 154–180 (2019)
2. Mueggler, E., Rebecq, H., Gallego, G., Delbruck, T., Scaramuzza, D.: The event-camera dataset and simulator: event-based data for pose estimation, visual odometry, and slam. Int. J. Robot. Res. **36**(2), 142–149 (2017)
3. Weikersdorfer, D., Adrian, D.B, Cremers, D., Conradt, J.: Event-based 3D slam with a depth-augmented dynamic vision sensor. In: 2014 IEEE International Conference on Robotics and Automation (ICRA), pp. 359–364. IEEE (2014)
4. Cannici, M., Ciccone, M., Romanoni, A., Matteucci, M.: Asynchronous convolutional networks for object detection in neuromorphic cameras. In: Proceedings of the IEEE/CVF Conference on Computer Vision and Pattern Recognition Workshops, pp. 1656–1665 (2019)
5. PROPHESEE. prophesee.ai website (2021)
6. Bochkovskiy, A., Wang, C.-Y., Liao, H.-Y.M: YOLOV4: optimal speed and accuracy of object detection. arXiv preprint arXiv:2004.10934 (2020)
7. Kalman, R.E.: A new approach to linear filtering and prediction problems. J. Basic Eng. **82**(1), 35–45 (1960)
8. Mills-Tettey, A., Stentz, A., Dias, M.B.: The dynamic Hungarian algorithm for the assignment problem with changing costs. Robotics Institute, Pittsburgh, PA, Technical report, CMU-RI-TR-07-27 (2007)
9. Boettiger, J.P.: A comparative evaluation of the detection and tracking capability between novel event-based and conventional frame-based sensors (2020)
10. Padala, V., Basu, A., Orchard, G.: A noise filtering algorithm for event-based asynchronous change detection image sensors on TrueNorth and its implementation on TrueNorth. Front. Neurosci. **12**, 118 (2018)
11. Bolelli, F., Cancilla, M., Baraldi, L., Grana, C.: Toward reliable experiments on the performance of connected components labeling algorithms. J. Real-Time Image Proc. **17**(2), 229–244 (2020)
12. Shapiro, L.G.: Connected component labeling and adjacency graph construction. Mach. Intell. Pattern Recogn. **19**, 1–30 (1996)
13. Lemaitre, F., Hennequin, A., Lacassagne, L.: Taming voting algorithms on GPUs for an efficient connected component analysis algorithm. In: ICASSP 2021–2021 IEEE International Conference on Acoustics, Speech and Signal Processing (ICASSP), pp. 7903–7907. IEEE (2021)
14. He, L., Ren, X., Gao, Q., Zhao, X., Yao, B., Chao, Y.: The connected-component labeling problem: a review of state-of-the-art algorithms. Pattern Recogn. **70**, 25–43 (2017)
15. Bolelli, F., Allegretti, S., Baraldi, L., Grana, C.: Spaghetti labeling: directed acyclic graphs for block-based connected components labeling. IEEE Trans. Image Process. **29**, 1999–2012 (2019)

16. Scholkopf, B.: Support vector machines: a practical consequence of learning theory. IEEE Intell. Syst. **13**, 18–28 (1998)
17. Freund, Y., Schapire, R.E.: A decision-theoretic generalization of on-line learning and an application to boosting. J. Comput. Syst. Sci. **55**(1), 119–139 (1997)
18. Zhao, Z.-Q., Zheng, P., Xu, S.T., Wu, X.: Object detection with deep learning: a review. IEEE Trans. Neural Netw. Learn. Syst. **30**, 3212–3232 (2018)
19. Redmon, J., Divvala, S., Girshick, R., Farhadi, A.: You only look once: unified, real-time object detection. In: Proceedings of the IEEE Conference on Computer Vision and Pattern Recognition, pp. 779–788 (2016)
20. Shafiee, M.J., Chywl, B., Li, F., Wong, A.: Fast Yolo: a fast you only look once system for real-time embedded object detection in video. arXiv preprint arXiv:1709.05943 (2017)
21. Redmon, J., Farhadi, A.: YOLO9000: better, faster, stronger. In: Proceedings - 30th IEEE Conference on Computer Vision and Pattern Recognition, CVPR, pp. 6517–6525 (2017)
22. Kuhn, H.W.: The Hungarian method for the assignment problem. Naval Res. Logistics Q. **2**(1–2), 83–97 (1955)
23. Mitrokhin, A., Fermüller, C., Parameshwara, C., Aloimonos, Y.: Event-based moving object detection and tracking. In: 2018 IEEE/RSJ International Conference on Intelligent Robots and Systems (IROS), pp. 1–9. IEEE (2018)
24. Brändli, C.: Event-based machine vision. Ph.D. thesis, ETH Zurich (2015)
25. Bagchi, S., Chin, T.-J.: Event-based star tracking via multiresolution progressive Hough transforms. In: Proceedings of the IEEE/CVF Winter Conference on Applications of Computer Vision, pp. 2143–2152 (2020)
26. Cannici, M., Ciccone, M., Romanoni, A., Matteucci, M.: A differentiable recurrent surface for asynchronous event-based data. In: Vedaldi, A., Bischof, H., Brox, T., Frahm, J.-M. (eds.) ECCV 2020. LNCS, vol. 12365, pp. 136–152. Springer, Cham (2020). https://doi.org/10.1007/978-3-030-58565-5_9
27. Suzuki, S., et al.: Topological structural analysis of digitized binary images by border following. Comput. Vis. Graph. Image Process. **30**(1), 32–46 (1985)
28. Roboflow. Roboflow website (2021)
29. Sanchez-Matilla, R., Poiesi, F., Cavallaro, A.: Online multi-target tracking with strong and weak detections. In: Hua, G., Jégou, H. (eds.) ECCV 2016. LNCS, vol. 9914, pp. 84–99. Springer, Cham (2016). https://doi.org/10.1007/978-3-319-48881-3_7

Quest for Speed: The Epic Saga of Record-Breaking on OpenCV Connected Components Extraction

Federico Bolelli, Stefano Allegretti, and Costantino Grana(✉)

Dipartimento di Ingegneria "Enzo Ferrari",
Università degli Studi di Modena e Reggio Emilia, Modena, Italy
{federico.bolelli,stefano.allegretti,costantino.grana}@unimore.it

Abstract. Connected Components Labeling (CCL) represents an essential part of many Image Processing and Computer Vision pipelines. Given its relevance on the field, it has been part of most cutting-edge Computer Vision libraries. In this paper, all the algorithms included in the OpenCV during the years are reviewed, from sequential to parallel/GPU-based implementations. Our goal is to provide a better understanding of what has changed and why one algorithm should be preferred to another both in terms of memory usage and execution speed.

Keywords: OpenCV · Connected Components Labeling

1 Introduction

OpenCV (Open Source Computer Vision Library) is a software library mainly aimed at real-time computer vision [37]. Originally developed by Intel, it was later supported by Willow Garage and then Itseez. The library is cross-platform and free for use under the open-source Apache 2 License. Starting with 2011, OpenCV features GPU acceleration for real-time operations.

A common basic task in image processing is to produce a description of the objects inside a binary image; this is often done by extracting its connected components. By considering the pixel lattice as a graph in which foreground pixels are nodes connected by edges to their foreground neighbors, a connected component on the graph corresponds to the common definition of an "object of interest". Based on the specific use case, two pixels can be considered connected or not, according to the definition of pixel connectivity: in 2D-images, pixels can be either *4-connected* (sides only) or *8-connected* (sides and corners). A possible solution to extract connected components (objects) is to use a Connected Components Labeling (CCL) algorithm: a procedure which generates a symbolic image in which each pixel of a single connected component is assigned a unique identifier.

The CCL algorithm has an exact output meaning that different algorithmic solutions should be mainly compared in term of speed and memory footprint.

P. L. Mazzeo et al. (Eds.): ICIAP 2022 Workshops, LNCS 13374, pp. 107–118, 2022.
https://doi.org/10.1007/978-3-031-13324-4_10

After the introduction of the task in 1966 [38], several proposals to optimize its computational load have been published for both sequential [7,9,14,15,20,22,24,25,28,31,42] and parallel architectures [1–4,29,35,43], taking into account also 3D volumes [8,27,40].

Connected Components Analysis, or CCA in short, extends CCL by computing some features of the connected components such as their bounding box, their area, or the first moments to compute center of gravity. CCA is basically a voting algorithm like histogram computation or Hough transform [30] and it is a mandatory step for many Computer Vision and Image Processing pipelines [5,13,17,18,23,32,34,36,41].

Connected Components extraction has been available since the early days of OpenCV and has evolved (in speed) with every release. Initially, the implementation available was based on the combination of two different functions: `FindContours` and `DrawContours`, respectively in charge of retrieving contours and the hierarchical information from binary images and drawing them. Since version *3.0.0*, `cv::connectedComponents` and `cv::connectedComponentsWithStats` APIs have been introduced, providing a major speed breakthrough for CCL computation within the library.

The goal of this paper is to review all the algorithms implemented in OpenCV during the years, thus providing the reader with a better understanding of what has changed and why one should choose one algorithm rather than another both in terms of memory usage and execution speed.

2 The First Approach

The extraction of Connected Components (CCs) from a binary image has been available since the first release of the OpenCV with the combination of `findContours` and `drawContours` functions (Listing 1.1).

`findContours` operates on a binary image by retrieving objects' contours. The function retrieves contours from the binary image using the algorithm described in [39]. The algorithm follows objects' borders with a sort of topological analysis capability. If one wants to convert a binary picture into the border representation, then they can extract the topological structure of the image with little additional effort by using this function. The information to be extracted

```
void cv::findContours (InputArray image, OutputArrayOfArrays contours,
    OutputArray hierarchy, int mode, int method, Point offset = Point())

void cv::drawContours (InputOutputArray image, InputArrayOfArrays
    contours, int contourIdx, const Scalar & color, int thickness = 1,
    int lineType = LINE_8, InputArray hierarchy = noArray(), int maxLevel
     = INT_MAX, Point offset = Point())
```

Listing 1.1. OpenCV C++ API for *findContours* and *drawContours* functions.

```
1 [...]
2 vector<vector<Point>> contours;
3 vector<Vec4i> hierarchy;
4 findContours(src, contours, hierarchy, RETR_CCOMP, CHAIN_APPROX_SIMPLE);
5 for (int idx = 0; idx >= 0; idx = hierarchy[idx][0]) {
6     Scalar color(rand() & 255, rand() & 255, rand() & 255);
7     drawContours(dst, contours, idx, color, FILLED, 8, hierarchy);
8 }
9 [...]
```

Listing 1.2. OpenCV example on how to retrieve connected components from a binary image and fill them with random colors. Tested on version *4.5.5*.

```
int cv::connectedComponents(InputArray image, OutputArray labels, int
    connectivity, int ltype)

int cv::connectedComponentsWithStats(InputArray image, OutputArray labels
    , OutputArray stats, OutputArray centroids, int connectivity, int
    ltype)
```

Listing 1.3. OpenCV C++ API for *connectedComponents* and *connectedComponentsWithStats* functions.

is the inclusion relation among the two types of borders: the *outer borders* and the *hole borders.* Since there exists one-to-one correspondence between an outer border and a 1-component, and between a hole border and a 0-component, the topological structure of a given binary image can be determined.

A topological representation can be mapped into connected components by filling the contours. An example is reported in Listing 1.2.

3 A Novel Interface

Unfortunately, finding the contours and flood filling them is not a smart way of performing CCL. For this reason, researchers and practitioners started using different implementations found online until the release of OpenCV 3.0.0, which introduced two new interfaces (Listing 1.3).

The `connectedComponents` function takes a binary image as input and produces an integer symbolic image in which all the pixels from the same object are assigned the same (unique) number. With the parameter `connectivity` the user can specify whether to use 4- or 8-connectivity to define pixel connectivity (i.e. considering pixel connected only if they share the same border, 4-connectivity, or also if they share vertexes, 8-connectivity). `ltype` specifies whether the output image should use 16- or 32-bit per pixel. The function returns the total number of labels $[0, N-1]$, where 0 represents the background label. In this version only the Scan Array-Based Union Find (SAUF) algorithm by Wu *et al.* was available.

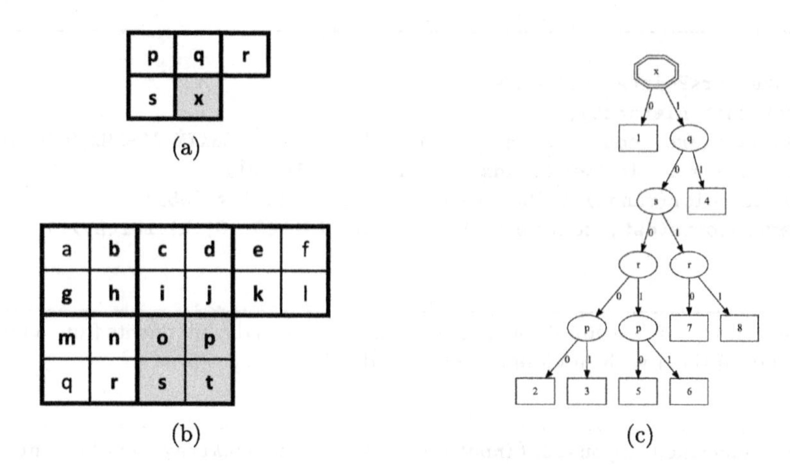

Fig. 1. (a) is the Rosenfeld mask used by SAUF to compute the label of pixel x during the first scan and (b) is the Grana mask used by BBDT to compute the label of pixels o, p, s and t. Finally, (c) is the optimal decision tree proposed in [42]. Internal nodes (ellipsis) represent the conditions to be checked, and leaves (rectangles) contain the actions to be performed, which are identified by integer numbers. The root of the tree, also a condition, is represented by a octagon. Action 1 represents *no action.* Action 2 is *new label.* Action 3 means $x \leftarrow p$, i.e. assign x the label of p. Action 4, 5, and 7 are respectively $x \leftarrow q$, $x \leftarrow r$, and $x \leftarrow s$. Finally, action 6 and 8 require merge between different label classes, specifically, $x \leftarrow r + p$, $x \leftarrow s + r$.

A `connectedComponentsWithStats` implementation is also available. This function allows to calculate at the same time the output symbolic image with labeled connected components and their statistics:

- the minimum bounding box containing the connected component;
- the area (in pixels) of the object;
- the centroids (x, y)-coordinates of connected components, including background.

All of this information is stored inside `stats` and *centroids* matrices. Also in this case, the SAUF algorithm is employed to identify connected objects.

The SAUF algorithm itself, introduced by Wu *et al.*, is based on two key elements:

- the use of the Union-Find algorithm to store and handle equivalences between pixel classes[1];

[1] The union-find data structure, first applied to CCL by Dillencourt *et al.* [19], provides two convenient procedures to deal with equivalence classes of labels: *Find*, which retrieves the representative label of an equivalence class, and *Union*, which merges two equivalence classes into one, ensuring that they share the same representative label.

- an optimal strategy, based on a manually identified decision tree, to reduce the average number of load/store operations during the first scan of the input image.

As most of the state-of-the-art algorithms for CCL, SAUF is based on a two scan (or two pass) approach. During the first scan of the image, the algorithm assigns temporary labels to pixels and records equivalences between classes. The second scan, instead, is meant to replace each temporary label with the representative of its equivalence class (usually the smallest one). The scanning approach is led by the Rosenfeld mask reported in Fig. 1a. Indeed, when labeling the current pixel, x, pixels p, q, r, and s are enough to determine the class which x belongs to. Moreover, if q is a foreground pixel, it is already connected with all the other foreground pixels in the "current" mask and this connectivity has already been recorded in the Union-Find data structure. This means that we can simply assign x the same class of q, saving three checks. When q is a background pixel, we can for example check pixel p. In this case, when p is foreground, r must be inspected also, to verify whether p and r are connected through x. If this is the case, a merge between the two classes have to be performed. Moving on with this reasoning, the decision tree depicted in Fig. 1c can be obtained. Other equivalently optimal[2] versions can be generated.

As said, combining the use of the optimal decision tree with the Union-Find algorithm optimized with path compression [42] translates into the SAUF algorithm. A similar approach can be applied to 4-connectivity producing this time a much simpler (and smaller) decision tree.

4 Going Faster with Blocks

In 2010, Grana *et al.* [22] introduced a major breakthrough, consisting in a 2×2 block-based approach denoted as *Block-Based with Decision Tree* algorithm (BBDT). The proposed algorithms make use of an optimal decision tree, generated upon the mask of Fig. 1, and the Union-Find algorithm implemented with Three Table Array (TTA) strategy proposed in [25].

The problem is modeled as a *command execution metaphor*: values of pixels in the scanning mask constitute a *rule* (binary string), which is associated to a set of equivalent actions in an *OR*-decision table. Given this decision table, an algorithm can simply read all the pixels inside the mask, identify the rule, and find the action to be performed in the corresponding column. A dynamic programming approach [24] is then used to convert the *OR*-decision table into an optimal binary decision trees. This approach allows to minimize the average number of conditions to be checked when choosing the correct action to be performed.

The possible actions are the same mentioned for SAUF algorithms, this time working with blocks: *no action* if the current block is background (i.e., all the

[2] Optimality is related to the number of accesses to the pixels in the scanning mask, *i.e.*, number of memory accesses.

```
int cv::connectedComponents(InputArray image, OutputArray labels, int
    connectivity, int ltype, int ccltype)

int cv::connectedComponentsWithStats(InputArray image, OutputArray labels
    , OutputArray stats, OutputArray centroids, int connectivity, int
    ltype, int ccltype)
```

Listing 1.4. OpenCV C++ API for *connectedComponents* and *connectedComponentsWithStats* functions.

pixels of the block are background), *new label* if it has no foreground neighbors, *assign* or *merge* based on the label of neighboring foreground pixels/blocks.

Since version *3.2.0*, the BBDT algorithm has been introduced in the OpenCV. Two new overloading functions, detailed in Listing 1.4, have been added to introduce the `ccltype` parameter while preserving the Application Binary Interface (ABI compatibility). This parameter makes the user able to select the algorithm to be used. Given that BBDT is only available for 8-connectivity, the SAUF version is always executed when labeling with 4-connectivity.

It is important to notice that, while the SAUF algorithm forces a row major ordering of labels, BBDT does not. This means that label ordering in the output *label* image may be different when executing the two algorithms, but with exactly the same semantic meaning.

5 Spaghetti for All

Many improvements have been proposed since the introduction of BBDT, and some of them introduced significantly novel ideas, in particular:

- realizing that it is possible to use a finite state machine to summarize the value of pixels already inspected by the horizontally moving scan mask [28];
- combining decision trees and configuration transitions in a decision forest, in which each previous pattern allows to "predict" some of the current configuration pixels values, thus allowing automatic code generation [20];
- switching from decision trees to Directed Rooted Acyclic Graphs (DRAGs), to reduce the machine code footprint and lessen its impact on the instruction cache [11].

Prediction, as introduced by He *et al.* [26], has proven to be one of the most useful additions, as it allows to exploit already available information, save expensive load/store operations, and reduce execution time consequently. When the scan mask is shifted along a row of the image it always contains some of the pixels it already contained in the previous step, though in different locations. If those pixels were indeed checked in the previous mask step, a second read of their value can be avoided by their removal from the decision process.

The procedure proposed in [20] was suitable to be automatized, but still a small mask was employed. The reason, in this case, was that the larger the mask

```
void cv::cuda::connectedComponents(InputArray image, OutputArray labels,
    int connectivity, int ltype, cv::cuda::
    ConnectedComponentsAlgorithmsTypes ccltype)
```

Listing 1.5. OpenCV C++ API for *connectedComponents* performed in CUDA.

is, the more decision trees will populate the resulting forest, and the higher every tree will be. The machine code that implements the algorithm resulting from the application of prediction to BBDT would be very large, and may have a negative impact on instruction cache. Therefore, despite load/store operations being less, the overall performance on real case datasets may be worse than that of the single tree variation. For this reason, all works on prediction chose to avoid the complexity of the BBDT mask, and simplified it in various ways.

In [7], the BBDT original mask and the *state prediction* paradigm are combined in the *Spaghetti Labeling* algorithm, by taking advantage of the code compression technique that converts a directed rooted tree into a DRAG [11]. The resulting process is modeled by a directed acyclic graph (DAG) with multiple entry points (roots), which correspond to the knowledge that can be inferred from the previous step. This guarantees a significant reduction of the machine code, even better than that achievable by a compiler, since it can leverage the presence of equivalent actions in the trees leaves, and compress not only equal subtrees, but also equivalent ones.

Spaghetti labeling has been included in OpenCV since version *4.5.2* and *3.4.14*. The signatures are the same as the previous ones, changing only the default value `ccltype = CCL_SPAGHETTI`.

The later introduction of GRAPHGEN [8], a technique for the automatic generation of decision DAGs inspired by Spaghetti, allowed, since version *4.5.5*, to also implement a 4-connected version of Spaghetti, making it the default algorithm for both 8- and 4-connectivity.

OpenCV aims at maximizing speed, thus parallelization is heavily employed throughout all library and a specifically developed framework is available. At the moment, following the embarrassingly parallel approach of [12], labeling algorithms are run on image stripes and a further joining stage is added. The parallel version of the algorithms is automatically employed if at least one of the allowed parallel frameworks is enabled and if the rows of the image are at least twice the number returned by `getNumberOfCPUs`.

6 GPU Implementation

Starting from the 4th major release of OpenCV, all CUDA modules are located in *opencv_contrib*,[3] an additional public repository containing extra modules that can be optionally added to the installation of the library. The CUDA version of

[3] https://github.com/opencv/opencv_contrib.

CCL has been recently added to *opencv_contrib*, and will be included in release *4.5.6*. Its signature, reported in Listing 1.5, was chosen to be as close as possible to the CPU version, with the only difference being the return type. This function, in fact, does not return the amount of labels assigned: the additional task of counting labels, which is trivial for most sequential algorithms, is instead considerably time consuming when performed in a massively parallel fashion, and for this reason it is excluded from the workload of CUDA CCL algorithms.

So far, the only available CUDA algorithm is Block-Based Komura Equivalence (BKE) [4], which takes advantages of both the Union-Find algorithm and the Block-Based approach and represents the current state of the art. In this proposal, the Union-Find structure is directly coded in the output image, in the sense that the provisional label assigned to each block doubles its meaning as the memory address of the parent in the Union-Find tree. This particular choice of provisional labels allows to avoid a specific data structure for the Union-Find.

Like all CUDA algorithms, BKE is composed of *kernels*, i.e. functions executed at the same time by a high number of threads. The kernels composing the algorithm are *Initialization*, *Compression*, *Reduction* and *FinalLabeling*, and are described in the following. Each uses a number of threads equal to the blocks in the image, so that each thread is responsible for labeling its own block, which will be referred to as X.

Initialization. Each thread looks at the neighborhood in order to find out which blocks are connected to X, then takes the smallest of their raster addresses and sets it as the initial label of X. From the Union-Find point of view, this means that X is assigned a parent in the forest. Finally, an information bitset detailing with pixels of the block are foreground and which blocks are connected to X is stored in the output image, along with the provisional label; it will be used again in subsequent kernels. In this case, the output image is used as a temporary buffer: this information bitset is only useful for the algorithm, and will not be present in the final output.

Compression. This kernel flattens the Union-Find trees coded in the image, by means of the *Find* operation: each thread reads the parent label of X, then the parent of the parent, and repeats the process until it reaches the root; then, it assigns the root label to X. After this compression, all trees have height 1.

Reduction. Each thread reads the information bitset stored in Initialization in order to find out which blocks are connected to X, and then proceeds to make sure that all of them are indeed in the same Union-Find tree. This is accomplished by means of the *Union* procedure, which takes two nodes as input, traces back their trees until the roots and finally links one root to the other. Of course, the neighbor blocks with the smallest address is excluded, since X has already been connected to it in *Initialization*. From the Union-Find point of view, the Reduction kernel completes the CCL task: each block is put in the same tree as all of its neighbor, and therefore each tree in the forest completely corresponds to a CC in the image. After Reduction, a second Compression is performed, again to flatten trees to height 1. This time, however, it also means that each block in the same tree has the same parent, and thus the same label.

FinalLabeling. The only remaining operation to perform at this point is to assign block labels to single pixels. Each thread reads the information bitset again, this time to check which pixels of the blocks are foreground; then, it assigns the label of X to all of them, and label 0 to the remaining pixels. This is the final rewriting of the output image, and overwrites the information bitset previously stored. After FinalLabeling, each pixel in the same CC has the same unique label, and thus the labeling task is completed.

7 Discussion

The inclusion of algorithms in OpenCV has been done after a careful and really open comparison of the execution times, evaluated using YACCLAB [11,21], a widely used [16,33] open source *C++* benchmarking framework for CCL algorithms. YACCLAB allows researchers to test state-of-the-art algorithms on real and synthetic generated datasets. The fairness of the comparison is guaranteed by compiling the algorithms with the same optimizations and by running them on the same data and over the same machine.

The algorithms provided by YACCLAB cover most of the paradigms for CCL explored in the past, along with a lower bound limit for all CCL algorithms over a specific dataset/image, obtained by reading once the input image and writing it on the output again.

The benchmark provides a template implementation of the algorithms over the labels solving strategy. Using different label solvers can significantly change the performance of a specific combination of dataset, algorithm and operating system.

The YACCLAB dataset covers most applications in which CCL may be useful, and features a significant variability in terms of resolution, image density, variance of density, and number of components. It includes six real-world datasets, and specifically: *3DPeS* [6], *Fingerprints*, *Medical*, *MIRflickr*, *Tobacco-800*, *XDOCS* [10].

A clear result is that, on average, Spaghetti Labeling is the optimal choice. In very specific corner cases, such as when the order of labels needs to be sorted by rows, or when the instruction cache is extremely small, other techniques could be employed. The combination of `FindContours` and `DrawContours` is a viable solution if your aim is to obtain the contours, because the connected components are an additional bonus. If you just need the connected components, these should be definitely avoided. The GPU version is now available and it makes sense if your images are already in GPU, allowing you to stay in GPU without moving back and forth from main memory to device memory. Even if the GPU version is faster than Spaghetti Labeling, the total amount of time required to move data between host and device plus the CCL procedure is higher than running in CPU directly.

8 Conclusion

With this paper we provided a review of the sequential and parallel implementation of CCL algorithms included in the OpenCV library. The open source nature of OpenCV allowed to spot numerous and subtle bugs, and it is always incredible how many small details may be overlooked in real world usage of code.

All the additions to OpenCV, not only for CCL, have been strongly motivated by independent performance evaluations, in terms of effectiveness, or (as for this specific case) speed. Every alternative proposal should be openly evaluated and the source code needs to be released publicly, in order to avoid contrasting claims of "I'm better than you". We want the user to git-pull our code and check if it really is the best for his use case, or not.

References

1. Allegretti, S., Bolelli, F., Cancilla, M., Grana, C.: Optimizing GPU-based connected components labeling algorithms. In: IPAS, pp. 175–180 (2018)
2. Allegretti, S., Bolelli, F., Cancilla, M., Grana, C.: A block-based union-find algorithm to label connected components on GPUs. In: Ricci, E., Rota Bulò, S., Snoek, C., Lanz, O., Messelodi, S., Sebe, N. (eds.) ICIAP 2019. LNCS, vol. 11752, pp. 271–281. Springer, Cham (2019). https://doi.org/10.1007/978-3-030-30645-8_25
3. Allegretti, S., Bolelli, F., Cancilla, M., Pollastri, F., Canalini, L., Grana, C.: How does connected components labeling with decision trees perform on GPUs? In: Vento, M., Percannella, G. (eds.) CAIP 2019. LNCS, vol. 11678, pp. 39–51. Springer, Cham (2019). https://doi.org/10.1007/978-3-030-29888-3_4
4. Allegretti, S., Bolelli, F., Grana, C.: Optimized block-based algorithms to label connected components on GPUs. IEEE Trans. Parallel Distrib. Syst. **31**, 423–438 (2019). https://doi.org/10.1109/TPDS.2019.2934683
5. Allegretti, S., Bolelli, F., Pollastri, F., Longhitano, S., Pellacani, G., Grana, C.: Supporting skin lesion diagnosis with content-based image retrieval. In: 2020 25th International Conference on Pattern Recognition (ICPR). IEEE, January 2021
6. Baltieri, D., Vezzani, R., Cucchiara, R.: 3DPeS: 3D people dataset for surveillance and forensics. In: Proceedings of the 2011 Joint ACM Workshop on Human Gesture and Behavior Understanding, pp. 59–64. ACM (2011)
7. Bolelli, F., Allegretti, S., Baraldi, L., Grana, C.: Spaghetti labeling: directed acyclic graphs for block-based connected components labeling. IEEE Trans. Image Process. **29**(1), 1999–2012 (2019)
8. Bolelli, F., Allegretti, S., Grana, C.: One DAG to rule them all. IEEE Trans. Pattern Anal. Mach. Intell., 1–12 (2021). https://doi.org/10.1109/TPAMI.2021.3055337
9. Bolelli, F., Baraldi, L., Cancilla, M., Grana, C.: Connected components labeling on DRAGs. In: 2018 24th International Conference on Pattern Recognition (ICPR), pp. 121–126 (2018)
10. Bolelli, F., Borghi, G., Grana, C.: XDOCS: an application to index historical documents. In: Serra, G., Tasso, C. (eds.) IRCDL 2018. CCIS, vol. 806, pp. 151–162. Springer, Cham (2018). https://doi.org/10.1007/978-3-319-73165-0_15
11. Bolelli, F., Cancilla, M., Baraldi, L., Grana, C.: Towards reliable experiments on the performance of Connected Components Labeling algorithms. J. Real-Time Image Proc. **17**(2), 229–244 (2018)

12. Bolelli, F., Cancilla, M., Grana, C.: Two more strategies to speed up connected components labeling algorithms. In: Battiato, S., Gallo, G., Schettini, R., Stanco, F. (eds.) ICIAP 2017. LNCS, vol. 10485, pp. 48–58. Springer, Cham (2017). https://doi.org/10.1007/978-3-319-68548-9_5
13. Canalini, L., Pollastri, F., Bolelli, F., Cancilla, M., Allegretti, S., Grana, C.: Skin lesion segmentation ensemble with diverse training strategies. In: Vento, M., Percannella, G. (eds.) CAIP 2019. LNCS, vol. 11678, pp. 89–101. Springer, Cham (2019). https://doi.org/10.1007/978-3-030-29888-3_8
14. Chang, W.Y., Chiu, C.C.: An efficient scan algorithm for block-based connected component labeling. In: 22nd Mediterranean Conference on Control and Automation, pp. 1008–1013 (2014)
15. Chang, W.Y., Chiu, C.C., Yang, J.H.: Block-based connected-component labeling algorithm using binary decision trees. Sensors **15**(9), 23763–23787 (2015)
16. Chen, J., Nonaka, K., Sankoh, H., Watanabe, R., Sabirin, H., Naito, S.: Efficient parallel connected component labeling with a coarse-to-fine strategy. IEEE Access **6**, 55731–55740 (2018)
17. Cipriano, M., et al.: Deep segmentation of the mandibular canal: a new 3D annotated dataset of CBCT volumes. IEEE Access **10**, 11500–11510 (2022)
18. Cipriano, M., Allegretti, S., Bolelli, F., Pollastri, F., Grana, C.: Improving segmentation of the inferior alveolar nerve through deep label propagation. In: Proceedings of the IEEE/CVF Conference on Computer Vision and Pattern Recognition (CVPR), pp. 1–10. IEEE (2022)
19. Dillencourt, M.B., Samet, H., Tamminen, M.: A general approach to connected-component labeling for arbitrary image representations. J. ACM **39**(2), 253–280 (1992)
20. Grana, C., Baraldi, L., Bolelli, F.: Optimized connected components labeling with pixel prediction. In: Blanc-Talon, J., Distante, C., Philips, W., Popescu, D., Scheunders, P. (eds.) ACIVS 2016. LNCS, vol. 10016, pp. 431–440. Springer, Cham (2016). https://doi.org/10.1007/978-3-319-48680-2_38
21. Grana, C., Bolelli, F., Baraldi, L., Vezzani, R.: YACCLAB - yet another connected components labeling benchmark. In: 2016 23rd International Conference on Pattern Recognition (ICPR), pp. 3109–3114 (2016)
22. Grana, C., Borghesani, D., Cucchiara, R.: Optimized block-based connected components labeling with decision trees. IEEE Trans. Image Process. **19**(6), 1596–1609 (2010)
23. Grana, C., Borghesani, D., Cucchiara, R.: Automatic segmentation of digitalized historical manuscripts. Multimedia Tools Appl. **55**(3), 483–506 (2011)
24. Grana, C., Montangero, M., Borghesani, D.: Optimal decision trees for local image processing algorithms. Pattern Recogn. Lett. **33**(16), 2302–2310 (2012)
25. He, L., Chao, Y., Suzuki, K.: A linear-time two-scan labeling algorithm. In: 2007 IEEE International Conference on Image Processing, pp. 241–244 (2007)
26. He, L., Chao, Y., Suzuki, K.: An efficient first-scan method for label-equivalence-based labeling algorithms. Pattern Recogn. Lett. **31**(1), 28–35 (2010)
27. He, L., Chao, Y., Suzuki, K.: Two efficient label-equivalence-based connected-component labeling algorithms for 3-D binary images. IEEE Trans. Image Process. **20**(8), 2122–2134 (2011)
28. He, L., Zhao, X., Chao, Y., Suzuki, K.: Configuration-transition-based connected-component labeling. IEEE Trans. Image Process. **23**(2), 943–951 (2014)

29. Hennequin, A., Lacassagne, L., Cabaret, L., Meunier, Q.: A new direct connected component labeling and analysis algorithms for GPUs. In: 2018 Conference on Design and Architectures for Signal and Image Processing (DASIP), pp. 76–81. IEEE (2018)
30. Illingworth, J., Kittler, J.: A survey of the Hough transform. Comput. Vis. Graph. Image Process. **44**(1), 87–116 (1988)
31. Lacassagne, L., Zavidovique, B.: Light speed labeling: efficient connected component labeling on RISC architectures. J. Real-Time Image Proc. **6**(2), 117–135 (2011). https://doi.org/10.1007/s11554-009-0134-0
32. Laradji, I.H., Rostamzadeh, N., Pinheiro, P.O., Vazquez, D., Schmidt, M.: Where are the Blobs: counting by localization with point supervision. In: Computer Vision – ECCV 2018, pp. 547–562 (2018)
33. Ma, D., Liu, S., Liao, Q.: Run-based connected components labeling using double-row scan. In: Zhao, Y., Kong, X., Taubman, D. (eds.) ICIG 2017. LNCS, vol. 10668, pp. 264–274. Springer, Cham (2017). https://doi.org/10.1007/978-3-319-71598-8_24
34. Pham, H.V., Bhaduri, B., Tangella, K., Best-Popescu, C., Popescu, G.: Real time blood testing using quantitative phase imaging. PLoS ONE **8**(2), e55676 (2013)
35. Playne, D., Hawick, K.: A new algorithm for parallel connected-component labelling on GPUs. IEEE Trans. Parallel Distrib. Syst. **29**(6), 1217–1230 (2018). https://doi.org/10.1109/TPDS.2018.2799216
36. Pollastri, F., Bolelli, F., Paredes, R., Grana, C.: Augmenting data with GANs to segment melanoma skin lesions. Multimedia Tools Appl. **79**(21–22), 15575–15592 (2019)
37. Pulli, K., Baksheev, A., Kornyakov, K., Eruhimov, V.: Realtime computer vision with OpenCV: mobile computer-vision technology will soon become as ubiquitous as touch interfaces. Queue **10**(4), 40–56 (2012). https://doi.org/10.1145/2181796.2206309
38. Rosenfeld, A., Pfaltz, J.L.: Sequential operations in digital picture processing. J. ACM **13**(4), 471–494 (1966)
39. Suzuki, S., Abe, K.: Topological structural analysis of digitized binary images by border following. Comput. Vis. Graph. Image Process. **30**(1), 32–46 (1985). https://doi.org/10.1016/0734-189X(85)90016-7
40. Söchting, M., Allegretti, S., Bolelli, F., Grana, C.: A heuristic-based decision tree for connected components labeling of 3D volumes. In: 2020 25th International Conference on Pattern Recognition (ICPR), pp. 7751–7758. IEEE, January 2021. https://doi.org/10.1109/ICPR48806.2021.9413096
41. Uslu, F., Bharath, A.A.: A recursive Bayesian approach to describe retinal vasculature geometry. Pattern Recogn. **87**, 157–169 (2019)
42. Wu, K., Otoo, E., Suzuki, K.: Two strategies to speed up connected component labeling algorithms. Pattern Analysis Application 0(LBNL-59102) (2005)
43. Zavalishin, S., Safonov, I., Bekhtin, Y., Kurilin, I.: Block equivalence algorithm for labeling 2D and 3D images on GPU. EI **2016**(2), 1–7 (2016)

An Efficient Run-Based Connected Component Labeling Algorithm for Processing Holes

Florian Lemaitre(✉), Nathan Maurice, and Lionel Lacassagne

LIP6, Sorbonne University, CNRS, Paris, France
florian.lemaitre@lip6.fr

Abstract. This article introduces a new connected component labeling and analysis algorithm framework that is able to compute in one pass the foreground and the background labels as well as the adjacency tree. The computation of features (bounding boxes, first statistical moments, Euler number) is done on-the-fly. The transitive closure enables an efficient hole processing that can be filled while their features are merged with the surrounding connected component without the need to rescan the image. A comparison with State-of-the-Art shows that this new algorithm can do all these computations faster than all existing algorithms processing foreground and background connected components or holes.

Keywords: Black & white processing · Connected component labeling and analysis · Euler number · Adjacency tree · Hole processing · Hole filling

1 Introduction and State-of-the-Art

Connected Component Labeling (CCL) is a fundamental algorithm in computer vision. It consists in assigning a unique number to each connected component of a binary image. Since Rosenfeld [26], many algorithms have been developed to accelerate its execution time on CPU [3,5,12], SIMD CPU [14,19], GPU [22] or FPGA [16].

In the same time, Connected Component Analysis (CCA) that consists in computing Connected Component (CC) features – like bounding-box to extract characters for OCR, or the first raw moments (S, S_x, S_y) for motion detection and tracking – has also risen [1,13,17,18,28]. Parallelized algorithms have also been designed [2,6,15]. The initial Union-Find algorithm [29] has been also analysed [30] and improved [7] with Decision Tree [31] and various path compression/modification algorithms [20,21].

Some other features – useful for pattern classification/recognition – are computed by another set of algorithms: the Euler number with Bit-Quads [8], the adjacency (also known as homotopy or inclusion) tree [25] and more recently, foreground (FG) and background (BG) labeling (also known as BW labeling) [10] and hole filling with also improvements in the last decade [23,32]. Hole filling

P. L. Mazzeo et al. (Eds.): ICIAP 2022 Workshops, LNCS 13374, pp. 119–131, 2022.
https://doi.org/10.1007/978-3-031-13324-4_11

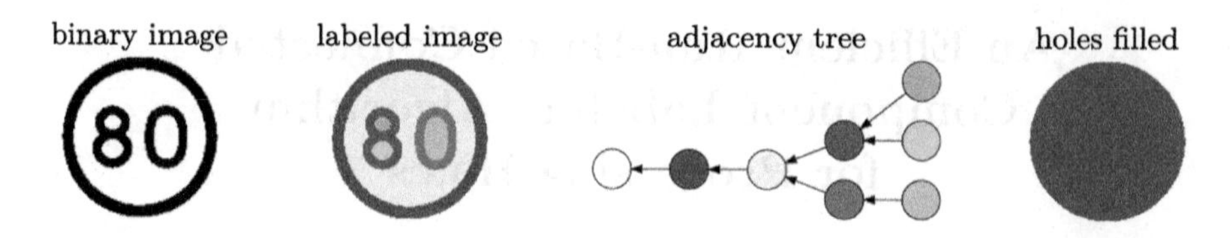

Fig. 1. Example of Black and White labeling with hole filling (FG in black)

is an important part of medical image processing [27,33]. An example of black and white labeling with adjacency tree and hole filling is shown in Fig. 1.

Our contribution is a fast algorithm framework to process holes in black and white images. It can compute features of the CCs, the adjacency tree or the euler number of the image, and can fill holes.

This paper is split into the three following parts: Sect. 2 provides an overview of our new CCL algorithm. The specificities of black & white and hole processing are detailed in Sect. 3. Section 4 presents a benchmark of existing algorithms and their analysis.

2 General Overview of Our New Algorithm

We chose to base our new black and white algorithm on the existing LSL algorithm [18] and especially its latest SIMD implementation, the FLSL algorithm [19]. The reason is two-fold: as the LSL is run-based (segment processing), it is able to compute features very quickly compared to pixel-based algorithms. The second reason is that FLSL is the fastest CCL algorithm currently available [19]. To be noted that FLSL does not explicitly support CCA, thus feature computation had to be back-ported from LSL to this new algorithm.

The LSL algorithm is a CCL/CCA algorithm based on Union-Find structure [29] to build the equivalence relationship between parts of the same connected component. The specificity of LSL is to be run-based: it first computes segments of same class pixels (either foreground of background), and then unifies intersecting segments from consecutive lines. This reduces both the number of temporary labels and the number of "Union" needed to process the image.

BW FLSL needs the following global tables:

- T: Equivalence table (Union-Find structure),
- F: Feature table, encodes the features of each label,
- I: Initial adjacency table, encodes the adjacency tree (explained later).

Figure 2 illustrates the LSL-related table usage on a simple example.

LSL is composed of four steps (Algorithm 1). During the first one, input pixels are read and grouped into segments of same class (foreground or background). This step computes the position of the segments (RLC_i) using semi-open intervals: $RLC_i[er]$ is the position of the first pixel of the er-th segment, whereas $RLC_i[er+1]$ is the position of the first pixel *after* the er-th segment. This step is taken verbatim from [19].

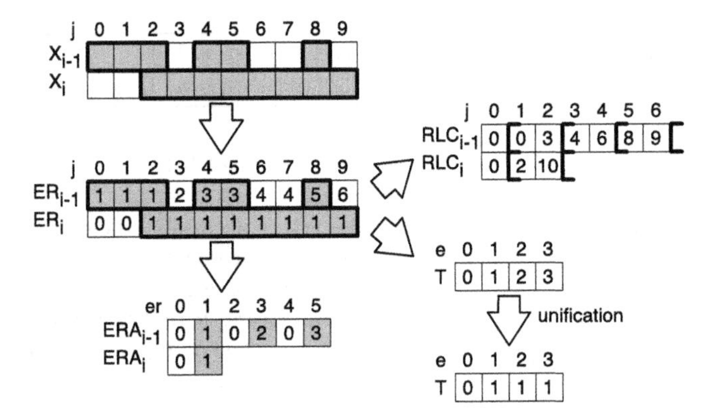

Fig. 2. Tables for the LSL (*ER* is actually not needed anymore).

Algorithm 1: New BW FLSL overview.
This paper contribution is highlighted with gray boxes.

```
1  ne ← 1            ▷ Reset number of labels
2  I[0] ← −1         ▷ Exterior component has no surrounding
3  F[0] ← ∅
4  for i = 0 to h − 1 do
5  |   RLE(i)        ▷ Step 1a: Detect segments
6  |_  Unify (i)     ▷ Step 1b: Merge labels from adjacent line segments
7  Close ()          ▷ Step 2: Transitively close the equivalence graph
8  if relabel then
9  |   for i = 0 to h − 1 do
10 |_  |_  Relabel (i)   ▷ Step 3: Write the label image
```

During the second step, temporary labels are assigned to segments, and segments from current line are "unified" with segments of the previous line. This is done by computing the intersection of current segments with the ones above, and mark them equivalent. The equivalences between labels are recorded in the equivalence table T. In addition, when a label is assigned to a segment, the features of this segment is computed and merged with the features of this label.

As for the FLSL, this step actually uses a Finite State Machine that works similarly to a merged sort where the segments of both consecutive lines are iterated together. The detailed implementation of this FSM can be found in Algorithm 2.

The third step is the transitive closure of the equivalence graph: it makes each temporary label point directly to the root. The equivalence trees are flattened. During this step, features from temporary labels are also merged into their root. As will be explained later in this paper (Sect. 3.2), the hole filling and the Euler number computation are also done during the transitive closure.

The final step is a relabeling step: it produces a labeled image where each pixel is assigned the final root label of its connected component. It is actually a

Algorithm 2: New BW unification (BW FLSL step 1b).
Black and White related processing is highlighted with gray boxes.

1 `init:`
2 $er \leftarrow 0$ ▷ Index of current line segment (relative label)
3 $er' \leftarrow 0$ ▷ Index of previous line segment
4 $a \leftarrow 0$ ▷ Label of current segment, the first is necessarily the exterior
5 $j_0 \leftarrow RLC_i[er]$ ▷ Starting position of current segment
6 $j_1 \leftarrow RLC_i[er+1]$ ▷ End position of current segment
7 $c_8 \leftarrow 0$ ▷ $c_8 = 1$ if current segment is 8-adjacent, here, BG is 4-adjacent
8 ▷ virtual segments allowing to avoid testing for the end of previous line
9 $S_0 \leftarrow RLC_{i-1}[ner_{i-1}]$ $S_1 \leftarrow RLC_{i-1}[ner_{i-1}+1]$ ▷ Save past-the-end
10 **if** ner_{i-1} **is** *odd* **then** $RLC_{i-1}[ner_{i-1}] \leftarrow w+1$
11 $RLC_{i-1}[ner_{i-1}+1] \leftarrow w+2$
12 **goto** `increment previous`
13 `new label:`
14 $T[ne] \leftarrow ne$ ▷ On-the-fly initialization of the equivalence table
15 $F[ne] \leftarrow \varnothing$ ▷ On-the-fly initialization of the feature table
16 $I[ne] \leftarrow a$ ▷ Initial adjacency: a is the label of previous segment
17 $a \leftarrow ne$
18 $ne \leftarrow ne+1$
19 `write label:`
20 $F[a] \leftarrow F[a] \cup$ `computeFeatures`(i, j_0, j_1)
21 $ERA_i[er] \leftarrow a$
22 `increment current:`
23 $er \leftarrow er+1$ ▷ Next segment of current line
24 $er' \leftarrow er'-1$ ▷ Previous segment of previous line intersects current segment
25 $c_8 \leftarrow c_8 \oplus 1$ ▷ Adjust adjacency for current component
26 $j_0 \leftarrow j_1$
27 $j_1 \leftarrow RLC_i[er+1]$
28 **if** $er = ner_i$ **then goto** `end`
29 **if** $RLC_{i-1}[er'] \geqslant j_1 + c_8$ **then goto** `new label`
30 `prolog:`
31 $a \leftarrow \text{Find}(T, ERA_{i-1}[er'])$
32 `increment previous:`
33 $er' \leftarrow er'+2$
34 **if** $RLC_{i-1}[er'] \geqslant j_1 + c_8$ **then goto** `write label`
35 `unify:`
36 $e \leftarrow \text{Find}(T, ERA_{i-1}[er'])$
37 ▷ Union of the two root labels e and a
38 **if** $e \neq a$ **then**
39   **if** $e < a$ **then** swap e, a
40   $T[e] \leftarrow a$
41 **goto** `increment previous`
42 `end:`
43 **if** ner_i **is** *odd* **and** $a \neq 0$ **then** $T[a] \leftarrow 0$
44 $RLC_{i-1}[ner_{i-1}] \leftarrow S_0$ $RLC_{i-1}[ner_{i-1}+1] \leftarrow S_1$ ▷ Restore past-the-end

Algorithm 3: New BW Transitive closure (BW FLSL step 2).
Black and White related processing is highlighted with gray boxes.

```
1  for e = 0 to ne − 1 do
2      a ← T[e]    ▷ ancestor
3      if Hole filling and e = a then    ▷ If label is root
4          i ← I[e]    ▷ label of the surrounding component
5          if T[i] > 0 then a ← i    ▷ e has a surrounding (is not the exterior)
6      if a < e then
7          r ← T[a]
8          T[e] ← r    ▷ Transitive Closure: r = T[T[e]]
9          F[r] ← F[r] ∪ F[e]    ▷ Feature merge
10     else    ▷ e is a root
11         I[e] ← T[I[e]]    ▷ point adjacency to root
12         if Euler number computation then E[e] ← 0
13 if Euler number computation then
14     for e = ne − 1 to 0 step −1 do
15         if T[e] = e then
16             i ← I[e]
17             E[i] ← E[i] + 1 − E[e]
```

Algorithm 4: New BW Relabeling (BW FLSL step 3).

```
1  j0 ← RLC_i[0]    ▷ j0 is 0
2  for er = 0 to ner_i − 1 step 1 do
3      e ← ERA_i[er]    ▷ provisional label
4      r ← T[e]    ▷ final label
5      j1 ← RLC_i[er + 1]
6      Y_i[j0, j1[ ← r    ▷ Memset
7      j0 ← j1    ▷ Register rotation
```

line by line RLE decoder. Like for FLSL, this algorithm is accelerated using an SIMD memset [19]. This step can be skipped if not required, for instance if one is interested only in the connected component features (CCA) without displaying the image of labels.

The two first steps (RLE encoder and segment unification) are done together and thus require only a single scan of the image. The transitive closure step does not scan the image, but requires to scan the equivalence table holding the relation between temporary labels. The relabeling step, when done, needs a second pass over the image to produce the image of labels.

3 Specificities of Black and White Labeling and Hole Processing

In the following, both BG and FG connected components are considered. For the sake of simplicity, a "component" designates a connected component.

3.1 Black and White Labeling

Classical LSL does not process background components, but thanks to its segment design and its semi-open interval encoding, it is easily adaptable. Indeed, the end of a foreground segment is the beginning of the following background one, and vice-versa. Thus, no modification to the RLC tables is required. The RLE encoder remains identical.

The unification step needs a few adjustments. First, it iterates over both odd (FG) and even (BG) segments instead of just the odd ones. This requires to adapt the FSM itself. Indeed, the classical FSM needs to handle cases where multiple segments should be skipped because of the lack of intersection. When processing both FG and BG segments, this cannot happen anymore as all segments (FG and BG) are iterated over sequentially. Consequently, we just need to check if the start of the segment on the previous line ($RLC_{i-1}[er']$) is after the end of the segment on current line ($RLC_i[er+1]$). This makes the new FSM actually *simpler* than the one used by the FLSL algorithm.

To process both FG and BG components, we need to consider complementary connectivity: either 8-adjacency for FG and 4-adjacency for BG, or the 4-adjacency for FG and 8-adjacency for the BG. This is done with the c_8 variable (Algorithm 2, line 7) that defines the current connectivity. It is equals to 1 when processing 8-adjacent component and 0 otherwise. Therefore, changing the background connectivity from 4 to 8 can be easily done by setting c_8 to 1 instead of 0 (Algorithm 2, line 7).

Labels are also assigned to background labels, thus, ERA_i does not necessarily have 0 at even indices. The first encoded segment of a line i is always a BG segment, but has zero length if the first pixel of the line is FG.

Like the unification, the relabeling also needs to iterate over both FG and BG segments.

3.2 Holes and Adjacency Tree Computation

Let us introduce some notations about holes. A component C_1 is surrounded by another component C_2 – written $C_1 \sqsubset C_2$ – *iif* all paths from C_1 to the exterior of the image contain at least one pixel from C_2. A hole in a foreground component W is a background component B that is surrounded by W.

The adjacency tree is encoded in a new table I. For a label e_1 associated to a component C_1, $e_2 = I[e_1]$ is one of the temporary labels of the unique component C_2 that is both adjacent to C_1 and surrounding C_1 ($C_1 \sqsubset C_2$). The label e_2 is not necessarily a root label during the execution of the algorithm.

$I[e_1]$ equals -1 if $e_1 = 0$, or if e_1 is not a root label ($T[e_1] \neq e_1$). In other words, the table I represents the adjacency tree whose edges are directed according to the surrounding relation. In the following, we consider for the sake of simplicity 8-adjacency for the FG and 4-adjacency for the BG.

We considered two methods to build the adjacency tree and the surrounding relation: detecting *closing pixels* [9], or looking at the adjacency at exterior pixels [23], and more precisely looking at the *initial adjacency*.

We chose to use the *initial adjacency* method as it saves one extra branch and one extra Find within the Unification compared to the closing pixel method. Moreover, the update of I when an adjacency is discarded is actually not necessary as I is accessed only for root labels whose initial adjacencies are kept by construction. While the adjacency is a local property, the surrounding is not and thus is defined and correct only when the component has been fully scanned. Consequently, initial adjacency builds a speculative I that is correct only at the end of the image scan and that cannot be worked on beforehand.

The *initial adjacency* method works as follow. Every time a new label is created, the label directly above the current pixel is recorded in I as its initial adjacency and speculative surrounding. It is actually simpler to look for the label on the left that is necessarily from the same component as above. When two root labels a and b (with $a < b$) are unified, the initial adjacency $I[b]$ is discarded in favor of $I[a]$ (and $T[b] \leftarrow a$). The order relation on labels implies that top pixels of a are higher than top pixels of b – or at least at the same height. It means that the higher initial adjacency and speculative surrounding is kept while the other is discarded. Once a component has been fully scanned, only the initial adjacency of the root label remains. The root label being by construction the label of top most pixels, its initial adjacency is necessarily on the exterior of the component. The remaining initial adjacency and speculative surrounding is thus necessarily a true surrounding.

Hole Filling is done during the transitive closure (Algorithm 3, lines 3–5). This is done by merging any component that is neither ⓪ nor directly surrounded by ⓪ with their surrounding component. The initial surrounding relation of such a component is transformed into an equivalence relation ($T[e] \leftarrow I[e]$). In fact, arbitrary connected operators can be implemented using the same principle. One would only need to change the criteria to merge a component into its surrounding in order to implement any connected operator.

Euler Number is the difference between the number of connected components and the number of holes [8]. Because we have labeled both BG and FG component, it is trivial to compute. In fact, thanks to the adjacency tree, we can even compute the euler number of a component without much effort (Algorithm 3, line 13–17).

3.3 Example

Figure 3 shows how our algorithm builds the equivalence table T and the adjacency tree I on a simple, yet complete, example. It shows the input image with initial labels and their speculative surrounding (FG in gray and BG in white), as well as a graph representing both the equivalence table T and the adjacency tree I.

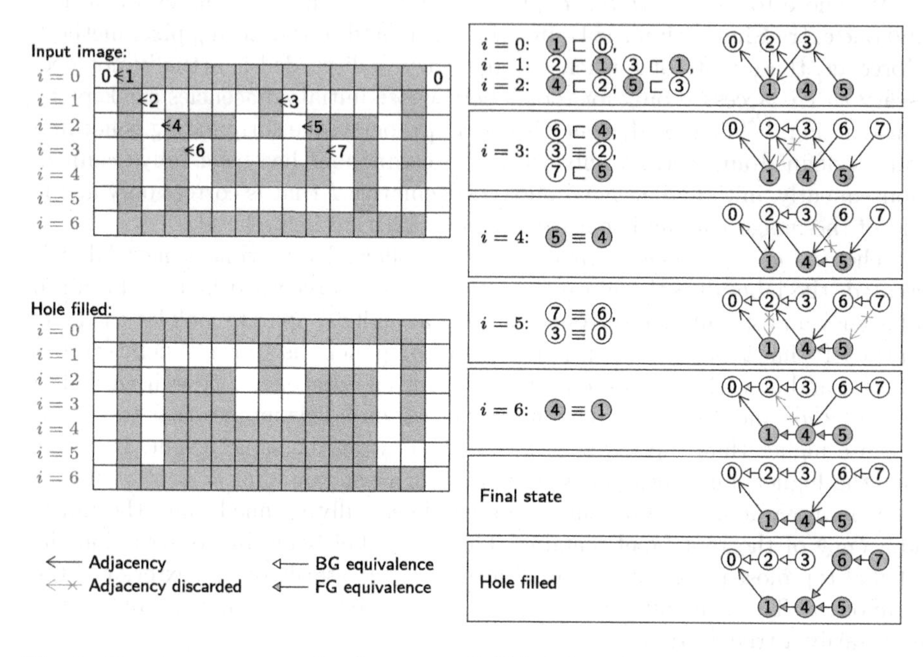

Fig. 3. Step by step example of our new BW labeling focusing on equivalence and adjacency computation.

On the first three lines ($i = 0$, $i = 1$ and $i = 2$), five new labels are created ❶, ②, ③, ❹ and ❺. Their initial adjacency is set as their speculative surrounding: ❶ $\sqsubset$ ⓪, ② $\sqsubset$ ❶, ③ $\sqsubset$ ❶, ❹ $\sqsubset$ ② and ❺ $\sqsubset$ ③.

At $i = 3$, two new labels are created with the following speculative surroundings: ⑥ $\sqsubset$ ❹ and ⑦ $\sqsubset$ ❺. In addition, ③ $\equiv$ ② is detected. Consequently, the speculative surrounding of ③ is discarded in favor of ② $\sqsubset$ ❶.

At $i = 4$, as ❺ $\equiv$ ❹, the speculative surrounding ❺ $\sqsubset$ ③ is discarded.

At $i = 5$, two new equivalences are detected: ② $\equiv$ ⓪ and ⑦ $\equiv$ ⑥. Therefore, the speculative surroundings of ② and ⑦ are dropped. The component ⓪ ② ③ has no more surrounding as ⓪ is the exterior of the image. While the algorithm is not capable to detect it, we can see that the surrounding ⑥ $\sqsubset$ ❹ is no more speculative and is actually definitive.

At $i = 6$, the last equivalence ❹ $\equiv$ ❶ is detected and the speculative surrounding ❹ $\sqsubset$ ② is discarded, and the surrounding ❶ $\sqsubset$ ⓪ is kept.

This leads to the final state before transitive closure where all remaining surroundings (⑥ ⊏ ④ and ① ⊏ ⓪) are no more speculative and are actually true surroundings. When holes are filled, the adjacency edge ⑥ ⊏ ④ is replaced by an equivalence edge ⑥ ≡ ④. Note that our algorithm actually fills hole during transitive closure and not beforehand.

4 Benchmark and Performance Analysis

We measured the performance of our algorithms using a protocol similar to [4]. All the algorithms are sequential and no multithreading is used. We tested randomly generated 2048 × 2048 images with varying density and granularity on a Skylake Gold 6126 Xeon @2.60 GHz. We focus our analysis on $g = 1$ as it is the worst case for run-based algorithms like FLSL. Grana's [3], Diaz' [24] and Lemaitre's [19] CCL algorithms have been ran and measured on this machine. The feature computation with FLSL has been back-ported from classical LSL and was not part of its paper. The other ones have been estimated from their paper. To have comparable results across machines, we give all the results in cycles per pixel (cpp) that is the execution time multiplied by the clock frequency and divided by the number of pixels. In addition, we tested multiple variants of our algorithm that computes a subset of *Euler number*, *hole filling*, *feature computation* and *relabel* in order to compare to existing algorithms that do not compute all of them. Especially, the Euler number computation has been implemented for the sole purpose of comparing our new algorithm with the State-of-the-Art. For CCA algorithms, the seven standard features are computed: the surface, the bounding box (x_{min}, x_{max}, y_{min}, y_{max}) and the first statistical raw moments (S_x, S_y).

Table 1. Processing time in cpp of the core part of our new BW FLSL as well as the extra processing time for extra computation. Minimal and maximal times are given for 2048 × 2048 random images. Min time reached for $d = 0\%$ and max time reached for $g = 1$ and $d \simeq 40\%$.

		Min	Max
BW + Adjacency	(BWA)	0.36	12.7
+Euler number	(E)	+ 0	+ 0.29
+Hole Filling	(H)	+ 0	+ 0.50
+Feature Computation	(F)	+ 0	+ 8.59
+Relabeling	(R)	+ 0.59	+ 3.66

Table 1 shows the minimal and maximal processing time of our new labeling algorithm. The first line corresponds to a *base* processing: foreground and background CC labeling and computing their adjacency tree (BWA). The next lines provide the extra times for extra computations like *Euler number* (E) or *hole filling* (H), *feature computation* (F) and *relabeling* (R). The extra times are the best (min) and worst (max) case we measured for doing these extra computations. One can then estimate the total processing time for a given combination of {E, H, F, R} just by adding the associated extra times.

On this table, we can observe that the minimal extra time for all computations but relabeling is 0. This is a property of run-based algorithms: those computation times depend on the number of segments – which is 1 per line for empty images.

In the worst case, Feature computation adds a large extra time because the seven features need to be written for each and every labels which highly increases the number of memory accesses. The minimal extra processing time for *relabeling* is non-zero as a second scan of the image is required to produce the output image of labels. Therefore, its computation should be avoided if not required. But thanks to the SIMD RLE decoder, this processing remains fast in the worst case. One can also notice that *Euler number* computation and *hole filling* are inexpensive using our approach.

Table 2. Performance comparison between State-of-the-Art algorithms and this work (BW FLSL). The "compute" column shows what is computed. Processing time in cycle/pixel for 2048 × 2048 random images at $g = 1$.

Algorithm		Compute	Min	Avg	Max
He bit-quad		E	2.87	14.0	23.7
He BW	(with R)	BWER	16.5	51.0	79.6
Diaz	(with R)	BWAR	18.4	36.8	59.0
Spaghetti(FG) + Spaghetti(BG)	(with R)	BWR	5.76	32.7	51.2
FLSL(FG+R) + FLSL(BG+R)	(with R)	BWR	2.34	17.0	24.4
FLSL(FG+F) + FLSL(BG+F)	(with F)	BWF	1.68	20.5	30.3
FLSL(BG) + FLSL(FG+F)	(with F)	WF**H**	1.89	19.8	33.7
BW FLSL+ER		BWAER	0.98	10.0	14.6
BW FLSL+F		BWAF	0.38	13.0	20.0
BW FLSL+FH		BWAF**H**	0.38	14.0	20.7

B : Black labeling (BG) A : Adjacency tree F : Feature Computation R : Relabel
W : White labeling (FG) E : Euler number **H** : Hole filling

In Table 2, each State-of-the-Art algorithm are compared to one configuration of our new algorithm that computes at least as much. Our base algorithm BW FLSL+ER that computes the adjacency tree, the Euler number relabels the output image is faster than any black and white CCL algorithm. In average it is 5.1× faster than He BW [11] and 3.6× faster than Diaz [24]. It is even faster than He bit-quad [32] whose sole purpose is to compute the Euler number of the image. This speed difference comes mainly from the efficient use of runs, the use of SIMD, and the low overhead computation of the adjacency tree. Even though a single execution of FLSL is faster than BW FLSL, FLSL process only a FG components. Thus, two executions of FLSL (and Spaghetti) are needed to compute any hole related property.

Therefore, BW FLSL is from 3.3× up to 5.9× faster than Spaghetti and from 1.4× up to 1.7× faster than FLSL to have both black and white labels or holes filled. In addition, BW FLSL computes the adjacency tree with no extra cost.

5 Conclusion

In this article, we have introduced a new connected component labeling and analysis algorithm that is able to do in one single pass of the image, both the Euler number computation but also a double foreground and background labeling with the adjacency tree computation. The modified transitive closure algorithm enables an efficient hole processing: holes can be filled and the surrounding connected components are updated on-the-fly whereas features are computed to take this change into account.

As far as we know our new algorithm outperforms all published algorithms for BW labeling and hole processing. In addition, it is easily tunable: its structure can be adapted to other connected operators like filtering out components based on their statistical features.

References

1. Bailey, D., Johnston, C.: Single pass connected component analysis. In: Image and Vision New Zeland (IVNZ), pp. 282–287 (2007)
2. Bailey, D.G., Klaiber, M.J.: Zig-zag based single-pass connected components analysis. J. Imaging **5**(45), 1–26 (2019)
3. Bolelli, F., Allegretti, S., Baraldi, L., Grana, C.: Spaghetti labeling: directed acyclic graphs for block-based connected components labeling. IEEE Trans. Image Process. **29**, 1999–2012 (2020)
4. Bolelli, F., Cancilla, M., Baraldi, L., Grana, C.: Toward reliable experiments on the performance of Connected Components Labeling algorithms. J. Real-Time Image Process. **17**(2), 229–244 (2018). https://doi.org/10.1007/s11554-018-0756-1
5. Cabaret, L., Lacassagne, L.: What is the world's fastest connected component labeling algorithm? In: IEEE International Workshop on Signal Processing Systems (SiPS), pp. 97–102 (2014)
6. Cabaret, L., Lacassagne, L., Etiemble, D.: Parallel light speed labeling for connected component analysis on multi-core processors. J. Real-Time Image Process. (JRTIP) **15**(1), 173–196 (2018)
7. Galil, Z., Italiano, G.: Data structures and algorithms for disjoint set union problems. ACM Comput. Surv. **23**(3), 319–344 (1991)
8. Gray, S.B.: Local properties of binary images in two dimensions. Trans. Comput. **20**(5), 551–561 (1971)

9. He, L., Chao, Y.: A very fast algorithm for simultaneously performing connected-component labeling and Euler number computing. Trans. Image Process. **24**(9), 2725–2735 (2017)
10. He, L., Chao, Y., Suzuki, K.: A new algorithm for labeling connected-components and calculating the Euler number, connected-component number, and hole number. In: International Conference on Pattern Recognition (ICPR), pp. 3099–3102 (2012)
11. He, L., Chao, Y., Suzuki, K.: An algorithm for connected-component labeling, hole labeling and Euler number computing. J. Comput. Sci. Technol. **28**(3), 468–478 (2013)
12. He, L., Ren, X., Gao, Q., Zhao, X., Yao, B., Chao, Y.: The connected-component labeling problem: a review of state-of-the-art algorithms. Pattern Recogn. **70**, 25–43 (2017)
13. He, L., Ren, X., Zhao, X., Yao, B., Kasuya, H., Chao, Y.: An efficient two-scan algorithm for computing basic shape features of objects in a binary image. J. Real-Time Image Proc. **16**(4), 1277–1287 (2016)
14. Hennequin, A., Masliah, I., Lacassagne, L.: Designing efficient SIMD algorithms for direct connected component labeling. In: ACM Workshop on Programming Models for SIMD/Vector Processing (PPoPP), pp. 1–8 (2019)
15. Hennequin, A., Meunier, Q.L., Lacassagne, L., Cabaret, L.: A new direct connected component labeling and analysis algorithm for GPUs. In: IEEE International Conference on Design and Architectures for Signal and Image Processing (DASIP), pp. 1–6 (2018)
16. Klaiber, M.J., Bailey, D.G., Simon, S.: A single-cycle parallel multi-slice connected components analysis hardware architecture. J. Real-Time Image Proc. **16**(4), 1165–1175 (2019)
17. Lacassagne, L., Zavidovique, A.B.: Light speed labeling for RISC architectures. In: IEEE International Conference on Image Analysis and Processing (ICIP) (2009)
18. Lacassagne, L., Zavidovique, B.: Light speed labeling: efficient connected component labeling on RISC architectures. J. Real-Time Image Process. (JRTIP) **6**(2), 117–135 (2011)
19. Lemaitre, F., Hennequin, A., Lacassagne, L.: How to speed connected component labeling up with SIMD RLE algorithms. In: ACM Workshop on Programming Models for SIMD/Vector Processing (PPoPP), pp. 1–8 (2020)
20. Manne, F., Patwary, M.M.A.: A scalable parallel union-find algorithm for distributed memory computers. In: Wyrzykowski, R., Dongarra, J., Karczewski, K., Wasniewski, J. (eds.) PPAM 2009. LNCS, vol. 6067, pp. 186–195. Springer, Heidelberg (2010). https://doi.org/10.1007/978-3-642-14390-8_20
21. Patwary, M.M.A., Blair, J., Manne, F.: Experiments on union-find algorithms for the disjoint-set data structure. In: Festa, P. (ed.) SEA 2010. LNCS, vol. 6049, pp. 411–423. Springer, Heidelberg (2010). https://doi.org/10.1007/978-3-642-13193-6_35
22. Playne, D.P., Hawick, K.: A new algorithm for parallel connected-component labelling on GPUs. IEEE Trans. Parallel Distrib. Syst. **29**(6), 1217–1230 (2018)
23. Díaz del Río, F., Molina-Abril, H., Real, P.: Computing the component-labeling and the adjacency tree of a binary digital image in near logarithmic-time. In: Marfil, R., Calderón, M., Díaz del Río, F., Real, P., Bandera, A. (eds.) CTIC 2019. LNCS, vol. 11382, pp. 82–95. Springer, Cham (2019). https://doi.org/10.1007/978-3-030-10828-1_7
24. del Rio, F.D., Sanchez-Cuevas, P., Molina-Abril, H., Real, P.: Parallel connected-component-labeling based on homotopy trees. Pattern Recogn. Lett. **131**, 71–78 (2020)

25. Rosenfeld, A.: Digital topology. Am. Math. Mon. **28**(8), 621–360 (1979)
26. Rosenfeld, A., Platz, J.: Sequential operator in digital pictures processing. J. ACM **13**(4), 471–494 (1966)
27. Somasundaram, K., Kalaiselvi, T.: A method for filling holes in objects of medical images using region labeling and run length encoding schemes. In: National Conference on Image Processing (NCIMP), pp. 110–115 (2010)
28. Tang, J.W., Shaikh-Husin, N., Sheikh, U.U., Marsono, M.N.: A linked list run-length-based single-pass connected component analysis for real-time embedded hardware. J. Real-Time Image Proc. **15**(1), 197–215 (2016)
29. Tarjan, R.: Efficiency of good but not linear set union algorithm. J. ACM **22**(2), 215–225 (1975)
30. Tarjan, R., Leeuwen, J.: Worst-case analysis of set union algorithms. J. ACM **31**, 245–281 (1984)
31. Wu, K., Otoo, E., Suzuki, K.: Optimizing two-pass connected-component labeling algorithms. Pattern Anal. Appl. **12**, 117–135 (2009)
32. Yao, B., He, L., Kang, S., Zhao, X., Chao, Y.: Bit-quad-based Euler number computing. IEICE Trans. Inf. Syst. **E100.D**(9), 2197–2204 (2017)
33. Zhao, H., Chen, Z.X.: A simple hole filling algorithm for binary cell images. Appl. Mech. Mater. **433–435**, 1715–1719 (2013)

LSL3D: A Run-Based Connected Component Labeling Algorithm for 3D Volumes

Nathan Maurice(✉), Florian Lemaitre, Julien Sopena, and Lionel Lacassagne

LIP6, Sorbonne University, CNRS, Paris, France
nathan.maurice@lip6.fr

Abstract. Connect Component Labeling (*CCL*) has been a fundamental operation in Computer Vision for decades. Most of the literature deals with 2D algorithms for applications like video surveillance or autonomous driving. Nonetheless, the need for 3D algorithms is rising, notably for medical imaging.

While 2D *CCL* algorithms already generate large amounts of memory accesses and comparisons, 3D ones are even worse. This is the *curse of dimensionality*. Designing an efficient algorithm should address this problem. This paper introduces a segment-based algorithm for 3D labeling that uses a new strategy to accelerate label equivalence processing to mitigate the impact of higher dimensions. We claim that this new algorithm outperforms State-of-the-Art algorithms by a factor from ×1.5 up to ×3.1 for usual medical datasets and random images.

1 Introduction

Connected Component Labeling (CCL) has been a fundamental algorithm in Computer Vision for decades [14,35,38]. It consists of finding connected components (sets of adjacent pixels) in a binary image and assigning them a unique identifier referred to as the *label*.

CCL is used in a wide array of applications, such as autonomous driving [10,39], video surveillance [20,36], medical applications [1,9,21,28,33] and other fields like [32] where a real-time implementation matters.

This article introduces a new 3D labeling algorithm named *LSL3D* and our contributions are twofold: 1) a new Finite State Machine (FSM) to efficiently process segments using Run-Length Encoding (RLE) and 2) a cache mechanism to re-use partial results and reduce computational complexity.

We claim that our new segment-based algorithm is 1.8× to 2.3× faster than State-of-The-Art algorithms for existing medical datasets. Moreover, for random 3D images, which are more stressing at low granularities, we claim that LSL3D is 1.5× to 3.1× faster.

The article is written as follows: Sect. 2 gives a background on *CCL* and details our benchmark protocol. Section 3 reviews existing literature. Then,

P. L. Mazzeo et al. (Eds.): ICIAP 2022 Workshops, LNCS 13374, pp. 132–142, 2022.
https://doi.org/10.1007/978-3-031-13324-4_12

Table 1. Average characteristics of 3D datasets.

Dataset	# subsets	Size	Density	Granularity	# runs	# CCs
mitochondria	3	1024 × 768 × 165	5.9	26.4	197,878	40
OASIS	373	256 × 256 × 128	19.8	4.2	236,718	3,200
Random	16	256 × 256 × 256	0–100	1–16		

Sect. 4 introduces three strategies for label equivalence management that attenuate the *curse of dimensionality*. Finally, Sect. 5 studies the impact of SIMD on our new *CCL* algorithms.

2 Classical Approaches to Connected Components Labeling and Their Evaluation

Fundamentally, *CCL* algorithms establish equivalences between foreground pixels if they are connected.

Only a single solution for the labeling of a given image exists. Qualitative compromises are therefore impossible, and research on *CCL* algorithms has been focused on the reduction of their execution time.

In this section, we will first detail the basis of modern algorithms, and the evaluation protocols: metrics, datasets and benchmark platform.

2.1 Main Principles of Modern *CCL* Algorithms

Modern algorithms are all derived from historical ones like those from Rosenfeld [35] or Haralick and Shapiro [14]. They are composed 3 steps:

1. a *provisional labeling*, that assigns a temporary label to each pixel and builds label equivalences,
2. label *equivalence solving*, that computes the *Transitive Closure* (TC) of the graph associated to the label equivalence table,
3. *final labeling*, to replace temporary labels with final labels (usually the smallest of each component).

Modern algorithms implement some algorithmic optimizations to accelerate these three steps. Since the bottleneck of these algorithms is usually their *control-flow* rather than memory accesses or calculations, datasets have a major impact on their performance.

2.2 Benchmarking Procedure and Datasets

To evaluate the algorithms' performance, two medical datasets have been used for the benchmarks: *mitochondria* [29] which includes 3 subsets and *OASIS* [30] which includes 373 subsets. Images from *mitochondria* contain large blobs and several small CCs. On the other hand, images from *OASIS* contain a hollow volume with complex shapes. This translates into a high and low granularity [7] for *mitochondria* and *OASIS*, respectively (Table 1).

On top of these medical datasets, we use random images, which have been generated using MT19937 [31] for reproducible results. The random images have a density $d \in [0\%; 100\%]$ and a granularity $g \in [1; 16]$ where the granularity describes the detail level of an image (size of individual blocks during generation). The algorithms have been evaluated on an Intel Xeon Gold 6126 using the *YACCLAB* [6] framework.

3 State-of-the-Art of 3D Algorithms

Literature on *CCL* algorithms is extensive and has been centered on 2D images. *CCL* on CPUs has been heavily studied and optimized [6,13,16,25]. On GPUs, after an early era of iterative algorithms [3,19,41], a new generation introduced by Komura [22] are now direct; a new way to manage equivalences and reduce memory accesses was introduced by Playne [34] and has become the basis of the fastest *CCL* algorithms [2,18].

CCL algorithms can be classified according to their neighborhood mask and how they process data: they can be pixel-based, block-based or segment-based.

3.1 Pixel-Based Algorithms

The extension of the *Rosenfeld* 2D algorithm to 3D is straightforward: the 9 adjacent pixels from the previous slice are added to the mask, for a total of 13 pixels.

The mask-based approach was improved by Wu [40] (*SAUF*). Wu realized that a decision could be taken without accessing all 4 pixels in the neighborhood for 2D images. A decision tree was proposed to access as few neighbors as possible. *SAUF* was later ported to 3D by Bolelli as *SAUF 3D* [4]. The decision tree was further optimized by He *et al.* with the *Label Equivalency Based* (*LEB*) algorithm for 2D [15] and 3D [27] images.

In [17], He noticed that the value of the previous pixel could simplify the following decision with fewer comparisons and introduced a graph of decision trees. This method was generalized by Grana with the *PRED* algorithm [11], which was later extended to 3D volumes by Bolelli with *PRED 3D* [4]. The introduction of *Direct Rooted Acyclic Graphs* (*DRAG*) by Bolelli [5] reduced the code footprint. *DRAG* were used to revisit existing algorithms, like with *SAUF++* and *PRED++* by Bolelli [4]. The same paper extended these new *SAUF++* and *PRED++* algorithms for 3D (*SAUF++ 3D* and *PRED++ 3D*).

3.2 Block-Based Algorithms

Grana [12] proposed a block-based approach (foreground pixels in the same 2×2 block are necessarily in the same component). The decision tree for block-based algorithms was then improved upon by Chabardes [8] with a forest of *decision trees*, and was later adapted to use a *DRAG* by Bolelli [5].

The block-based approach was extended to 3D volumes by Sochting [37] with the *Entropy Partition Decision Tree*. *EPDT* algorithms handle pixels by blocks of either $2 \times 1 \times 1$ (*EDPT_19c* and *EDPT_22c*) or 2×2 (*EDPT_26c*).

3.3 Segment-Based Algorithms

While block-based approaches have been shown to be efficient, pixels can also be regrouped as segments. For a given line, it iterates over each column and aggregates consecutive pixels into segments using a *Run Length Encoding* (interval encoding indices). Then, it checks for segment adjacency (overlaps between current segment and segment of the previous line) and performs unions when needed.

A *Run Based Two Scans* strategy (*RBTS*) for 2D images was used by He [26] and later extended to 3D volumes [27]. The segment-based approach has also been proposed by Lacassagne with the *Light Speed Labeling* (*LSL*) [23,24] for 2D labeling. *LSL* also uses a RLE but adds a *line-relative labeling* (ER tables) combined with a table of segments (RLC) to accelerate adjacency detection and equivalence building.

4 LSL3D and Efficient Unification Strategies for 3D Volumes

This section presents, step-by-step, the improvement and the transformation of the classical 2D LSL algorithm into an optimized 3D version. Step zero is the extension of the 2D version to 3D, keeping the line-relative labeling (version named `LSL_ER`). It can be viewed as a legacy version for comparison [24]. The first step is the replacement of the ER tables by a Finite State Machine (FSM) (`LSL_FSM`). The second improvement is a cache-reuse mechanism to perform unions/unifications with double-lines (`LSL_FSM_DOUBLE`).

Our successive *LSL* implementations have been tested according to the benchmark in Sect. 2.2. They are compared to 7 algorithms from the State-of-the-Art: *LEB*, *RBTS*, *PRED++ 3D*, *SAUF++ 3D*, *EDPT_19c*, *EDPT_22c* and *EDPT_26c*. Among the *EPDT* algorithms, we only present the best one (*EDPT_22c*). The results are shown on Figs. 2 and 3 and will be evaluated throughout the following sections.

4.1 Extension of the Segment-Based Unification for 3D Volumes

In order to find overlapping segments without iterating several times on the current line, the first *LSL* algorithm [24] finds overlaps by accessing the ER table. In 2D, two ER tables (current and previous line) are necessary at any given time. On the other hand, due to the raster scan, two planes are required in 3D (current and previous plane). This can degrade performance on large images:

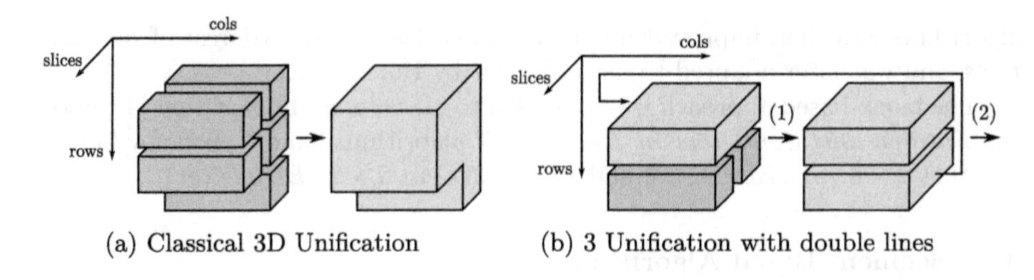

(a) Classical 3D Unification (b) 3 Unification with double lines

Fig. 1. Segment-based unification for 3D volumes. (Color figure online)

for instance, on *mitochondria*, the ER tables use a total of 3.1 MB of memory and thus do not fit within the 1.0 MB L2 cache of the Xeon.

The performance of our `LSL_ER` implementation in 3D can be seen on Fig. 2. `LSL_ER` becomes faster at $g > 4$ on random images and widens the gap at higher granularities (up to $\times 1.3$ for $g = 16$).

For medical images, `LSL_ER` is overall faster than the state of the art (Fig. 3): while *RBTS* is $\times 1.1$ faster on *mitochondria*, it is also slower by a factor of $\times 1.3$ on *OASIS*. Similarly, *PRED++ 3D* is as fast on *OASIS* but slower by a factor of $\times 1.3$ on *mitochondria*.

The limits of *LSL* can be explained by the duration of the RLE step, especially on large images (between 60% and 70% of the execution time on *mitochondria*). Not only does it create an array of segments (RLC table) but the initialization of the ER table is costly as it contains one element per pixel.

4.2 A Finite-State Machine-Based Unification

Overlapping segments between lines can also be found without ER using a Finite-State Machine (FSM). In the 2D unification [25], each state of the 2D FSM encodes segment configurations between the current and previous lines. Merging two lines involves iterating over both at the same time: a new label is created for each isolated segment, whereas the components of two overlapping segments are merged together.

While the FSM-based implementation of *LSL3D* is efficient on simple images, this is not the case for more complex images. Indeed, Fig. 2 shows that `LSL_FSM` is $\times 1.3$ faster than ER for high-granularity images ($g = 16$), but slower by a factor of $\times 0.61$ for $g = 1$.

The execution time of `LSL_FSM` follows a similar trend on medical images, as can be seen on Fig. 3: *mitochondria* the FSM-based algorithm improves the execution time compared to the ER-based unification by a factor $\times 1.1$. On the other hand, for *OASIS*, it is slower by a factor of $\times 0.95$.

The overhead of the unification phase explains the results. Despite a lack of ER tables and a faster RLE step, the complexity of the 3D FSM (27 states and 55 transitions in 3D, vs. 8 states and 14 transitions for its 2D counterpart) degrades the accuracy of the branch predictor. This is particularly penalizing on complex images with many segments and a wider diversity of segments configurations (and thus, with more state transitions being performed).

4.3 Computational Reuse of Merged Lines

The complexity of the FSM has led to performance limitations on complex images. To overcome these limitations, we have redesigned the FSM to store and reuse partial results. On top of factorizing calculations, this also simplifies the FSM (from 27 states and 55 transitions to just 9 states and 18 transitions).

More precisely, as shown on Fig. 1, two consecutive iterations on the same slice process three lines several times: the current line (in red) and two neighboring lines (in blue) will be re-processed in the following fusion. This redundancy can be removed by caching partial results (in green) into a double-line array.

In order to simplify the computational reuse, a 2×2 mask is used for lines, as displayed on Fig. 1b. Two phases are required: a first step (1) unifies the current line (red line) and its neighbor (blue line) from the previous slice. It produces a temporary line (green) that contains overlaps from both lines. Then a second step (2) unifies the double-line with the one produced in the previous step. The newly-created double-line is re-used in the next unification to avoid redundant processing. The former double-line is discarded, and its memory is recycled for the next iteration.

The performance of `LSL_DOUBLE` on random images (Fig. 2) shows that `LSL_DOUBLE` is on average better than `LSL_ER` and `LSL_FSM` for all granularities: It is indeed $\times 1.3 - 1.5$ faster than the best algorithm for $g = 4$ and $g = 16$ and only $\times 0.94$ the speed of the best.

Besides these good results, Fig. 2 also shows that `LSL_DOUBLE` is more resistant to increasing densities (gap between green a purple lines, beyond 25% density). Indeed, the number of segments at these densities is statistically high, which slows down segments-based algorithms. In `LSL_DOUBLE`, the phenomenon is reduced by the fusion of segments within double-lines: more segments implies more overlaps, which leads to more fusions and fewer segments in the double-line. This makes the double-line strategy particularly relevant for complex images. On *mitochondria* (Fig. 3a), `LSL_DOUBLE` is as fast as `LSL_FSM`. However, unlike `LSL_FSM`, it is as fast as `LSL_ER` on *OASIS*. These results make the double-line algorithm at least as fast as the *best* algorithm on both *OASIS* and *mitochondria* (Fig. 3).

5 Architecture-Specific Optimizations of Run-Length Encoding on 3D Images

As seen in the previous section, the double-line unification reduces the execution time of *LSL3D*: the unification (which does not need extra ER tables) becomes highly optimized. The RLE and relabeling steps thus become the main bottlenecks. ($\approx 90\%$ and $\approx 70\%$ of the execution time for *mitochondria* and *OASIS*).

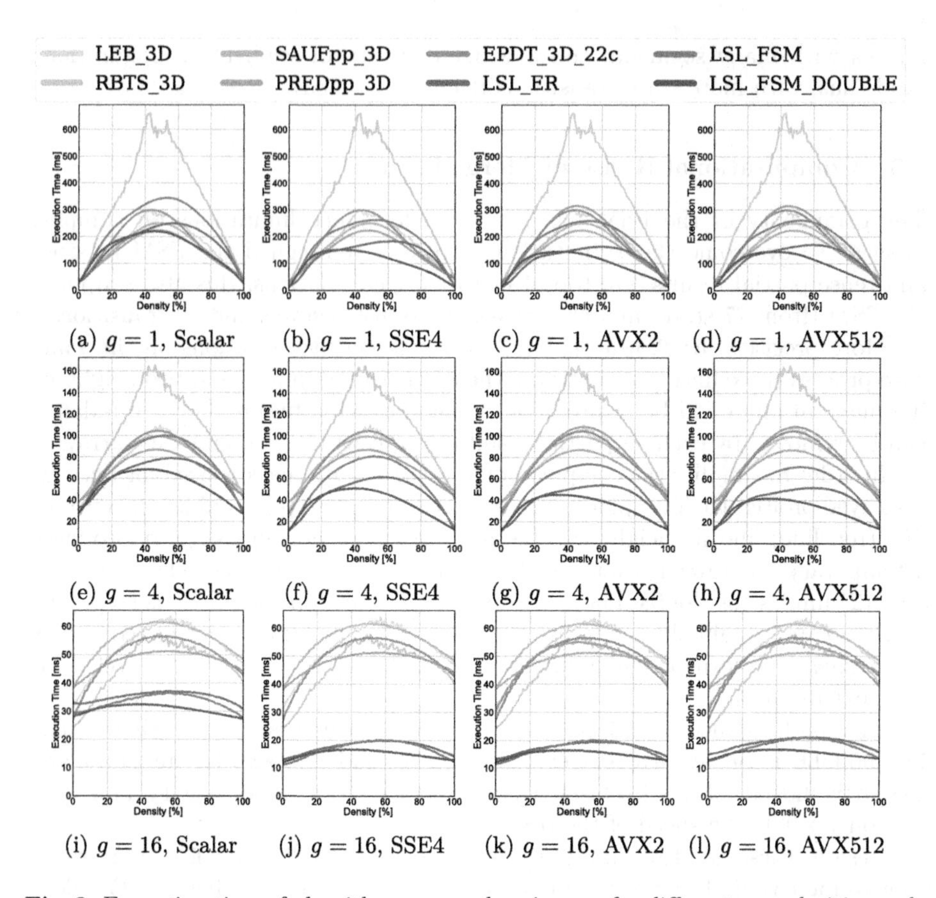

(a) $g = 1$, Scalar (b) $g = 1$, SSE4 (c) $g = 1$, AVX2 (d) $g = 1$, AVX512

(e) $g = 4$, Scalar (f) $g = 4$, SSE4 (g) $g = 4$, AVX2 (h) $g = 4$, AVX512

(i) $g = 16$, Scalar (j) $g = 16$, SSE4 (k) $g = 16$, AVX2 (l) $g = 16$, AVX512

Fig. 2. Execution time of algorithms on random images for different granularities and densities on Xeon (Color figure online)

Fortunately, these steps lend themselves well to instruction level parallelism with SIMD [25]. Several SIMD implementations of the RLE and relabeling have been tested: SSE4, AVX2 and AVX512.

As can be seen on Fig. 2, the use of SIMD accelerates the execution of all *LSL* versions on random images. In fact, SSE4 alone is enough to make LSL_ER faster than State-of-the-Art algorithms on random images by a factor of ×1.1 to ×2.6 (Fig. 2). This is also true for medical images, with a speedup of ×1.4 to ×2.0 compared to the State-of-the-Art algorithm on *mitochondria* and *OASIS* (Fig. 3).

The use of more complex instruction sets such as AVX2 or AVX512 do not provide additional speedups over SSE4 on simple images such as *mitochondria* ((Fig. 3a), but nonetheless improves execution times by 10% on complex images (*OASIS*). For the AVX2 version, the lack of dedicated *compress* instructions makes the speedup constrained by look-up table accesses in the RLE step. On the other hand, the AVX512 *compress* instructions on the Xeon are only

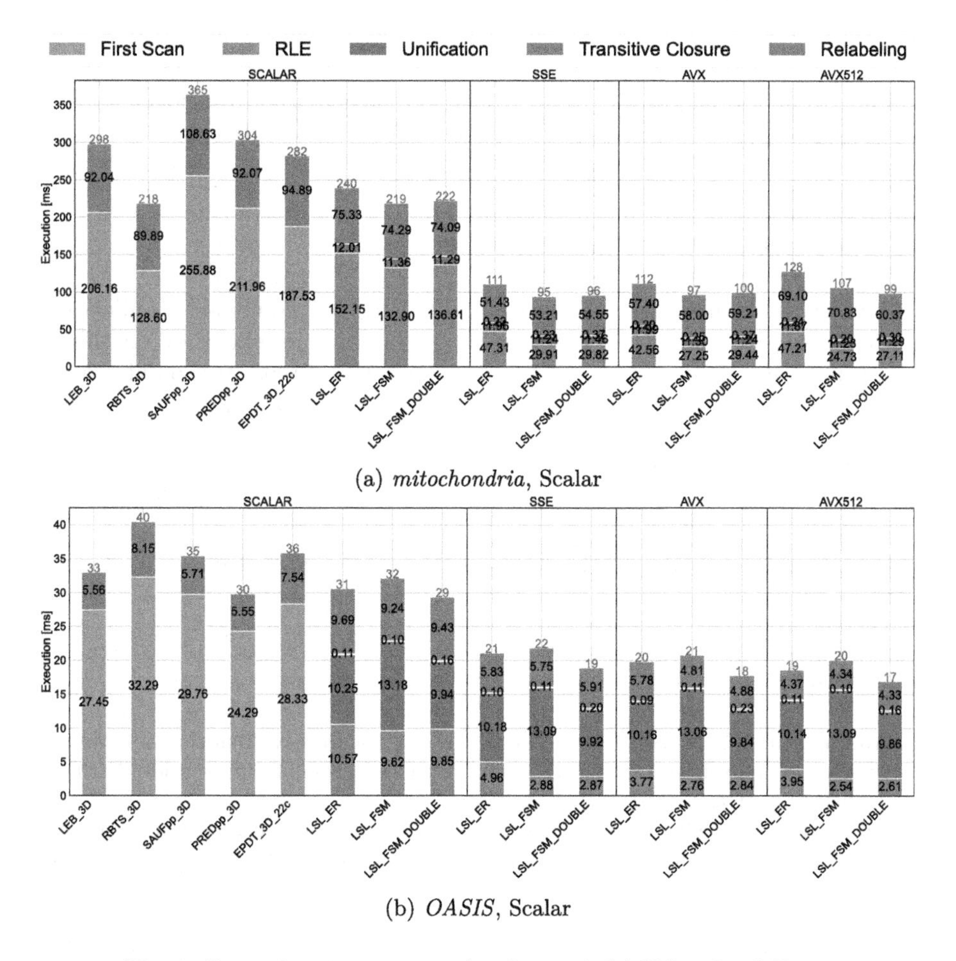

(a) *mitochondria*, Scalar

(b) *OASIS*, Scalar

Fig. 3. Execution time on *mitochondria* and *OASIS* on Intel Xeon

available for 32-bits elements. A conversion of elements to 16-bits for processing the segments (RLC table encodes segments using 16-bits numbers) is thus required and adds an overhead.

The combination of SIMD with the double-line mechanism gives an even greater acceleration: on random images, `LSL_DOUBLE+AVX512` is on average $\times 1.5$ to $\times 3.0$ faster than the best algorithm from the State-of-the-Art, whereas it is faster by a factor $\times 1.7$ and $\times 2.2$ on natural images.

The speedup of the best *LSL3D* version compared to the best State-of-the-Art can be found on Table 2.

Table 2. Best speedups of LSL3D versus best State-of-the-Art algorithms

Dataset	Mitochondria	Oasis	Random $g=1$	Random $g=4$	Random $g=16$
Speedup	$\times 2.3$	$\times 1.8$	$\times 1.5$	$\times 2.2$	$\times 3.1$

6 Conclusion and Future Work

This article introduces a new algorithm, *LSL3D*, that combines a unification approach based on a finite state machine to improve its efficiency on simple images and a *double*-line mechanism and computational reuse for complex ones. On top of a scalar extension, we use SIMD (`SSE4`, `AVX2`, `AVX512`) instructions to accelerate the RLE compression and decompression steps.

Evaluation of performances on medical datasets and random images shows that *LSL3D* outperforms State-of-the-Art algorithms by a factor $\times 1.5$ up to a factor $\times 3.1$, on an Intel Xeon.

Future works will address the parallelization of this algorithm for multi-core CPUs and GPUs.

References

1. Abuzaghleh, O., Barkana, B.D., Faezipour, M.: Noninvasive real-time automated skin lesion analysis system for melanoma early detection and prevention. IEEE J. Transl. Eng. Health Med. **3**, 1–12 (2015)
2. Allegretti, S., Bolelli, F., Grana, C.: Optimized block-based algorithms to label connected components on GPUs. IEEE Trans. Parallel Distrib. Syst. **31**(2), 423–438 (2020)
3. Barnat, J., Bauch, P., Brim, L., Češka, M.: Computing strongly connected components in parallel on CUDA. In: 2011 IEEE International Parallel Distributed Processing Symposium, pp. 544–555 (2011)
4. Bolelli, F., Allegretti, S., Grana, C.: One DAG to rule them all. IEEE Trans. Pattern Anal. Mach. Intell. (2021)
5. Bolelli, F., Baraldi, L., Cancilla, M., Grana, C.: Connected components labeling on DRAGs. In: 2018 24th International Conference on Pattern Recognition (ICPR), Beijing, August 2018, IEEE, pp. 121–126 (2018)
6. Bolelli, F., Cancilla, M., Baraldi, L., Grana, C.: Toward reliable experiments on the performance of connected components labeling algorithms. J. Real-Time Image Process. (JRTIP) **17**, 229–244 (2018). https://doi.org/10.1007/s11554-018-0756-1
7. Breen, E.J., Jones, R.: Attribute openings, thinnings, and granulometries. Comput. Vis. Image Underst. **64**(3), 377–389 (1996)
8. Chabardès, T., Dokládal, P., Bilodeau, M.: A labeling algorithm based on a forest of decision trees. J. Real-Time Image Proc. **17**(5), 1527–1545 (2019). https://doi.org/10.1007/s11554-019-00912-8
9. Chen, W., Giger, M.L., Bick, U.: A fuzzy C-means (FCM)-based approach for computerized segmentation of breast lesions in dynamic contrast-enhanced MR images. Acad. Radiol. **13**(1), 63–72 (2006)

10. Farhat, W., Faiedh, H., Souani, C., Besbes, K.: Real-time embedded system for traffic sign recognition based on ZedBoard. J. Real-Time Image Proc. **16**(5), 1813–1823 (2017). https://doi.org/10.1007/s11554-017-0689-0
11. Grana, C., Baraldi, L., Bolelli, F.: Optimized connected components labeling with pixel prediction. In: Blanc-Talon, J., Distante, C., Philips, W., Popescu, D., Scheunders, P. (eds.) ACIVS 2016. LNCS, vol. 10016, pp. 431–440. Springer, Cham (2016). https://doi.org/10.1007/978-3-319-48680-2_38
12. Grana, C., Borghesani, D., Cucchiara, R.: Optimized block-based connected components labeling with decision trees. Trans. Image Process. **19**(6), 1596–1609 (2010)
13. Gupta, S., Palsetia, D., Patwary, M.A., Agrawal, A., Choudhary, A.: A new parallel algorithm for two-pass connected component labeling. In: Parallel & Distributed Processing Symposium Workshops (IPDPSW), pp. 1355–1362. IEEE (2014)
14. Haralick, R.: Some neighborhood operations. In: Real-Time Parallel Computing Image Analysis, pp. 11–35. Plenum Press (1981)
15. He, L., Chao, Y., Suzuki, K.: A linear-time two-scan labeling algorithm. In: 2007 IEEE International Conference on Image Processing (San Antonio, TX, USA, 2007), pp. V - 241–V - 244. IEEE (2007)
16. He, L., Ren, X., Gao, Q., Zhao, X., Yao, B., Chao, Y.: The connected-component labeling problem: a review of state-of-the-art algorithms. Pattern Recogn. **70**, 25–43 (2017)
17. He, L., Zhao, X., Chao, Y., Suzuki, K.: Configuration-transition-based connected-component labeling. IEEE Trans. Image Process. **23**(2), 943–951 (2014)
18. Hennequin, A., Lacassagne, L., Cabaret, L., Meunier, Q.: A new direct connected component labeling and analysis algorithms for GPUs. In: 2018 Conference on Design and Architectures for Signal and Image Processing (DASIP), pp. 76–81. IEEE (2018)
19. Hwu, W.W. (ed.): GPU Computing Gems. Morgan Kaufman (2001). Chap 35: Connected Component Labeling in CUDA
20. Joshi, K.A., Thakore, D.G.: A survey on moving object detection and tracking in video surveillance system. Int. J. Soft Comput. Eng. **2**(3), 44–48 (2012)
21. Khan, N., et al.: Automatic segmentation of liver & lesion detection using H-minima transform and connecting component labeling. Multimedia Tools Appl. **79**(13), 8459–8481 (2019). https://doi.org/10.1007/s11042-019-7347-4
22. Komura, Y.: GPU-based cluster-labeling algorithm without the use of conventional iteration: application to Swendsen-Wang multi-cluster spin flip algorithm. Comput. Phys. Commun. **194**, 54–58 (2015)
23. Lacassagne, L., Zavidovique, A.B.: Light speed labeling for RISC architectures. In: IEEE International Conference on Image Analysis and Processing (ICIP) (2009)
24. Lacassagne, L., Zavidovique, B.: Light speed labeling: efficient connected component labeling on RISC architectures. J. Real-Time Image Proc. **6**(2), 117–135 (2011)
25. Lemaitre, F., Hennequin, A., Lacassagne, L.: How to speed connected component labeling up with SIMD RLE algorithms. In: Proceedings of the ACM 2020 Sixth Workshop on Programming Models for SIMD/Vector Processing (San Diego, CA, USA, February 2020), pp. 1–8. ACM (2020)
26. He, L., Chao, Y., Suzuki, K.: A run-based two-scan labeling algorithm. IEEE Trans. Image Process. **17**(5), 749–756 (2008)
27. He, L., Chao, Y., Suzuki, K.: Two efficient label-equivalence-based connected-component labeling algorithms for 3-D binary images. IEEE Trans. Image Process. **20**(8), 2122–2134 (2011)

28. Litjens, G., et al.: Deep learning as a tool for increased accuracy and efficiency of histopathological diagnosis. Sci. Rep. **6**, 26286 (2016)
29. Lucchi, A., Li, Y., Fua, P.: Learning for structured prediction using approximate subgradient descent with working sets. In: 2013 IEEE Conference on Computer Vision and Pattern Recognition (Portland, OR, USA, June 2013), pp. 1987–1994. IEEE (2013)
30. Marcus, D.S., Fotenos, A.F., Csernansky, J.G., Morris, J.C., Buckner, R.L.: Open access series of imaging studies: longitudinal MRI data in nondemented and demented older adults. J. Cogn. Neurosci. **22**(12), 2677–2684 (2010)
31. Matsumoto, M., Nishimura, T.: Mersenne twister web page. http://www.math.sci.hiroshima-u.ac.jp/~m-mat/MT/emt.html
32. Millet, M., Rambaux, N., Petreto, A., Lemaitre, F., Lacassagne, L.: A new processing chain for detection and tracking of meteors from space. In: International Meteor Conference, September 2021
33. Nazlibilek, S., Karacor, D., Ercan, T., Sazli, M.H., Kalender, O., Ege, Y.: Automatic segmentation, counting, size determination and classification of white blood cells. Measurement **55**, 58–65 (2014)
34. Playne, D.P., Hawick, K.: A new algorithm for parallel connected-component labelling on GPUs. IEEE Trans. Parallel Distrib. Syst. **29**(6), 1217–1230 (2018)
35. Rosenfeld, A., Platz, J.: Sequential operator in digital pictures processing. J. ACM **13**(4), 471–494 (1966)
36. Salau, J., Krieter, J.: Analysing the space-usage-pattern of a cow herd using video surveillance and automated motion detection. Biosys. Eng. **197**, 122–134 (2020)
37. Bolelli, F., Allegretti, S., Grana, C.: A heuristic-based decision tree for connected components labeling of 3D volumes: implementation and reproducibility notes. In: Kerautret, B., Colom, M., Krähenbühl, A., Lopresti, D., Monasse, P., Talbot, H. (eds.) RRPR 2021. LNCS, vol. 12636, pp. 139–145. Springer, Cham (2021). https://doi.org/10.1007/978-3-030-76423-4_9
38. Veillon, F.: One pass computation of morphological and geometrical properties of objects in digital pictures. Signal Process. **1**(3), 175–179 (1979)
39. Weng, H.-M., Chiu, C.-T.: Resource efficient hardware implementation for real-time traffic sign recognition. In: 2018 IEEE International Conference on Acoustics, Speech and Signal Processing (ICASSP), pp. 1120–1124. IEEE (2018)
40. Wu, K., Otoo, E., Shoshani, A.: Optimizing connected component labeling algorithms. In: Fitzpatrick, J.M., Reinhardt, J.M. (eds.) Medical Imaging (San Diego, CA, April 2005), p. 1965 (2005)
41. Ziegler, G., Rasmusson, A.: Efficient volume segmentation on the GPU. In: GPU Technology Conference, pp. 1–44, Nvidia (2010)

Artificial Intelligence for Preterm Infants' HealthCare - AI-Care

Deep-Learning Architectures for Placenta Vessel Segmentation in TTTS Fetoscopic Images

Alessandro Casella[1,2(✉)], Sara Moccia[3,4], Ilaria Anita Cintorrino[2], Gaia Romana De Paolis[2], Alexa Bicelli[2], Dario Paladini[5], Elena De Momi[2], and Leonardo S. Mattos[1]

[1] Department of Advanced Robotics, Istituto Italiano di Tecnologia, Genoa, Italy
alessandro.casella@polimi.it

[2] Department of Electronics, Information and Bioengineering, Politecnico di Milano, Milan, Italy

[3] The BioRobotics Institute, Scuola Superiore SantAnna, Pisa, Italy

[4] Department of Excellence in Robotics and AI, Scuola Superiore SantAnna, Pisa, Italy

[5] Department of Fetal and Perinatal Medicine, Istituto "Giannina Gaslini", Genoa, Italy

Abstract. Twin-to-Twin Transfusion Syndrome (TTTS) is a rare pregnancy pathology affecting identical twins, which share both the placenta and a network of blood vessels. Sharing blood vessels implies an unbalanced oxygen and nutrients supply between one twin (the donor) and the other (the recipient). Endoscopic laser ablation, a fetoscopic minimally invasive procedure, is performed to treat TTTS by restoring a physiological blood supply to both twins lowering mortality and morbidity rates. TTTS is a challenging procedure, where the surgeons have to recognize and ablate pathological vessels having a very limited view of the surgical size. To provide TTTS surgeons with context awareness, in this work, we investigate the problem of automatic vessel segmentation in fetoscopic images. We evaluated different deep-learning models currently available in the literature, including U-Net, U-Net++ and Feature Pyramid Networks (FPN). We tested several backbones (i.e. ResNet, DenseNet and DPN), for a total of 9 experiments. With a comprehensive evaluation on a novel dataset of 18 videos (1800 frames) from 18 different TTTS surgeries, we obtained a mean intersection-over-union of 0.63 ± 0.19 using U-Net++ model with DPN backbone. Such results suggest that deep-learning may be a valuable tool for supporting surgeons in vessel identification during TTTS.

Keywords: TTTS · Vessels segmentation · Computer-assisted intervention · Deep learning

P. L. Mazzeo et al. (Eds.): ICIAP 2022 Workshops, LNCS 13374, pp. 145–153, 2022.
https://doi.org/10.1007/978-3-031-13324-4_13

1 Introduction

Twin-to-Twin Transfusion Syndrome (TTTS) occurs in the 10–15% of monochorionic pregnancies (i.e. twin pregnancies with shared placenta). TTTS imply unequal blood flow along placental blood vessels, due to the presence of abnormal anastomoses [3]. If not treated, TTTS can have serious consequences for both twins, with the risk of perinatal mortality of one or both foetuses that can exceed the 90% of cases [5].

TTTS is commonly treated with fetoscopic minimally invasive surgery, where surgeons treat the pathological anastomoses via laser photocoagulation. The identification of pathological anastomoses is a challenging task due to different factors, such as limited Field-of-View (FoV) on the surgical site, turbidity of amniontic fluid, large variability in the illumination level, noise in endoscopic images and occlusions caused by surgical instruments and fetuses. In anterior placental procedures, where the 30° fetoscope is used, the FoV is further reduced due to the view angle between camera and placenta surface. These factors impair surgeon's ability to remain oriented during the procedure, which often results in increased procedural duration and incomplete ablation of anastomoses [4].

Intra-operative automatic vessel segmentation may be a valuable tool to provide surgeons with context awareness for identifying abnormal anastomoses. Moreover, vessel segmentation was proven to be a strong prior for other computer-assisted surgery algorithms, including mosaiking for FoV expansion [2]. However, few attempts have been devoted to address the problem of vessel segmentation. These attempts include [15], which is the first to investigate the use of deep learning for fetoscopy vessel segmentation using U-Net, and [2], which compares U-Net architectures using different encoders (i.e., VGG-16, ResNet-50 and ResNet-101). The work in [15] achieves DSC of 0.55 ± 0.22 on a dataset of 345 frames from 10 TTTS procedures. The work in [2] presents a dataset of 483 intraoperative frames from 6 different in vivo TTTS surgeries. U-Net with ResNet-101 backbone is used, achieving mean Dice Similarity Coefficient (DSC) of 0.78 ± 0.13.

It emerged from the literature survey that no common benchmark exists on vessel segmentation and more studies need to be performed to translate the research methodologies into the clinical practice. Obtaining an accurate segmentation is, indeed, a challenging task. This can be explained considering (i) the intrinsic challenges of fetoscopy videos, which also hampers surgeons' context awareness, and (ii) the lack of large publicly available datasets for algorithm training and testing [7,8]. In this work, we investigate convolutional neural network (CNN) architectures for placental vessel segmentation in fetoscopic images acquired in the actual surgical practice.

The contribution of this work can be summarized as follows:

- Comprehensive study of CNN segmentation architectures, highlighting the impact of specific combinations of architectures and backbones in vessel segmentation performance

- Validation on a new in-vivo dataset acquired during the actual surgical practice. With 1800 frames from 18 different TTTS procedures, the dataset is the larger in the field.

2 Materials and Methods

CNNs for segmentation usually consist of a backbone for feature extraction, also known as encoder (Sect. 2.1), and a decoder (Sect. 2.2) to process the extracted features and compute the segmentation mask. We here evaluated three of the most used backbones, i.e. Residual (ResNet), Densely Connected (DenseNet), and Dual Path (DPN), along with decoder topology from the three most popular architectures for medical image segmentation (i.e., U-Net, Feature Pyramid Network (FPN) and U-Net++).

2.1 Backbones

Since the introduction of U-Net, researchers have implemented skip connections in multiple ways to foster features extraction and travelling through CNN models to tackle several challenges in segmentation tasks.

Increasing the depth of the segmentation network showed to improve segmentation performance but could lead to the vanishing gradient problem and high memory footprint. In ResNet, short skip connections enable features to be reused, travelling from initial to deeper layers, avoiding gradient vanishing with a limited overhead since no additional parameters are required [10]. Although different versions of ResNet have been presented so far, in this work, we use ResNet-50 (23 million parameters) as reference with previous works from state of the art for vessel segmentation [2,13].

The same idea underlying ResNet is further extended in the dense blocks of DenseNet. In each dense block, the features from each layer are iteratively concatenated with those of previous layers. In addition, short skip connections that realise the dense connectivity between layers enable the reuse of feature maps from previous layers. The introduction of dense connectivity further increases efficiency and enables new feature exploration, pushing even further the depth of the backbone [11]. Dense connectivity was found to be also effective in inter-fetal membrane segmentation from fetoscopic images [8]. We used the DenseNet backbone with 169 layers in our tests (12 million parameters).

DPN take the best of both worlds combining residual and dense connectivity. This design allows features in common to be shared among different layers keeping the flexibility to explore new features through the dual path. [9]. We used the DPN backbone with 68 layers in our tests (11 million parameters).

2.2 Decoder Architectures

In U-Net, the encoder is connected to the decoder through long skip connections. Skip connections enable U-Net to compute the segmentation mask using fine-grained features learned at the encoder level [14].

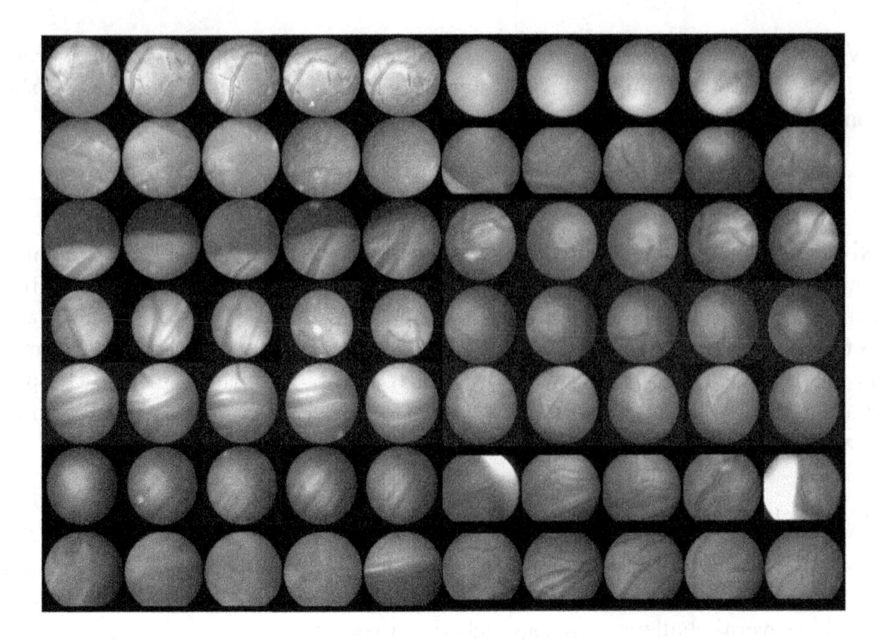

Fig. 1. Samples images from the proposed dataset.

In FPN, the long skip connections between encoder and decoder are maintained with the addition of 1×1 convolutions to improve features semantics. The final output is obtained by combining predictions at different scales. The decoder, or top-down pathway in FPN, upsample spatially coarser but semantically stronger feature maps from higher pyramid levels. With lower-levelRz semantics but higher spatial accuracy, encoder feature maps are merged to those from the decoder through the skip connections [12].

The findings in dense connectivity are shifted to encoder-decoder connectivity in U-Net++. This architecture redesigns skip pathways (i.e. how the feature maps travel through the network) with dense connections to fuse in the decoder features from different encoder stages [17]. In our tests we used U-Net++ without deep supervisions.

3 Experimental Protocol

Our evaluation dataset was collected at Department of Fetal and Perinatal Medicine, Istituto Giannina Gaslini, Genoa (Italy) and University College Hospital, London (United Kingdom). The dataset characteristics are summarized in Table 1. Sample images are shown in Fig. 1. The dataset consists of 1800 frames extracted from 18 videos of 18 different surgeries. A total of 100 frames was extracted from each video to have a balanced number of frames among the different surgeries. Frames with only visible vessels were annotated using the Pixel Annotation Tool [6] under the supervision of expert surgeons. This dataset is a subset of the dataset presented in [1] where only vessels are annotated.

Table 1. Summary of the challenge dataset. For each video, image resolution, number of annotated frames and ratio between vessel and background (BG) pixels.

# Video	Frame resolution (pixels)	# frames	BG/Vessel (Avg. ratio)
1	470 × 470	100	11.13%
2	540 × 540	100	6.32%
3	550 × 550	100	10.93%
4	480 × 480	100	7.22%
5	500 × 500	100	17.03%
6	450 × 450	100	17.24%
7	640 × 640	100	11.11%
8	720 × 720	100	5.84%
9	660 × 660	100	20.00%
10	380 × 380	100	6.86%
11	680 × 680	100	8.90%
12	720 × 720	100	9.29%
13	380 × 380	100	8.83%
14	400 × 400	100	5.16%
15	400 × 400	100	12.08%
16	720 × 720	100	8.47%
17	400 × 400	100	15.83%
18	320 × 320	100	6.63%
All videos		**1800**	**10.49%**

We perform data augmentation on the dataset to increase the generalization capability of each tested network. The data augmentation introduced photometric distortions (i.e. translations, rotation, shear) to simulate additional views, change in contrast, brightness and blur to add several challenging characteristics of fetoscopic images.

All the architectures are implemented in PyTorch and trained on a NVIDIA RTX 2080Ti with 12 GB of memory, using SGD optimiser with a sinusoidal learning rate ranging from 10^{-2} to10^{-4} to optimise the combined-loss function (CL):

$$CL = L_{DSC} + CE + \lambda \sum w_i^2 \tag{1}$$

where L_{DSC} is defined as:

$$DSC = \frac{2TP}{2TP + FP + FN} \tag{2}$$

being TP the number of vessel pixels correctly identified, and FP and FN the background and vessel pixels misclassified, and CE is the cross-entropy loss and the last term implements the L_2 weights regularisation.

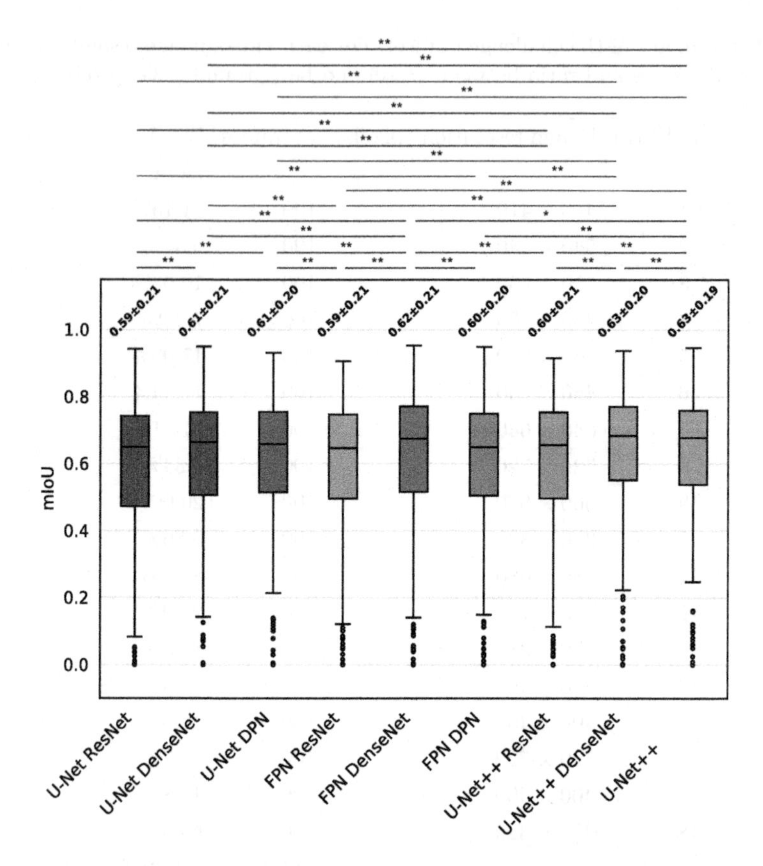

Fig. 2. Boxplot of performance comparison between all the tested architectures in terms of mean Intersection-over-Union (IoU). Statistical significance was tested using Friedman test with post-hoc analysis. (p-value: $* < 0.05$, $** < 0.01$, $*** < 0.001$)

The combined loss function aims to jointly maximize the overlap and likelihood of the predicted segmentation with the groundtruth to improve generalisation while the regularisation prevents model overfitting [16]. All the architectures were trained with a batch size of 64.

In order to assess the robustness of the segmentation, 6-fold cross-validation was performed using 3 patients per fold.

The Friedman test on mean IoU imposing a significance level (p) equal to 0.05, were used to assess whether or not remarkable differences existed between the tested architectures.

4 Results

The best mean IoU among all folds was achieved by U-Net++ with DenseNet and DPN backbone (mean $IoU = 0.63 \pm 0.20$ and mean $IoU = 0.63 \pm 0.19$ respectively).

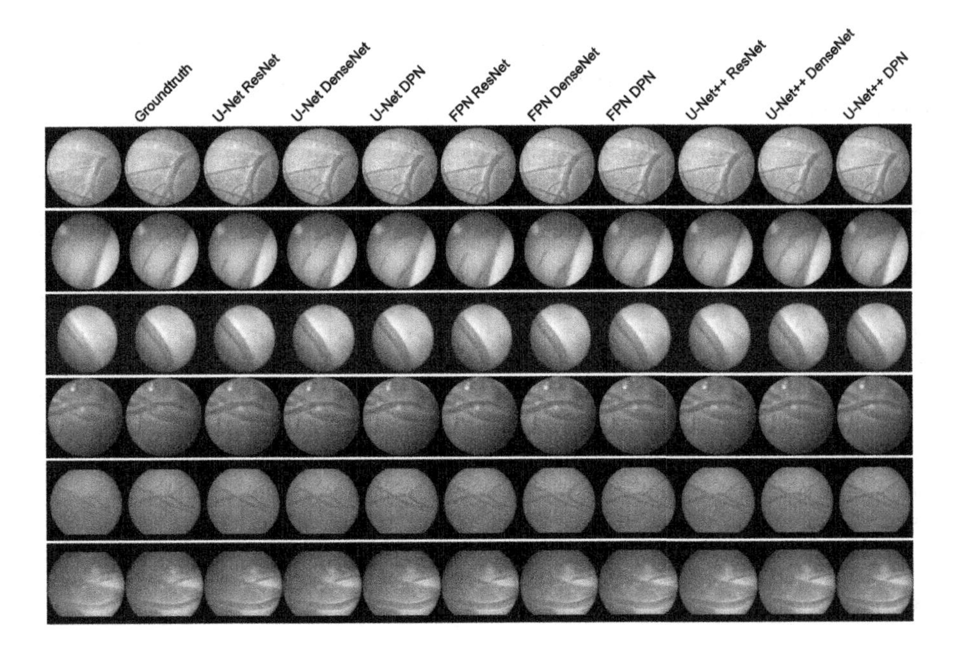

Fig. 3. Sample images along with the manual annotation (groundtruth) and predicted output for the tested architectures. Vessels are superimposed in red. (Color figure online)

The boxplot in Fig. 2, shows the performance in terms of IoU of the tested networks. Mean value and standard deviation are reported on top of each method.

The U-Net ResNet presented in [2] and FPN ResNet achieved the worst results among all folds with a mean IoU of 0.59 ± 0.21. FPN DPN and U-Net++ ResNet showed comparable result with a mean IoU of 0.60 ± 0.20 and 0.60 ± 0.21 respectively. The U-Net DenseNet and FPN DenseNet achieved a slight improvement in terms of average IoU (0.61 ± 0.21 and 0.62 ± 0.21 respectively), while U-Net DPN had a slight reduction of the standard deviation (0.61 ± 0.20).

Visual samples of the predicted segmentations by the tested CNNs are shown in Fig. 3. Each row shows the original image, the groundtruth and segmentation results of a sample frame extracted from the testing videos for each fold.

5 Discussion and Conclusions

In this work, we compared different CNN architectures and backbones for placenta vessels segmentation in in-vivo fetoscopy video. The evaluation was performed on a new dataset of 1800 manually annotated fetoscopic frames from 18 different surgeries.

All the tested architectures achieved comparable performance on our dataset. Despite the low difference in mean IoU between all the tested methods, the use of dense connectivity quantitatively shown to improve the performance of vessel segmentation in fetoscopic images. These results confirm the findings also reported in [8].

The presence of dense connectivity in both backbone and decoder that can be seen in U-Net++, DenseNet and DPN contributes to increasing the segmentation performance and robustness by increasing the mean IoU and reducing its variability. The post-hoc analysis on the Friedman test confirms that these architectures perform better than the others.

A limitation of the experimental protocol may be seen in the dataset size and annotated classes, which could be enriched to conduct a more in-depth analysis on multi-class segmentation.

To conclude, the achieved results suggest that CNNs, and in particular those which implement dense connectivity, may be an effective tool in supporting surgeons in the identification of placenta vessels from fetoscopic videos. This may have a positive impact on TTTS surgery, by lowering the surgery duration and, as a consequence, by reducing surgeons' mental workload and patients' risks.

References

1. Bano, S., et al.: FetReg: placental vessel segmentation and registration in fetoscopy challenge dataset. arXiv preprint arXiv:2106.05923 (2021)
2. Bano, S., et al.: Deep placental vessel segmentation for Fetoscopic Mosaicking. In: Martel, A.L., et al. (eds.) MICCAI 2020. LNCS, vol. 12263, pp. 763–773. Springer, Cham (2020). https://doi.org/10.1007/978-3-030-59716-0_73
3. Baschat, A., et al.: Twin-to-twin transfusion syndrome (TTTS). J. Perinatal Med. **39**(2) (2011)
4. Beck, V., Lewi, P., Gucciardo, L., Devlieger, R.: Preterm prelabor rupture of membranes and fetal survival after minimally invasive fetal surgery: a systematic review of the literature. Fetal Diagn. Therapy **31**(1), 1–9 (2012)
5. Bolch, C., et al.: Twin-to-twin transfusion syndrome neurodevelopmental follow-up study (neurodevelopmental outcomes for children whose twin-to-twin transfusion syndrome was treated with placental laser photocoagulation). BMC Pediatrics **18**(1), 1–11 (2018)
6. Breheret, A.: Pixel Annotation Tool (2017). github.com/abreheret/PixelAnnotationTool
7. Casella, A., Moccia, S., Frontoni, E., Paladini, D., De Momi, E., Mattos, L.S.: Inter-foetus membrane segmentation for TTTS using adversarial networks. Ann. Biomed. Eng. **48**(2), 848–859 (2019). https://doi.org/10.1007/s10439-019-02424-9
8. Casella, A., Moccia, S., Paladini, D., Frontoni, E., De Momi, E., Mattos, L.S.: A shape-constraint adversarial framework with instance-normalized spatio-temporal features for inter-fetal membrane segmentation. Med. Image Anal. **70**, 102008 (2021)
9. Chen, Y., Li, J., Xiao, H., Jin, X., Yan, S., Feng, J.: Dual path networks. In: Advances in Neural Information Processing Systems, vol. 30 (2017)
10. He, K., Zhang, X., Ren, S., Sun, J.: Deep residual learning for image recognition. In: Proceedings of the IEEE Conference on Computer Vision and Pattern Recognition, pp. 770–778 (2016)

11. Jégou, S., Drozdzal, M., Vazquez, D., Romero, A., Bengio, Y.: The one hundred layers Tiramisu: fully convolutional DenseNets for semantic segmentation. In: Proceedings of the IEEE Conference on Computer Vision and Pattern Recognition Workshops, pp. 11–19 (2017)
12. Lin, T.Y., Dollár, P., Girshick, R., He, K., Hariharan, B., Belongie, S.: Feature pyramid networks for object detection. In: Proceedings of the IEEE Conference on Computer Vision and Pattern Recognition, pp. 2117–2125 (2017)
13. Moccia, S., De Momi, E., El Hadji, S., Mattos, L.S.: Blood vessel segmentation algorithms review of methods, datasets and evaluation metrics. Comput. Methods Prog. Biomed. **158**, 71–91 (2018)
14. Ronneberger, O., Fischer, P., Brox, T.: U-Net: convolutional networks for biomedical image segmentation. In: Navab, N., Hornegger, J., Wells, W.M., Frangi, A.F. (eds.) MICCAI 2015. LNCS, vol. 9351, pp. 234–241. Springer, Cham (2015). https://doi.org/10.1007/978-3-319-24574-4_28
15. Sadda, P., Imamoglu, M., Dombrowski, M., Papademetris, X., Bahtiyar, M.O., Onofrey, J.: Deep-learned placental vessel segmentation for intraoperative video enhancement in fetoscopic surgery. Int. J. Comput. Assist. Radiol. Surg. **14**(2), 227–235 (2018). https://doi.org/10.1007/s11548-018-1886-4
16. Van Laarhoven, T.: L2 regularization versus batch and weight normalization. arXiv preprint arXiv:1706.05350 (2017)
17. Zhou, Z., Rahman Siddiquee, M.M., Tajbakhsh, N., Liang, J.: UNet++: a nested U-Net architecture for medical image segmentation. In: Stoyanov, D., et al. (eds.) DLMIA/ML-CDS -2018. LNCS, vol. 11045, pp. 3–11. Springer, Cham (2018). https://doi.org/10.1007/978-3-030-00889-5_1

An Advanced Tool for Semi-automatic Annotation for Early Screening of Neurodevelopmental Disorders

Giuseppe Massimo Bernava[1(✉)], Marco Leo[2], Pierluigi Carcagnì[2], and Cosimo Distante[2]

[1] Institute for Chemical-Physical Processes (IPCF), National Research Council of Italy, Viale Ferdinando Stagno d'Alcontres 37, 98158 Messina, Italy
giuseppe.bernava@cnr.it

[2] Institute of Applied Sciences and Intelligent Systems, National Research Council of Italy, Via Monteroni snc, 73100 Lecce, Italy

Abstract. Non-invasive solutions (no sensors nor markers) appear the most appealing for assessment of body movements and facial dynamics in order to predict Neurodevelopmental disorders (NDD) even in the first days of life. To this aim, recent advances in machine learning applied could be effectively exploited on visual data framing the children, but they suffer from the scarcity of annotated data for training the algorithms. In order to fill this gap, in this paper, a semi-automatic tool specifically designed for labelling videos of children in cribs is introduced. It consists of a Graphical User Interface allowing to select: 1) videos, or static images, to be processed and 2) the desired annotation goal achieved by state-of-the-art deep learning based neural architectures.

1 Introduction

Neurodevelopmental disorders (NDDs) are defined as a group of conditions with onset in the developmental period, inducing deficits that produce impairments of functioning. NDDs are characterized by an inability to reach cognitive, emotional, and motor developmental milestones. NDDs comprise intellectual disability (ID); Communication Disorders; Autism Spectrum Disorder (ASD); Attention-Deficit/Hyperactivity Disorder (ADHD); Neurodevelopmental Motor Disorders, including Tic Disorders (sudden twitches, movements, or sounds that people do repeatedly); and Specific Learning Disorders [20]. In addition to neonatal magnetic resonance imaging, NDDs can be predicted by using movements assessment [10]. Facial recognition is another possible way to diagnose a patient, for example, because of distinct attributes in ASD children. Indeed, scientists found that children diagnosed with autism share common facial feature distinctions from children who are not diagnosed with the disease Children with autism have an unusually broad upper face, including wide-set eyes. They also have a shorter middle region of the face, including the cheeks and nose [12]. Eye movement data can be also distinctive traits of autism [8]. On the other hand, within

P. L. Mazzeo et al. (Eds.): ICIAP 2022 Workshops, LNCS 13374, pp. 154–164, 2022.
https://doi.org/10.1007/978-3-031-13324-4_14

the ADHD field, responses to stimuli, in terms of motor and effective, have been commonly used as predictive signs which lie early in life [25]. It urges to increase efforts in producing systematic reviews on early behavioral markers for each NDD [18].

The causes of NDDs are multiple, both genetic and environmental but, the exact causes driving atypical neurodevelopment remain poorly understood. Even if NDDs are considered highly heritable it is quite difficult to plan efficient screening programs. Besides, ethical, economic, legal and social aspects should be carefully considered. Traditional screening programs are based on human observation and then they have low-throughput and high costs. Hence, screenings are often limited to very high-risk children and the majority of cases emerge later in life when clinical interventions become less effective and the quality of life of individuals is irremediably compromised. Besides, differences in assessments of raters with various levels of experience introduce a bias in the diagnosis and assessment steps [22]. In order to overcome the aforementioned drawbacks, and kin the light of the aforementioned clinical findings and prompted by recent advances in hardware and software technologies, several researchers tried to introduce automatic systems to analyse baby's behavioral features, even in cribs. The use of physical sensors is discouraged by the sparsity of spatial data, difficulties to get consistent positioning and possible modifications of the behaviours to be observed. Alternatively, active/passive visual markers can be positioned on the children and acquired by optical devices. Their use is discouraged by the difficulties to get consistent positioning and then by long set-up times [9] making them not suitable for at-home usages.

Hence, non-invasive solutions (no sensors nor markers) appear the most appealing. Under this premise, recent advances in machine learning applied could be effectively exploited on visual data framing the children [19], but they suffer from the scarcity of annotated data for training the algorithms. Data privacy adherence and compliance is the main issue but it can be properly following ethical features in Governments' guidances. Another relevant issue concerns the labelling procedure which is not just time-consuming but labour-intensive as well. Manually labelling thousands of frames is a very hard task and for this reason, there are a few (and scarcely populated) annotated datasets of children that could feed AI algorithm for early NDDs diagnosis and assessment [14].

In order to fill this gap, in this paper, a semi-automatic tool for labelling videos of children is provided. It consists of a Graphical User Interface allowing to select videos, or static images, to be processed and the desired annotation goal. The provided annotation types allow the user to automatically point out hands and feet, the whole skeleton and the facial dynamics. Each annotation type relies on state-of-the-art deep learning methods. To the best of our knowledge, this is the first tool providing such a large range of annotation choices (existing ones concentrated only on a specific feature, e.g. pose) with very recent and effective algorithms, even working on mobile platforms (thanks to lightweight architectures). The rest of the paper is organized as follows: Sect. 2 reports related work and Sect. 3 describes the main technical features proposed annotation tool. Its main deep learning components are then detailed in Sect. 3.2 and then, in Sect. 4 the dataset of children in cribs used to train and test the algorithms is briefly described. Section 5 reports some outcomes of the annotations performed on videos in the children dataset and finally Sect. 6 concludes the paper.

2 Related Work

2.1 Marker-Less AI Tools for Children's Motion Analysis

It is quite difficult to find tools properly designed and set up for video-based analysis of children's motor performance [23]. In particular, two systems are described in the following. In chronological order, the first one is the AVIM system [21], a monitoring system developed in C# language using the OpenCV image processing library and specifically designed for an objective analysis of infants from 10 days to the 24th week of age. It acquires and records images and signals from a webcam and a microphone but also allows users to perform both audio and video editing. Very useful functionalities are the possibility of adding notes during the recording and to play/cut/copy and assess on-the-fly the sequences of interest. Besides, it extracts from the image the 2D position of the body segments to help the study of the movements according to amplitude, average speed and acceleration. The body analysis can concentrate either on the lower body, based on three points only, or on the full body by taking into account 8 points (right shoulder, left shoulder, left hand/wrist, base of the sternum, pubis/genitals, tight foot/ankle, left foot/ankle). It is worth noting that in both modalities (lower body or full body), all the points are manually placed and then tracked over time in order to extract some motion parameters according to the clinical literature are automatically extracted [13]. Some acoustic parameters (and related statistics) are automatically estimated as well (fundamental frequency, first two resonance frequencies of the vocal tract, kurtosis, skewness and time duration of each cry unit).

The second device deserving a mention is MOVIDEA [2] which has been designed for semi-automatic video-based analysis of infants' motor performance. It includes a camera placed 50 cm above the child, at chest height, and software designed to extract kinematic features of limbs of a newborn (up to 24 weeks old) at home while lying on a bed, upon a green blanket. A Graphical User Interface completes the system and it allows the software operator to interact with the system. At first, the operator has to identify the limb by selecting the central point of the region of interest (i.e., hand, foot). The system then tracks the selected point frame by frame using the Kanade-Lucas-Tomasi algorithm [26] and movement features of extracted trajectories are compared with the reference ones for the identification of pathological motion patterns [5].

2.2 Existing Tools for Face Childrens' Analysis

A semi-automatic annotation methodology for annotating massive face datasets has been firstly proposed in [24]. It relied on generative models such as Active

Orientation Models (AOMs) that are a variant of Active Appearance Models [6]. The accompanying facial landmark annotations consist of a set of 68 points for images in the range −45°:+45° and 39 points for profile images. The tool was used only for annotating adult faces. To the best of our knowledge, there exists no tool specifically designed for annotating children's faces. Furthermore, even works introducing algorithms for analyzing children's faces are few. Examples are the tools in [1,16] aiming at facial analysis for diagnoses and assessment of children with neurocognitive disorders.

3 The Proposed Annotation Tool

The proposed tool has been developed by using Proteo framework [4]. Proteo is a framework designed to give therapists and researchers without specific skills in software development the capability of defining telerehabilitation service customization and developing rehabilitation-oriented serious games, starting from game templates. Moreover, due to its features Proteo is also suitable for the development of other types of applications. Proteo natively works in client-server mode. The client can be installed on a computer, on a tablet or even run on the browser. All scripts are stored on the server instead and they are divided into 3 macro categories: scripts, plugins and libraries. During the login phase the client downloads a script. This script is executed by the client and contains the whole user interface. The client will interface with the plugins on the server through a REST interface. All permission management is handled by the framework.

To the aim of developing the proposed annotation tool, the "movelab.lua" script has been developed. The administration scripts (admin.lua, editor.lua) with their plugins, image analysis plugins (deepcrimson.lua) and data management plugins (deepindigo.lua, nfs.lua and proteo.lua) exploited in this paper were already part of the Proteo framework. Besides, several support libraries have been developed to allow advanced features exploiting deep learning both on the server and client sides.

Proteo allows to call these libraries both locally and remotely. In the following we will analyze the structure of the main scripts according to the features of the software.

3.1 The Graphical User Interface

For the development of the GUI, the native Proteo API was used but a pure Lua library was developed for the creation of the graphs. The interface is divided into three parts. On the left side there is a list of available files for processing. You can list all the files in a folder (through a filter based on the file name) or you can load the list of files on the remote server. That way you can work on a set of videos without the need to download it. Once you click "Open" the entire video is loaded into memory using a structure that allows users to store different information for each frame.

The central part of the interface shows the current frame, the title and a set of controls to move around within the video. It also displays over imposed information, such as skeleton, landmarks and so on, which can integrated with bounding box and motions tracks trough the two check boxes at the bottom of the GUI. User can also draw a line on the video to calculate the distance between two points.

The rightmost of the interface provides the possibility of selecting a type of analysis to be performed among the available options in the list provided by clicking on 'select model' function. Options are:

1. DLNet18
2. DLNet50
3. BlazePose
4. BlazeFace
5. Holistic
6. Selfie Segmentation
7. DeepLabv3
8. Facial Dynamics

After selecting the model, the user can select whether to apply the analysis to a single frame or to the whole video. In the second case each frame is analyzed separately starting from the current one. Below the buttons there are 4 bottoms allowing to graph and save information (quantity of motion, ground truth, extracted features) about the selected video. In Fig. 1 the GUI while running the BlazePose algorithm is shown.

3.2 Deep Learning Components

If the required analysis is *DLNet18* or *DLNet50 type*, two modified residual neural network (ResNet) [11] are used. A residual network is formed by stacking several residual blocks together. A residual block is a stack of layers set in such a way that the output of a layer is taken and added to another layer deeper in the block. The non-linearity is then applied after adding it together with the output of the corresponding layer in the main path. This by-pass connection is known as the shortcut or the skip-connection. This has been demonstrated very effective to overcome the vanishing/exploding gradients and performance degradation problem, i.e. as the network depth increases, the accuracy saturates and then degrades rapidly. In the proposed tool DLNet18 and DLNet50 types refer to a processing by a Residual Network having 18 and 50 layers respectively. Both networks were fine tuned, trained on ImageNet [7], using video from the dataset described in Sect. 4. Both nets were trained to point out hands and foot central point in each image containing a human body.

If *Blaze pose type* is selected for analysis the selected videos are processed by the neural architecture named BlazePose introduced in [3]. BlazePose is a lightweight convolutional neural network architecture for human pose estimation that is tailored for real-time inference on mobile devices. During inference, the

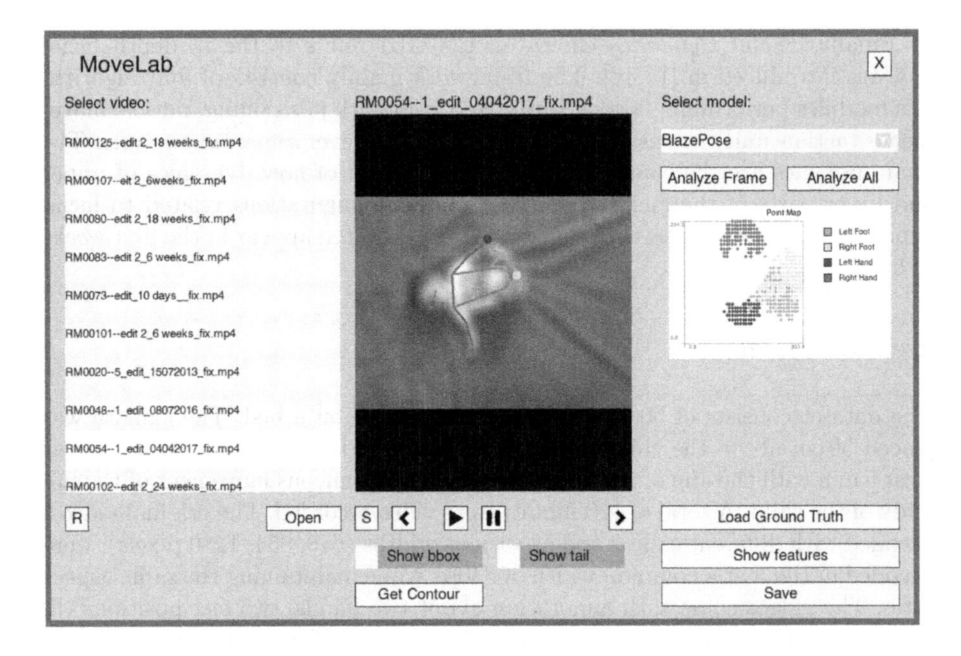

Fig. 1. The GUI while exploiting BlazePose

network produces 33 body keypoints for a single person and runs at over 30 frames per second on off-the shelf mobile phones. It uses both heatmaps and regression to keypoint coordinates.

Holistic type refers to the MediaPipe Holistic pipeline[1] instead. MediaPipe is a framework aimed to build prototypes by combining existing perception components, to advance them to polished cross-platform applications and measure system performance and resource consumption on target platforms [17]. In the deployed configuration, holistic type simultaneously performs human pose, face landmarks, and hand tracking. MediaPipe Holistic utilizes the pose, face and hand landmark models in MediaPipe Pose, MediaPipe Face Mesh and MediaPipe Hands respectively to generate a total of 543 landmarks (33 pose landmarks, 468 face landmarks, and 21 hand landmarks per hand). It can run in real-time on mobile devices and provides higher accuracy by integrating separate models each of which are optimized for their particular domain (i.e. pose, face and hands).

MediaPipe is also used for Segmentation type. It processes selected videos and images using MediaPipe Selfie Segmentation,[2] that segments the prominent humans in the scene. It can run in real-time on both smartphones and laptops. Underlying models are based on MobileNetV3, with modifications to make them more efficient.

---

[1] https://google.github.io/mediapipe/solutions/holistic.

[2] https://google.github.io/mediapipe/solutions/selfie_segmentation.html.

Finally, *Facial Dynamics* choice on the GUI refers to the in depth facial analysis introduced in [15,16]. The framework mainly consists of four algorithmic modules performing: face detection and landmark positioning, muscle movements (action unit) intensity estimation and high-level semantic analysis. The final outcomes are the computational quantification of how dynamics of upper and lower parts of the facial are similar to the configurations related to facial expressions (both volunteers or non-volunteers, as often appear in the first weeks of life).

4 The Dataset

The dataset consists of 600 videos of children lying on a bed. The camera was placed 50 cm above the child, at chest height. The recording took place for at least 5 min with the aim of acquiring images of spontaneous movement of the full body of the child. A total of 150 infants were video recorded. The original videos, encoded with different codecs and at various widths (640, 854, 1280 pixels), were encoded in H264 at a common width of 640px while maintaining the same aspect ratio. The videos come with annotation about two hands, two feet positions (in each frame), body length and face size (only in the first frame). For an additional level of privacy it is possible to set the software to always show the blurred videos (preserving the analysis on the original frames), the screen shots of this paper were made using this setting.

5 Experimental Results

In this paragraph we will present some types of analysis that can be performed with Movelab, without addressing the clinical aspects that will be analyzed in a future work. One of the main aspects of children's motion analysis is the automatic determination of hand and foot positions. As stated in Sect. 3 the annotation tool makes available two different models (DL50 and BlazePose) for human motion analysis based on ResNet50 and the version of BlazePose provided with the MediaPipe tool respectively. The first experiments aimed at qualitatively and quantitatively compare the two available algorithmic choices on childrens videos. Using a small portion of the test videos (not used during ResNet50 training), a Percentage-of-Correct-Keypoints(PCK) for individual limbs was computed. The face size present in the dataset was used as a reference to calculate PCK. Results are reported in Fig. 2. On the x-axis there are the 4 different regions used for computing PCK: LF (left foot), RF (right foot), LH (left hand) and RH (right hand). The small number of tests does not allow to draw general conclusions but, it is quite evident that ResNet50 works more accurately than BlazePose. However, it is worth noting that BlazePose turns out to be a very good motion analysis system also in this context, considering that the analysis time is a tenth of that of ResNet50 and that BlazePose returns many more points and not only the 4 analyzed.

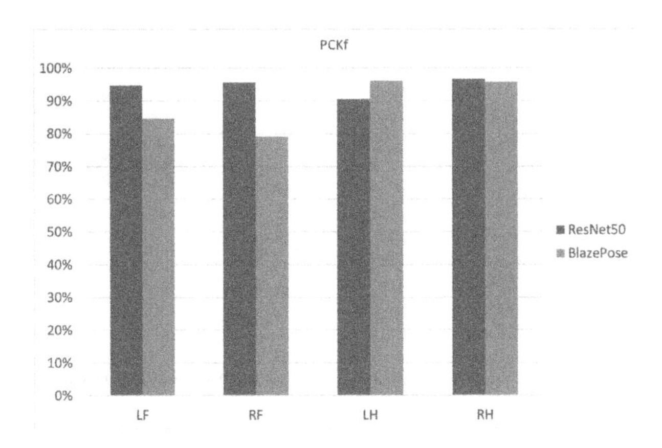

Fig. 2. PCKF comparison between DL50 and BlazePose options in the tool. See the text for explanation.

The temporal trend of the outcomes of both algorithms can be displayed by the GUI (see Fig. 3). This allows clinicians to quickly observe how each limb of the child moved in the video.

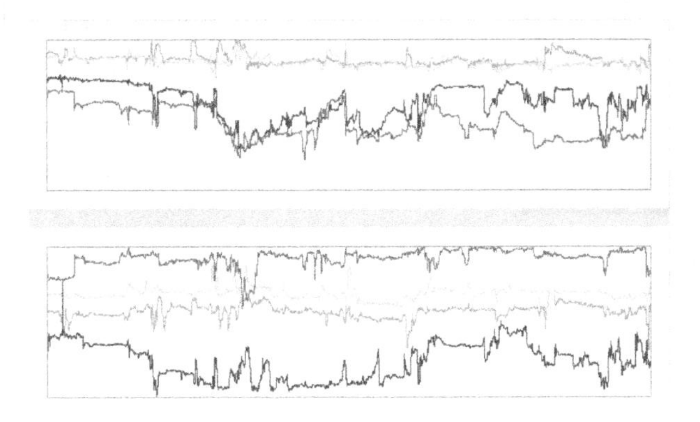

Fig. 3. Motion of the limbs in a video of the dataset (Left Foot green, Right Foot yellow, Left Hand blu, Right Hand magenta). Upper graph refers to x coordinates of hands and feet (4 graphs). Lower graphs refers to y-coordinates instead. (Color figure online)

It is also possible to quickly show the time evolution of the limbs with a graph that also calculates the main kinematic features such as speed and acceleration, cross-correlations and periodicity of movements. In addition the global quantity of motion can be plotted by considering the pixels changes between consecutive frames in the area segmented as belonging to the child by 'Selfie Segmentation' option in the GUI (see Fig. 3) (Fig. 4).

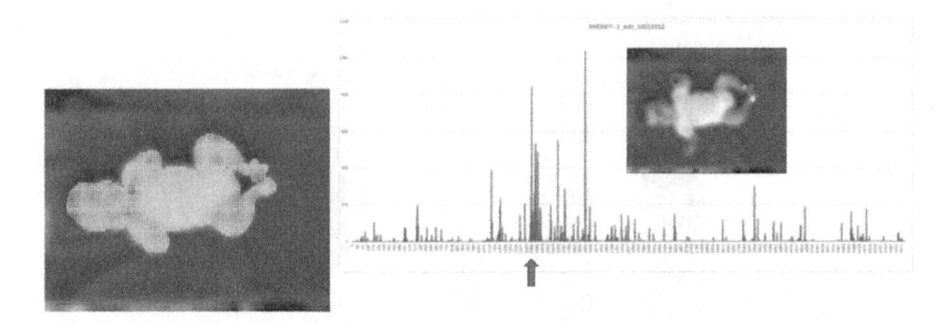

Fig. 4. Selfie-segmentation of the child's body in a frame (left) and plot of the quantity of motion along a video (right).

Concerning provided option for facial dynamics, some outcomes of the tool on a video of the dataset are reported in Fig. 5.

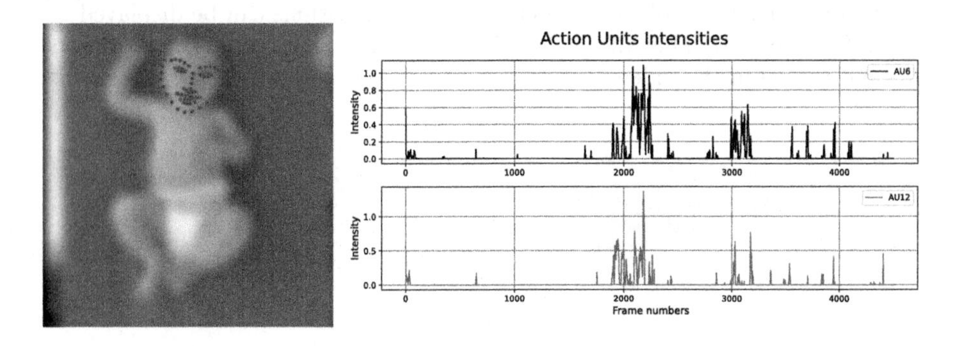

Fig. 5. Face analysis trough the tools and action units plots.

6 Conclusions

Early diagnosis and assessment of neurodevelopmental disorders from videos is a growing research field. Observing several functional features, such as motor and facial dynamics of baby in crib along several hours of videos, might be crucial for quickly taking important clinical decisions. Unfortunately this is a time demanding task leveraging on the availability of trained personnel. Recent deep learning techniques are proving their worth to overcome these drawback, but in many cases being able to use them requires specific technical skills which are often not present in a clinically oriented research team. Therefore, being able to apply, trough a user-friendly GUI, different techniques to the same video, being able to quickly compare results, and also being able to modify the software's functionality quickly can be key to making research in this field faster. To further enhance the research in this field, in this paper the Movelab script, built through the Proteo framework, has been introduced. Proteo has proven to be a powerful tool for building cross-platform applications capable of harnessing the power of libraries such as OpenCV, TensorFlowLite and SDL in a simple and intuitive way. The software allows to compute features chosen according to their reported relevance in the literature and the occurrence of neurodevelopmental disorders. Specifically, three different classes of features were studied for the description of movements in infants: features extracted from the analysis of limb trajectories, features extracted from the analysis of movement of full body and features inherent to the facial dynamics. For the first class, an ad-hoc network was trained, and a pre-trained network (BlazePose) capable of obtaining comparable results was used. As future work the results extracted from the network will be analyzed for a more accurate comparison, taking into account the features of BlazePose: higher number of keypoints and possibility to work in real time also on mobile devices. For the second class it was introduced the possibility to obtain the metrics through the analysis of the data obtained from a segmentation analysis (SelfieSegmentation). For the third class algorithms capable of detecting and positioning facial landmarks, estimating muscle movement intensities (action unit) and extracting semantic information were used. As future work, algorithms capable of partitioning the body into different zones will be explored and further measurements on individual zones will be performed. Besides, the possibility to analyze gaze tracks of the child will be investigated.

Acknowledgment. The support, ideas, and data collected by researchers Maria Luisa Scattoni and Angela Caruso of the Istituto Superiore di Sanitá were essential to carry out this work.

References

1. Akter, T., et al.: Improved transfer-learning-based facial recognition framework to detect autistic children at an early stage. Brain Sci. **11**(6), 734 (2021)
2. Baccinelli, W., et al.: Movidea: a software package for automatic video analysis of movements in infants at risk for neurodevelopmental disorders. Brain Sci. **10**(4), 203 (2020)
3. Bazarevsky, V., Grishchenko, I., Raveendran, K., Zhu, T., Zhang, F., Grundmann, M.: BlazePose: on-device real-time body pose tracking. arXiv preprint arXiv:2006.10204 (2020)
4. Bernava, G., Nucita, A., Iannizzotto, G., Caprì, T., Fabio, R.A.: Proteo: a framework for serious games in telerehabilitation. Appl. Sci. **11**(13), 5935 (2021)
5. Caruso, A., et al.: Early motor development predicts clinical outcomes of siblings at high-risk for autism: insight from an innovative motion-tracking technology. Brain Sci. **10**(6), 379 (2020)
6. Cootes, T.F., Edwards, G.J., Taylor, C.J.: Active appearance models. IEEE Trans. Pattern Anal. Mach. Intell. **23**(6), 681–685 (2001)
7. Deng, J., Dong, W., Socher, R., Li, L.J., Li, K., Fei-Fei, L.: ImageNet: a large-scale hierarchical image database. In: CVPR09 (2009)
8. Duan, H., Min, X., Fang, Y., Fan, L., Yang, X., Zhai, G.: Visual attention analysis and prediction on human faces for children with autism spectrum disorder. ACM Trans. Multimedia Comput. Commun. Appl. (TOMM) **15**(3s), 1–23 (2019)

9. Ghazi, M.A., Ding, L., Fagg, A.H., Kolobe, T.H., Miller, D.P.: Vision-based motion capture system for tracking crawling motions of infants. In: 2017 IEEE International Conference on Mechatronics and Automation (ICMA), pp. 1549–1555. IEEE (2017)
10. Hadders-Algra, M.: Early diagnostics and early intervention in neurodevelopmental disorders-age-dependent challenges and opportunities. J. Clin. Med. **10**(4), 861 (2021)
11. He, K., Zhang, X., Ren, S., Sun, J.: Deep residual learning for image recognition. In: Proceedings of the IEEE Conference on Computer Vision and Pattern Recognition, pp. 770–778 (2016)
12. Hosseini, M.P., Beary, M., Hadsell, A., Messersmith, R., Soltanian-Zadeh, H.: Deep learning for autism diagnosis and facial analysis in children. Front. Comput. Neurosci. **15** (2021)
13. Kanemaru, N., et al.: Specific characteristics of spontaneous movements in preterm infants at term age are associated with developmental delays at age 3 years. Dev. Med. Child Neurol. **55**(8), 713–721 (2013)
14. Leo, M., Bernava, G.M., Carcagnì, P., Distante, C.: Video-based automatic baby motion analysis for early neurological disorder diagnosis: state of the art and future directions. Sensors **22**(3), 866 (2022)
15. Leo, M., et al.: Computational analysis of deep visual data for quantifying facial expression production. Appl. Sci. **9**(21), 4542 (2019)
16. Leo, M., et al.: Computational assessment of facial expression production in ASD children. Sensors **18**(11), 3993 (2018)
17. Lugaresi, C., et al.: MediaPipe: a framework for building perception pipelines. arXiv preprint arXiv:1906.08172 (2019)
18. Micai, M., Fulceri, F., Caruso, A., Guzzetta, A., Gila, L., Scattoni, M.L.: Early behavioral markers for neurodevelopmental disorders in the first 3 years of life: an overview of systematic reviews. Neurosci. Biobehav. Rev. **116**, 183–201 (2020)
19. Moccia, S., Migliorelli, L., Carnielli, V., Frontoni, E.: Preterm infant's pose estimation with spatio-temporal features. IEEE Trans. Biomed. Eng. **67**(8), 2370–2380 (2019)
20. Morris-Rosendahl, D.J., Crocq, M.A.: Neurodevelopmental disorders-the history and future of a diagnostic concept. Dialogues Clin. Neurosci. **22**(1), 65 (2020)
21. Orlandi, S., et al.: AVIM-a contactless system for infant data acquisition and analysis: software architecture and first results. Biomed. Signal Process. Control **20**, 85–99 (2015)
22. Peyton, C., et al.: Inter-observer reliability using the general movement assessment is influenced by rater experience. Early Human Dev. **161**, 105436 (2021)
23. Raghuram, K., et al.: Automated movement recognition to predict motor impairment in high-risk infants: a systematic review of diagnostic test accuracy and meta-analysis. Dev. Med. Child Neurol. **63**(6), 637–648 (2021). https://doi.org/10.1111/dmcn.14800
24. Sagonas, C., Tzimiropoulos, G., Zafeiriou, S., Pantic, M.: A semi-automatic methodology for facial landmark annotation. In: Proceedings of the IEEE Conference on Computer Vision and Pattern Recognition (CVPR) Workshops, June 2013
25. Shephard, E., et al.: Early developmental pathways to childhood symptoms of attention-deficit hyperactivity disorder, anxiety and autism spectrum disorder. J. Child Psychol. Psychiatry **60**(9), 963–974 (2019)
26. Tomasi, C., Detection, T.K.: Tracking of point features. Int. J. Comput. Vis. **9**, 137–154 (1991)

Some Ethical Remarks on Deep Learning-Based Movements Monitoring for Preterm Infants: Green AI or Red AI?

Alessandro Cacciatore[1](✉), Lucia Migliorelli[2], Daniele Berardini[2], Simona Tiribelli[3], Stefano Pigliapoco[1], and Sara Moccia[4]

[1] Department of Humanities - Languages, Language Liaison, History, Arts, Philosophy, Universitá degli Studi di Macerata, Macerata, Italy
a.cacciatore1@unimc.it

[2] Department of Information Engineering, Universitá Politecnica delle Marche, Ancona, Italy

[3] Department of Political Sciences, Communication, and International Relations, Universitá degli Studi di Macerata, Macerata, Italy

[4] Department of Excellence in Robotics and AI, The BioRobotics Institute, Scuola Superiore Sant'Anna, Pisa, Italy

Abstract. Preterm infants' spontaneous movements monitoring is a valuable ally to early recognise neuro-motor impairments, especially common in infants born before term. Currently, highly-specialized clinicians assess the movements quality on the basis of subjective, discontinuous, and time-consuming observations. To support clinicians, automatic monitoring systems have been developed, among which Deep Learning algorithms (mainly Convolutional Neural Networks (CNNs)) are up-to-date the most suitable and less invasive ones. Indeed, research in this field has devised highly reliable models, but has shown a tendency to neglect their computational costs. In fact, these models usually require massive computations, which, in turn, require expensive hardware and are environmentally unsustainable. As a consequence, the costs of these models risk to make their application to the actual clinical practice a privilege. However, the ultimate goal of research, especially in healthcare, should be designing technologies that are fairly accessible to as many people as possible. In light of this, this work analyzes three CNNs for preterm infants' movements monitoring on the basis of their computational requirements. The two best-performing networks achieve very similar accuracy (Dice Similarity Coefficient around 0.88) although one of them, which was designed by us following the principles of Green AI, requires half as many Floating Point Operations (47×10^9 vs 101×10^9). Our research show that it is possible to design highly-performing and cost-efficient Convolutional Neural Networks for clinical applications.

Keywords: Preterm infants · Deep Learning · Green AI

P. L. Mazzeo et al. (Eds.): ICIAP 2022 Workshops, LNCS 13374, pp. 165–175, 2022.
https://doi.org/10.1007/978-3-031-13324-4_15

1 Introduction

The World Health Organization[1] estimates that 11% infants are born prematurely, *i.e.*, before the 37th week of gestation, every year. This average datum does not consider the huge variations of premature birth rates occurring across time and space. In Italy, this rate is "normally" around 7%, but it has increased up to 11% in the post-COVID-19 era (years 2020–2021), as reported by the Italian Society of Neonatology.[2] Other reports [2,4] show how sensitive this phenomenon is to social, economic, and genetic factors. This suggests that the rate of premature births in a population can increase unpredictably. Although, in the last decades and in developed countries, survival rates among preterm infants have increased [27], this does not mean that a premature birth comes without sequelae. In fact, preterm infants happen to be more likely to develop neuromotor diseases and deficits [27].

Early detection of potential neuro-motor impairments caused by a preterm birth can allow for a targeted intervention that exploits brain cells' neural plasticity while it is at its highest levels, i.e., during early childhood [12,15]. In the past decades, new methods and techniques have gained ground in this field, and the General Movements Assessment (GMA) [19,25] stands out for its precision in early identifying possible future impairments. GMA requires a visual analysis of the infant's movements by clinicians to rate movements quality.

Although on the one hand GMA is a powerful technique, on the other hand it relies uniquely on the clinician's experience, discretion, and expertise, which makes it a highly subjective technique, prone to intra- and inter-operator variability. Moreover, GMA is a discontinuous analysis because the clinician clearly cannot constantly observe the infant. Because of these downsides, research has pushed to develop automatic methods to support clinicians with more objective and stable evaluations. This was done, at first, by tracking infant's movements via sensors such as accelerometers [20,26]. Since these sensors must be directly taped to the limbs, they are accurate in tracking displacements, but quite unsuitable for this clinical application, because of preterm infants' small and fragile bodies. In fact, accelerometers might jeopardize infants' spontaneous movements or irritate their skin [20]. For this reason, the focus was shifted to contact-less methods, like RGB-D cameras, which let infants, operators, and parents move freely. In order to be able to analyze movements in the three dimensions, and to preserve subjects' privacy, depth cameras are to be preferred to RGB ones. Another advantage of using depth cameras is that the resulting videos are robust to light and illumination shifts. However, light shifts are not the only source of variability in the context of Neonatal Intensive Care Units (NICUs), which host infants of different ages and who might have to wear oxygen masks, plasters, or other medical devices. Therefore, given the high variability of the environment at issue, the use of images (or videos) requires the design of Deep Learning

[1] https://www.who.int/news-room/fact-sheets/detail/preterm-birth.

[2] https://www.sin-neonatologia.it/covid-neonati-dati-registro-sin-confermano-aumento-nascite-premature/.

(DL) algorithms, typically Convolutional Neural Networks (CNNs), to detect (or track) the infant's limbs movements. These kinds of approaches, which our research focuses on, were used by the authors in [14,17,21,22], and [23]. They all designed CNNs to process images (either depth frames or RGB) with the aim of estimating the pose of infants, as it is a prior for GMA [20] and for infants' limb movements monitoring.

Over the past few decades, DL research has devised increasingly more powerful and reliable architectures. The introduction of these algorithms in the healthcare sector could really pave the way to a more personalized medicine. However, the diffusion of automatic DL-based monitoring applications would be hindered by a main drawback of these technologies. In fact, excellent results can be achieved, only by using computational-intensive DL models, which require specific (*i.e.*, expensive) hardware to run. Especially when dealing with such a sensitive and often needy field like healthcare, one must always keep in mind that different clinical contexts across the world may (and most probably do) have access to different funds. Researchers who intend to develop new biomedical DL frameworks must pursue, from the very design of their projects, the ideal of a wise and clever use of potentially scarce financial resources. It is inherently pointless to devise a biomedical tool, like a CNN-based framework, that borders on perfection but requires specific and expensive hardware, because most of the human population would not have the chance to enjoy its benefits. As biomedical researchers, to pay attention to this means to consider the ethical dimension of *fairness*[3] in the design of the technologies that we develop, in order to make their benefits fairly enjoyable by more people across countries. Besides, there is a deeper issue that DL research has been neglecting: the carbon footprint that comes from the large-scale deployment of cumbersome CNNs, *i.e.*, their *sustainability*. In fact, a DL-based technology is expensive when it requires much computation, and much computation requires much electrical energy, most of which is still produced with fossil fuels like oil and coal [1]. Therefore, these two ethical problems, the fairness and the sustainability of DL-based technologies, go hand in hand.

This study is intended to show that paying special attention to the computational efficiency, and, therefore, to the fairness of DL algorithms, does not prevent the design of highly-performant technologies. This is done by comparing three different CNN architectures used in the field of preterm infants' limbs detection from depth images acquired in the clinical practice.

2 State of the Art: Green AI and Red AI

The impressive performances reached by CNNs have already been mentioned in Sect. 1 and are well known among DL researchers. However, Schwartz et al. [24] point out that these performances are essentially "bought" at the cost of massive computations. This behavior has led DL to be, first, environmentally unfriendly and, second, a niche field for new researchers and end-users because they can

[3] Throughout the paper, the word "fairness" will always refer to *distributive fairness* in the discussed algorithms. To expand different dimensions of fairness that can be promoted via AI technology, see G. Tiribelli (2022) [6].

access it only if they can afford expensive computers and expensive computations. Schwartz defines this as Red AI: striving for higher performance accuracy without considering the economic and environmental costs of the model. On the contrary, Green AI endorses a new DL model, only if it achieves similar or higher performances than the state-of-the-art competitors and its energy consumption is lower. In sum, Green AI supports models that are performing and sustainable.

As a matter of fact, edge computing (EC) has been focusing on developing lightweight models that can be deployed on small devices like smartphones. MobileNets [9] are probably the clearest example of this. Although this research opens up the possibility of making DL-based technologies accessible to a large portion of the population, we consider it improper to include EC in the broader field of Green AI. In fact, EC-research is compelled to devise small and efficient CNNs, because the end-user devices for which they are intended (mainly smartphones) are not meant for massive and long computations, and the deployment of computational-demanding models would be impossible. On the contrary, Green AI chooses to use only a portion of the available resources, either economic and computational, because other geographical contexts might not be able to access them. This choice is driven by the ethical reasons described in Sect. 1: fairness and affordability of technologies, as well as their environmental sustainability.

3 Methods

3.1 Considered Deep Learning Architectures

Three networks (visible in Fig. 1) are going to be discussed in this study: *BabyPoseNet*, the detection network introduced in [17], *EDANet*, introduced in [11], and *TwinEDA*, a CNN that we have specifically designed to be accurate and cost-efficient.

BabyPoseNet. This CNN designed by Moccia et al. [17] was devised to detect the positions of limbs and limbs' connections in depth-video frames of preterm infants. The architecture, perfectly symmetrical, is that of an encoder-decoder framework. Along the encoding path, the data are down-sampled through a series of convolutional blocks outputting an increasing number of feature maps. The dimensional reduction lessens the computational costs, whereas more and more information is extracted as more kernels are used. After that, the decoding path consists of a mirrored series of transposed convolutions that restore the data original dimensions, while reducing the number of kernels up to 20, the total number of output maps (See Sect. 3.2).

By looking at Fig. 1 (right), the most apparent feature of this architecture is its total parallelism: at each encoding or decoding stage, the data are processed by two identical, parallel paths which concatenate right after. This approach is justified by the fact that the network was especially designed to detect two different sets of entities, joints and joint-connections, that, however, belong to the same human body (hence the concatenation after each stage).

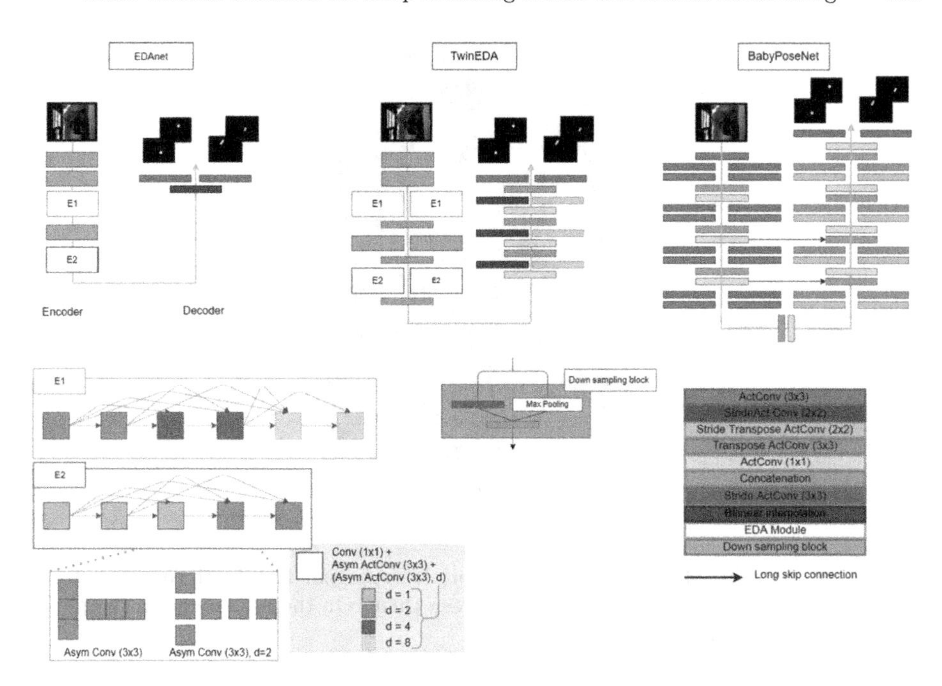

Fig. 1. The architectures of EDANet, TwinEDA, and BabyPoseNet, from left to right. EDANet was put in the middle in order to clearly show the similarities it shares with the two other CNNs. The left branch of each network represents the encoding path, whereas the decoder is shown on the right. On the bottom part of the figure is the legend of all the colors and boxes used to represent the architectures. *ActConv* means that the Convolutional layer is followed by a Batch Normalization and an ReLU Activation layers, whereas *Stride* specifies that the convolutional kernel is strided.

EDANet. The architecture introduced in [11] was meant to perform real-time image segmentation. Real-time image analysis requires very low computational times, achievable only by highly-efficient CNNs. Indeed, the authors state from the very beginning that their aim is to create an efficient and accurate network. Figure 1 (left) displays the architecture of EDANet, which follows an encoder-decoder scheme, too. However, unlike BabyPoseNet, EDANet's scheme is not symmetrical. The encoding path contains efficient structural blocks, like dilated [29] and asymmetric [28] convolutions. These layers approximate a standard convolution, achieving comparable performances and requiring much less computation than the standard operation. Along the decoding path, the original dimensions of the data are restored through a simple interpolation, which does not introduce trainable parameters. This choice is consistent with the aim of devising an efficient CNN, but reduces the learning capacity of the architecture.

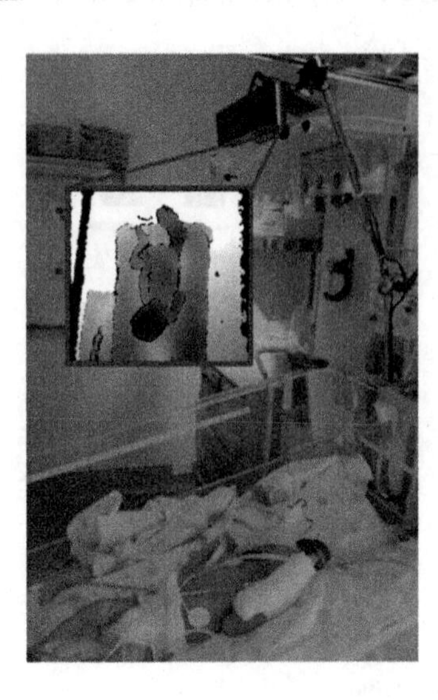

Fig. 2. A crib in the NICU in the Salesi Hospital in Ancona. The depth camera is fixed to the crib to avoid relative movements between them. On the foreground, a sample of depth frame is shown.

TwinEDA. Inspired by BabyPoseNet's high accuracy and by TwinEDA's fast predictions, we designed TwinEDA, a CNN that could perform similarly to BabyPoseNet while being more efficient. In this sense, this research was guided by the main principles of Green AI. As shown in Fig. 1 (middle), TwinEDA's architecture was built by combining the blocks that characterize the two baseline CNNs: the lightweight asymmetric and dilated convolutions, and the computationally expensive bi-branch structure.

3.2 The BabyPose Dataset

Since the GMA relies on the observation of limb movements, the infant's body was modeled as a set of 12 joints (shoulders, elbows, and wrists; hips, knees, and ankles) and the eight joint-connections between them. All the CNNs discussed in this work were trained on the expanded version of the babyPose Dataset [16], which now contains 1000 depth-video frames from 27 preterm infants (for a total of 27000 annotated frames). The videos were acquired inside the Salesi Hospital in Ancona, Italy, and Fig. 2 shows the camera-crib setting. The expected output from the CNNs is a set of 20 binary maps, each of which shows the position of one of the human-body entities (joints or joint-connections) described above.

3.3 Evaluation Metrics

In compliance with the definition of Green AI [24], the CNNs that are going to be analyzed will be evaluated in terms of detection quality and computational efficiency.

Efficacy Metrics. CNNs detection quality are going to be evaluated in terms of Dice Similarity Coefficient (DSC) and Recall (Rec), defined as follows:

$$DSC = \frac{2 \times TP}{2 \times TP + FP + FN} \quad (1)$$

$$Rec = \frac{TP}{TP + FN} \quad (2)$$

where TP and FP are the true joint (or joint-connection) and background pixels detected as joints. FN are the pixels belonging to a joint (or joint-connection), but ascribed as background.

Efficiency Metrics. Less "direct" metrics will be used to assess how efficient a CNN is. As stated in [24] and [13], some of the most intuitive metrics (*elapsed real time* or *number of parameters*) are not suitable for a general assessment of a CNN's efficiency, especially when comparing different networks. For instance, two different CNNs can use the same number of parameters but manage them in different ways, which results in different time and energy consumption. Or again, the elapsed real time depends on the used hardware (CPUs, GPUs, or TPUs) or on the depth of the model (*i.e.*, the number of layers defining it).

Eventually, Schwartz et al. [24] only endorse the use of the *total number of floating point operations*[4] (*FPO*, also referred to as FLOPs), as universal measure of how much computation a specific model requires. Ma et al. [13] suggest to use the model's memory requirements (*Size*), too, as it influences the choice of the required hardware. In light of this, *FPO* and the *Size* are going to be reported for each model. Additionally, the *elapsed real time*, or *Inference time* (t_{inf}), was chosen as efficiency metric too, in order to compare different CNNs on the basis of their speed (*i.e.*, the efficiency) on the same hardware.

[4] The FLOPs were computed with a dedicated Python package, available at https://github.com/tokusumi/keras-flops.

Table 1. The measures of efficacy and efficiency are presented for each convolutional neural network (CNN). The efficacy metrics are median values over the 8 joints and 12 connections to be detected by the CNNs. Floating Point Operations (*FPO*) are expressed in Giga (10^9 FLOPs) and are computed on a batch composed by 16 elements. Model *Size* is in megabytes, whereas Inference speed (t_{inf}) is in milliseconds needed to process a single frame.

Network	Efficacy		Efficiency		
	DSC	*Rec*	*FPO*	Size	t_{inf}
BabyPoseNet	0.89	0.87	101	182	10
TwinEDA	0.88	0.84	47	45	20
EDANet	0.8	0.69	6.7	9	1

4 Results and Discussion

As can be seen from Table 1, the efficacy metrics decrease from top (i.e., BabyPoseNet) to bottom (i.e., EDANet), whereas the trend for the efficiency ones is exactly the opposite. The trend for efficacy metrics is an expected behaviour as more complex network architectures (e.g., BabyPoseNet) can generally capture more complex relationships between data. However the loss in efficiency does not always match the gain in terms of detection quality, and typically the benefit is not worth the cost. This is a documented tendency, too [3].

If compared to EDANet, TwinEDA features a bi-branch structure and a more complex decoder, which undoubtedly requires more computational power. The thus-obtained structure is more similar to that of BabyPoseNet, but, unlike it, the most computational-requiring operations in TwinEDA (such as convolutions) are replaced with less demanding approximations (e.g., asymmetric and dilated convolutions). Because of this, TwinEDA performs similarly to BabyPoseNet (see values of *DSC* and *Rec*) while being way less energy, time, and memory-demanding. Therefore, TwinEDA has the same requirements as BabyPoseNet to be used in the actual clinical practice to monitor preterm infants' movements, and it could be deployed on a larger scale thanks to its lower costs.

TwinEDA matches the requirements of GreenAI [24] and it is an example of how computationally lightweight architectural blocks can make the whole CNN more efficient. In order to promote the principle of fairness, researchers may follow two possible paths to an efficient CNN. The first one (the one we have followed) requires an initial cost-agnostic design, followed by a minimization and optimization process that makes use of lighter architectural blocks. However, this method cannot always ensure good results, since to change the architecture of a CNN is to change its generalization capability, and the quality loss consequent to these substitutions cannot be a priori quantified. The second way to devise an efficient architecture is to pay particular attention to its efficiency from the very design of the model. Efficiency metrics can be evaluated even before the networks is trained, so it is very easy to keep track of their changes. The problem that this method entails is that the researcher has limited choices in the design phase.

A good practice in this respect and the future direction of this work is to use model compression [3] to reduce the size of a CNN.[5] Model compression can be achieved via network pruning [7,10], parameter quantization [18], or knowledge distillation [8]. All these techniques require to train a cumbersome model and to convert it into a lighter and more efficient one with a relatively small loss in terms of efficacy. This means to obtain an effective model that can be deployed even on cost-effective devices.

There are many ways to devise accessible and fair DL technologies and many more will hopefully be found out. We, as researchers, have the duty and the responsibility to try them all.

5 Conclusion

In this work, we have tried to provide measures to enhance fairness and, hence, sustainability of three CNNs used to monitor preterm infants' spontaneous movements. The focus on these themes is justified by the fact that the power of DL algorithms to support clinicians risks to be hindered by their computational and energetic costs, affordable only in limited contexts.

Currently, advancements in the DL domain heavily rely and depend on a huge consumption of resources, and so does the deployment of the designed DL models. This practice, aside from being environmentally and economically unfriendly (as explained in Sect. 1), makes research itself unapproachable to new researchers and, ultimately, usable only in economically-wealthy contexts. For this reason we consider that, even when the economic resources are plentiful, the costs of a newly designed DL algorithm should always be taken into account and minimized when possible. Without this, DL would only serve some privileged sectors and countries, and, ultimately, increase social inequality. In this way, research risks (if this has not happened yet) to stumble in one of the main downsides of what Galimberti [5] defines as "the age of technology", *i.e.* an era in which technological development is no longer carried out for the sake of humans but for its own. We believe that it is possible to obtain high performances from lightweight DL models, and that researchers are now standing at a crossroads. The first path would lead to rethinking research in the DL domain, by embracing the ethical paradigm of fairness in DL algorithms. The second one would mean to rethink our whole society, because it relies every day more heavily on an increasingly expensive and less sustainable DL.

Efficiency can guarantee sustainability, and sustainability can enable fairness. Adopting a resource-friendly and responsible behavior when designing DL frameworks is the only way to make them available worldwide and to let all mankind fully enjoy their benefits.

[5] Model compression can be applied on any Artificial Neural Network, but for the sake of consistency we only refer CNNs.

References

1. Agency, I.E.: Key world energy statistics 2021. IEA, Paris (2021). https://www.iea.org/reports/key-world-energy-statistics-2021
2. Anum, E.A., Springel, E.H., Shriver, M.D., Strauss, J.F.: Genetic contributions to disparities in preterm birth. Pediatr. Res. **65**(1), 1–9 (2009)
3. Buciluă, C., Caruana, R., Niculescu-Mizil, A.: Model compression. In: Proceedings of the 12th ACM SIGKDD International Conference on Knowledge Discovery and Data Mining, pp. 535–541 (2006)
4. Culhane, J.F., Goldenberg, R.L.: Racial disparities in preterm birth. In: Seminars in Perinatology, vol. 35, pp. 234–239. Elsevier (2011)
5. Galimberti, U.: Man in the age of technology. J. Anal. Psychol. **54**(1), 3–17 (2009)
6. Giovanola, B., Tiribelli, S.: Weapons of moral construction? On the value of fairness in algorithmic decision-making. Ethics Inf. Technol. **24**(1), 1–13 (2022)
7. Han, S., et al.: DSD: dense-sparse-dense training for deep neural networks. arXiv preprint arXiv:1607.04381 (2016)
8. Hinton, G., Vinyals, O., Dean, J., et al.: Distilling the knowledge in a neural network, vol. 2, no. 7. arXiv preprint arXiv:1503.02531 (2015)
9. Howard, A.G., et al.: MobileNets: efficient convolutional neural networks for mobile vision applications. arXiv preprint arXiv:1704.04861 (2017)
10. LeCun, Y., Denker, J., Solla, S.: Optimal brain damage. In: Advances in Neural Information Processing Systems, vol. 2 (1989)
11. Lo, S.Y., Hang, H.M., Chan, S.W., Lin, J.J.: Efficient dense modules of asymmetric convolution for real-time semantic segmentation. In: Proceedings of the ACM Multimedia Asia, pp. 1–6 (2019)
12. Luby, J.L., Baram, T.Z., Rogers, C.E., Barch, D.M.: Neurodevelopmental optimization after early-life adversity: cross-species studies to elucidate sensitive periods and brain mechanisms to inform early intervention. Trends Neurosci. **43**(10), 744–751 (2020)
13. Ma, N., Zhang, X., Zheng, H.T., Sun, J.: ShuffleNet V2: practical guidelines for efficient CNN architecture design. In: Proceedings of the European Conference on Computer Vision (ECCV), pp. 116–131 (2018)
14. McCay, K.D., Ho, E.S., Shum, H.P., Fehringer, G., Marcroft, C., Embleton, N.D.: Abnormal infant movements classification with deep learning on pose-based features. IEEE Access **8**, 51582–51592 (2020)
15. Meisels, S.J., Shonkoff, J.P.: Early childhood intervention: a continuing evolution (2000)
16. Migliorelli, L., Moccia, S., Pietrini, R., Carnielli, V.P., Frontoni, E.: The babypose dataset. Data Brief **33**, 106329 (2020)
17. Moccia, S., Migliorelli, L., Pietrini, R., Frontoni, E.: Preterm infants' limb-pose estimation from depth images using convolutional neural networks. In: 2019 IEEE Conference on Computational Intelligence in Bioinformatics and Computational Biology (CIBCB), pp. 1–7. IEEE (2019)
18. Polino, A., Pascanu, R., Alistarh, D.: Model compression via distillation and quantization. arXiv preprint arXiv:1802.05668 (2018)
19. Prechtl, H.F.: State of the art of a new functional assessment of the young nervous system. An early predictor of cerebral palsy. Early Hum. Dev. **50**(1), 1–11 (1997)
20. Raghuram, K., et al.: Automated movement recognition to predict motor impairment in high-risk infants: a systematic review of diagnostic test accuracy and meta-analysis. Dev. Med. Child Neurol. **63**(6), 637–648 (2021)

21. Reich, S., et al.: Novel AI driven approach to classify infant motor functions. Sci. Rep. **11**(1), 1–13 (2021)
22. Sakkos, D., Mccay, K.D., Marcroft, C., Embleton, N.D., Chattopadhyay, S., Ho, E.S.: Identification of abnormal movements in infants: a deep neural network for body part-based prediction of cerebral palsy. IEEE Access **9**, 94281–94292 (2021)
23. Schmidt, W., Regan, M., Fahey, M., Paplinski, A.: General movement assessment by machine learning: why is it so difficult. J. Med. Artif. Intell. **2** (2019). https://jmai.amegroups.com/article/view/5058. ISSN = 2617-2496
24. Schwartz, R., Dodge, J., Smith, N.A., Etzioni, O.: Green AI. Commun. ACM **63**(12), 54–63 (2020)
25. Touwen, B.: Variability and stereotypy in normal and deviant development. Clin. Dev. Med. **67**, 99–110 (1978)
26. Viganò, A.: Design and development of a device for the functional evaluation of newborns nervous system in clinical practice (2015)
27. Wardlaw, T., You, D., Hug, L., Amouzou, A., Newby, H.: UNICEF report: enormous progress in child survival but greater focus on newborns urgently needed. Reprod. Health **11**(1), 1–4 (2014)
28. Yang, H., et al.: Asymmetric 3D convolutional neural networks for action recognition. Pattern Recogn. **85**, 1–12 (2019)
29. Yu, F., Koltun, V.: Multi-scale context aggregation by dilated convolutions. arXiv preprint arXiv:1511.07122 (2015)

Towards a Complete Analysis of People: From Face and Body to Clothes - T-CAP

Effect of Gender, Pose and Camera Distance on Human Body Dimensions Estimation

Yansel González Tejeda(✉) and Helmut A. Mayer

Department of Artificial Intelligence and Human Interfaces, Paris Lodron University of Salzburg, Salzburg, Austria
yansel.gonzalez-tejeda@stud.sbg.ac.at, helmut@cs.sbg.ac.at

Abstract. Human Body Dimensions Estimation (HBDE) is a task that an intelligent agent can perform to attempt to determine human body information from images (2D) or point clouds or meshes (3D). More specifically, if we define the HBDE problem as inferring human body measurements from images, then HBDE is a difficult, inverse, multi-task regression problem that can be tackled with machine learning techniques, particularly convolutional neural networks (CNN). Despite the community's tremendous effort to advance human shape analysis, there is a lack of systematic experiments to assess CNNs estimation of human body dimensions from images. Our contribution lies in assessing a CNN estimation performance in a series of controlled experiments. To that end, we augment our recently published neural anthropometer dataset by rendering images with different camera distance. We evaluate the network inference absolute and relative mean error between the estimated and actual HBDs. We train and evaluate the CNN in four scenarios: (1) training with subjects of a specific gender, (2) in a specific pose, (3) sparse camera distance and (4) dense camera distance. Not only our experiments demonstrate that the network can perform the task successfully, but also reveal a number of relevant facts that contribute to better understand the task of HBDE.

Keywords: Human body dimensions estimation · Human body measurements · Deep learning

1 Introduction

Human Body Dimensions Estimation (HBDE) is a task that an intelligent can perform to attempt to determine human body information from images (2D) or point clouds or meshes (3D). For instance, estimating the height and the shoulder width of a person from a picture or a 3D mesh. Being humans in the center of society, one would expect that intelligent agents should be able to perceive the shape of a person and reason about it from an anthropometric perspective, i.e., be capable of accurately estimating her human body measurements.

P. L. Mazzeo et al. (Eds.): ICIAP 2022 Workshops, LNCS 13374, pp. 179–190, 2022.
https://doi.org/10.1007/978-3-031-13324-4_16

This problem can be characterized by specifying the intelligent agent's perceptual input. If the HBDE problem is circumscribed to inferring human body measurements from images, then HBDE is, theoretically, a difficult, inverse, multi-task regression problem.

Practically, HBDE from images is a compelling problem, as well. HBDE plays an important role in several areas ranging from digital sizing [28], thought ergonomics [6] and computational forensics [29], to virtual try-on [20] and even fashion design and intelligent automatic door systems [17]. Moreover, since accurately estimating a person's body measurements would decrease the probability that the person returns clothes acquired online, HBDE has gained attention as an important step toward a more individual-oriented clothes manufacture.

Inverse problems such as HBDE can be tackled with convolutional neural networks (CNN). However, most studies in the field of HBDE have only focused on investigating to what extent CNNs can predict body measurements. A number of factors can affect this prediction, but researchers have not treated them in depth. What is not yet clear is the impact of the person's gender, pose, and camera distance with respect to the subject, on the estimation performance.

In this paper, we investigate these dependencies with a series of experiments. Despite the tremendous effort from researchers to attempt to better understand HBDE, there is lack of this kind of experiment in the literature. We believe that our contribution will shed light on how a CNN estimate HBDs. Upon publication, we will make our code publicly available for research purposes.[1]

2 The Problem of Human Body Dimensions Estimation

As stated above, CNNs can be employed to approach the HBDE problem. However, supervised learning methods demand large amounts of data. Unfortunately, this kind of data is extremely difficult to collect. For the network input, several persons must be photographed with the same camera under equal lighting conditions. Further, in order to study the effect of pose, the subjects must adopt several poses; and to study the effect of camera distance, they would have to be again photographed. The supervision signal is even more challenging and costly: these same subjects must be accurately measured with identical methods to acquire their body dimensions. This is **the data scarcity problem in HBDE**.

A possible solution is to generate realistic 3D human meshes and calculate HBDs from these meshes. But the HBD calculation is by no means a trivial task. Properly defining HBDs suffers from two issues: inconsistency and uncertainty.

HBDs definitions differ depending on their intended purpose. To just mention one example, health studies measure waist circumference at the midpoint between the inferior margin of the last rib and the iliac crest [7]. However, while investigating the height of the waist for clothing pattern design, [12] found seven different waist definitions and [11] directly enunciated that not all body measurements defined by 3D scanning technologies are valid for clothing pattern. This

[1] Code under https://github.com/neoglez/gpcamdis_hbde.

multiplicity of definitions complicates consistent conceptualization for machine learning.

Furthermore, HBDs are defined based on skeletal joins and/or body landmarks. These reference criteria are highly uncertain and depend on the person performing the measurement. A single HBD may exhibit important variability due to observer or instrument error. Also, researchers and practitioners base their analysis on HBD by presenting a figure of a thin subject with the measurements depicted by segments without further elucidation. This approach hinders the HBDs calculation reproducibility.

Formally, the HBDE problem has been defined by [13] as a deep regression problem. Given an image $\mathcal{I}$ from a 3D human body with HBD D, the goal is to return a set $\hat{D}$ of estimated human body dimensions, that is

$$\hat{D} = \mathcal{M}(\mathcal{I}(D)). \quad (1)$$

The dataset is assumed to be drawn from a generating distribution and the deep neural network $\mathcal{M}$ minimizes the prediction error.

3 Related Work

Obviously, human body dimensions are determined by human shape. In the field of Human Shape estimation (HSE), shape has been ambiguously presented either as a parametric model acting as proxy to a 3D mesh or directly as a triangular mesh. In a community effort to be more precise, the task of shape estimation has been currently sharper defined as human mesh recovery, estimation or reconstruction. Additionally, pose estimation has been established as inferring the location of skeleton joints, albeit these not being anatomically correct. In the last five years, the body of work in these two fields has exploded. Since human mesh and pose estimation are barely indirectly related to our work, we will not discuss them here. In contrast, we focus on end-to-end adults HBDE from images, i.e., the model input are images of adult subjects and the output are human body measurements.

Undoubtedly, anthropometry has contributed most to human shape analysis. Important surveys such as CAESAR (1999) [24], ANSUR I and II (2017) [1] and NHANES (1999–2021) [2], have collected HBDs. However, they did not take images of the subjects. This makes unclear how the CNN input could be obtained. Recently, other datasets have been released for specific tasks, e.g., [23] propose a dataset with images and seven HBDs for estimation in the automotive context.

Of all these compendiums, CAESAR is probably the most convenient data in terms of realism. It contains rigorously recorded human body dimensions and 3D scans, from which realistic images could be synthesized. The project costed six million USD (see [24] executive summary). Consequently, this data is highly expensive. Alternatively, we employ a generative model derived from real humans, capable of producing thousands of 3D meshes from which we can calculate and visualize the HBDs.

Certainly, height is the HBD that has been investigated the most [5,8,14–16, 19,21,27,29]. Very early work [16] investigated the effect of gender and inverted pictures when humans estimate height from images. They quantified estimation performance using Pearson's Correlation Coefficient and established that the estimated and ground truth height where highly correlated. This fact has been confirmed recently by [19], which also concluded that humans estimate height inaccurately. Other HBDs have been explored, e.g., waist [12] but, in general, they have received significantly less attention.

Strongly related to our work are studies using or generating synthetic data and calculating or manually collecting HBDs [3,9,10,26,30–32]. None of these works investigated the effect of gender, pose or camera distance in the estimation performance. Here, we explore these interactions.

Recently, [4] proposed a baseline for HBDE given height and weight. They claimed that linear regression estimates accurately HBDs when the inputs are height and weight. Like we, this method use ground truth derived from the SMPL model [18]. Despite their input being different to ours, we will use this work for comparison.

A *neural anthropometer* (NeuralAnthro) was introduced by [13]. The CNN was trained on grayscale synthetic images of moderate complexity, i.e., no background, limited human poses, and fix camera perspective and distance. In this work, we go further and increase the image complexity, making the input more challenging to the intelligent agent conducting HBDE.

4 Material and Methods

We now detail the dataset and CNN (model) of the supervised learning approach that governs our experiments.

4.1 Dataset

We start with the NeuralAnthro synthetic dataset. The reason to use a synthetic dataset is the cost and effort that collecting "real" data would imply. Since coherent pose variability is more difficult to find in real datasets, another important aspect is the possibility to vary the subject posture to experiment with different poses. While we did not collect our data from physical humans, we use the SMPL model, which is derived from real humans. SMPL is the most employed model in academia and industry for its realism and simplicity [25].

Input. Figure 1 depicts our dataset. We obtained the 3D meshes, 6000 female and male subjects in pose zero and pose one (total 12000 meshes), from the neural anthropometer dataset [13].

Using the current standard method to employ a render engine to produce the mesh corresponding images, we simulate the cinematographic technique of *tracking back* (sparsely and densely varying the camera distance to the mesh), as follows.

Fig. 1. Our curated dataset. We augment the NeuralAnthro dataset, containing images of female (left) and male (right) subjects in pose zero (arms stretched to the sides) and pose one (arms lowered) taken with a camera at a fix distance, by rendering photos with sparse and dense camera distances (center). Note that the subjects appear nearer or farther in the images. All instances are 200×200 pixels grayscale images displaying a single subject.

Sparse camera distance: back tracking by placing the camera at distances 4 m, 5 m and 6 m.
Dense camera distance: back tracking by randomly placing the camera between distances 4,2 m and 7,2 m.

In total, we synthesize 72000 pictures from the 12000 meshes. The images correspond to meshes of a specific gender and a definite pose, taken at specific camera distance with respect to the subject.

Supervision Signal. While the data scarcity problem is the major challenge in HBDE, another problem is measurement inconsistency. There is no consensus regarding the correct manner to define a specific measurement, let alone several of them. The problem arises even when HBDs are automatically computed by 3D scanning technologies [20], making manually corrections unavoidable. The united method introduced by [13] with Sharmeam (**Sh**oulder width, right and left **arm**s length and ins**eam**) and Calvis (**C**hest, w**a**ist and pe**lvis** circumference plus height) allows us to resolve the inconsistency issue because it provides a proper method to calculate eight HBDs. Additionally, it agrees, to a large extent, with anthropometry and tailoring.

4.2 Neural Anthropometer

The NeuralAnthro is a small, easily deployable CNN that we use to conduct our experiments. We use the same experimental setting as in the original paper [13], i.e., we train for 20 epochs and use mini-batches of size 100. We report results based on 5-fold cross-validation. We minimize the mean squared error between the actual and the estimated HBDs using stochastic gradient descent with a momentum 0.9; the learning rate is set to 0.01.

5 Results and Discussion

For the presentation of the results we use the following abbreviations: shoulder width (SW), right arm length (RAL), left arm length (LAL), inseam (a.k.a. crotch height) (I), chest circumference (CC), waist circumference (WC), pelvis circumference (PC) and height (H). Average MAD (AMAD) and Average RPE (ARPE) are both represented by a capital A. The figures we present are interesting in several ways. Due to space restrictions we can not discuss exhaustively all their aspects. Therefore, we examine the most salient results.

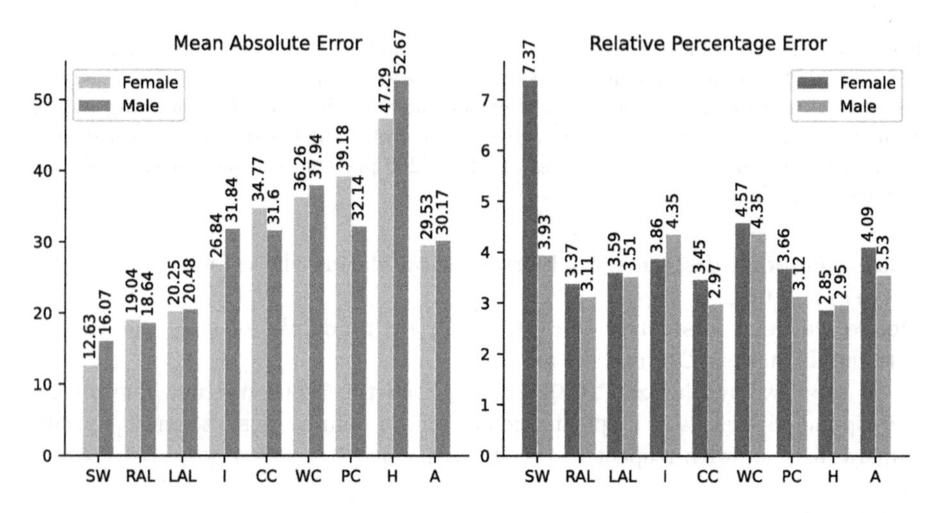

Fig. 2. Effect of gender on HBDE. Left: we display Mean Absolute Error (MAE) in *mm*; right: Relative Percentage Error (RPE).

5.1 Effect of Gender

We start our discussion by evaluating the network performance when the input are images from humans of a specific gender in pose zero or one. We define training with two gender as *unisex training* and *gender training* when the input are subjects of a specific gender. Figure 2 shows the results.

Like [8], we observe that height estimation is more accurate in unisex training, compared to gender training (RPE 1.58 unisex training reported in [13] vs. gender training 2.85 female and 2.95 male).

For the network, it is considerably more difficult to estimate female gender training SW than male gender training SW. Although female gender training SW MAE is lower than male gender training (12.63 mm vs. 16.07 mm), the inverse relation can be observed, when considering RPE (7.37 vs. 3.93).

Curiously, regarding the effect of gender, the CNN and humans appear to estimate height differently. Unlike [16]'s results, Fig. 2 shows that the female height estimation error (RPE 2.85) is lower than male (RPE 2.95). Perhaps it is not surprisingly, that this relation holds for inseam as well (RPE 3.86 vs. 4.35). With the exceptions of these two HBDs, the RPE of estimating other HBDs is larger for female as for male subjects.

5.2 Effect of Pose

Figure 3 presents the breakdown of the estimation error when we train the network individually with images of humans in pose zero and pose one (*multi-pose training*). Surprisingly, the network estimated shoulder width more poorly when the subject was in pose one as in pose zero (RPE 6.4 vs. 6.0). One would expect that estimating SW would be easier when the subject is in pose one, because the arms are lowered, and, therefore, the shoulder joints could be easier recognized.

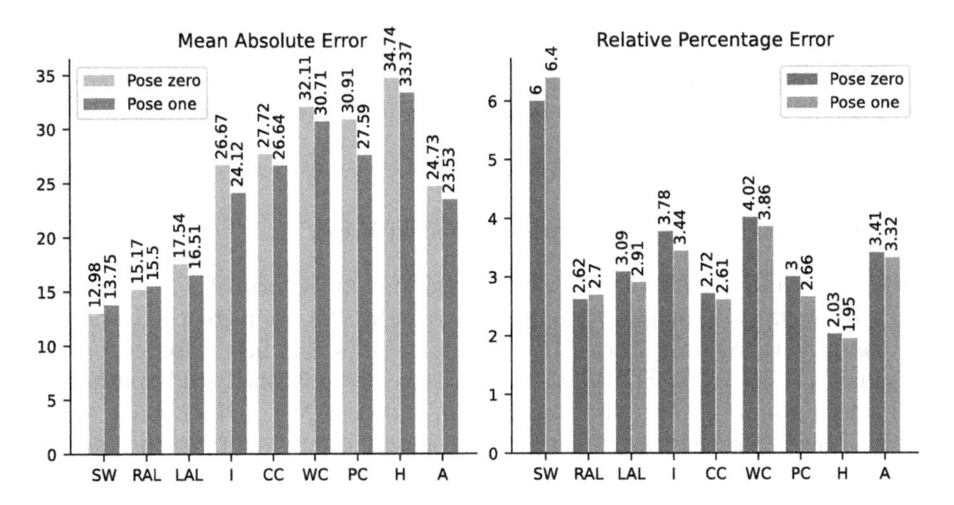

Fig. 3. Effect of pose on HBDE. Left: we display Mean Absolute Error (MAE) in *mm*; right: Relative Percentage Error (RPE).

5.3 Effect of Camera Distance

The most interesting finding was that the network is able to accurately estimate all HBDs independently of the camera distance to the person (ARPE 3.04, 3.03, 2.96, 3.57 and 3.11), when training with sparse camera distance 4 m, 5 m and 6 m and randomly chosen camera distance respectively. This fact challenges intuition, e.g., contradicts current research claiming that the network can only correctly estimate height if the evaluation is performed for a particular camera distance [4]. But this finding is in accordance to when humans estimating height as reported in preliminary work [16] (Fig. 4).

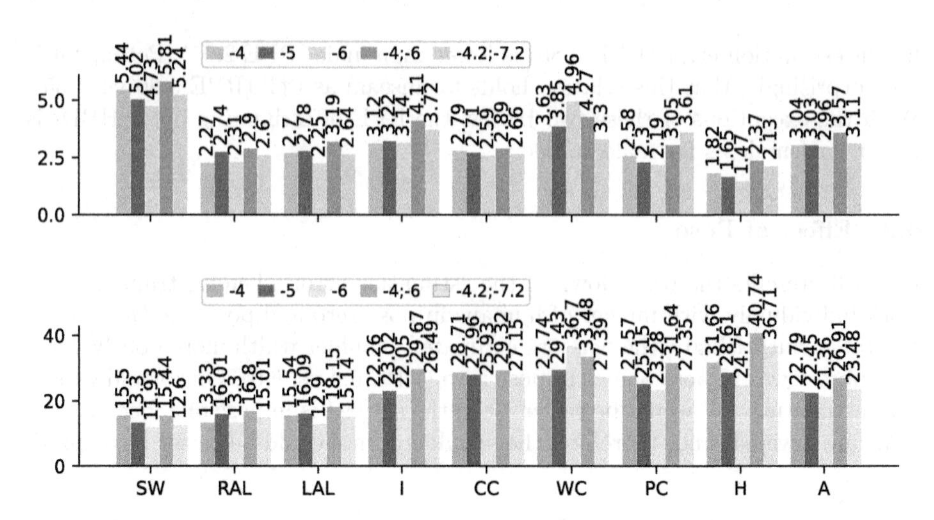

Fig. 4. Effect of camera distance on HBDE. We placed the camera at distances -4 m, -5 m and -6 m, and randomly distances sampled from $-4{,}2$ m to $-7{,}2$ m with respect to the subject. Top: Relative Percentage Error (RPE); bottom: Mean Absolute Error (MAE) in mm.

5.4 Quantitative Comparison to Related Work

Although we did not aim to present a method that outperform SOTA estimation methods, we discuss comparative quantitative results for completeness. Basically, we compare to NeuralAnthro's [13] original results, the best baseline results (Baseline I = 2) on ANSUR data in [4], a recently published study on height estimation from real images by humans [19], and the ANSUR II allowable error.

We have been eminently cautious in comparing our results in the task of human body dimension estimation. Several reasons hinder a fair comparison and constitute a major obstacle to advance the field.

First, in the literature, Mean Absolute Error (MAE) and Mean Absolute Difference (MAD) refer to the same error quantity. Also, Relative Percentage Error (RPE) has not been consistently reported. RPE is important because human body dimensions are not in the same scale. For instance, probably $40mm$ MAE, say, account for a lower height estimation error (better performance) as for head circumference error (worst performance). Besides MAD and RPE, estimation performance has been reported by Mean ± Std. Dev and *success rate* [31], seemingly *Expert ratio*. This inconsistency in reporting results complicates significantly the comparison with other research. Second, most method's input are 3D, therefore, inadequate for comparison to 2D methods.

Third, we require that methods' result has been reported persistently in the literature. Neither we compare to results reported online that are not longer available, nor to results that has been used for comparison but we were not able to locate in the original cited paper. Last, in the literature, different datasets have been used for comparing. This might render previous and this compari-

Table 1. Comparison to four related methods. We compare estimation performance in terms of MAD error in mm. We do not present HBDs that are not comparable (na: not applicable). Minimal errors are bold and we emphasized ANSUR II allowable error. Additionally, we enclosed in parenthesis our experiment setting that achieved best estimation results.

Method	SW	RAL	LAL	I	CC	WC	PC	H
Baseline (I = 2) [4]	na	na	na	na	29.1	37.9	**21.6**	na
NeuralAnthro [13]	12.54	**12.98**	13.48	**12.17**	**25.22**	27.53	25.85	27.34
Our experiments	**11.93** (6 m)	13.30 (6 m)	**12.9** (6 m)	22.05 (6 m)	25.93 (6 m)	**27.39** (−4.2; −7.2) m	23.28 (6 m)	**24.75** (6 m)
Humans observing real images [19]	na	na	na	na	na	na	na	64.0
Humans, ANSUR (Allowable error ANSUR II) [22]	na	na	na	***10.0***	***14.0***	***12.0***	***12.0***	***6.0***

son counterproductive. For example, see ANSUR [22] App. G for an extensive account on comparability limitations.

Table 1 shows the comparison. The input to Baseline is not images (like ours) but height and weight. However, that research does establish a conceptual baseline: HBDE methods should estimate body measurements with higher accuracy compared to regression. This statement should not be categorically interpreted. Methods requiring images as input without any other information are more challenging and, therefore, might exhibit less accuracy. As it can be seen, NeuralAntro estimates more precisely RAL, I and CC as the regression baseline, when applicable, and all of our experiment settings. This might happen because NeuralAnthro was trained and evaluated with fixed camera distance. The network probably found more difficult learning when trained with three different camera distances. Nevertheless, being SW the most difficult HBD to estimate, our experiment with one camera distance at $6m$ manifests the best estimation performance. Moreover, our experiment setting with randomly selected camera distances shows the best WC estimation performance.

As the authors indicate in [19], height estimation by humans exhibits poor performance. The cause is, probably, that the persons estimated the HBD from real images, which is the input with highest complexity, compare to synthetic controlled data.

Estimation error of all HBD lies over the ANSUR II allowable error, but the fact that the NeuralAnthro is a small CNN could indicate, that by incrementing the size of the network, the estimation performance could be improved as well.

6 Conclusions and Future Work

In this paper, we assessed the performance of a neural network employed to estimate human body measurements from images. To that end, we augmented our recently published dataset containing images of female and male subjects in two poses, with images of these subjects synthesized using different camera distances with respect to the subjects. We trained a CNN with two genders, two different poses and sparse and dense camera distances. After training we evaluated the network performance in terms of MAE and RPE.

The CNN estimated HBDs of male subjects more accurately than those of females. The shoulder width predictions exhibit a surprising pose dependency. The width is estimated more correctly for subjects with arms spread out to the side (compared to subjects with lowered arms, where the contours of the shoulders are more pronounced). In contrast to our expectations, network performance decreases only slightly when perceiving humans from a range of (camera) distances instead of a fixed distance; given that the person is completely visible in the image. In general, shoulder width is the most difficult HBD to estimate.

6.1 Future Work

An important question that needs to be answered is why the estimation is, in general, highly accurate (errors are reported in mm). Exploring to what extent synthetic data is representative of the real HBDs would contribute to understand this phenomenon. Increasing the level of realism of the images would probably have the strongest effect in HBDE. Also, investigating the minimum amount of data for conducting HBDs with reasonable accuracy, would help determining bounds to collect a plausible real dataset, therefore, alleviating the data scarcity problem.

References

1. The anthropometric survey of US army personnel. https://www.openlab.psu.edu/ansur2/
2. NHANES questionnaires, datasets, and related documentation. https://wwwn.cdc.gov/nchs/nhanes/continuousnhanes/default.aspx
3. Ashmawi, S., Alharbi, M., Almaghrabi, A., Alhothali, A.: Fitme: body measurement estimations using machine learning method. Procedia Comput. Sci. **163**, 209–217 (2019). https://doi.org/10.1016/j.procs.2019.12.102, https://www.sciencedirect.com/science/article/pii/S1877050919321416. 16th Learning and Technology Conference 2019Artificial Intelligence and Machine Learning: Embedding the Intelligence
4. Bartol, K., Bojanić, D., Petković, T., Peharec, S., Pribanić, T.: Linear regression vs. deep learning: a simple yet effective baseline for human body measurement. Sensors **22**(5) (2022). https://doi.org/10.3390/s22051885
5. BenAbdelkader, C., Yacoob, Y.: Statistical body height estimation from a single image. In: 8th IEEE International Conference on Automatic Face and Gesture Recognition (FG 2008), Amsterdam, The Netherlands, 17–19 September 2008, pp. 1–7 (2008). https://doi.org/10.1109/AFGR.2008.4813453

6. Brito, M.F., Ramos, A.L., Carneiro, P., Gonçalves, M.A.: Ergonomic analysis in lean manufacturing and Industry 4.0—a systematic review. In: Alves, A.C., Kahlen, F.-J., Flumerfelt, S., Siriban-Manalang, A.B. (eds.) Lean Engineering for Global Development, pp. 95–127. Springer, Cham (2019). https://doi.org/10.1007/978-3-030-13515-7_4
7. Canhada, S.L., et al.: Ultra-processed foods, incident overweight and obesity, and longitudinal changes in weight and waist circumference: the Brazilian Longitudinal Study of Adult Health (ELSA-Brasil). Public Health Nutr. **23**(6), 1076–1086 (2020). https://doi.org/10.1017/S1368980019002854
8. Dey, R., Nangia, M., Ross, K.W., Liu, Y.: Estimating heights from photo collections: a data-driven approach, pp. 227–238. Association for Computing Machinery, New York (2014). https://doi.org/10.1145/2660460.2660466
9. Dibra, E., Jain, H., Öztireli, C., Ziegler, R., Gross, M.: HS-Nets: estimating human body shape from silhouettes with convolutional neural networks. In: 2016 Fourth International Conference on 3D Vision (3DV), pp. 108–117. IEEE (2016). https://doi.org/10.1109/3DV.2016.19
10. Dibra, E., Öztireli, C., Ziegler, R., Gross, M.: Shape from selfies: human body shape estimation using CCA regression forests. In: Leibe, B., Matas, J., Sebe, N., Welling, M. (eds.) ECCV 2016. LNCS, vol. 9908, pp. 88–104. Springer, Cham (2016). https://doi.org/10.1007/978-3-319-46493-0_6
11. Gill, S., Ahmed, M., Parker, C., Hayes, S.: Not all body scanning measurements are valid: perspectives from pattern practice. In: 8th International Conference and Exhibition on 3D Body Scanning and Processing Technologies, Conference date: 10-10-2017 Through 11–10-2017, October 2017. https://doi.org/10.15221/17.043
12. Gill, S., Parker, C.J., Hayes, S., Brownbridge, K., Wren, P., Panchenko, A.: The true height of the waist: explorations of automated body scanner waist definitions of the TC2 scanner. In: Proceedings of the 5th International Conference on 3D Body Scanning Technologies, Lugano, Switzerland, 21–22 October 2014. Hometrica Consulting - Dr. Nicola D'Apuzzo, October 2014. https://doi.org/10.15221/14.055
13. Gonzalez Tejeda, Y., Mayer, H.A.: A neural anthropometer learning from body dimensions computed on human 3D meshes. In: 2021 IEEE Symposium Series on Computational Intelligence (SSCI), pp. 1–8 (2021).https://doi.org/10.1109/SSCI50451.2021.9660069
14. Guan, Y.P., et al.: Unsupervised human height estimation from a single image. J. Biomed. Sci. Eng. **2**(06), 425 (2009)
15. Günel, S., Rhodin, H., Fua, P.: What face and body shapes can tell us about height. In: ICCV Workshops, pp. 1819–1827 (2019). https://doi.org/10.1109/ICCVW.2019.00226
16. Kato, K., Higashiyama, A.: Estimation of height for persons in pictures. Percept. Psychophys. **60**(8), 1318–1328 (1998)
17. Kiru, M.U., Belaton, B., Mohamad, S.M.S., Usman, G.M., Kazaure, A.A.: Intelligent automatic door system based on supervised learning. In: 2020 IEEE Conference on Open Systems (ICOS), pp. 43–47. IEEE (2020). https://doi.org/10.1109/ICOS50156.2020.9293673
18. Loper, M., Mahmood, N., Romero, J., Pons-Moll, G., Black, M.J.: SMPL: a skinned multi-person linear model. ACM Trans. Graph. **34**, 248:1–248:16 (2015)
19. Martynov, K., Garimella, K., West, R.: Human biases in body measurement estimation, October 2020. https://epjdatascience.springeropen.com/articles/10.1140/epjds/s13688-020-00250-x

20. Michael, E., Antje, C., Silke, S.: Online shopping featuring "my customized avatar" - generating customized avatars for a sustainable shopping experience in e-commerce. In: 12th International Conference and Exhibition on 3D Body Scanning and Processing Technologies (2021). 10.15221/21
21. Momeni-k, M., Diamantas, S.C., Ruggiero, F., Siciliano, B.: Height estimation from a single camera view. In: VISAPP, no. 1, pp. 358–364 (2012)
22. Paquette, S.: Anthropometric Survey (ANSUR) II Pilot Study: Methods and Summary Statistics. Anthrotch, US Army Natick Soldier Research, Development and Engineering Center (2009)
23. Pini, S., D'Eusanio, A., Borghi, G., Vezzani, R., Cucchiara, R.: Baracca: a multimodal dataset for anthropometric measurements in automotive. In: 2020 IEEE International Joint Conference on Biometrics (IJCB), pp. 1–7. IEEE (2020)
24. Robinette, K.M., Daanen, H., Paquet, E.: The CAESAR project: a 3-D surface anthropometry survey. In: Second International Conference on 3-D Digital Imaging and Modeling (Cat. No. PR00062), pp. 380–386. IEEE (1999)
25. Shi, F., et al.: Review of artificial intelligence techniques in imaging data acquisition, segmentation, and diagnosis for COVID-19. IEEE Rev. Biomed. Eng. **14**, 4–15 (2020)
26. Škorvánková, D., Riečický, A., Madaras, M.: Automatic estimation of anthropometric human body measurements. arXiv preprint arXiv:2112.11992 (2021)
27. Sriharsha, K.V., Alphonse, P.J.A.: Anthropometric based Real Height Estimation using multi layer peceptron ANN architecture in surveillance areas. In: 2019 10th International Conference on Computing, Communication and Networking Technologies (ICCCNT), pp. 1–6 (2019)
28. Tanasa, D., Cojocea, E.: ESENCA - a fit predictor built for the 21st century. In: Proceedings of of 3DBODY.TECH 2021, pp. 83–90. IEEE (2021). 10.15221/21
29. Thakkar, N., Farid, H.: On the feasibility of 3D model-based forensic height and weight estimation. In: Proceedings of the IEEE/CVF Conference on Computer Vision and Pattern Recognition, pp. 953–961 (2021). https://doi.org/10.1109/CVPRW53098.2021.00106
30. Yan, S., Kämäräinen, J.K.: Learning anthropometry from rendered humans. arXiv preprint arXiv:2101.02515 (2021)
31. Yan, S., Wirta, J., Kämäräinen, J.K.: Anthropometric clothing measurements from 3D body scans. Mach. Vis. Appl. **31**, 1–11 (2020)
32. Yan, S., Wirta, J., Kämäräinen, J.K.: Silhouette body measurement benchmarks. In: 2020 25th International Conference on Pattern Recognition (ICPR), pp. 7804–7809 (2021). https://doi.org/10.1109/ICPR48806.2021.9412708

StyleTrendGAN: A Deep Learning Generative Framework for Fashion Bag Generation

Laura Della Sciucca[1,3], Emanuele Balloni[1,2(✉)], Marco Mameli[1], Emanuele Frontoni[1,2], Primo Zingaretti[1], and Marina Paolanti[1,2]

[1] VRAI Vision Robotics and Artificial Intelligence Lab, Dipartimento di Ingegneria dell'Informazione (DII), Università Politecnica delle Marche, Via Brecce Bianche 12, 60131 Ancona, Italy
m.mameli@pm.univpm.it, p.zingaretti@univpm.it
[2] Department of Political Sciences, Communication and International Relations, University of Macerata, 62100 Macerata, Italy
{emanuele.balloni,emanuele.frontoni,marina.paolanti}@unimc.it
[3] Department of Humanities, University of Macerata, 62100 Macerata, Italy
laura.dellasciucca@unimc.it

Abstract. Dealing with fashion multimedia big data with Artificial Intelligence (AI) algorithms has become an appealing challenge for computer scientists, since it can serve as inspiration for fashion designers and can also allow to predict the next trendy items in the fashion industry. Moreover, with the global spread of COVID-19 pandemic, social media contents have achieved an increasingly crucial factor in driving retail purchase decisions, thus it has become mandatory for fashion brand analysing social media pictures. In this light, this paper aims at presenting StyleTrendGAN, a novel custom deep learning framework that has the ability to generate fashion items. StyleTrendGAN combines a Dense Extreme Inception Network (DexiNed) for sketches extraction and Pix2Pix for the transformation of the input sketches into the new handbag models. StyleTrendGAN increases the efficiency and accuracy of the creation of new fashion models compared to previous ones and to the classic human approach; it aims to stimulate the creativity of designers and the visualization of the results of a production process without actually putting it into practice. The approach was applied and tested on a newly collected dataset, "MADAME" (iMage fAshion Dataset sociAl MEdia) of images collected from Instagram. The experiments yield high accuracy, demonstrating the effectiveness and suitability of the proposed approach.

1 Introduction

Fashion industry has grown into one of the biggest segments of the economy in the world, evaluated at 3 trillion dollars as of 2018, characterizing two percent of global GDP.[1] Fashion brands face a regularly changing consumer inclination

[1] https://fashionunited.com/global-fashion-industry-statistics/.

P. L. Mazzeo et al. (Eds.): ICIAP 2022 Workshops, LNCS 13374, pp. 191–202, 2022.
https://doi.org/10.1007/978-3-031-13324-4_17

in fashion as more fashionistas and teenagers explore new outfits to refill their wardrobe and improve their style. This prompts for the need for fashion designers to create up-to-date, trendy designs and patterns to satisfy the arbitrary preferences of the consumers. Social media use has been constantly growing over the recent years and has become a decisive element of the online shopping and clothing trend tracking experience. In fact, nowadays, approximately 74% of customer use social media to make purchase decisions [4]. This aspect is also due to the fact that influencers achieved higher network engagement than celebrities, thanks to their 'subject' expertise for related fashion product categories. Designers sometimes take inspirations from different sources to conceive their unique works, such as from their own personal life experiences, variation of mix-and-match styles, and creative works by other designers. Notwithstanding, to maintain with the continuous change consumer preference in fashion and style, the creation of more fashionable outfits and designs would be essential.

With the current advancement of Artificial Intelligence (AI) and with its deep learning (DL) subset algorithms, domain involving creativity and art can be strengthened by these advanced models. In fact, the traditional 3D rendering pipeline can produce beautiful and realistic imagery, but only in the hands of trained artists. The idea of shortening the traditional 3D modeling and rendering pipeline started from image-based rendering techniques. These techniques focus on re-using image content from a pool of training pictures. For a limited range of image synthesis and editing scenarios, these non-parametric techniques allow non-experts to author photorealistic imagery. In recent years, the idea of direct image synthesis (without using the traditional rendering pipeline) has gotten significant interest because of promising results from DL architectures such as Variational Autoencoders (VAEs) and Generative Adversarial Networks (GANs).

Regarding this context, in this work StyleTrendGAN is presented, a novel custom deep learning framework able to generate new fashion items. The goal of StyletrendGAN is not to replace designers but to help them by providing more references and inspiration, and by creating adequate models that can ease their work and provide customers with delightful and sophisticated designs. StyleTrendGAN consists of two phases: firstly, a Dense Extreme Inception Network (DexiNed) [17] has been chosen to process the images in order to obtain their edges. The resulting images were then separated into ground-truth images and their relative sketches. Then, they are divided into train and test sets and fed to the Pix2Pix network [9], to obtain new bag models with the application of textures.

StyleTrendGAN has been evaluated on a newly labelled dataset of images containing handbags collected from Instagram. "MADAME" Dataset comprises of 55.087 pictures which show handbags. The dataset has been manually labelled by human annotators in 10 classes of bags, thus providing a more precise dataset.

All in all, we can thus summarize the main contributions of this paper as follows:

- development of a generative framework that improves user's shopping experiences by developing an AI framework that can assist fashion designers in their work;
- Our proposed model by creating new items is a way for effective future trend prediction;
- a new challenging dataset of images containing 10 different bag classes collected by Instagram, hand-labelled with ground truth and publicly available to the research community.

This paper is structured as follows: Sect. 2 gives an overview of the background knowledge of related studies; in Sect. 3 there is a detailed description of the methods and the entire workflow together with the required materials; Sect. 4 presents an explanation of the results obtained accompanied by conclusions in Sect. 5.

2 State-of-the-Art

Many studies have been conducted in recent years in the fashion field as well as in its trend over time [1], given the massive production of fashion items and its increasing global relevance. An overview on datasets and DL methods available in the fashion world is provided in [14].

Different approaches have been developed, mainly with the implementation of different types of Generative Adversarial Networks [5]. GarmentGAN [18] uses two GANs to transfer garments with the purpose of try-on. It uses a two-stage framework consisting of a shape transfer network that learns to generate a semantic map (given the image of the person and desired fashion item), and an appearance transfer network that synthesizes a realistic image of the person wearing the garment while preserving finer semantic details. In 2017, Jetchev et al. introduced Conditional Analogy Generative Adversarial Network (CAGAN) [10], which was one of the first works to allow to learn the relation between paired images present in training data, and then generalize and generate images that correspond to the relation, but were never seen in the training set. To discern the changes in color, texture and shape from one another, Yildirim et al. [23] proposed an improvement on traditional GAN approaches, by customizing conditional GANs with consistency loss functions. Namboodiri et al. [15] improved CAGAN by reworking the U-Net generator architecture, along with the usage of a combination of ReLu and Leaky ReLu activation functions. Another challenging task which regards the use of GANs in the fashion fields is about overlaying an item on a model with an arbitrary pose. The approach followed by Pandey, N. et al. [16] introduced Poly-GAN, a new conditional GAN which directly generates garments on the human pose instead of adapting it to the body shape. Xintong Han et al. [6] worked on the same application by following a different procedure. In particular, they analyzed the limits of GANs in relation to their ability of applying realistic deformations to the garment. For this reason, they focused their attention on the development of VITON (virtual try-on network). Related works for the virtual try-on approach include MG-VTON [3], which uses a Warp-GAN to warp clothes to human structure and CP-VTON, which implements a Geometric Matching Module and a Try-On Module for clothes fitting. Other articles presented a novel application of GANs to generate new clothing starting from requirements specified by the customers [20] or by a simple description [25]. Similarly, the work of Texture GAN developed in 2017 by W. Xian et al. [21] followed the same objective. Users "drag" one or more example textures onto sketched objects and the network realistically applies these textures to the indicated objects. Obviously, these approaches require large texture datasets and even user intervention. FashionGAN [24] uses cGAN to allow users to input a fashion sketch and a fabric image to receive an image of the fabric applied on the garment from the sketch.

A different GANs application is the one followed by Liu et al. [12] whose aim was to study the combination of clothing attributes for outfit matches to generate outfit composition. Hsiao et al. [8] instead, proposed an outfit improvement challenge by making minimal edits without changing the properties of the fashion model with the aim of making the existing outfit more fashionable. Recently, a work by Jiang et al. [11] proposed a new method to create new clothing styles from an image and a single or multiple style images; they proposed two fashion style generator frameworks: FashionG for single-style generation and a spatially constrained FashionG (SC-FashionG) framework for mix-and-match style generation.

In 2015, S. Xie and Z. Tu [22] developed a new edge detection algorithm, Holistically-Nested Edge Detection (HED), which performs image-to-image prediction by means of a deep learning model that leverages fully convolutional neural networks and deeply-supervised nets. HED automatically learns rich hierarchical representations (guided by deep supervision on side responses), important to resolve the challenging ambiguity in edge and object boundary detection. They implemented this framework using the publicly available Caffe Library, building it on top of the publicly available implementations of FCN and DSN. Later, in 2018 S. Niklaus gave a personal re-implementation of Holistically-Nested Edge Detection using PyTorch.[2] This approach has proven to give back thick and poor quality edge detection, even with blur. Furthermore, some edges were left out, so it could be considered not so useful for the aim of this work. Dense Extreme Inception Network (DexiNed) was implemented in 2020 by X. Soria et al. [17]; this is the network of choice for this work. It is an improvement over HED, as it can generate thin edge-maps, avoiding missed edges at the same time.

Considering the state of art in this context, StyleTrendGAN comprises a Dense Extreme Inception Network (DexiNed) [17] and the Pix2Pix network [9], to obtain new bag models with the application of textures. With respect to the above mentioned state-of-the-art works, our approach has been applied on MADAME, a newly labelled dataset of images containing bag coming from social media, specifically on Instagram.

[2] https://github.com/sniklaus/pytorch-hed.

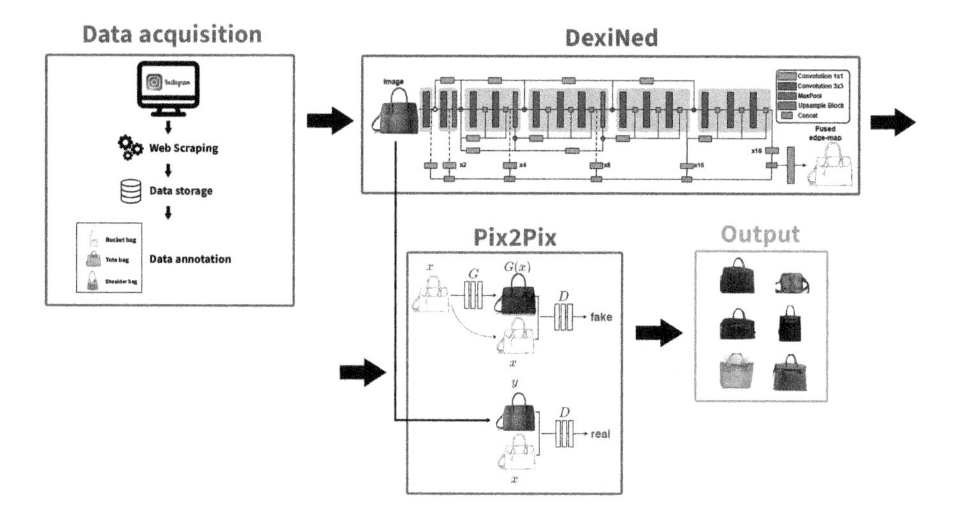

Fig. 1. StyleTrendGAN workflow.

3 Materials and Methods

This section introduces the overall framework of StyleTrendGAN as well as the dataset used for evaluation. The framework is depicted in Fig. 1 and comprises two main components: DexiNed [17] for sketches extraction and Pix2Pix [9] for the transformation of the input sketches into the new handbag models. Further details are given in the following subsections. StyleTrendGAN is comprehensively evaluated on MADAME, a publicly available dataset collected for this work. The details of the data collection and ground truth labeling are discussed in Subsect. 3.2.

3.1 Deep Generative Models

As already stated, DexiNed [17] is chosen as the edge detection network. It introduces a new Convolutional Neural Network (CNN) architecture for edge detection, capable of generating thin edge-maps, without missing edges. Unlike the other state-of-the-art CNN based edge detectors, this model has a single training stage, but it is still able to outperform previous models in the usage of edge detection datasets. Moreover, DexiNed does not need pretrained weights, and it is trained from scratch with fewer parameters tuning. The result of edge extraction is given as input to the Generative Adversarial Network in order to obtain the desired output. In particular, the Pix2Pix network [9] is adopted, as it follows a totally random and automatic approach, often obtaining surprising results. Figure 2 shows an example of sketches extraction has been implemented with the DexiNed.

The generations of new handbag models has been implemented with a Pix2Pix network [9]. It is basically a Conditional GAN to map edges to photo.

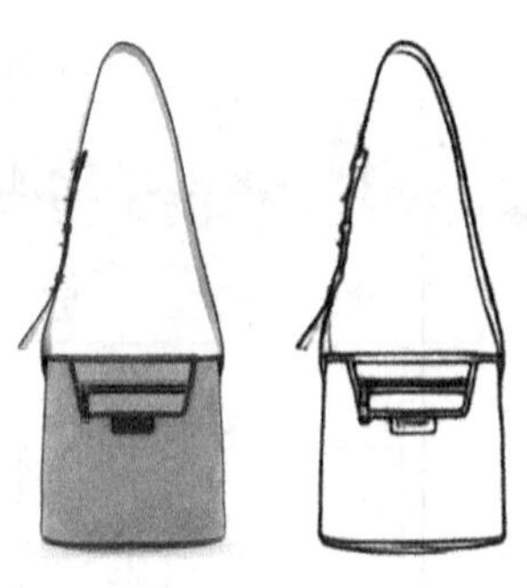

Fig. 2. An example of sketches extraction with DexiNed.

Unlike an unconditional GAN, both the generator and discriminator observe the input edge map. This method differs from prior works in several architectural choices for the generator and discriminator: for the generator a "U-Net"-based architecture has been used and for the discriminator a convolutional "PatchGAN" classifier has been used, which only penalizes structure at the scale of image patches.

U-Net is an encoder-decoder architecture with skip connections between mirrored layers in the encoder and decoder stacks. This approach has been followed because, for many image translation problems, there is a great amount of low-level information shared between input and output and it would be desirable to shuttle this information directly across the net. For example, in the case of image colorization, the input and output share the location of prominent edges. PatchGAN tries to classify if each $N \times N$ patch in an image is real or fake. We run this discriminator convolutionally across the image, averaging all responses to provide the ultimate output of D. It has been demonstrated that N can be much smaller than the full size of the image and still produce high quality results. This is advantageous because a smaller PatchGAN has fewer parameters, runs faster, and can be applied to arbitrarily large images.

3.2 MADAME Dataset

The "MADAME" dataset (iMage fAshion Dataset sociAl MEdia) consists of 55.087 images collected from Instagram. To collect the images, it is implemented a social media crawler algorithm that is able to search images starting from chosen keywords or hashtag. Since our aim was to generate new bags, our social crawler retrieves raw information from Instagram. The classes defined are: backpack, Belt bags, Bucket bags, Clutch bags, Mini bags, Cross-body bags, Shoulder bags, Tote bags, pochette and bowling bag.

Figure 3 shows examples of MADAME dataset and Table 1 reports the number of pictures for each class.

Fig. 3. Example of two handbags (clutch and bowling bag) in the dataset.

Table 1. Number of images for each class of MADAME dataset.

Class	# of images
Backpack	4778
Belt bags	3673
Bucket bags	3689
Clutch bags	6890
Mini bags	5239
Cross-body bags	5798
Shoulder bags	7096
Tote bags	6553
Pochette	3267
Bowling bag	8104

The dataset was then fed to the DexiNed network, to obtain the corresponding sketches. At this point the sketches were separated from the ground-truth images. Both groups were then divided into 80% for the training set and 20% for the test set, preparing them for the usage with Pix2Pix GAN, maintaining the sketches-to-original images correspondence.

3.3 Performance Metrics

The objective of a Conditional GAN such as Pix2Pix can be expressed as:

$$L = E_{x,y}[logD(x, y)] + E_{x,z}[log(1 - D(x, G(x, z)))] \tag{1}$$

The generator tries to minimize this objective against an adversarial discriminator that tries to maximize it.

4 Results and Discussion

In this section, the results of the experiments conducted on MADAME Dataset are reported. Two types of Learning Rate Scheduler have been tested: Linear and Cosine. The former makes the LR decay linearly as the epochs pass while the latter is a type of scheduler widely used in computer vision that generally gives better results [13]. Its function is so defined:

$$n_t = n_T + \frac{n_0 - n_T}{2}(1 + cos(\frac{\pi}{T})), \tag{2}$$

where $t \in [0.T]$, n_T is the target rate at time T and n_0 is the initial learning rate.

Three different configurations for the hyper-parameters were set, as reported in Table 2.

Table 2. Experiments settings

Hyper-parameters	First configuration	Second configuration	Third configuration
N° Epochs	350	500	500
Batch size	8	8	8
LR	0.0002	0.0002	0.0002
LR scheduler	Cosine	Cosine	Linear

Graphics of the loss trend during training with the first Fig. 4, second Fig. 5 and third Fig. 6 configuration of hyper-parameters are reported. It can be observed that the loss trend of the generator is decreasing in all the three experiments, as it should be. In the first graphic it could be seen that at the 350th epoch the generator loss (G_L1) value is over the value of 10. Instead, in the other two graphics, it could be seen that, as the number of epochs increases, the trend tends to decrease until it reaches values below 10, particularly with the linear LR scheduler.

On the other hand, the discriminator loss is always close to zero. This is due to the fact that the generator has not yet learned to fool him correctly.

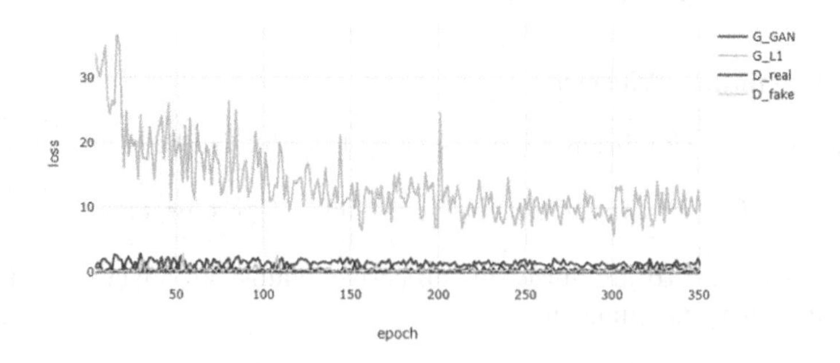

Fig. 4. Loss over time of the first training

To assess the goodness of the proposed method, three metrics have been used: Inception Score (IS) [19], Fréchet Inception Distance (FID) [7] and Kernel Inception Distance (KID) [2]. Results are shown in Table 3.

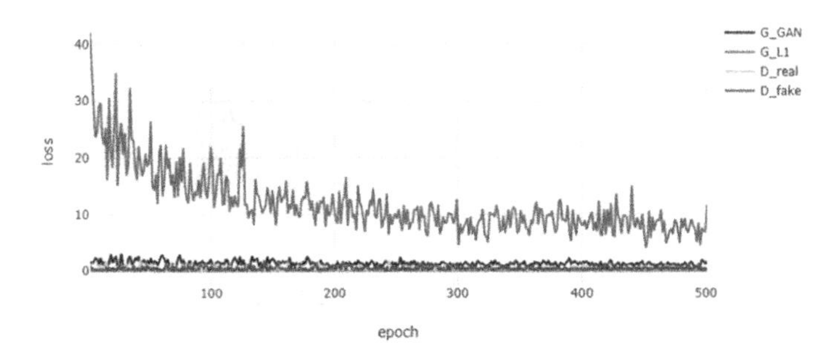

Fig. 5. Loss over time of the second training

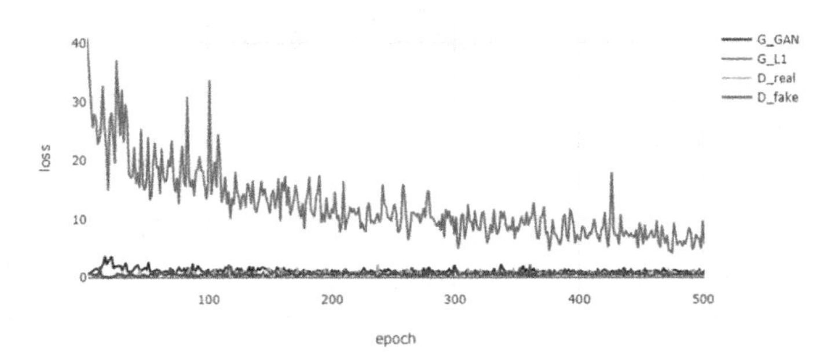

Fig. 6. Loss over time of the third training

Table 3. Experiments results

Metrics	First configuration	Second configuration	Third configuration
IS	0.178	0.877	1.209
FID	186.938	17.8	1.68
KID	2.87	1.773	0.0938

The results reveal how the third configuration (500 epochs and Linear LR scheduler) has proven to be the most effective in terms of all the metrics used. The only underwhelming result is the Inception Score, which is pretty low for all the configurations; it has to be noted, though, that is not a reliable metric in this case because, as pointed out in [2], it uses ImageNet dataset for score evaluation, so it has problems when the dataset domain is not present in the ImageNet classes, as it is in this case.

Even though the results are good, some tests showed noisy and blurred outputs. This is due to the limited number of images of MADAME dataset. By increasing the cardinality of the dataset and the number of epochs and looking for a better combination between the starting learning rate and his scheduler, even better results could be obtained with this method.

Fig. 7. Qualitative results of StyleTrendGAN

Some qualitative results of StyleTrendGAN are shown in Fig. 7. Starting from the edge-oriented images obtained as output from the DexiNed, Pix2Pix takes them as input and generates new bags as final output, with surprising results.

5 Conclusions and Future Works

This paper aimed at improving the state-of-the-art models in the fashion field, by proposing StyleTrendGAN a deep generative framework that achieves good results in new handbag models generation. As a matter of fact, DexiNed proved to be the first DL-based approach able to generate thin edge-maps from handbags images. Experimental results and comparisons with previous approaches shows the validity of DexiNed, outperforming them when evaluated in other edge oriented datasets. The results obtained with Pix2Pix suggest that Conditional adversarial networks are a promising approach for edges-to-photo translation tasks. In addition, these networks proved to be able to adapt to the task and data at hand, which makes them applicable in a wide variety of settings. During the development of the work, this approach turned out to match well sketches with the original handbags from which they were extracted.

Obviously, this paper could be considered as a starting point for the development of multiple future works. A possible improvement challenge could be trying to obtain good results even when the match between sketches and original images is not guaranteed or when the dataset images increase in complexity. In fact, it could be interesting to improve this work by considering more complex images with a different background or even with the presence of people or other objects depicted in them.

Other future works could see the implementation of such a model into a Decision Support System; furthermore, it could be interesting to extend the scope of the model outside the handbags area.

References

1. Al-Halah, Z., Stiefelhagen, R., Grauman, K.: Fashion forward: forecasting visual style in fashion. arXiv:1705.06394 (2017)
2. Bińkowski, M., Sutherland, D.J., Arbel, M., Gretton, A.: Demystifying MMD GANs. arXiv:1801.01401 (2018)
3. Dong, H., Liang, X., Wang, B., Lai, H., Zhu, J., Yin, J.: Towards multi-pose guided virtual try-on network. arXiv:1902.11026 (2019)
4. Gekombe, C., Tumsifu, E., Jani, D.: Social media use among small and medium enterprises: a case of fashion industry growth. Univ. Dar es Salaam Libr. J. **14**(2), 3–18 (2019)
5. Goodfellow, I., et al.: Generative adversarial nets. In: Ghahramani, Z., Welling, M., Cortes, C., Lawrence, N., Weinberger, K.Q. (eds.) Advances in Neural Information Processing Systems. vol. 27. Curran Associates, Inc. (2014). https://proceedings.neurips.cc/paper/2014/file/5ca3e9b122f61f8f06494c97b1afccf3-Paper.pdf
6. Han, X., Wu, Z., Wu, Z., Yu, R., Davis, L.S.: VITON: an image-based virtual try-on network. CoRR arXiv:1711.08447 (2017)
7. Heusel, M., Ramsauer, H., Unterthiner, T., Nessler, B., Klambauer, G., Hochreiter, S.: GANs trained by a two time-scale update rule converge to a Nash equilibrium. CoRR arXiv:1706.08500 (2017)
8. Hsiao, W.L., Katsman, I., Wu, C.Y., Parikh, D., Grauman, K.: Fashion++: minimal edits for outfit improvement, pp. 5046–5055 (2019). https://doi.org/10.1109/ICCV.2019.00515
9. Isola, P., Zhu, J.Y., Zhou, T., Efros, A.A.: Image-to-image translation with conditional adversarial networks. In: Proceedings of the IEEE Conference on Computer Vision and Pattern Recognition, pp. 1125–1134 (2017)
10. Jetchev, N., Bergmann, U.: The conditional analogy GAN: swapping fashion articles on people images. arXiv:1709.04695 (2017)
11. Jiang, S., Li, J., Fu, Y.: Deep learning for fashion style generation. IEEE Trans. Neural Netw. Learn. Syst., 1–13 (2021). https://doi.org/10.1109/TNNLS.2021.3057892
12. Liu, L., Zhang, H., Ji, Y., Wu, Q.M.J.: Toward AI fashion design: an attribute-GAN model for clothing match. Neurocomputing **341**, 156–167 (2019). https://doi.org/10.1016/j.neucom.2019.03.011
13. Loshchilov, I., Hutter, F.: SGDR: stochastic gradient descent with warm restarts arXiv:1608.03983 (2016)
14. Mameli, M., Paolanti, M., Pietrini, R., Pazzaglia, G., Frontoni, E., Zingaretti, P.: Deep learning approaches for fashion knowledge extraction from social media: a review. IEEE Access **10**, 1545–1576 (2021). https://doi.org/10.1109/ACCESS.2021.3137893
15. Namboodiri, R., Singla, K., Kulkarni, V.: GAN based try-on system: improving CAGAN towards commercial viability. In: 2021 12th International Conference on Computing Communication and Networking Technologies (ICCCNT), pp. 1–6 (2021). https://doi.org/10.1109/ICCCNT51525.2021.9579703
16. Pandey, N., Savakis, A.E.: Poly-GAN: multi-conditioned GAN for fashion synthesis. Neurocomputing **414**, 356–364 (2020). CoRR arXiv:1909.02165
17. Poma, X.S., Riba, E., Sappa, A.: Dense extreme inception network: towards a robust CNN model for edge detection. In: Proceedings of the IEEE/CVF Winter Conference on Applications of Computer Vision, pp. 1923–1932 (2020)

18. Raffiee, A.H., Sollami, M.: GarmentGAN: photo-realistic adversarial fashion transfer. arXiv:2003.01894 (2020)
19. Salimans, T., et al.: Improved techniques for training GANs. In: Lee, D., Sugiyama, M., Luxburg, U., Guyon, I., Garnett, R. (eds.) Advances in Neural Information Processing Systems, vol. 29. Curran Associates, Inc. (2016). https://proceedings.neurips.cc/paper/2016/file/8a3363abe792db2d8761d6403605aeb7-Paper.pdf
20. Shastri, H., Lodhavia, D., Purohit, P., Kaoshik, R., Batra, N.: Vastr-GAN: versatile apparel synthesised from text using a robust generative adversarial network. In: 5th Joint International Conference on Data Science and Management of Data (9th ACM IKDD CODS and 27th COMAD), pp. 222–226. Association for Computing Machinery, New York (2022). https://doi.org/10.1145/3493700.3493721
21. Xian, W., Sangkloy, P., Lu, J., Fang, C., Yu, F., Hays, J.: TextureGAN: controlling deep image synthesis with texture patches. CoRR arXiv:1706.02823 (2017)
22. Xie, S., Tu, Z.: Holistically-nested edge detection. arXiv:1504.06375 (2015)
23. Yildirim, G., Seward, C., Bergmann, U.: Disentangling multiple conditional inputs in GANs. arXiv:1806.07819 (2018)
24. Yirui, C., Liu, Q., Gao, C., Su, Z.: FashionGAN: display your fashion design using conditional generative adversarial nets. In: Computer Graphics Forum, vol. 37, pp. 109–119 (2018). https://doi.org/10.1111/cgf.13552
25. Zhu, S., Fidler, S., Urtasun, R., Lin, D., Loy, C.C.: Be your own Prada: fashion synthesis with structural coherence. CoRR arXiv:1710.07346 (2017)

Gender Recognition from 3D Shape Parameters

Giulia Martinelli(✉), Nicola Garau, and Nicola Conci

University of Trento, Via Sommarive, 9, 38123 Povo, Trento, TN, Italy
{giulia.martinelli-2,nicola.garau,nicola.conci}@unitn.it

Abstract. Gender recognition from images is generally approached by extracting the salient visual features of the observed subject, either focusing on the facial appearance or by analyzing the full body. In real-world scenarios, image-based gender recognition approaches tend to fail, providing unreliable results. Face-based methods are compromised by environmental conditions, occlusions (presence of glasses, masks, hair), and poor resolution. Using a full-body perspective leads to other downsides: clothing and hairstyle may not be discriminative enough for classification, and background cluttering could be problematic. We propose a novel approach for body-shape-based gender classification. Our contribution consists in introducing the so-called Skinned Multi-Person Linear model (SMPL) as 3D human mesh. The proposed solution is robust to poor image resolution and the number of features for the classification is limited, making the recognition task computationally affordable, especially in the classification stage, where less complex learning architectures can be easily trained. The obtained information is fed to an SVM classifier, trained and tested using three different datasets, namely (i) FVG, containing videos of walking subjects (ii) AMASS, collected by converting MOCAP data of people performing different activities into realistic 3D human meshes, and (iii) SURREAL, characterized by synthetic human body models. Additionally, we demonstrate that our approach leads to reliable results even when the parametric 3D mesh is extracted from a single image. Considering the lack of benchmarks in this area, we trained and tested the FVG dataset with a pre-trained Resnet50, for comparing our model-based method with an image-based approach.

Keywords: Gender recognition · Body shape · Parametric human body model

1 Introduction

Gender recognition has a wide range of application areas, ranging from human-computer interaction to surveillance systems, as well as commercial developments with particular attention to retail analytics. For this task, the observation of the face is generally considered amongst the most relevant element of the body. However, there exists a large set of additional cues, which can be analyzed so as to infer the gender information. This includes, for example hairstyle, body shape, clothing, eyebrows, posture and gait, as well as vocal traits, based on the voice pitch. Such additional features allow for the recognition through a multi-modal observation, exploring different dimensions, such as appearance, motion, and sound.

P. L. Mazzeo et al. (Eds.): ICIAP 2022 Workshops, LNCS 13374, pp. 203–214, 2022.
https://doi.org/10.1007/978-3-031-13324-4_18

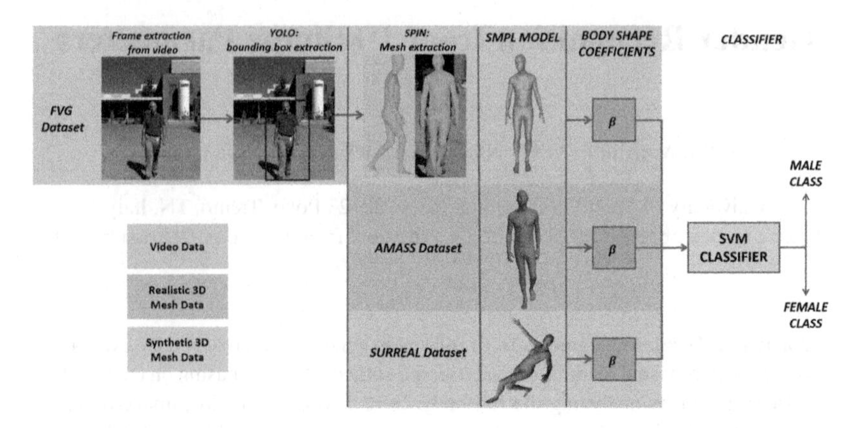

Fig. 1. Overview of the proposed pipeline. AMASS [12] and SURREAL [24] are characterized by SMPL [11] parametric mesh; the body shapes are therefore given. The FVG [26] dataset consists of videos of walking subjects. The parametric mesh is extracted using the SPIN [10] algorithm, from which the SMPL body shape coefficients are extracted and fed to the SVM classifier.

According to the information used for the classification, the existing gender recognition literature can be divided into two main categories: appearance and non-appearance-based approaches. The former leverages the features extracted from human physical appearance. These features can be static, denoting characteristics that are always present in an individual [6] (face, eyebrows, hand geometry), or dynamic [9], as body movement, activity recognition, or apparel information, like the detection of clothing and jewelry. The literature has also explored the analysis of other non-appearance-based features, extracting for example daily social network data [4]: information such as daily activities, logging emails, blogs, and handwriting can be used as features for classification. Such studies, however, are out of the scope of this work.

We propose a novel model-based approach for gender recognition that consists in extracting the parametric 3D human body model. The use of a model-based solution helps resolving the potential ambiguities that might arise when looking at aesthetic and appearance-based features only. In fact, the key goal of our work consists of using the SMPL [11] body-shapes parameters, which are invariant to clothing, hairstyle or other parameters commonly associated to one particular gender. In this way, we ensure that the model is sufficiently simple and reflects a standardized representation. In literature, only a few works address this problem using the body-shape information and, to the best of our knowledge, none of them use the parametric human body model SMPL [11]. In addition, most of the existing works use 3D human mesh vertices as features, significantly increasing the computational complexity, since the feature space that needs to be investigated is very large. In our case, the number of features is shrunk to only ten features. We also implement a CNN for comparison purposes, to evaluate our work against image-based methods. To do so, we use a pre-trained Resnet50 [7] and we perform training and testing on the FVG [26] dataset, comparing the results against the ones obtained by the SVM, fed with the mesh parameters extracted from the video sequence.

The main technical contributions of this work can be summarized as follows:

- We propose an effective descriptor using the SMPL body shape parameter for gender recognition via a 3D human model. We prove that this type of classification is suitable for those datasets that are composed of 3D meshes, as well as videos, exhibiting the potential for the application in a wide set of use-cases, including video surveillance, robotics, and biometrics.
- We show how our classifier, with a reduced feature space, improves the results obtained by other model-based solutions proposed in the literature.

2 Related Work

Gender Recognition from Body Shapes. In literature, only a few works address the problem of gender classification using the body shape information. In fact, while 2D image data can be often misleading due to camera view point and image resolution, 3D shape models offer a more comprehensible description of the observed object (subject) at a negligible incremental cost. The authors in [21] propose a gender recognition solution based on 3D human body shapes obtained with laser scanning. The paper does not consider the full body, and the authors use multiple features extracted from the subjects' chest and torso. Furthermore, the authors assert in the conclusion that their approach fails in classifying overweight or fully dressed individuals. More recently, other works focus on the 3D mesh of the human body. The same authors present another research on gender classification in [22], where they perform the recognition task by considering the shape landmarks of 3D human body model. The work proposed in [25] considers the body shape as feature, and the classification relies on the geodesic distance on the mesh. They discover that the most relevant features are the geodesic distance between the chest and the wrist, as well as the one between the lower back and the face. The approach proposed in [16], introduces a 2D-vertex-based gender recognition model. The authors compare the performance of two classifiers, Support Vector Machines (SVMs) and Extremely Randomized Trees (ERTs). They obtain the most remarkable results by using as input feature the vertices of 3D mesh and the SVM as classifier, with an accuracy of 78%. Using a 3D vertex-based methods makes the feature space of the classifier very large. Originally, their meshes contained between 67290 and 68300 vertices; this required a re-sampling (using a uniform probability distribution), to the bottom side, namely 67290 vertices. Since this number was still very large to be processed, they extracted the most relevant features by using Principal Component Analysis (PCA), resulting overall in 350 components.

Gender Recognition from Full-Body Images. In computer vision, gender recognition from whole-body images is a challenging task because the features extracted may not be discriminative enough for the classification and because background cluttering may be problematic. Gender classification has recently been addressed using convolutional neural networks. In [13], a CNN is trained considering the whole person body (Global CNN), the upper and then the lower portion of the human body (Local CNN). The Local CNN of the upper body achieves the highest accuracy because the face of a person is more discriminative than the rest of the body. This is supported by a feature visualization method that shows where the CNN extracts the features on the image. When the face is not visible, features are concentrated in the rest of the body. In this case, the information is achieved from clothing, hairstyle, and body shapes information. Sometimes these features are not enough for accurate classification. This is confirmed also by Raza et al. in [18], where they propose an appearance gender recognition method where a deep neural network is used to extract the silhouette of the pedestrian image. The silhouette is then used as a binary mask to remove the background from the image. The outcome is fed into a stacked sparse autoencoder (SSAE). The gender is classified considering three different camera views (frontal, back, and mixed) and they obtain the lowest accuracy score, as expected, on the back view. The mixed views obtain an accuracy slightly lower than the one in the front. The frontal view is in fact more distinctive, as it contains information extracted not only from the body but also from the face. This proves that the body features may not be discriminative enough to reach the accuracy of face features. Ng et al. [13] show that by combining the Global and Local CNN from the upper part of the body it is possible to obtain a better model, outperforming the state-of-the-art methods.

Human Mesh Recovery from Natural Images. Model-based human pose estimation can be faced following two different approaches. Optimization-based methods iteratively fit a parametric human body model, e.g. SMPL [11], to estimate the body pose and shape of the 2D observations, usually 2D joints locations. This solution has been presented as an alternative to preexisting models coming from the scans of different bodies in a varied set of poses. With this model, Loper et al. [11] created realistic animated human bodies that represent different body shapes that deform naturally with pose and exhibit soft-tissue motions like those of real humans. In contrast, regression based methods use a deep network to directly estimate the model parameters from pixels. Both methods have some pros and cons. Optimization based methods tend to be very slow and sensitive to initialization. Regression based methods, instead of taking only a sparse set of 2D location, take into account all pixels values; at the same time, this leads to a mediocre image-model alignment, and a large quantity of data is usually necessary for training. Regarding the first approach, SMPLify [3] has been the first method that automatically estimates the 3D pose and shape of human body. The most recent works have focused on regression; in fact, since there is a deficiency of images with full 3D shape ground truth, alternative supervision signals to train the deep networks are searched. The majority of the solutions uses 2D annotations including 2D keypoints, silhouettes, or part segmentation. This information can be used as input [23], intermediate representation [14,17], and supervision [8,14,17,20,23]. In this context, the SPIN algorithm [10], acronym of **SMPL oPtimization IN the loop**, presents a novel way of tackling the problem, finding a way to use the two methods in a collaborative fashion.

3 Datasets

Front View Gait Dataset (FVG). The FVG dataset [26] contains videos of 226 walking subjects, annotated by gender. It focuses only on the frontal view with three different near frontal-view angles towards the camera and other variations in terms of walking, speed, carrying, clothing, cluttered background and time. The 226 subjects walk along a straight line of 16 m toward the camera. The resolution of the video is full HD and the height of the person ranges from 101 to 909 pixels. For every subject, 12 videos have been captured, with different inclination of the camera (-45°, 0, 45°) and four variations of walking pace.

Archive of Motion Capture as Surface Shapes Dataset (AMASS). The AMASS dataset [12] consists of a collection of 15 MoCap datasets with gender annotation, represented with a common framework and parameterization. This has been achieved by converting the MoCap data into realistic 3D human meshes represented by a rigged SMPL body model, via the Mosh++ method.

Synthetic hUman foR REAL Dataset (SURREAL). The SURREAL dataset [24] contains 6 million frames of synthetic humans with ground truth pose, depth maps, segmentation masks, and gender information. The synthetic bodies are created using the SMPL body model. The SMPL parameters are fitted using the MoSh method from raw 3D MoCap marker data. The synthetic data has been generated rendering the following pieces of information: (i) a 3D human body model, whose pose was estimated with a motion capture system (ii) a frame using background image (iii) a texture map on the body, together with lightning and camera position. All these data are combined together in order to increase the diversity of the dataset.

4 Approach

The processing pipeline we propose consists of two stages: (i) extraction and preparation of the features, and (ii) classification. Since we are considering three different datasets, the pipeline slightly differs depending on which one is being used (see Fig. 1). In particular, AMASS and SURREAL are characterized by parametric SMPL models; therefore the body shape parameters are given. For the FVG dataset, instead, an additional processing stage for features extraction is needed. This is performed by using the SPIN [10] algorithm, as follows:

- The parameters of the SMPL human parametric model are regressed with a deep network.
- These regressed values are used by an iterative fitting in order to align the model to the 2D keypoints.
- The fitted model is used as supervision for the network, closing the loop between the regression and optimization method.

The SMPL body model provides a function $\mathcal{M}\left(\vec{\beta},\vec{\theta}\right)$, that takes as input the body shape parameters $\vec{\beta}$ and the pose parameters $\vec{\theta}$, and gives as output the body mesh $M \in \mathbb{R}^{3N}$ with $N = 6890$ the number of vertices. The body pose is defined by a standard skeletal rig, composed by $K = 23$ joints; the pose is then defined by $|\theta| = 3 \times 23 + 3 = 72$ parameters (3 for each joint plus 3 for the root orientation). The body shapes of different people are represented by the function:

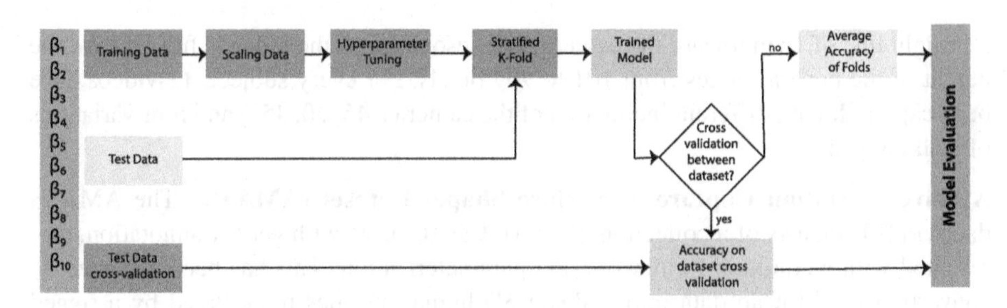

Fig. 2. Classification Pipeline. The training data are scaled and split in folds. On the training data the SVM Hyper-parameters are tuned. If the training and testing set belong to the same dataset, the accuracy of the model is the average accuracy over splits. Otherwise, the accuracy is calculated on the new testing set.

$$B_S\left(\vec{\beta}; \mathcal{S}\right) = \sum_{n=1}^{|\vec{\beta}|} \beta_n \mathbf{S}_n \tag{1}$$

where $\vec{\beta} = [\beta_1, ..., \beta_{|\vec{\beta}|}]^T$, $|\vec{\beta}|$ is the number of linear shape coefficients, and the $\mathbf{S}_n \in \mathbb{R}^{3N}$ represents the orthonormal principal components of shape displacements. In the end, the body shape parameters are only ten and they can be defined as the principal components of the shape variation learned from 3D scans of thousands of people.

In summary, the main steps of the proposed methodology are listed hereafter:

1. The image is cropped, extracting the bounding box around the person using YOLO [19] as a detector. A bounding box is required by the SPIN algorithm, as it assumes that the person is centered in the image;
2. The cropped image is passed to the SPIN algorithm that extracts the body shape and pose coefficients;
3. The ten body shape coefficients are used as features for the classification, and split into training and test samples, following a cross-validation approach;
4. The training data is scaled and the tuning of hyper-parameters is performed;
5. Finally, the trained model accuracy is calculated.

4.1 Model Selection

In machine learning, we know that tuning the hyper-parameters is a key step, which allows building a robust and accurate model, preventing over/underfitting. In our implementation, we use the Grid Search method. We tune two grids: simple linear kernel, $\langle x, x' \rangle$ with five possible values of the regularization parameter, C and a RBF kernel with five different values of γ and four values of C. Since the chosen datasets exhibit a severe class imbalance we divide the data following a **stratified k-fold cross validation**; it consists of a variation of the k-fold method, where each fold is composed approximately by the same percentage of samples belonging to both classes. This allows us to mitigate the possible effects of gender classification, due to the gender unbalance. The tuning of the parameters has been done in two steps. First, we calculate the most suitable number of splits dividing each dataset in a range of 3 to 10, and, for each of them, performing a model fitting. The final choice has been done by considering the number of splits that returns the highest accuracy. The second step consists in tuning the hyper-parameters and we choose the combination of parameters with the highest accuracy obtained from the confusion matrix. Finally, we proceed with model training over our different datasets with the hyper-parameters found. The accuracy of the model is calculated by averaging the accuracy of each split if the training and testing set belong to the same dataset, while with cross testing among different datasets the accuracy is calculated on the new testing set. The classification pipeline is illustrated in Fig. 2. All the experiments have been conducted on a NVIDIA RTX 3090, using Pytorch for the network implementation and Scikit-Learn for the SVM implementation.

Table 1. Cross Validation Results. The experiments are conducted with different data splits: for example [FVG + A] - [A] means that the classifier has been trained on FVG and AMASS, and tested on AMASS. We also tested the algorithm adding progressively a larger amount of synthetic samples to AMASS and FVG. For example, [A + Sn] - [A]: n is the ID of the training set ($n = \{1, 2, 3, 4\}$); [A + Sn] - [A] means training on AMASS and SURREAL, and testing on AMASS. A larger ID number corresponds to a larger amount of SURREAL data.

Experiment	# Train	# Test	# Female	# Male	Accuracy (%)
[A]-[A]	317	159	68	91	84.23
[S]-[S]	3800	1900	977	923	99.94
[FVG]-[FVG]	5650	1130	415	715	87.38
[FVG + A] - [FVG]	214	79	22	57	83.5
[FVG + A] - [A]	214	104	58	46	95.2
[FVG + S1] - [FVG]	154	79	22	57	83.5
[FVG + S2] - [FVG]	183	79	22	57	86.1
[FVG + S3] - [FVG]	220	79	22	57	83.5
[FVG + S4] - [FVG]	294	79	22	57	84.8
[A + S1] - [A]	383	111	61	50	81.1
[A + S2] - [A]	455	111	61	50	82.9
[A + S3] - [A]	547	111	61	50	84.7
[A + S4] - [A]	730	111	61	50	86.5

5 Results

In this section, we describe the conducted experiments and the corresponding results, to validate the effectiveness of our classifier using the SMPL body shapes parameters for gender classification. We perform cross training and testing on three different types of dataset: synstetic, real and registration scans. In this way we want to demonstrate the effectiveness of the classifier on different type of data. We investigate the accuracy of the model when combining real and synthetic data from different datasets. The experimental results are listed in Table 1. The highest accuracy is obtained by the SURREAL dataset, as we expected; SURREAL is a synthetic dataset and the body shape parameters have a perfect distribution between −5 and 5, making it a rather simple dataset to work with. As far as the FVG dataset is concerned, the returned accuracy is 87.38%. When we train and test on the AMASS dataset, the accuracy of classification decreases; so, even if this dataset is made of real registration scans, it consists of subject with a strong diversity in body shape. Instead the FVG dataset consists of real data, but its accuracy is higher than the AMASS dataset because it is characterized by subjects that do not strongly vary their body shapes. As far as the wrongly classified samples in AMASS is concerned, we assume that the performance decreases because the subjects are characterized by a sparse diversity in body shape. The failures in FVG occur generally when the subject is very far from the camera, namely exhibiting a reduced number of relevant pixels. This makes the extraction of the SMPL parameters with SPIN not sufficiently reliable. We then train and test the AMASS and FVG dataset adding in the training phase a progressively larger amount of synthetic data (from SURREAL): as we expected the accuracy increases when the synthetic data grow. As mentioned previously this is motivated by the fact that the synthetic samples are less subject to variations, making the classification easier and less prone to be adopted as substitutes for the real ones in this specific task.

5.1 Comparison with Previous Body Shape-Based Methods

Since the novelty of our work consists in using the SMPL meshes, we could not find in the literature other works for a straightforward comparison. The available state-of-the-art papers [21,22,25] use the CAESAR dataset [2] characterized by meshes extracted through a laser scanner. We could not apply our method to these datasets because they are not characterized by SMPL mesh. Nevertheless, we still try to provide a fair comparison, although the differences between the meshes affect the features extracted for the classification. These features consist of the Geodesic Distances (GD) [25] between landmarks, which corresponds to the length of shortest path between two points constrained on the shapes, Normal Distributions (ND) [21] on the chest region, mesh Vertices Coordinates (VC) [16] and Landmarks Positions [22] (LP). Looking

Table 2. Comparison with previous body shape-based methods. The term **RegS** stands for Registration Scans, **S** for Synthetic Shapes and **RealD** are Real Data (i.e. video data). **Dataset** indicates the train/test data, **Method** and **Features** the method and features used for the classification respectively. The results of our solution are listed in the last four rows.

Dataset	Type	Method	Features	Train	Test	Accuracy(%)	Pre-processing	Feature space	Landmarks
CAESAR	RegS	[25]	GD	500	500	96.1	✗	11	✓
CAESAR	RegS	[21]	ND	1224	1224	96	✓	100	✗
CAESAR	RegS	[22]	LP	1224	1224	98.9	✗	219	✓
POSER [1]	RegS-S	[16]	VC	450	140	75	✗	350	✗
AMASS	RegS	**Ours**	SMPL	317	159	**84.23**	✗	10	✗
FVG	RealD	**Ours**	SMPL	5650	1130	**87.38**	✗	10	✗
AMASS - SURREAL	RegS-S	**Ours**	SMPL	5146	1030	**97.8**	✗	10	✗
FVG - SURREAL	RealD-S	**Ours**	SMPL	8987	1123	**92**	✗	10	✗

at the methodology more in detail (see Table 2), the previous solutions require landmark detection or a pre-alignment process. They also have a much larger feature space. Instead, our method does not need any landmark or pre-processing step; furthermore, it has a much smaller feature space, resulting in a much faster computation. It is worth mentioning that our method can be effective also when using small datasets for training and testing, when compared to the ones used by the competing solutions. In addition, the proposed method uses a SMPL mesh that can also be extracted from a single image, thus it can be applied even when a laser scanner [21,22,25] is not available (e.g. surveillance), giving the solution generalization and scalability properties A fairer comparison can be conducted looking at the 3D vertex-based method presented in [16]. The authors achieve an accuracy of 75% using a very large number of features, even after feature reduction. With this respect, our method attains an accuracy of 87.38%, avoiding any feature reduction processes (e.g. PCA) since the SMPL mesh shrinks the feature space to only ten parameters.

5.2 Comparison with Image-Based Methods

In order to prove the effectiveness of our solution, also when compared to image-based methods, we use a pre-trained Resnet50 and we train and test the architecture on the FVG dataset. The comparison is summarized in Table 3. The CNN reaches an average accuracy of 80% in the validation phase. When using the same dataset, our proposed method reaches 87%. This proves the peculiarity of the body-shape features used in this work with respect to the common features used by a simple CNN. This is also proved in [13], where the highest results is obtained when the face of the subject is visible (80.8%). In [18] they obtained an accuracy of 82.9% on frontal views and 82.4% on mixed views. In Fig. 3 we can see examples of misclassified subjects by the CNN but correctly classified by our method considering the body shape parameters extracted from the 3D mesh. In fact, our solution does not rely on visual features and only considers the body shape information for gender classification, thus making it robust to camera pose changes, face appearance, and clothing. The last three columns report error in classification for both CNN and our method, possibly due to light conditions. For this reason, we made a use of a neutral body model for the incorrect classification samples only for visualization purpose, since the body model does not alter the values of body shape parameters.

Table 3. Comparison with image-based methods. We compare our proposed method against previous image-based works, as well as against the benchmark CNN we have implemented.

Method	Dataset	View	Accuracy (%)
Ng. et al. [13]	MIT [15]+APiS [27]	Upper frontal body part	80.8
Ng. et al. [13]	MIT+APiS	Global + Upper frontal parts	82.5
Raza et al. [18]	MIT+PETA [5]	Frontal	82.9
Raza et al. [18]	MIT+PETA [5]	Mixed	82.4
Our CNN	FVG	Frontal all body	80
Our method	FVG	Frontal all body	**87.38**

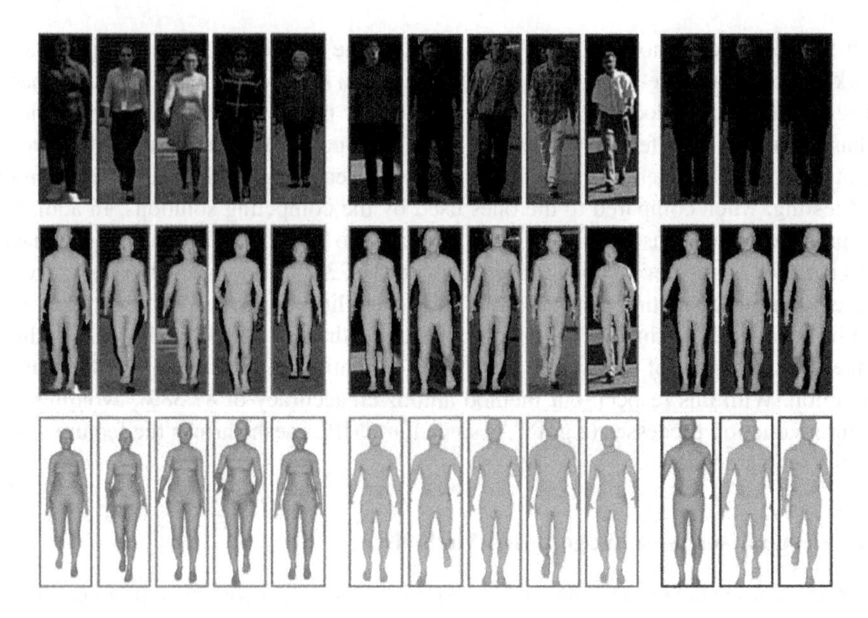

Fig. 3. Examples of classification output. The first row is characterized by subjects misclassified by the CNN. The second and third rows represent the classification output of our method. The red and green borders indicate respectively wrong and good classification output. The correct gender is indicated by the color of meshes in the third row. (Color figure online)

6 Conclusions

We propose a novel approach for gender classification using SMPL body shapes parameters. This is suitable for all those datasets that are characterized by 3D meshes, as well as videos, exhibiting the potential for the application in a wide set of use-cases, as video surveillance, robotics, biometrics. Considering the low-dimensionality of the feature space that allows for fast computation, the proposed approach obtains satisfactory results, yet adding desirable properties, such as the use of a parametric mesh that provides a simple and a standard representation, with a number of vertices that is lower than the one used by competing methods. Our approach outperforms also the results of image-based competing methods, since the features we adopt do not depend on camera view, and they are robust to face occlusion. In the future, our goal is to create a new dataset characterized by SMPL parametric shapes for gender recognition, as well as the recognition of additional attributes as, for example, age. We plan to use the DMPL [11] model, that has the same advantages of SMPL model but it considers the body deformations produced by the body movements and impact forces with the ground.

References

1. Poser - 3d character art and animation software. https://www.posersoftware.com/
2. Blackwell, S., et al.: Civilian american and european surface anthropometry resource (caesar), Descriptions, vol. 2, p. 192, June 2002

3. Bogo, F., Kanazawa, A., Lassner, C., Gehler, P., Romero, J., Black, M.J.: Keep it SMPL: automatic estimation of 3d human pose and shape from a single image. In: Leibe, B., Matas, J., Sebe, N., Welling, M. (eds.) ECCV 2016. LNCS, vol. 9909, pp. 561–578. Springer, Cham (2016). https://doi.org/10.1007/978-3-319-46454-1_34
4. Burger, J., Henderson, J., Kim, G., Zarrella, G.: Discriminating gender on twitter, pp. 1301–1309, January 2011
5. Deng, Y., Luo, P., Loy, C.C., Tang, X.: Pedestrian attribute recognition at far distance, pp. 789–792, November 2014. https://doi.org/10.1145/2647868.2654966
6. Dhomne, A., Kumar, R., Bhan, V.: Gender recognition through face using deep learning. Proc. Comput. Sci. **132**, 2–10 (2018). https://doi.org/10.1016/j.procs.2018.05.053
7. He, K., Zhang, X., Ren, S., Sun, J.: Deep residual learning for image recognition (2015)
8. Kanazawa, A., Black, M.J., Jacobs, D.W., Malik, J.: End-to-end recovery of human shape and pose (2018)
9. Kastaniotis, D., Theodorakopoulos, I., Economou, G., Fotopoulos, S.: Gait-based gender recognition using pose information for real time applications. In: 2013 18th International Conference on Digital Signal Processing (DSP), pp. 1–6 (2013). https://doi.org/10.1109/ICDSP.2013.6622766
10. Kolotouros, N., Pavlakos, G., Black, M.J., Daniilidis, K.: Learning to reconstruct 3d human pose and shape via model-fitting in the loop. In: Proceedings of the IEEE International Conference on Computer Vision, pp. 1–10 (2019)
11. Loper, M., Mahmood, N., Romero, J., Pons-Moll, G., Black, M.J.: SMPL: a skinned multi-person linear model. ACM Trans. Graphics (Proc. SIGGRAPH Asia) **34**(6), 248:1–248:16 (2015)
12. Mahmood, N., Ghorbani, N., Troje, N.F., Pons-Moll, G., Black, M.J.: AMASS: Archive of motion capture as surface shapes. In: International Conference on Computer Vision, pp. 5442–5451, Oct 2019
13. Ng, C.-B., Tay, Y.-H., Goi, B.-M.: Pedestrian gender classification using combined global and local parts-based convolutional neural networks. Pattern Anal. Appl. **22**(4), 1469–1480 (2018). https://doi.org/10.1007/s10044-018-0725-0
14. Omran, M., Lassner, C., Pons-Moll, G., Gehler, P.V., Schiele, B.: Neural body fitting: Unifying deep learning and model-based human pose and shape estimation (2018)
15. Oren, M., Papageorgiou, C., Sinha, P., Osuna, E., Poggio, T.: Pedestrian detection using wavelet templates, pp. 193–199, July 1997. https://doi.org/10.1109/CVPR.1997.609319
16. Pablo, N., Bruno, P., Celia, C., Virginia, R., Rolando, G.J., Claudio, D.: Gender recognition using 3d human body scans. In: 2018 IEEE Biennial Congress of Argentina (ARGENCON), pp. 1–6 (2018). https://doi.org/10.1109/ARGENCON.2018.8646293
17. Pavlakos, G., Zhu, L., Zhou, X., Daniilidis, K.: Learning to estimate 3d human pose and shape from a single color image (2018)
18. Raza, M., Sharif, M., Yasmin, M., Khan, M., Saba, T., Fernandes, S.: Appearance based pedestrians' gender recognition by employing stacked auto encoders in deep learning. Future Gener. Comput. Syst. **88**, 28–39 (2018). https://doi.org/10.1016/j.future.2018.05.002
19. Redmon, J., Divvala, S., Girshick, R., Farhadi, A.: You only look once: unified, real-time object detection (2016)
20. Tan, V., Budvytis, I., Cipolla, R.: Indirect deep structured learning for 3d human body shape and pose prediction. In: BMVC (2017)
21. Tang, J., Liu, X., Cheng, H., Robinette, K.M.: Gender recognition using 3-d human body shapes. IEEE Trans. Syst. Man Cybern. Part C (Appli. Rev.) **41**(6), 898–908 (2011). https://doi.org/10.1109/TSMCC.2011.2104950
22. Tang, J., Liu, X., Cheng, H., Robinette, K.M.: Gender recognition with limited feature points from 3-d human body shapes. In: 2012 IEEE International Conference on Systems, Man, and Cybernetics (SMC), pp. 2481–2484 (2012). https://doi.org/10.1109/ICSMC.2012.6378116

23. Tung, H.Y.F., Tung, H.W., Yumer, E., Fragkiadaki, K.: Self-supervised learning of motion capture (2017)
24. Varol, G., et al.: Learning from synthetic humans. In: CVPR (2017)
25. Wuhrer, S., Shu, C., Rioux, M.: Posture invariant gender classification for 3d human models. In: 2009 IEEE Computer Society Conference on Computer Vision and Pattern Recognition Workshops, pp. 33–38 (2009). https://doi.org/10.1109/CVPRW.2009.5204295
26. Zhang, Z., et al.: Gait recognition via disentangled representation learning. In: Proceeding of IEEE Computer Vision and Pattern Recognition, Long Beach, CA, June 2019
27. Zhu, J., Liao, S., Lei, Z., Yi, D., Li, S.: Pedestrian attribute classification in surveillance: database and evaluation, pp. 331–338, Dec 2013. https://doi.org/10.1109/ICCVW.2013.51

Recognition of Complex Gestures for Real-Time Emoji Assignment

Rosa Zuccarà, Alessandro Ortis(✉), and Sebastiano Battiato

Department of Mathematics and Computer Science, University of Catania, Catania, Italy
{ortis,battiato}@dmi.unict.it

Abstract. Gesture recognition allows humans to interface and interact naturally with the machine. This paper presents analytical and algebraic methods to recognize specific combinations of facial expressions and hand gestures, including interactions between hands and face. The methodologies for extracting the features for both faces and hands were implemented starting from landmarks identified in real-time by the MediaPipe framework. To benchmark our approach, we selected a large set of emoji and designed a system capable of associating chosen emoji to facial expressions and/or hand gestures recognized. Complex poses and gestures combinations have been selected and assigned to specific emoji to be recognized by the system. Furthermore, the Web Application we created demonstrates that our system is able to quickly recognize facial expressions and complex poses from a video sequence from standard camera. The experimental results show that our proposed methods are generalizable, robust and achieve on average 99,25% of recognition accuracy.

Keywords: Facial expressions · Hand gestures · Real time recognition · Emoji

1 Introduction and Motivations

Human communication is often a complex combination of facial expressions, hand gestures and speech, all of which contribute significantly to a spoken message [1]. Facial expression is one of the main ways by which human beings communicate their intentions and emotions. For this reason, Facial Expression Recognition (FER) is very important and can be used in many applications such as driver safety, healthcare, video conferencing, virtual reality, and Human-Machine Interface (HMI) [2–4]. The state of the art of the FER shows that most of the designed systems perform three phases: image pre-processing, feature extraction, expression classification. The calculation of the feature vector describing a facial expression is often based on the landmarks estimated by face detection algorithms. Classification is usually done by training a learning model. The method proposed in [5] uses an estimator of 68 facial landmarks and calculates a feature vector of distance by a selected set of landmarks around mouth

P. L. Mazzeo et al. (Eds.): ICIAP 2022 Workshops, LNCS 13374, pp. 215–227, 2022.
https://doi.org/10.1007/978-3-031-13324-4_19

area and eyes since muscles in those areas change with facial expressions. The difference between emotional and neutral feature vectors is considered as final features to identify emotions using Random Forest Classifier. In the work in [6] feature vector is constructed using all 68 estimated landmarks and considering for each: coordinates, the distance from a fixed point and the direction of the direct line towards the fixed point. Facial expression recognition was achieved using Support Vector Classifier (SVC). The paper [7] presents the FER method based on Convolution Neural Network (CNN). Considering the complexity of the system, for CNN it is necessary to collect a huge amount of data, so that the trained network has better generalization performance and can reduces overfitting. Hand detection plays an important role in hand gesture recognition for applications such as sign language recognition, Human-Computer Interaction (HCI), driver hand behaviour monitoring and virtual/augmented reality interaction [8,9]. Gupta et al. [10] proposed a method for the "static hand gesture" recognition based on 15 local Gabor filters, to reduce the complexity, followed to a combination of Principal Component Analysis (PCA) and Linear Discriminant Analysis (LDA) for make system invariant to scale and rotation problem. Classification of gestures is done with the help of a one-against-one multiclass Support Vector Machine (SVM). Chen et al. [11] introduced a "hand dynamic gesture" recognition system from 2D video. First is applied a real-time hand tracking algorithm, then is calculated the vector of spatial and temporal features that is as the input to the Hidden Markov Model (HMM) based recognition system. Koh [12] designs a two-stage hand gesture recognition system (trajectory-based, shape-based) for distinguishes 15 user-defined hand gestures that are highly representative to Visual Communication Markers (VCMs) such as emoji. After different steps of image pre-processing and feature extraction, the classification is performed by Machine Learning (ML) algorithms. Some research efforts focus on integrating more aspects of communication, such as facial expression, hand gestures or speech recognition activities [13]. The researchers dealt with both the extraction phase of the characteristics and the phase of recognition of the expression or hand gestures using ML algorithms. The drawbacks of the solutions based on Deep Learning (DL) models are often related to high computational requirements, or specific hardware needs to perform the inferences. Indeed, research solutions are often time and computational expensive. This represents the main gap between research state of the art and its applications for the deployment of services and feasible solutions. Therefore, in our research on the recognition of complex combinations of facial expressions and gestures, the recognition phase was carried out using the characteristics calculated by analytical and algebraic methods. The development of an efficient recognition system must overcome challenges in the face or hand detection phase such as: the segmentation, in the presence of complex backgrounds in which there are many objects in an image, the representation of the local form of the hand [14], in the representation of the global configuration of the body and the modelling of the gestural sequence [15]. Well-performing face and hand detection models that also guarantee to be performed quickly in real-time and on mobile devices are respectively: MediaPipe Face Mesh [16] and MediaPipe Hands [17], both available in MediaPipe, an open-source framework for building multimodal and cross-platform applied ML pipelines. The ML pipeline of the two solutions: "MediaPipe Face Mesh" and "MediaPipe Hands" consist of two real-time deep neural network models that work together: a detector that operates on the full image and computes face or hand locations and a 3D face landmark model or 3D hand landmark model that operates on those locations and predicts the approximate surface geometry via regression. MediaPipe Hands infers 21 3D landmarks of a hand, while MediaPipe Face Mesh estimates 468 3D face landmarks. With the aim to recognize complex gestures, we defined a pool of proper analytical geometry methods on the landmarks estimated by the MediaPipe models avoiding to lean on further DL models for classification. This allowed the implementation of a real-time complex gesture recognition system, which can be executed without high computational or specific hardware requirements.

2 Proposed Method

This paper aims to recognize complex combinations of facial expressions and hand gestures, including interactions that could occur between the face and hands. To this end, we defined a set of complex gestures associated to emoji and implemented a gesture-to-emoji recognition system based on analytical and algebraic methods able to process a video flow in real time. Emojis are pictograms or ideograms used to express emotions through facial images or to describe concepts through images of objects, places, activities, foods, plants, animals. They are usually introduced to emphasize the message in digital conversations and social media post sharing. They are used in text communications to emulate visual cues such as facial expressions, poses and gestures [18]. To recognize hand gestures, we have taken into consideration: if it is the right or left hand, the orientation of the hand (vertical or horizontal), the region (palm or back), the position of the hand in the plane, the bending of the hand, the poses of each finger (closing, folding) and finally the reciprocal position between the fingers (dilation, proximity (touch) or crossing). Whereas features extracted from eyebrows, eyes and mouth have been considered for the facial expression.

2.1 Methods for Recognizing Facial Expressions

To estimate the state of the eye opening (closed, open or wide open) or of the mouth (closed, ajar, open, or wide open) the most significant landmarks are selected and the corresponding 2D coordinate values are used to compute the following AR_1 (i.e., Aspect Ratio) metric:

$$AR_1 = \frac{\sum_{i=1}^{3} d_i}{2d_{C_{12}}} = \frac{\sum_{i=1}^{3} \sqrt{(P_i^u.x - P_i^l.x)^2 + (P_i^u.y - P_i^l.y)^2}}{2\sqrt{(C_1.x - C_2.x)^2 + (C_1.y - C_2.y)^2}} \tag{1}$$

In particular, we extracted the same number of landmarks around the mouth and each eye and applied the same equation for the estimation of the closing degree. Figure 1 shows the landmarks of eyelids, whereas Fig. 2 shows the landmarks extracted around the mouth. The metric (Eq. 1) is partially insensitive to the small variations in the proportions of the eyes or mouth that occur between different individuals, it is invariant with respect to uniform resizing of the image and rotation of the face in the plane. For the eyes this invariance is obtained by normalizing the sum of the distances between the contours of the eyelids with respect to the distance between the corners of the eye (see Fig. 1). Whereas for the mouth, the landmarks taken into consideration are those relating to the inner edge and the corners of the vermilion (see Fig. 2). A similar feature was suggested in [19] to measure the eye blink and for correct recognition a classifier that takes a larger temporal window of a frame into account is trained. Instead, in our approach we have empirically established the threshold values used to determine the appropriate state of opening the mouth or eyes, collecting the values of the "Aspect Ratio" extracted from several acquisition sessions with different participants. It has been observed that the value of the metric increases with increasing eye or mouth opening. The separation values between the states are determined as the average between two averages of the adjacent states. Intervals of values that define the eye-opening states are determined as follows: less than 0.5 for closed eyes, between 0.5 and 0.7 for open eyes and greater than 0.7 for wide eyes. Intervals of values that define the mouth-opening states are determined as follows: less than 0.22 for closed mouth, between 0.22 and 0.55 for ajar mouth, between 0.55 and 0.9 for open mouth and greater than 0.9 for wide open mouth. Oral commissures are the points where the upper and lower lips meet, usually known as the corners of the mouth. Happiness and sadness are detected comparing the positions of the landmarks of the oral commissures with the are compared with those of the chin and cheek, as detailed in the following Definition 1 and Definition 2 (see Fig. 3), respectively.

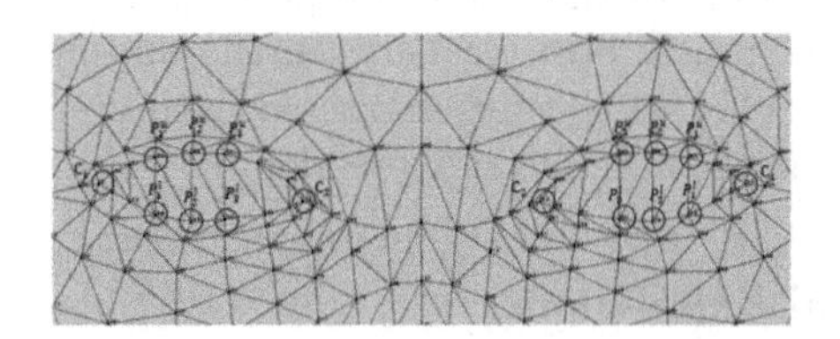

Fig. 1. Landmarks of eyelids (P_i^k) and of the left and right eye corners (C_j).

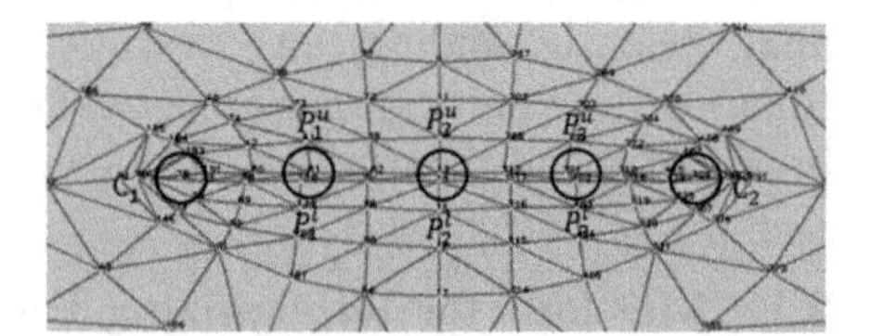

Fig. 2. Landmarks of the upper and lower inner border (P_i^k) and the right and left angles of the mouth vermilion (C_j).

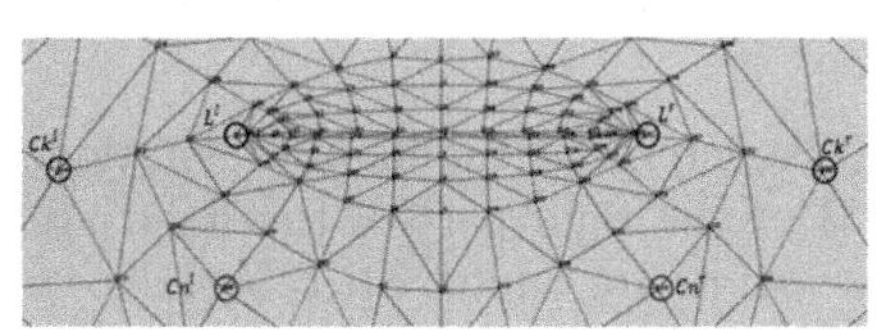

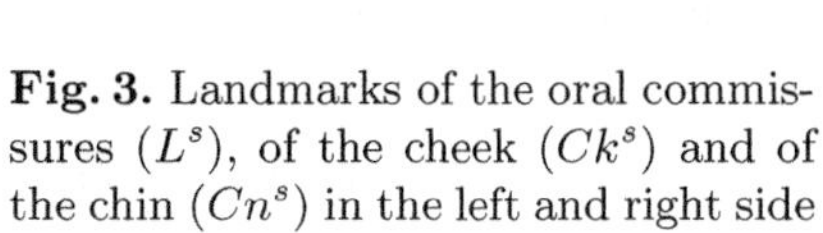

Fig. 3. Landmarks of the oral commissures (L^s), of the cheek (Ck^s) and of the chin (Cn^s) in the left and right side of the face.

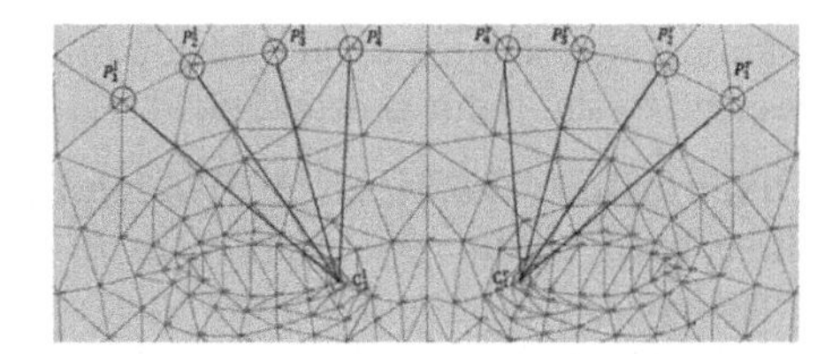

Fig. 4. Landmarks (C_1^s) of the inner corner of the right and left eye and landmarks of the lower part of the forehead (P_i^s).

Definition 1 (Smiling lips). *$L^l.y < Ck^l.y$ and $L^l.x < Cn^l.x$, $L^r.y < Ck^r.y$ and $L^r.x > Cn^r.x$*

Definition 2 (Sad lips). *$L^l.y > Ck^l.y$ and $L^l.x > Cn^l.x$, $L^r.y > Ck^r.y$ and $L^r.x < Cn^r.x$*

To determine whether a subject has raised or lowered eyebrows, landmarks of the inner corners of the eyes were selected because these points are stable in the two eyebrow poses. In addition, landmarks of the lower forehead were selected because their position varies with the occurrence of the muscular movements of the eyebrows (see Fig. 4). We have built a metric AR_2 robust enough to be invariant under different factors such as face size or the distance between the face and the camera. Indeed, as shown in Eq. 2, the sum of the distances between inner corner eye and points of the forehead is normalized by the distance between the landmarks C_1^l and P_4^l (left side) or between C_1^r and P_4^r (right side).

$$AR_2^s = \frac{\sum_{i=1}^{4} d_i^s}{2d_4^s} = \frac{\sum_{i=1}^{4} \sqrt{(P_i^s.x - C_1^s.x)^2 + (P_i^s.y - C_1^s.y)^2}}{2\sqrt{(P_4^s.x - C_1^s.x)^2 + (P_4^s.y - C_1^s.y)^2}} \quad s = l \text{ (left side) or } s = r \text{ (right side)} \tag{2}$$

A similar metric is proposed by the authors in [5] who have developed a method to compute, normalize and extract a feature vector which represent facial emotions that are classified using the random forest-based classification technique. While in our approach the threshold values are established to determine the states that indicate the positions of the eyebrows ("normal", "raised", "lowered") and these are set using the same procedure seen for the other components of the face, described above.

2.2 Methods for Recognizing Hand Gestures

For each hand detected by the MediaPipe Hand detector, the characteristics that describe its configuration are extracted. Then, approaches are proposed to determine the orientation, the fold, the position in the plane of the hand and to identify the palm or the back. Furthermore, for each finger, the closure, the folding, possible reciprocal position between the other fingers and alignment with respect to the palm is established.

Orientation: Vertical or Horizontal: To determine the hand orientation, three fixed landmarks (marked as 0, 9, 13) of the rigid part of the hand are considered and the coordinate values of these points are compared. Table 1 shows the various conditions for establishing the orientation of the hand.

Table 1. Conditions for establishing the orientation of the hand. The notation "0.y" stands for: coordinate y of the point marked as 0.

Vertical orientation with fingers pointing up	Horizontal orientation with fingers pointing to the right	Horizontal orientation with fingers pointing to the left	Vertical orientation with fingers pointing down
Orientation: 1	Orientation: 2 Slice: top	Orientation: 2 Slice: down	Orientation: 3
$0.y \geq 9.y$ and $0.y > 13.y$	$9.y \leq 0.y$ and $0.y \leq 13.y$	$13.y \leq 0.y$ and $0.y < 9.y$	$0.y < 9.y$ and $0.y < 13.y$

Hand Region: Palm or Back: The MediaPipe hand detector can distinguish the right hand from the left one. This information, estimated by the detector, together with the orientation, assigned by our algorithm, is exploited to determine whether the palm or the back of the hand is shown. In particular, we observed that in each orientation of the left or right hand, the point to be examined corresponds to the metacarpal of the thumb (i.e., landmark 1). We thus found relationships between the coordinates of this point and the coordinates of the landmarks of the rigid part of the hand, marked as 0, 5, 9, 13, and 17. We have noticed that these relationships vary within certain regions of the plane, thus allowing it to be divided into "slices". In this way, when the hand is positioned in the XY plane of the "world" with a certain inclination, the algorithm assigns to it the appropriate "slice" in which it lies. In particular, for "Orientation 1" and "Orientation 3", the x (or y) coordinate of landmark 1 is compared with the maximum or minimum computed from the set of x (or y) coordinate values of the metacarpal landmarks, marked as 1, 5, 9, 13, 17. Furthermore, a similar comparison is performed for the landmark 0, corresponding to the wrist of the hand[1]. Figure 5 shows an example of the conditions for the palm of the left hand positioned in "Orientation 1". For the hand positioned in "Orientation 2", to establish the region (palm or back) the coordinates of the landmarks 0 and 1 are compared, while to determine thumb-up or thumb-down the comparison is performed between the landmarks 0, 9, and 13. Figure 6 shows an example of the conditions for the back of the right hand positioned horizontally with the thumb up or down.

[1] $max_x = \max\{1.x,\ 5.x,\ 9.x,\ 13.x,\ 17.x\}$, $min_x = \min\{1.x,\ 5.x,\ 9.x,\ 13.x,\ 17.x\}$, $max_y = \max\{1.y,\ 5.y,\ 9.y,\ 13.y,\ 17.y\}$, $min_y = \min\{1.y,\ 5.y,\ 9.y,\ 13.y,\ 17.y\}$, $\mathrm{Max} = \max\{0.y, 1.y, 5.y, 9.y, 13.y, 17.y\}$, $\mathrm{Min} = \min\{0.y, 1.y, 5.y, 9.y, 13.y, 17.y\}$.

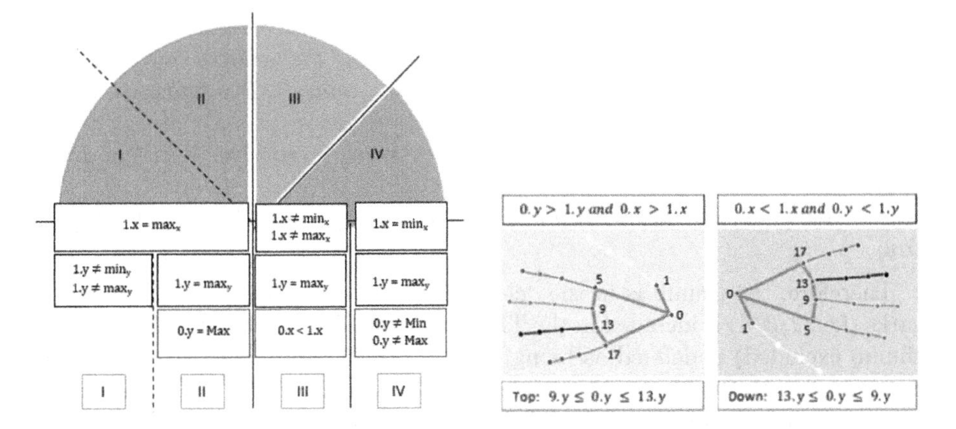

Fig. 5. Conditions for identifying the palm of the left hand positioned in "Orientation 1".

Fig. 6. Conditions that determine the back of the right hand positioned in "Orientation 2".

Alignment of the Finger with Respect to the Metacarpus: The direction of the fingers with respect to the metacarpus was analysed to establish the alignment of the fingers to the palm. Then we calculate the normal to the plane of the palm formed by two vectors having as extremes, respectively the landmarks 0 and 5 and the landmarks 0 and 17. For each of the four fingers, the unit direction vectors of the straight lines passing through the reference points corresponding to MCP and tip were determined. The alignment was examined by calculating the angle of inclination of the unit vector of the normal with respect to the unit vector of direction of the finger. We define the hand in the "straight state" when the four fingers have an inclination of no less than 70°.

Recognizing the Fingers in the "Closed" State: The thumb is defined to be in the "closed" state when at least one of its parts (i.e., TIP, PIP, MCP) reaches the top of the palm region. Whereas the remaining four fingers are defined in the "closed" state when their tip reaches the palm of the hand. To simulate this mechanism, the following two Regions of Interest (RoI) have been defined:

Definition 3 (Thumb Closing RoI). *Thumb Closing RoI is the region defined around the landmarks 5, 9, 13, and 17 and having the vertices in the points:* $A = (min_x,\ min_y)$, $B = (max_x,\ min_y)$, $C = (min_x,\ max_y)$, $D = (max_x,\ max_y)$. *Where:*
$min_x = min\{5.x,\ 9.x,\ 13.x,\ 17.x\} - offset$, $min_y = min\{5.y,\ 9.y,\ 13.y,\ 17.y\} - offset$,
$max_x = max\{5.x,\ 9.x,\ 13.x,\ 17.x\} + offset$, $max_y = max\{5.y,\ 9.y,\ 13.y,\ 7.y\} + offset$.

Definition 4 (Finger Closing RoI). *Finger Closing RoI is the region defined around the landmarks 1, 5, 9, 13, and 17 and having the vertices in the points:* $A = (min_x, min_y)$, $B = (max_x, min_y)$, $C = (min_x, max_y)$, $D = (max_x, max_y)$. *Where:*
$min_x = min\{1.x, 5.x, 9.x, 13.x, 17.x\} - offset$, $min_y = min\{1.y, 5.y, 9.y, 13.y, 17.y\}$,
$max_x = max\{1.x, 5.x, 9.x, 13.x, 17.x\} + offset$, $max_y = max\{1.y, 5.y, 9.y, 13.y, 17.y\}$.

Therefore, the thumb is in the "closed" state when at least one of its landmarks (i.e., 2, 3, 4) is included in the Thumb Closing RoI. Each of the four fingers (thumb excluded) is defined as being in the "closed" state when the landmark corresponding to the "tip" is included in the Finger Closing RoI. The proposed approach is invariant to the size, different positions and inclinations of the hand, and to the distance between the hand and the camera.

Recognizing Fingers in the "Bent" State: Another movement considered for the fingers is the bending of the three phalanges. To describe this behaviour, a similar approach to the one described above has been implemented. In fact, for each finger three regions have been determined: one which includes the landmarks corresponding to three interphalangeal joints (i.e., MCP, PIP, DIP), the second innermost which includes two of the joints (i.e., PIP, MCP) and the third which presents the MCP joint. To define the finger in the "bent" state, at least one of the three conditions must occur:

- Condition 1[2]: $min_x \leq tip.x \leq max_x$ and $min_y \leq tip.y \leq max_y$
- Condition 2[3]: $min_x \leq dip.x \leq max_x$ and $min_y \leq dip.y \leq max_y$
- Condition 3[4]: $min_x \leq pip.x \leq max_x$ and $min_y \leq pip.y \leq max_y$

Reciprocal Position Between the Fingers: Dilation, proximity and crossing are the reciprocal position between the fingers. To recognize when the fingers are two by two dilated, the distance of the reference point MCP_i from the line passing through the points MCP_{i+1} and TPC_{i+1} of the adjacent finger was compared with the distance of the reference point TCP_i from the same line. If the distance calculated from the TCP_i landmark is greater than that calculated by the MCP_i landmark, then the pair of fingers is defined in the "dilated" state. To describe the junction of two fingertips, a region (RoI) has been defined around one of the two fingertips and then a check is executed to see if the tip of the other finger belongs to that region. To represent the two crossed fingers, a region (RoI) around the three interphalangeal joints (TIP, DIP, PIP) was defined on one of the two fingers, and then it was checked whether the landmark DIP of the other finger belongs to that region.

---

[2] $max_x = \max\{\text{mcp.x, pip.x, dip.x}\} + \text{offset}$, $min_x = \min\{\text{mcp.x, pip.x, dip.x}\} - \text{offset}$,
$max_y = \max\{\text{mcp.y, pip.y, dip.y}\} + \text{offest}$, $min_y = \min\{\text{mcp.y, pip.y, dip.y}\} - \text{offset}$.

[3] $max_x = \max\{\text{mcp.x, pip.x}\} + \text{offest}$, $min_x = \min\{\text{mcp.x, pip.x}\} - \text{offset}$,
$max_y = \max\{\text{mcp.y, pip.y}\} + \text{offset}$, $min_y = \min\{\text{mcp.y, pip.y}\} - \text{offest}$.

[4] $max_x = \{\text{mcp.x}\} + \text{offset2}$, $\text{min}_x = \{\text{mcp.x}\} - \text{offset2}$, $max_y = \{\text{mcp.y}\} + \text{offest2}$, $min_y = \{\text{mcp.y}\} - \text{offset2}$.

Interaction Between Face and Hands: the method proposed to analyse the interaction between the face and the hands is to trace regions (RoI-s) around specific landmarks of the face, and then map the coordinates of the vertices of the RoI-s on the image plane. So, when significant landmarks of the hand, also mapped on the image plane, satisfy the conditions of belonging to these regions, then a "hand-face" interaction is considered to have taken place. This method has the characteristic of being invariant to the scale and inclinations of poses.

Head Tilt: the inclination of the head was determined by calculating the angle (α) between the x-axis of the image plane and the straight line passing through the two landmarks corresponding to the vertex of the head and the vertex of the chin. Condition to define the head in the "inclined" state is $|\alpha| \leq \frac{\pi}{2.5}$.

3 Real-Time Emoji Assignment

The above detailed methods have been implemented in a real-time video analysis process. At each frame, hands and face landmarks are extracted and related features are stored, considering the structures detailed in Table 2 and Table 3 for face (eyebrows, eyes, mouth) and hands respectively.

Table 2. Dictionary for facial components.

	Keys	Values
eyes_dict	Left	close, open, wide open
	Right	close, open, wide open
	State	closeLeft, closeRight, closeTwo, openTwo, wideLeft, wideRight, wideTwo
mouth_dict	State	close, ajar, open, wide
	State_Lips	smile, sad
eyebrows_dict	Left	normal, raised, lowered
	Right	normal, raised, lowered
	State	normalTwo, upperLeft, upperRight, upperTwo, upLow, lowerLeft, lowerRight, upperTwo, lowUp

Table 3. Dictionary for hand gestures.

Objects	Keys	Subkeys	Values
Left Right	Orientation		1, 2, 3
	Slice		1, 2, 3, top, down
	Part		palm, back
	Finger_Closed	thumb, first, second, third, fourth	True, False
	Finger_Bent	first, second, third, fourth	True, False
	Dilatate		True, False
	Folded		true, false
	Tag_gesture		All_Closed, 4_No_Closed, All_Bent, All_Stretched, All_Dilatate, All_Together, Hand_Straight

Were selected a large set of emojis and defined the task of assigning the correct emoji to recognized move. In particular, thirteen facial expression emojis were chosen, thirteen types of hand gestures, collecting 208 hand emojis to represent the right or left hand, palm or back, and the possible positions. Furthermore, emojis can be displayed according to the inclination of the hand[5].

[5] For further details refer to the complete set of encoded emojis and related gestures reported in the supplementary material available at the following link.

To show that the system is able to recognize the possible interactions between hands and face, 17 moves have also been chosen to be represented with emoji (see Table 4).

Table 4. Complex emoji and detailed gesture description.

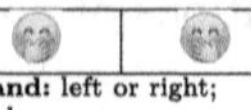 **Hand:** left or right; back **Fingers:** dilated **Tips of:** index finger and middle finger touch chin **Mouth:** closed **Lips:** smile **Eyes:** closed	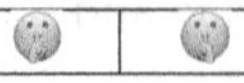 **Hand:** left or right; palm **Index gesture:** touches nose **Eyes:** open	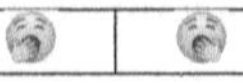 **Hand:** left or right; back. **Fingers:** dilated **Tips of:** index finger, middle finger touch chin **Mouth:** wide open **Eyes:** closed	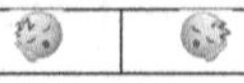 **Head:** tilted left or right **Hand:** right or left straight **Hand:** touches sphenoid and cheek **Mouth:** not closed **Eyes:** closed	**Eyes:** wide open **Hands:** left and right **Tips of:** index finger, middle finger, and ring finger touch cheeks
			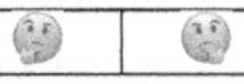	
Hands: left, right; back **Hand orientation:** vertical upwards **Index finger, middle finger, ring finger, little finger:** not closed **Middle-Index, Middle-Ring:** not dilated **Thumb:** closed **Tips of:** index finger, middle finger touch mouth **Eyes:** open	**Hands:** left, right; back **Hand orientation:** vertical with fingers up **Left hand:** inclined to the right **Right hand:** inclined to the left **Left hand:** touches left eye **Right hand:** touches right eye **Mouth:** closed **Lips:** smile	**Hands:** left and right **Hand orientation:** vertical upwards **Fingers:** not bent **Tips of:** little finger, ring finger, and middle finger touch sphenoid **Eyes:** open **Mouth:** closed **Lips:** smile	**Hand:** left or right; back. **Thumb-Index:** dilated **Middle finger, ring finger, little finger:** closed **Thumb:** touches cheek **Index finger:** touches chin **Eyes:** open **Mouth:** closed or ajar **Lips:** normal or sad	**Hands:** left, right; back **Index gesture:** touch cheeks **Mouth:** closed or **Lips:** sad **Eyes:** closed
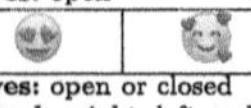				
Eyes: open or closed **Hands:** right, left; palm **Heart:** touch of the fingertips of two hands **Mouth:** closed **Lips:** smile				

4 Experimental Results

Through our Web Application we tested our solution on real-time videos acquired by a common webcam, involving 15 people varying age, gender, with different facial features or proportions of the hands. During the testing each user is asked to perform multiple sequences of complex gestures selected at random and corresponding to five Categories of Emoji: a single facial expression (Cat.1), a single hand gesture (Cat.2), two hands (Cat.3), a face accompanied by at least one hand (Cat.4) and finally an emoji representing the interaction between face and hands (Cat.5). The time available to carry out each pose is 120 s. If the required pose is correctly translated to corresponding emoji within this time the matching has success, otherwise it is considered as failed[6]. For each Category of Emoji, the time needed by the participants to achieve success were collected and the statistics are shown in Table 5. The recorded times and accuracy reflect the level of complexity of the move to be represented and the skill of the subject in simulating emoji. The user-independent recognition time is very low. Figure 7 shows for each Category of Emoji six examples of test involving different subjects, with different background, lighting conditions, camera distance and orientation. It is

[6] The video related to a complete test is available on the supplementary material.

possible to observe how the system is able to recognize complex gestures even with a very large range of variabilities in real-time. Experiments shown that our methods for the recognition of expressions and gestures are generalizable, robust, and achieve on average 99,25% of accuracy.

Table 5. Time performance and accuracy of the recognition system for Categories of Emoji.

	Min (sec.)	Max (sec.)	Mean (sec.)	Accuracy (%)
Cat.1	0,010969	0,87204	1,295784	100
Cat.2	0,020056	8,657423	0,49301	100
Cat.3	0,015962	46,93312	2,268474	98,76
Cat.4	0,03921	48,07147	3,415971	97,53
Cat.5	0,019948	21,2524	1,014331	100
Avg.				99,25

Fig. 7. Examples of complex gesture recognition in real-time for Categories of Emoji.

5 Conclusion

The proposed recognition method is based on the detection models offered by MediaPipe and consists in recognizing complex facial expressions, hand gestures, and interactions between face and hands in real-time. With the aim of recognizing complex gestures, we analysed the components of the face and hands in the various configurations assumed in the different moves. We paid attention to some peculiarities of the hand, being able to distinguish the palm from the back, the orientation and the direction of the hand. We established relation between the landmarks estimated by the detectors and defined a pool of analytical and algebraic methods. This allowed the implementation of a system for recognizing complex gestures in real time, avoiding computational or time expensive approaches. To show the validity of our approach, we have selected a large set of emojis and defined the task of assigning the correct emoji to the recognized gestures. Test results obtained from the real-time video analysis process show that complex gesture recognition system is fast, generalizable to various people and robust to a large set of variabilities. This work represents a first step toward a generalizable real-time complex gesture recognition. In the future we could define a more rigorous evaluation protocol and carry out large-scale experiments.

References

1. Clough, S., Duff, M.C.: The role of gesture in communication and cognition: implications for understanding and treating neurogenic communication disorders. Front. Hum. Neurosci. **14**, 1–22 (2020)
2. Battiato, S., Conoci, S., Leotta, R., Ortis, A., Rundo, F., Trenta, F.: Benchmarking of computer vision algorithms for driver monitoring on automotive-grade devices. In 2020 AEIT International Conference of Electrical and Electronic Technologies for Automotive (AEIT AUTOMOTIVE), pp. 1–6. IEEE (November 2020)
3. Altameem, T., Altameem, A.: Facial expression recognition using human machine interaction and multi-modal visualization analysis for healthcare applications. Image Vis. Comput. **103**, 104044 (2020)
4. Dey, S., Laha, A., Paul, A., Roy, S., Paul, S.: Facial expression recognition in video call. Int. J. Eng. Res. Technol. (IJERT) **09**(11), 159–161 (2021). NCETER - 2021
5. Munasinghe, M.I.N.P.: Facial expression recognition using facial landmarks and random forest classifier. In 2018 IEEE/ACIS 17th International Conference on Computer and Information Science (ICIS), pp. 423–427. IEEE (June 2018)
6. Rohith Raj, S., Pratiba, D., Ramakanth Kumar, P.: Facial expression recognition using facial landmarks: a novel approach. ASETS J. **5**, 24–28 (2020)
7. Wang, M., Tan, P., Zhang, X., Kang, Yu., Jin, C., Cao, J.: Facial expression recognition based on CNN. J. Phys: Conf. Ser. **1601**(5), 052027 (2020)
8. Matos, A., Filipe, V., Couto, P.: Human-computer interaction based on facial expression recognition: a case study in degenerative neuromuscular disease. In: Proceedings of the 7th International Conference on Software Development and Technologies for Enhancing Accessibility and Fighting Info-exclusion, pp. 8–12 (December 2016)
9. Borghi, G., Frigieri, E., Vezzani, R., Cucchiara, R.: Hands on the wheel: a dataset for driver hand detection and tracking. In 2018 13th IEEE International Conference on Automatic Face & Gesture Recognition, FG 2018, pp. 564–570. IEEE (May 2018)
10. Gupta, S., Jaafar, J., Ahmad, W.F.W.: Static hand gesture recognition using local Gabor filter. Procedia Eng. **41**, 827–832 (2012)
11. Chen, F.S., Fu, C.M., Huang, C.L.: Hand gesture recognition using a real-time tracking method and hidden Markov models. Image Vis. Comput. **21**(8), 745–758 (2003)

12. Koh, J.I.: Developing a hand gesture recognition system for mapping symbolic hand gestures to analogous emoji in computer-mediated communications. In: Proceedings of the 25th International Conference on Intelligent User Interfaces Companion (2020)
13. Song, N., Yang, H., Wu, P.: A gesture-to-emotional speech conversion by combining gesture recognition and facial expression recognition. In 2018 1st Asian Conference on Affective Computing and Intelligent Interaction (ACII Asia), pp. 1–6. IEEE (May 2018)
14. Liu, N., Lovell, B.C.: Hand gesture extraction by active shape models. In: Digital Image Computing: Techniques and Applications, DICTA 2005, p. 10. IEEE (December 2005)
15. Elmezain, M., Al-Hamadi, A., Pathan, S.S., Michaelis, B.: Spatio-temporal feature extraction-based hand gesture recognition for isolated American sign language and Arabic numbers. In: 2009 Proceedings of 6th International Symposium on Image and Signal Processing and Analysis, pp. 254–259. IEEE (September 2009)
16. Kartynnik, Y., Ablavatski, A., Grishchenko, I., Grundmann, M.: Real-time facial surface geometry from monocular video on mobile GPUs. arXiv preprint arXiv:1907.06724 (2019)
17. Zhang, F., et al.: MediaPipe hands: on-device real-time hand tracking. arXiv preprint arXiv:2006.10214 (2020)
18. Ortis, A., Farinella, G.M., Battiato, S.: Survey on visual sentiment analysis. IET Image Process. **14**(8), 1440–1456 (2020)
19. Soukupova, T., Cech, J.: Eye blink detection using facial landmarks. In: 21st Computer Vision Winter Workshop, Rimske Toplice, Slovenia (February 2016)

Generating High-Resolution 3D Faces Using VQ-VAE-2 with PixelSNAIL Networks

Alessio Gallucci[1,2](✉), Dmitry Znamenskiy[2], Nicola Pezzotti[1,2], and Milan Petkovic[1,2]

[1] Eindhoven University of Technology, Eindhoven, The Netherlands
a.gallucci@tue.nl, alessio.gallucci@philips.com
[2] Philips Research, Eindhoven, The Netherlands

Abstract. The realistic generation of synthetic 3D faces is an open challenge due to the complexity of the geometry and the lack of large and diverse publicly available datasets. Generative models based on convolutional neural networks (CNNs) have recently demonstrated great ability to produce novel synthetic high-resolution images indistinguishable from the original pictures by an expert human observer. However, applying them to non-grid-like data like 3D meshes presents many challenges. In our work, we overcome the challenges by first reducing the face mesh to a 2D regular image representation and then exploiting one prominent state-of-the-art generative approach. The approach uses a Vector Quantized Variational Autoencoder VQ-VAE-2 to learn a latent discrete representation of the 2D images. Then, the 3D synthesis is achieved by fitting the latent space and sampling it with an autoregressive model, PixelSNAIL. The quantitative and qualitative evaluation demonstrate that synthetic faces generated with our method are statistically closer to the real faces when compared to a classical synthesis approach based on Principal Component Analysis (PCA).

Keywords: 3D face synthesis · Generative modeling · 2D regular representation

1 Introduction

In the last two decades, the applications of virtual 3D models have risen exponentially. Two factors behind the rise are the growth in computational power and the economic benefit derived by simulating physical phenomena employing 3D models. Today, 3D face models serve many fields including animation of faces [1–3], recognition of expression [4], and face recognition [5]. For example, realistic faces generation is important for crowd generation in virtual reality environments [6]. However, the applications are often limited since face data contains privacy and sensitive information that reduces or blocks data sharing and aggregation from multiple sources. This poses a limitation to the generation of realistic 3D faces which can be used in several contexts.

To overcome such limitations, we propose to replace the original dataset with a synthetic replica and present a compelling solution to the generation of synthetic 3D faces using a machine learning approach. As a first step, we follow the seminal work of Blanz and Vetter [7] and register a common reference 3D template into every scan bringing all

P. L. Mazzeo et al. (Eds.): ICIAP 2022 Workshops, LNCS 13374, pp. 228–239, 2022.
https://doi.org/10.1007/978-3-031-13324-4_20

raw scans in full correspondence with a common parametrization. The parametrization is insufficient to generate synthetic scans since the registered template has thousands of highly correlated vertices. Generation methods should consider this correlation by finding a low-dimensional decorrelated surface representation. Thus, Blanz and Vetter [7] reduced the vertex coordinates to a small number of decorrelated scores with a data-driven approach using Principal Component Analysis (PCA). Sampling new scans with PCA is then straightforward; however, interpolating in the reduced PCA subspace will not always result in natural human shapes due to the linear nature of the method. To overcome the drawbacks of PCA and similar linear methods, deep generative models based on Convolutional Neural Networks (CNNs) are employed to capture more complex non-linear interactions in the data. The current state of the art advances in the field of geometric deep learning [8] leverage the power of CNNs by adapting them to work on meshes [9, 10]. Graph convolutions, however, restrict the resolution, and therefore, the accuracy and amount of surface details of the 3D template.

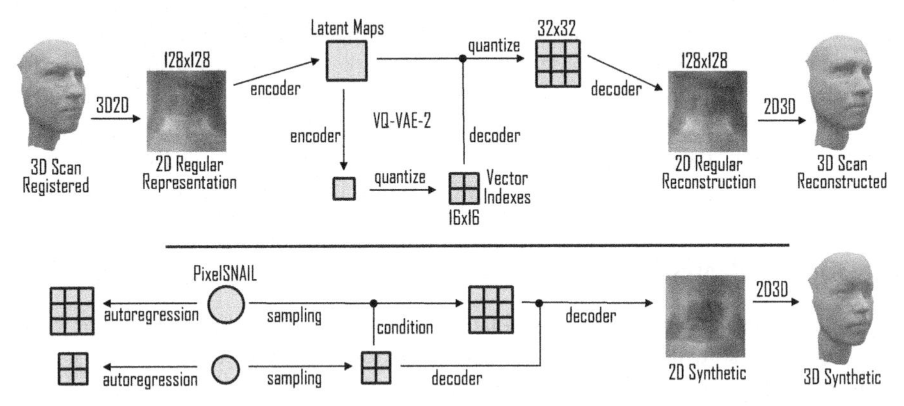

Fig. 1. Method's flowchart. The registered 3D scans are first converted into a regular 2D image to feed the VQ-VAE-2 autoencoder. The PixelSNAIL later learns a prior over the latent space, sample novel synthetics codes, decode them into the new geometric images, and, subsequently, to 3D scans.

By choosing the 3D mesh template with vertices connected as a 2D grid, our approach processes 3D meshes as 2D images. This makes the 3D2D mapping straightforward, enables 2D image synthesis methods, and avoids the challenges of graph convolutions [11]. Figure 1 shows schematically the solution we adopted, which is defined by two phases: in the first phase, see the top part of the figure, the method learns a discrete latent image representation given by a two layer quantized variational autoencoder VQ-VAE-2 [12]; then, in a later stage shown in the bottom of the figure, a powerful autoregressive network PixelCNN [13, 14] with self-attention [15], called as PixelSNAIL [16] is used to fit the latent space and sample from it. By employing such a novel pipeline, we empirically found that our method gives more natural 3D shapes compared to the PCA-based one. Due to the high variability of plausible 3D human shapes, the subjective evaluation is not enough to properly assess the quality and diversity of the generated face. A major challenge is defining a proper surrogate measure that evaluate how "human" is a 3D scan. In the literature, two metrics are commonly used to evaluate 3D scans: specificity [17] measures how close a scan is to the (training) set and diversity [18] measures the difference between a pairs of scans. In our work, instead of reporting a single number generated by such metrics, we compare their empirical distributions of the synthesized scans versus a test set of real faces. This allows us to measure the realism and diversity of the generated faces in terms of a previously unseen test dataset. The remaining paper content is structured as follows: the next section gives an overview of the prior art, Sect. 3 describes the approach for the synthetic generation, in the fourth section, we present the experiments and the relative quantitative evaluation. Lastly, we conclude and give acknowledgments.

2 Related Works

In the following, we present various approaches to synthetic head generation. Many works of research still rely on linear models [19] or multilinear models [20] due to their simplicity and due to the expansion of 3D Morphable Models [21]. Tran *et al.* [22] proposed a robust CNN-based approach to regress the PCA scores from pictures for face recognition and discrimination. In another work, the multilinear models are used to transfer facial expression and have the ability to animate faces [20]. While being simple and easy to train, they do not consider the input geometry. Additionally, a review of current methods regressing and sampling PCA scores is beyond the scope of the paper.

2.1 3D to 2D Representations

Many 2D representation methods originate from the solution of the rendering problem which relies on so called UV maps to map 2D texture image on a 3D object. The UV maps, by definition, provide a bijective mapping from the 3D mesh triangles to their images on the texture image. In 2002, Gu *et al.* [23] showed how to optimally cut a surface and sample it over a regular 2D square grid generating the so-called geometry image. The problem might be more straightforward for a face geometry since the UV maps can be created by warping of the 3D templates with a regular grid. Booth *et al.* [24] presented a list of possible optimal implementations. Figure 2 shows the selected regular template geometry for this paper on the left, the geometry image derived from the grid and a test texture on the middle, and the texture rendered on the template on the right. As shown in the picture, the main drawbacks of such methods are the artifacts around the cuts or borders. However, in this example, such artifacts do not conflict with our requirements for an accurate face model and not a full-head one.

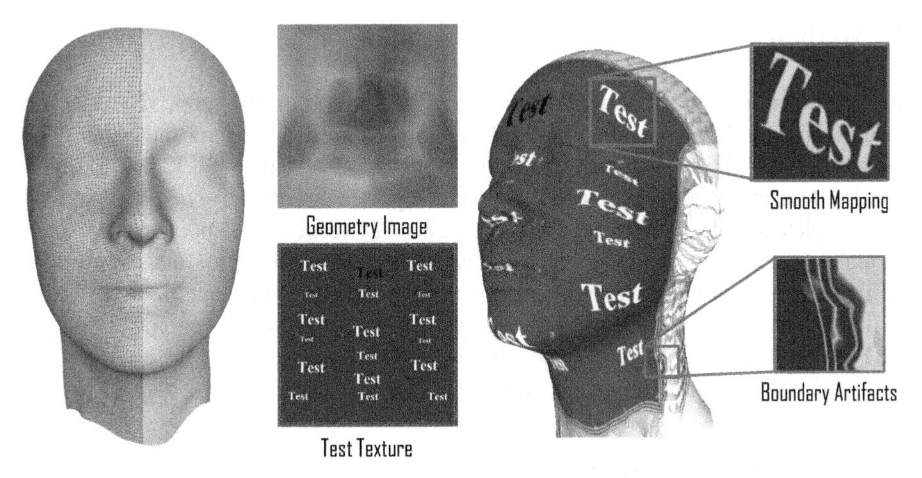

Fig. 2. Template. The face template on the left has a regular (triangular) mesh grid – for visualization purposes the template is half rendered surfaces and half mesh. On the middle top column, the so-called 128×128 *geometry image* for the facial template, which is naturally derived from the regular structures of the template mesh, and a test texture to visualize the smoothness of the UV map. On the right the test visualization with example of boundary artifacts of the UV map.

2.2 3D Face Generation with GANs

The most common generative models employ Generative Adversarial Networks (GANs) [25]. Abrevaya *et al.* [18], investigated the use of Wasserstein GAN [26] to generate novel 3D faces with the ability to control and modify their expression. However, in our work, we directly map the input surface into a geometry image, avoiding the need for a fully connected generator which might not efficiently handle the complexity of the shapes. Slossber *et al.* [27] and the extension work in Shamai *et al.* [28], similarly to our work, converted the 3D into a 2D regular representation through non-rigid registration techniques. In Moschoglou *et al.* [29], the template was mapped using a cylindrical unwrapping as introduced by Booth and Zafeiriou [24]. While using similar concepts, our work does not use adversarial training, a key difference that makes our method easier to train and less affected by the so-called *mode collapse* which affects GAN architectures.

2.3 3D Face Generation with Autoencoders

Apart from generative GANs models, recently, many works have overcome the linear modeling limitations by using VAEs [30]. For example, Bagautdinov *et al.* [31] modeled the face using a multiscale approach for different frequencies of details. Fernandez Abrevaya *et al.* [32] exploited the power of CNN-based encoder by coupling it with a multilinear decoder. In Li, K. *et al.* [33], a multi-column graph convolutional networks is designed to synthesize 3D surfaces. They first applied a spectral decomposition of the meshes and then trained multiple columns of graph convolutional networks. While these methods are similar to our approach, they also differ as no one uses the quantized autoencoder with an autoregressive network. Moreover, they do not convert the data into 2D geometry images but directly feed the registered 3D scans.

3 Method

The definition and registration of the face template are out of the scope of the current paper. Conceptually, we have followed the method explained in Blanz and Vetter [7, 6] and have morphed all scans by means of non-rigid registration methods [34, 35]. A more detailed description of our parametric models is reported in Gallucci *et al.* [36]. Since the face template already has a grid structure of 128×128 vertices, we apply a vertex-based normalization to map the range of input values into interval [0, 1], and therefore, to facilitate the follow up processing with the neural networks. The mapping to [0, 1] also facilitates the normalized *xyz* facial data visualization of the as *rgb* (geometry-) images, see Fig. 2 for an example. The range parameters for each grid vertex were retained for denormalizing the synthetic images into 3D shapes.

3.1 VQ-VAE-2 with PixelSNAIL

The VQ-VAE model is introduced in [37] and it builds on top of the Variational AutoEncoder (VAE) [30, 38] by generalizing ideas from classical image compression methods like jpeg. Given a dataset of observations $\{\mathbf{x}^{(1)}, \mathbf{x}^{(2)}, \ldots, \mathbf{x}^{(N)}\}$, the goal of a VAE is to learn, without supervision, a lower dimensional representation in terms of latent variables $\mathbf{z}$. It is composed by an encoder E, which map the input image into latent variables, and a decoder, which reconstruct the image from the compressed representation. In other words, the decoder network models the joint distribution $p(\mathbf{x}|\mathbf{z})p(\mathbf{z})$ while the encoder models the posterior distribution $q(\mathbf{z}|\mathbf{x})$.

In the VQ-VAE framework, the prior distribution is based on K prototype latent vectors $\{\boldsymbol{e}^{(1)}, \boldsymbol{e}^{(2)}, \ldots, \boldsymbol{e}^{(K)}\}$ of dimension D which quantize the latent maps $E(\mathbf{x})$, generated by the encoder. There are exactly K different latent vectors to choose from, so each pixel on the latent maps is represented with the nearest quantizing vector. In Razavi *et al.* [12] the two layer autoencoder VQ-VAE-2 is presented and trained on ImageNet [39]. The autoencoder is the upgrade of the VQ-VAE to include multiple hierarchical layers which provide different quantized codebooks at different hierarchies. The decoder then reconstructs the image using the latent maps conditioning the higher levels, which have a smaller resolution, to the bottom ones. In the original setup the input 24-bit image with resolution 256×256 was reduced to 64×64 bottom map and 32×32 top map with $K = 512 = 2^9$ different quantizing vectors of dimension $D = 64$.

For new data generation we apply the autoregressive model, PixelCNN [13, 14] with self-attention [15], called PixelSNAIL [16]. In this setup the autoregressive model can efficiently model the prior distribution of the latent codes, creating photo-realistic synthetic images. The idea behind the PixelCNN model is to learn the conditional distribution of the given sequence of random variables. When applied to the latent space, the latent codes of the whole image are sorted from top left to bottom right to predict the next code value, which is a discrete probability distribution over the K codes, in an autoregressive fashion. In our example, the autoregressive model learns the joint distributions of the latent codes on the top layer and then the distribution of the bottom codes conditioned on the top codes.

3.2 Metrics for Quantitative Evaluation

Our goal is to provide always realistic synthetic samples, and to achieve it, we visually inspected the generated scans and selected suitable metrics to prove this statement. The main idea is to prove that synthetic scans are statistically indistinguishable from a test set of original scans excluded from training. Before computing the metrics, the scans need to have identical parametrization corresponding to the 3D template. The identical parametrization enables a simple distance metric between a pair of scans, defined as the Root Mean Squared Distance between the corresponding pairs of vertices, after the rigid alignment of one scan to another [40].

We have employed two derivative metrics used in the literature to evaluate synthetic scans. The first metric is called diversity and has been introduced by Abrevaya et al. [18] with the aim to produce a single number measuring the heterogeneity of a set of scans. The diversity is defined as a distance between a random pair of synthetic scans. In our work, we compare the empirical distribution of the diversity of 250 generated scans with the empirical distribution of diversity in 250 original scans from the test dataset. The second selected metric is called specificity and is defined in Davies et al. [17] for a scan as the minimal distance to the scans in the training dataset. Similar to the diversity distribution, we evaluate the empirical distribution of the specificity over 250 synthetic scans and compare it with the empirical distribution of specificity in 250 original scans from the test dataset.

4 Experiments

Within this work, we have considered two datasets of registered scans already available at Philips Research: the *SizeChina* dataset of 3D head scans [41], and the *CAESAR* of full body scans [42, 43] where only the head was extracted. The above data gave us more than 5000 registered 3D face templates. We augment the face dataset by performing a symmetric reflection over the y-axis. While in this application we assume that asymmetries are normally distributed on the left and on the right of face we do not know if this is true. Nonetheless, we still believe this augmentation does not hamper the results of the approach. The dataset was split stratified according to participant id into train 90%, validation 5%, and test set 5%. We tested and computed the metrics only on the test set without considering the augmentations. For the sake of experimental reproducibility, we did not perform any other augmentation neither in training nor in test time. However, we believe further realistic augmentations would impact and consolidate the results. Nevertheless, we also notice that the current set of scans is enough to achieve the desired outcome of statistical indistinguishability from the test set.

In our experiments we focus on a two layers VQ-VAE hierarchy with input grid resolution of 128×128 and relative latent maps of dimension 32×32 and 16×16. We follow the approach described in Gallucci *et al.* [44] to find the best combination of $K = [64, 128, 256512]$, $D = [2, 4, 8, 16, 32, 64]$ and found out that, according to reconstruction error, $K = 512$ is always better than smaller values. Vice versa for big enough K we notice that smaller dimension of D provides the best outcomes. Hence, we used $D = 2$ for our final VQ-VAE-2 model. We also reduced the batch size to 32 compared to the original implementation for both the autoencoder and the autoregressive model the constraints of our computational resources. The reconstruction root mean squared weighted error per participant is 0.29 mm compared to 0.97 mm for PCA and is mostly accumulated in the areas with higher curvature or with low weights such as eyes, mouth, nostril, and neck, as shown in Fig. 3. We use vertex weights to improve the results in three different situations: to facilitate the registration of the raw scans, to maximize the PCA encoding energy in the face area of interest, and within the quantitative metrics, reported in the next section, to focus the attention of the metrics on more important facial areas of the model. Figure 3 shows the errors maps per vertex across the test set population: as expected the neural network reconstruction outperforms the smaller and linear PCA model – where we encoded the registered scans in "only" 200 principal components following Gallucci *et al.* [36]. To sample a new scan with PCA, we decode a 3D scan from the random PCA scores, where each score was sampled independently from the respective marginal empirical Cumulative Density Function (eCDF) computed over all PCA encoded scans. The above procedure guarantees that the synthetic PCA scans inherit the eCDF from the original data. Concerning the PixelSNAIL autoregressive model, we used the original configuration for ImageNet apart from the batch size, 32 in our example, and total number of epochs, 420 for both top and bottom hierarchy. The autoregressive models' validation accuracies in predicting the latent codes after 420 epochs are 0.87 for the top space and 0.91 for the bottom one. All the models were trained on PyTorch [45] with the same hyperparameters as in the original implementation (excluding the one explicitly mentioned above).

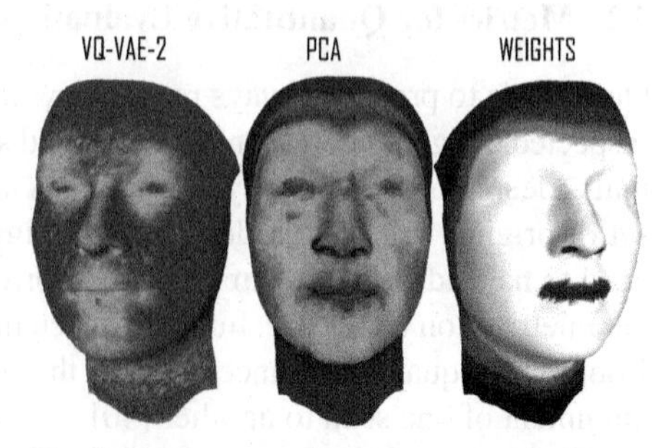

Fig. 3. Test set reconstruction mean absolute weighted error on the left using VQ-VAE-2 and on the middle using PCA. The meshes jet color-code ranges from blue 0.0 mm error map to red 1.0mm. On the right the used vertex weights are shown with color-code that ranges from 0.0 black to 1.0 white. (Color figure online)

Comparing the scans by visual inspection is not a trivial task and often not an objective metric. However, we believe that it is possible to spot some differences in the shape distribution between the different sets by looking at the overall scan's appearance. We present some scans randomly selected in Fig. 4: the top right shows registered scans with the relative PCA encoded version on the top left; the bottom scans are synthetic and generated with our approach on the left and with PCA on the right. The PCA ones present more variability, or, in other words, more shapes differences compared to the other sets. The synthetic scans generated with our approach, from a visual inspection, present similar shape variability to the original scans compared to the PCA synthetic. Nevertheless, this is not enough since our approach may simply replicate or clone the original training data. We test these hypotheses in the following quantitative analysis proving that the synthetic scans are novel and different from the original training ones.

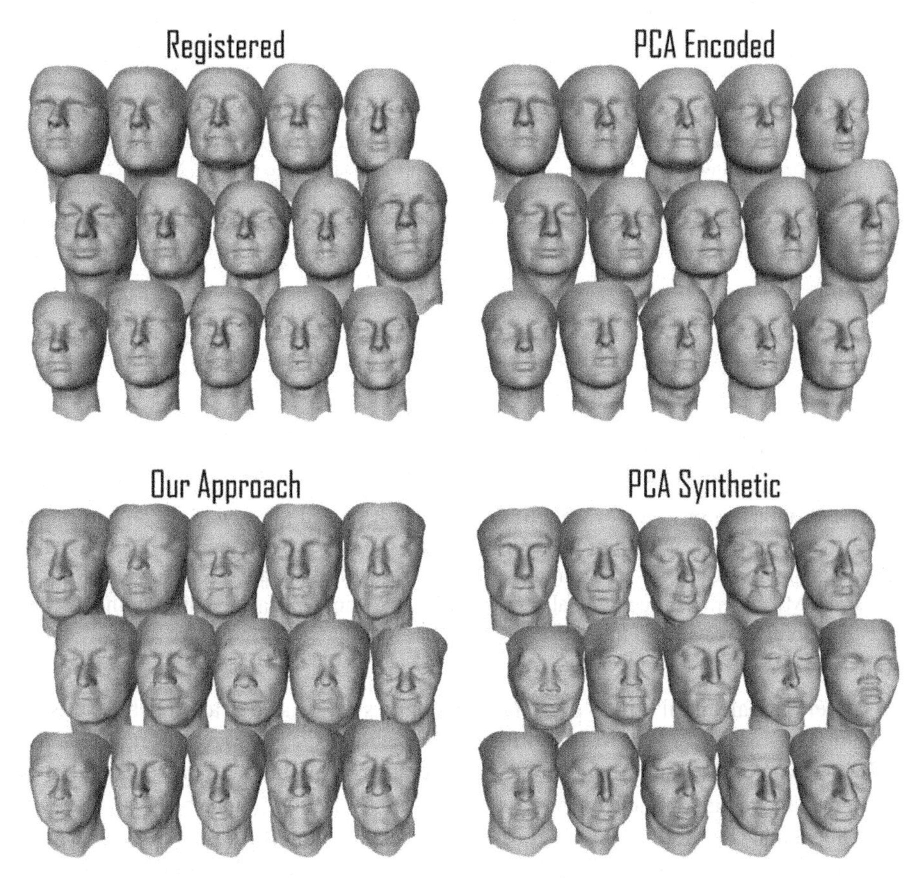

Fig. 4. Example of facial scans (without selection). A batch of registered scans (top left), same scans encoded (top right), synthetic scans generated with our approach (bottom left), and PCA synthetic scans (bottom right).

4.1 Quantitative Evaluation

We have analyzed 2D representations for registered raw versus PCA-encoded vertices and computed empirical distributions for specificity and diversity metrics. Given V the vertices of a synthetic scan its specificity S is defined has.

$$S(V) = \min_{t \in T} \left[\frac{\sum_{i=1}^{N} w_i \|v_i - v_i^t\|_2^2}{\sum_{i=1}^{N} w_i} \right]^{1/2}$$

where t is the index of the training set, $N = 128 \times 128 = 16384$ the total number of vertices $v_i \in V$, $w_i \in W$ are the weights for the ith vertex as shown on the right of Fig. 3. The diversity D of a pair of scans with vertices $v_i^1 \in V^1, v_i^2 \in V^2$ is defined as

$$D\left(V^1, V^2\right) = \left[\frac{\sum_{i=1}^{N} w_i \|v_i^1 - v_i^2\|_2^2}{\sum_{i=1}^{N} w_i}\right]^{\frac{1}{2}}.$$

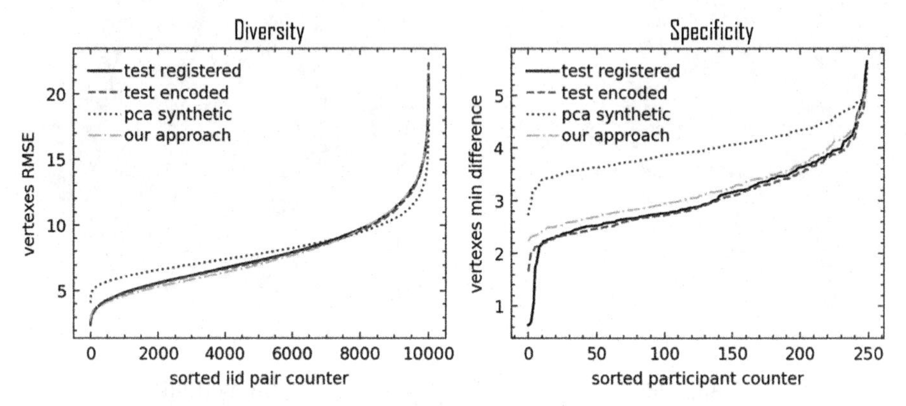

Fig. 5. EQuantitative metrics. On the left, the diversity is plotted for each i.i.d. pair and shows that the PCA distribution is "flatter" as excepted by the linear method. On the right, the specificity (the minimum distance versus the training set is kept) shows that our approach is much closer to the test set. Moreover, the specificity also proves that we do not replicate the input training scans since the minimum distance is markedly above 0 mm – with our approach above 2 mm.

The empirical distributions presented in Fig. 5 show that our approach results in synthetic faces which are statistically close to the original scans in the test set, unlike the PCA-based method which shows a flattened diversity distribution and higher specificity. The figure shows that our approach closely follows the distribution of the test-registered and -encoded scans, both in terms of diversity and specificity distributions. Moreover, since the specificity is computed against the training scans, we demonstrate that the faces generated by our approach are diverse from the training set since they do not collapse to zero and have a minimum distance above 2 mm. The higher specificity of the PCA-synthetic scans also confirms the qualitative evaluation, see example scans in Fig. 4, that PCA-based faces have more extreme characteristics.

5 Discussion and Conclusion

We presented a novel approach that can generate high-resolution synthetic 3D scans that combine traditional 3D parameterization approaches with the recent VQ-VAE-2 and PixelSNAIL deep learning based generative models. Our approach does not require the parametrization of the 3D face model and can be directly applied to registered templates, hence, allowing for a richer generation domain since synthetic scans can be outside the PCA linear sub-space. However, the major contribution of our work is that our method strictly outperforms the linear PCA classical approach and generates realistic high-resolution scans. We consider this only a first step in proving the validity of this approach – future work will perform a benchmark versus other state-of-the-art generative models. One main challenge is the lack of a clear quantitative metric to judge whether a scan belongs to the "real" class since the proposed diversity and specificity metrics may not be enough to capture all the relevant shape information. Additionally, while we believe the two selected metrics are suited for the current evaluation of realistic human faces, different metrics can be developed in the future. A natural extension of our approach that can partially solve the metrics problem could combine the 3D shape synthesis with the photo-realistic texture synthesis adding the *rgb* to the *xyz* channels within the 2D representation.

Acknowledgments. We thank Philips Research for providing access to the datasets of facial scans and software resources to manage the high-resolution parametric models.

References

1. Liu, S.-L., Liu, Y., Dong, L.-F., Tong, X.: RAS: a data-driven rigidity-aware skinning model for 3D facial animation. In: Computer Graphics Forum, pp. 581–594 (2020)
2. Carrigan, E., Zell, E., Guiard, C., McDonnell, R.: Expression packing: as-few-as-possible training expressions for blendshape transfer. In: Computer Graphics Forum, pp. 219–233 (2020)
3. Li, T., Bolkart, T., Black, M.J., Li, H., Romero, J.: Learning a model of facial shape and expression from 4D scans. ACM Trans. Graph. **36**, 191–194 (2017)
4. Valev, H., Gallucci, A., Leufkens, T., Westerink, J., Sas, C.: Applying delaunay triangulation augmentation for deep learning facial expression generation and recognition. In: Del Bimbo, A., et al. (eds.) ICPR 2021. LNCS, vol. 12663, pp. 730–740. Springer, Cham (2021). https://doi.org/10.1007/978-3-030-68796-0_53
5. Taigman, Y., Yang, M., Ranzato, M., Wolf, L.: DeepFace: closing the gap to human-level performance in face verification. In: Proceedings of the IEEE Conference on Computer Vision and Pattern Recognition, pp. 1701–1708 (2014)
6. Varol, G., et al.: Learning from synthetic humans. In: Proceedings of the IEEE Conference on Computer Vision and Pattern Recognition, pp. 109–117 (2017)
7. Blanz, V., Vetter, T.: A morphable model for the synthesis of 3D faces. In: Proceedings 26th Annual Conference on Computer Graphics and Interactive Techniques, pp. 187–194 (1999)
8. Bronstein, M.M., Bruna, J., LeCun, Y., Szlam, A., Vandergheynst, P.: Geometric deep learning: going beyond euclidean data. IEEE Signal Process. Mag. **34**, 18–42 (2017)
9. Ranjan, A., Bolkart, T., Sanyal, S., Black, M.J.: Generating 3D faces using convolutional mesh autoencoders. In: Ferrari, V., Hebert, M., Sminchisescu, C., Weiss, Y. (eds.) ECCV 2018. LNCS, vol. 11207, pp. 725–741. Springer, Cham (2018). https://doi.org/10.1007/978-3-030-01219-9_43
10. De Haan, P., Weiler, M., Cohen, T., Welling, M.: Gauge equivariant mesh CNNs: anisotropic convolutions on geometric graphs. arXiv Prepr. arXiv2003.05425 (2020)
11. Zhang, S., Tong, H., Xu, J., Maciejewski, R.: Graph convolutional networks: a comprehensive review. Comput. Soc. Netw. **6**(1), 1–23 (2019). https://doi.org/10.1186/s40649-019-0069-y
12. Razavi, A., van den Oord, A., Vinyals, O.: Generating diverse high-fidelity images with VQ-VAE-2. In: Advances in Neural Information Processing Systems, pp. 14837–14847 (2019)

13. Van Oord, A., Kalchbrenner, N., Kavukcuoglu, K.: Pixel recurrent neural networks. In: International Conference on Machine Learning, pp. 1747–1756 (2016)
14. den Oord, A., Kalchbrenner, N., Espeholt, L., Vinyals, O., Graves, A., et al.: Conditional image generation with pixelcnn decoders. In: Advances in Neural Information Processing Systems, pp. 4790–4798 (2016)
15. Vaswani, A., e al.: Attention is all you need. In: Advances in Neural Information Processing Systems, pp. 5998–6008 (2017)
16. Chen, X., Mishra, N., Rohaninejad, M., Abbeel, P.: PixelSNAIL: an improved autoregressive generative model. In: 35th International Conference on Machine Learning ICML 2018, vol. 2, pp. 1364–1372 (2018)
17. Davies, R., Twining, C., Taylor, C.: Statistical Models of Shape: Optimisation and Evaluation. Springer, London (2008). https://doi.org/10.1007/978-1-84800-138-1
18. Abrevaya, V.F., Boukhayma, A., Wuhrer, S., Boyer, E.: A decoupled 3D facial shape model by adversarial training. In: Proceedings of the IEEE/CVF International Conference on Computer Vision, pp. 9419–9428 (2019)
19. Thies, J., Zollhofer, M., Stamminger, M., Theobalt, C., Nießner, M.: Face2face: real-time face capture and reenactment of RGB videos. In: Proceedings of the IEEE Conference on Computer Vision and Pattern Recognition, pp. 2387–2395 (2016)
20. Vlasic, D., Brand, M., Pfister, H., Popovic, J.: Face transfer with multilinear models. In: ACM SIGGRAPH 2006 Courses, pp. 24–es (2006)
21. Booth, J., Roussos, A., Zafeiriou, S., Ponniah, A., Dunaway, D.: A 3D morphable model learnt from 10,000 faces. In: Proceedings of the IEEE Conference on Computer Vision and Pattern Recognition, pp. 5543–5552 (2016)
22. Tuan Tran, A., Hassner, T., Masi, I., Medioni, G.: Regressing robust and discriminative 3D morphable models with a very deep neural network. In: Proceedings of the IEEE Conference on Computer Vision and Pattern Recognition, pp. 5163–5172 (2017)
23. Gu, X., Gortler, S.J., Hoppe, H.: Geometry images. In: Proceedings of the 29th Annual Conference on Computer Graphics and Interactive Techniques, pp. 355–361 (2002)
24. Booth, J., Zafeiriou, S.: Optimal UV spaces for facial morphable model construction. In: 2014 IEEE International Conference on Image Processing (ICIP), pp. 4672–4676 (2014)
25. Goodfellow, I., et al.: Generative adversarial nets. In: Advances in Neural Information Processing Systems, pp. 2672–2680 (2014)
26. Arjovsky, M., Chintala, S., Bottou, L.: Wasserstein generative adversarial networks. In: International Conference on Machine Learning, pp. 214–223 (2017)
27. Slossberg, R., Shamai, G., Kimmel, R.: High quality facial surface and texture synthesis via generative adversarial networks. In: Leal-Taixé, L., Roth, S. (eds.) ECCV 2018. LNCS, vol. 11131, pp. 498–513. Springer, Cham (2019). https://doi.org/10.1007/978-3-030-11015-4_36
28. Shamai, G., Slossberg, R., Kimmel, R.: Synthesizing facial photometries and corresponding geometries using generative adversarial networks. ACM Trans. Multimedia Comput. Commun. Appl. **15**, 1–24 (2019)
29. Moschoglou, S., Ploumpis, S., Nicolaou, M.A., Papaioannou, A., Zafeiriou, S.: 3DFaceGAN: adversarial nets for 3D face representation, generation, and translation. Int. J. Comput. Vis. **128**, 2534–2551 (2020)
30. Kingma, D.P., Welling, M.: Auto-encoding variational Bayes. In: 2nd International Conference on Learning Representations ICLR 2014 - Conference Track Proceedings, pp. 1–14 (2014)
31. Bagautdinov, T., Wu, C., Saragih, J., Fua, P., Sheikh, Y.: Modeling facial geometry using compositional VAEs. In: Proceedings of the IEEE Conference on Computer Vision and Pattern Recognition, pp. 3877–3886 (2018)

32. Abrevaya, V.F., Wuhrer, S., Boyer, E.: Multilinear autoencoder for 3D face model learning. In: 2018 IEEE Winter Conference on Applications of Computer Vision (WACV), pp. 1–9 (2018)
33. Li, K., Liu, J., Lai, Y.-K., Yang, J.: Generating 3D faces using multi-column graph convolutional networks. In: Computer Graphics Forum, pp. 215–224 (2019)
34. Tam, G.K.L.L., et al.: Registration of 3D point clouds and meshes: a survey from rigid to Nonrigid. IEEE Trans. Vis. Comput. Graph. **19**, 1199–1217 (2013)
35. van Kaick, O., Zhang, H., Hamarneh, G., Cohen-Or, D.: A survey on shape correspondence. In: Eurographics Symposium on Geometry Processing (2011)
36. Gallucci, A., Znamenskiy, D., Petkovic, M.: Prediction of 3D body parts from face shape and anthropometric measurements. J. Image Graph. **8**, 67–77 (2020)
37. van den Oord, A., Vinyals, O., et al.: Neural discrete representation learning. In: Advances in Neural Information Processing Systems, pp. 6306–6315 (2017)
38. Kingma, D.P., Welling, M.: An introduction to variational autoencoders. arXiv Prepr. arXiv1906.02691 (2019)
39. Russakovsky, O., et al.: ImageNet large scale visual recognition challenge. Int. J. Comput. Vision **115**(3), 211–252 (2015). https://doi.org/10.1007/s11263-015-0816-y
40. Kabsch, W.: A solution for the best rotation to relate two sets of vectors. Acta Crystallogr. Sect. A Cryst. Phys. Diffr. Theor. Gen. Crystallogr. **32**, 922–923 (1976)
41. Ball, R., Molenbroek, J.F.M.: Measuring Chinese heads and faces. In: Proceedings of the 9th International Congress of Physiological Anthropology, Human Diversity Design for Life, pp. 150–155 (2008)
42. Robinette, K.M., Daanen, H., Paquet, E.: The CAESAR project: a 3-D surface anthropometry survey. In: Second International Conference on 3-D Digital Imaging and Modeling (Cat. No.PR00062), pp. 380–386 (1999)
43. Robinette, K.M., Daanen, H.: Lessons learned from CAESAR: a 3-D anthropometric survey, 5 (2003)
44. Gallucci, A., Pezzotti, N., Znamenskiy, D., Petkovic, M.: A latent space exploration for microscopic skin lesion augmentations with VQ-VAE-2 and PixelSNAIL. In: SPIE Medical Imaging Proceedings (2021)
45. Paszke, A., et al.: Automatic differentiation in PyTorch (2017)

Artificial Intelligence for Digital Humanities - AI4DH

The Morra Game: Developing an Automatic Gesture Recognition System to Interface Human and Artificial Players

Franco Delogu[1](✉), Francesco De Bartolomeo[2], Sergio Solinas[2], Carla Meloni[3], Beniamina Mercante[2], Paolo Enrico[2], Rachele Fanari[3], and Antonello Zizi[3]

[1] Lawrence Technological University, Southfield, MI 48075, USA
fdelogu@ltu.edu

[2] University of Sassari, Piazza Università, 21, 07100 Sassari, SS, Italy

[3] University of Cagliari, Via Università, 40, 09124 Cagliari, CA, Italy

Abstract. Morra is an ancient hand game still played nowadays. In its more popular variant, two players simultaneously extend one hand in front of the opponent to show a number of fingers, while uttering a number from 2 to 10. The player who successfully guesses the total number of fingers scores a point. Morra can be defined as a serious game, as it has the potential to positively affect cognition and to improve cognitive and perceptual skills. Moreover, with its involvement of many perceptual, cognitive and motor skills, morra is ideal to test several cognitive processes. This paper describes aspects of Gavina 2121, an artificial Morra player that successfully predicts the numbers of human opponents taking advantage of the limited ability of humans in random sequence generation. This study focuses on automatic gesture recognition. We developed and tested a system to allow Gavina 2121 to detect and count in real time the number of extended fingers in a human hand. The system is based on the open source MediaPipe Hand framework developed by Google. Our tests indicate that the system is able to accurately recognize the number of fingers extended by a human hand in real time, both in prone and supine positions. The system is still imprecise in semi-naturalistic conditions of an actual morra game, where the fingers of two hands need to be computed simultaneously. Our test, still in its pilot phase, shows promising results towards a flexible implementation of an artificial morra player that can sensibly expand the educational, rehabilitation and research applications of Morra.

1 Introduction

Gaming is an ubiquitous activity that has characterized human behavior in every part of the world at any historical period. In the last 20 years, there was a growing interest in the study of beneficial influences of videogames on cognition and emotion [1]. Several games, often defined as "serious games" are currently developed with the specific purpose of positively affecting behavior and cognition [2]. While the most of interest is dedicated to cognitive development and rehabilitation contexts, there is also an emerging interest in the effect of gaming in the general adult population. In a systematic review and meta-analysis, Pallavicini and colleagues [3] found support to the hypothesis of a beneficial

P. L. Mazzeo et al. (Eds.): ICIAP 2022 Workshops, LNCS 13374, pp. 243–253, 2022.
https://doi.org/10.1007/978-3-031-13324-4_21

role of video gaming on healthy adult in multiple domains of cognition, processing speed, response time, memory, task-switching, mental spatial rotation and emotion.

There is evidence about positive influences on cognition of non-digital gaming in general. For example, many studies on board games like Chess, Shogi and Go indicate that they are effective in cognitive rehabilitation [4] and in improving cognitive and perceptual skills [5]. It is important to note that the distinction between videogames and in-person games is often blurred and, by consequence, similar effects can be obtained in digital and non-digital versions of the same games. For example, most of the popular board and card games nowadays have digital versions available to the market. Furthermore, there is evidence that serious games have compatible effects in the non-digital and in digital formats (see for example [6]). Recently, we have seen the emergence of exergames, which combine the involvement of sensorimotor skills typical of physical exercise with the power of digital settings [7]. Finally, hand games have been used in many "serious" settings, like computational thinking education [8] and early childhood cognitive development [9, 10].

A subset of non-digital games is represented by hand games. Hand games have been greatly popular in history, perhaps because of their simplicity. In fact, hand games do not require any particular setting, devices or apparatuses to be played. The most studied hand game is Roshambo (Rock-Paper-Scissors, RPS), which is a non-cooperative strategic game that has been used as a non-computerized exergame in cognitive declining elderly adults [11], to investigate cognitive strategies in schizophrenia [12], to understand strategic interactions between healthy adults [13]. A game similar to RPS, yet more complex, is Morra, an ancient hand game still played nowadays. In its more popular variant, two players simultaneously extend one arm in front of the opponent to show a number of fingers, while uttering a number from 2 to 10. The player who successfully guesses the total number of fingers shown by the two hands scores a point. From a cognitive point of view, Morra is a complex activity which involves, and possibly integrates, many perceptual, cognitive and motor processes. During Morra, while listening and seeing the numbers presented by the opponent, a player needs to select two numbers, one to be shown with the fingers and one to be spoken. In order to be successful, a player should select those numbers in a very careful way: the to-be-shown number should be difficult to predict and the to-be-said number should be selected in order to target the number the opponent will show. This requires memory of previously shown and said numbers by the player and by his/her opponent. Moreover, this involves executive functions [14] to inhibit the numerals uttered which must always be greater than the numbers of shown fingers, and a dual-task attentional performance, to simultaneously detect and process visual (fingers) and verbal (spoken numbers) information. The task also requires an integration of visual information, with an automatic recall of the arithmetical fact [15] and the verbal information, which concerns the numbers said by both players which are compared with the arithmetical fact to decide who makes the point. All these operations, summarized in the diagram in Fig. 1, are conducted in a very small amount of time (more than one round per second).

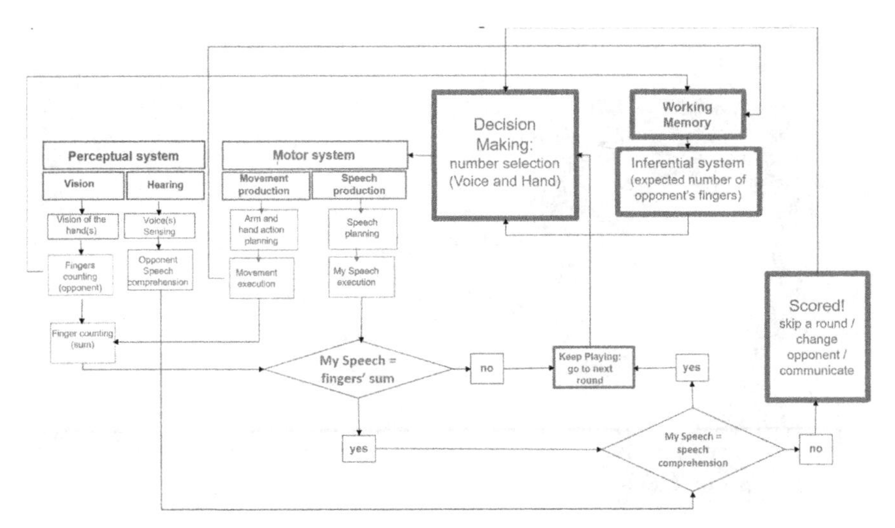

Fig. 1. A speculative model of the processes involved in each round of Morra playing

Morra analysis can provide a new approach to study the interaction between several cognitive functions in an ecological setting [16]. Considering its complexity, Morra is also a good candidate to be included in the category of serious games. The development of an artificial agent able to play Morra at different levels of expertise against human opponents is an important tool that can serve several goals in education, rehabilitation, cognitive training in healthy adults and basic research. In this paper we will focus on the development of an artificial agent able to play the Morra game against humans.

Previous studies have been published on the development of robots able to play hand games. In particular, several robots have been developed to play RPS against humans [17–21].

Morra and RPS are similar in many aspects as they are both zero-sum competitive games requiring the integration of sensorimotor skills, executive functions, attention and decision making. However, the two games also differ in many respects. Morra has a more complicated structure, having a much larger set of combinations to remember, involving the integration of two sensory modalities to receive the inputs, and requiring more advanced defensive and attack strategies to master the game [16].

In recent years, Zizi developed the Morra system Gavin 1.0, an artificial agent able to autonomously and successfully play Morra against a human opponent [22]. Recently, Gavina 2121, a new implementation based on the experience acquired from Gavin 1.0 has been developed. Gavina 2121 (see Fig. 2) has the ability to play Morra against human players and also allows the analysis of the behavior of human players in terms of numeric sequences produced during games.

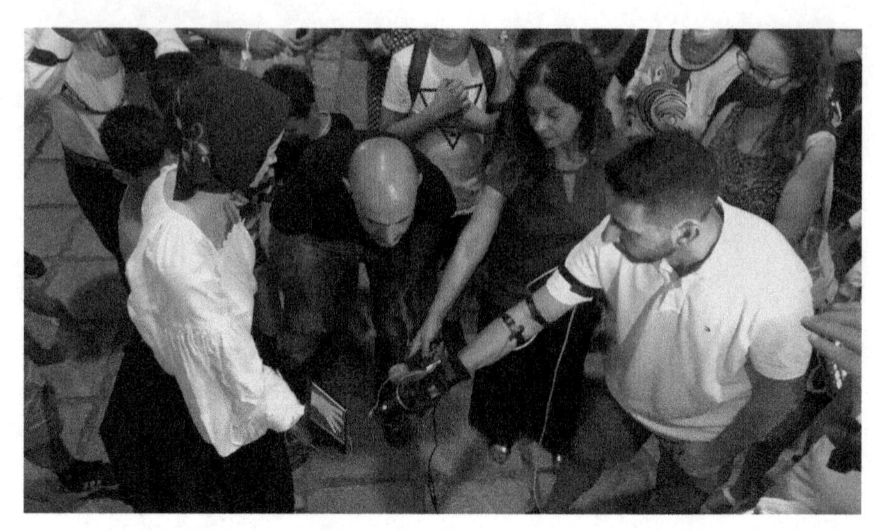

Fig. 2. Gavina 2121 plays against a human opponent in a public square in Bitti (Sardinia)

In each round, Gavina moves its robotic right arm synchronously with the opponent's movement, and shows a certain number of fingers. Simultaneously, Gavina, like its human counterpart, tries to guess the sum of all displayed fingers, the ones shown by the opponent and by Gavina itself. Gavina's main objective is to defeat its opponent. For this purpose, the system tends to show random sequences of numbers while trying to detect repetition of non-random sequences of numbers shown by its opponents. Gavina achieves this goal by using a machine learning (ML) system based on a fully automated Bayesian network, which converges to progressively more accurate predictions. The fact that Gavina systematically outperforms human competitors supports the theory that humans are bad randomizers of sequences [23].

The first version, Gavin 1.0, worked as a black box and did not allow the extractions of the strategies used to win. As Gavina2121 has the additional scope of helping the analysis of numeric sequences produced by humans, we developed a hybrid system, which is able to provide information on how the game estimates are produced, a characteristic of the so-called expert systems. The most recent implementation of Gavina uses a nondeterministic version of the n-gram model through a Bayesian network implemented with probability hypercubes, in a similar way to the Markov model. The system consists of 5 predictors and an arbiter who decides which predictor is likely to have the most successful choice in relation to the measurement of support and confidence. Each predictor focuses on human number sequences of different lengths and calculates the probability of the repetition of patterns of 1, 2, 3, 4 and 5 numbers respectively. For example, if the system uses sequences of 2 numbers to accurately predict the next number of a specific human player, it means that, in general, that player tend to repeat a certain pair of numbers, say 2 followed by a 5, and that using this information is for the system the best way to predict the next number outcome. The choices of the arbiter are recorded and can be used to interpret human performances.

The development of gesture and voice recognition is a very important topic in the development of multimodal interfaces [24]. A fundamental component of a hand game robotic system able to play against human opponents is the gesture recognition system used to recognize the hand configuration of the human player in real time. This is not a simple problem, as human hands during hand games need to be tracked at high speed with non-blurred images stable enough to allow gesture recognition. In this article we will mostly focus on the sensing characteristics and how they are integrated with the computational core and with the actuator skills of the robot.

Previously developed systems [25] used a high-speed vision system (500 fps) to actively track and recognize the human hand gestures, processing single frames to identify fingertips located outside a predetermined circular boundary centered on the hand palm. More recently, a similar system used the Leap motion device and two separate machine learning architectures to evaluate kinematic hand data on-the-fly to recognize and segment human motion activity and to classify hand gestures [20]. However, both these implementations suffer from the use of costly capture devices and do not ensure sufficient accuracy in finger counts.

In previous implementations of Gavina, we tried different solutions to the gesture recognition problem in Morra. Our first solution was to use five Hall effect sensors [26] positioned on each fingertip of the human player and a small magnet positioned in the center of the palm of the same hand. When the player extends a certain number of fingers, it creates an unambiguous pattern of magnetic activation in the Hall sensor system. Specifically, the measurement of the variations of the magnetic fields detected by each of the sensors returns the exact number played. This method, which in the experimentation phase proved to be extremely accurate, fails to detect the correct number of fingers when the human player produces ambiguous bending or extending of fingers.

Other solutions have been provided through the use of systems dedicated to motion detection, such as kinect and leap motion. However, such attempts suffer from several limitations due to the dynamics of the game. Specifically, players often rotate their wrist and tend not to bend or extend fingers completely. Also, the presence of other hands in the frame, the extremely variable lighting conditions, the variability of hands' position in the playing space, would make motion detection technology unreliable in the specific context of Morra.

Considering the previous attempts, we are currently working on developing a recognition system that can univocally return the number of extended fingers of a human hand in real time. In this study we describe the pilot implementation of MediaPipe Hands [27] within our artificial Morra agent Gavina 2121. Our scope is to demonstrate that MediaPipe is a robust, reliable, flexible and easy to implement automatic gesture recognition system in Gavina 2121.

2 Methods

The hand tracking solutions previously implemented in Gavina required cumbersome apparatuses or devices. In fact, the magnetic solution described above required a glove in which to install a magnet and sensors, and the use of additional sensors placed on the forearm to signal, with the extension of the arm, the temporal proximity of a new measurement. Analogously, the use of motion detection devices required the integration of different devices that are not easy to implement in quasi-naturalistic settings, like morra tournaments. Therefore, we decided to test a different approach in which the hand position is captured and tracked without utilizing additional devices like magnetic systems or motion capture devices. To accomplish this task we used the MediaPipe hand tracking framework [27]. MediaPipe Hands can predict landmarks on an image or a video sequence using a pretrained convolutional neural network (CNN) and represent its prediction on the hand detected by drawing the hand landmarks frame by frame. In detail, a visual object is passed through a machine learning pipeline that involves two subsequent models: the Palm Detection Model makes the system able to draw an initial bounding box of the palm that becomes the next input of the Hand Landmark Model. Finally, the latter model traces the 21 hand landmarks of the detected hand(s) using a regressor to decide their positions and draw them on the visual content combined with a prediction (and a label) of the detected hand(s).

On a practical level, we invoked the constructor of the object-class Hands by passing the parameters necessary to define the specifics of the model. As the MediaPipe reference source suggests, this framework can accept five parameters. From these five, four were crucial for our purposes during the testing phase of the algorithm: model_complexity provided the chance to opt for a more complex convolutional network structure by passing the value 1, max_num_hands allowed us to decide the maximum number of hands that framework must track. Finally, min_detection_confidence describes the minimum probability to detect the hand(s) in the scene and min_tracking_confidence makes the user able to set up a value that indicates the probability threshold for tracking successfully the hand(s). The choice of passing those values to the model was motivated by the outcomes of the preliminary phases of the algorithm. In fact, we didn't observe any substantial differences in the accuracy from changing the parameters.

2.1 Apparatus

For our tests we used a HP ProBook 455 G2 laptop, a built-in 708879-3C2 Webcam module for image acquisition. We developed our code in Python 3.9 programming language. For image processing we used real time image acquisition, prerecorded morra from tobii glasses 2 eye tracker, a sony HDR MV1 video camera and several commercially available models of smartphone cameras.

2.2 Procedure

Using the framework of Mediapipe hand, which provides 21 landmarks of the 3-d position of a hand in real time, we developed an algorithm for finger counting. In the testing phase we assessed the reliability of automatic finger counting in different conditions. We concentrated on hand position variability, counting fingers in real time and counting fingers from recorded videos.

Hand Position Variability: Our first step was to test whether the system could detect the position of the fingers of a human hand both when the hand is in a prone or supine position. This test was necessary because Morra players in actual Morra games often alternate supine and prone hand positions when extending their fingers in front of their opponent.

Counting Fingers in Real Time: The second step was to allow the system to count fingers while displaying random numbers of fingers in real time, with an approximate frequency of a number per second. This test was necessary to simulate the way Gavina detects and recognizes numbers of extended fingers in a Morra game against a human opponent. Specifically, to test if a finger is extended, the system compares the Y coordinates of the distal phalanx (tip of the finger) and the proximal phalanx (the one connected to the methacarp) from the hand landmarks received from MediaPipe. If the Y coordinate of the distal phalanx is greater than the Y coordinate of the proximal phalanx, the system will assign the status of "extended" to the finger in analysis. Finally, the system will count how many extended statuses are present, determining the number presented by the human player.

Counting Fingers in Images from Pre-recorded Videos: Finally, we tested if the system was able to detect and recognize the number of extended fingers on pre recorded videos of Morra games. This step was important to assess the possibility to automatically tabulate data of actual Morra games between two or four human contenders.

3 Results

Hand Position Variability: We tested the ability of the algorithm to detect fingers' position in prone and supine hand positions (Fig. 3). As in the first tests the software was unable to detect the numbers in prone hands, we modified the original function and split the supine/prone hand cases by considering the landmarks of the wrist and the base of the middle finger and comparing their y-coordinates. MediaPipe demonstrates an accuracy of 95.7% in palm position detection. Indeed, in our final test, results indicate that our apparatus is able to correctly detect palm position and fingers with high accuracy that reflect high scores in finger counts (see next paragraph for statistics).

Counting Fingers in Real Time: Our system counts the number of fingers by comparing Y coordinates of the distal and proximal phalanxes of each of the five fingers and then counting the number of extended vs. non-extended statuses. Our test indicates that our algorithm, in a total of 30 trials for prone and supine hand, could achieve 86% and 93% of accuracy, respectively.

Counting Fingers in Images from Pre-recorded Videos: The quality and the modality of the Morra game recordings was very variable. Specifically, they used smartphones, video cameras and a mobile eye tracking system to record Morra games from a sample of college students at Lawrence Technological University. Moreover, Morra games in an ecological setting have the two hands of the opponents in close spatial proximity with one another. This makes it very hard for an autonomous recognition system to distinguish, select and process the two hands in separation. During our tests, both the variability of the videos and the simultaneous presence of two hands in the same frame,

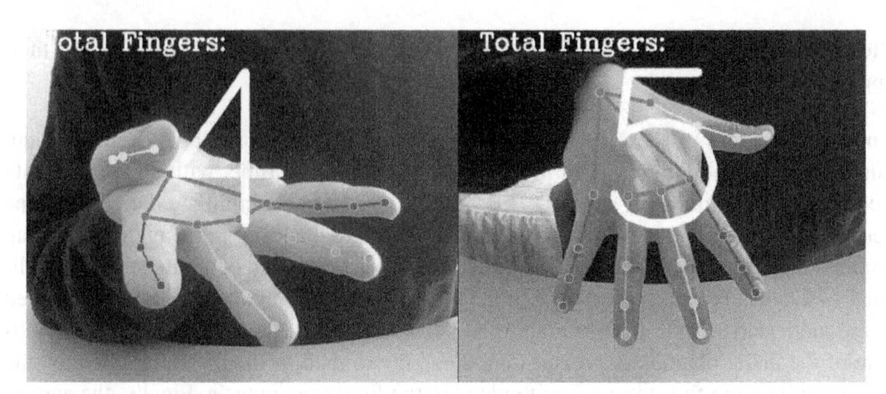

Fig. 3. The system shows high accuracy in counting the number of extended fingers in real time from hands in both supine and prone positions

made automatic finger counting very challenging to the system. The main issues consist in unsteady camera recording, broad scene focus, poor frame angles. For these reasons, we splitted the videos to test the model over consistent game sequences where the frames reproduced a clear choice of the player. By reducing the exposure to those issues we were able to investigate how we can improve the further recordings to prevent the system from possible distractors and isolate the recordings containing only frames of hands involved in the game (see Fig. 4).

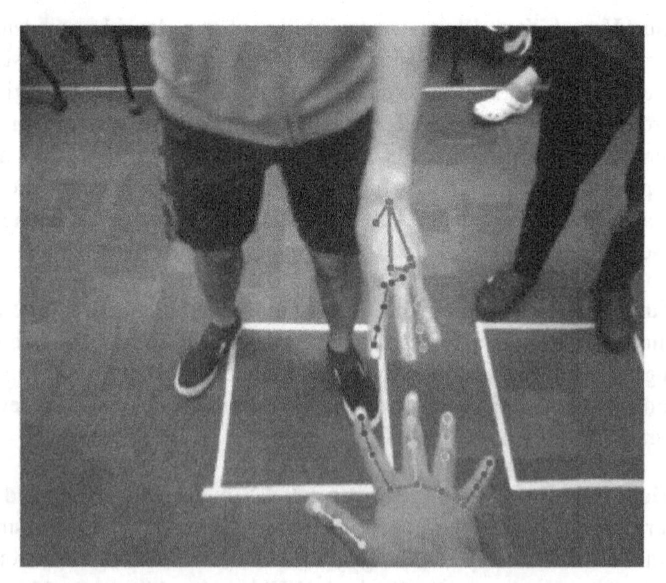

Fig. 4. Simultaneous detection and tracking of landmarks from two hands in a prerecorded video

4 Discussion

In this study we tested the robustness, reliability, flexibility and simplicity of implementation of MediaPipe Hand [27] as an automatic gesture recognition system for our artificial Morra agent Gavina 2121. Specifically, we assessed the accuracy of automatic recognition of the number of extended fingers of a human hand by MediaPipe in different settings: prone and supine hands in real time camera acquisition and with pre recorded videos.

Our results indicate that MediaPipe is able to count the number of extended fingers of a human hand with good precision both in supine and prone hand positions making it a good candidate for implementation in Gavina2121. However, the accuracy of recognition is reduced when the system is detecting finger position from pre recorded videos. In this case, the presence of two hands in the same frame, the variability of the quality of the videos and the always different dynamic of the motor behavior of human players makes the automatic gesture recognition very challenging.

Reliable gesture recognition is of vital importance for our Morra study as it is applicable to several research contexts and experimental paradigms. For example, playing Morra against an artificial agent allows the setting of a flexible training environment in which the user can employ different levels of difficulty with which to customize the robot's skills. Also, accurate gesture recognition allows telemorra, in which two human opponents can play Morra against each other online in a virtual setting and have Gavina as a point counter and referee. Telemorra can be applied to pedagogical and research contexts especially in cognitive development and rehabilitation settings. Our Morra agent can be flexibly employed in many contexts, from schools, rehabilitation centers, experimental psychology and cognitive neuroscience laboratories.

Like other serious games [4, 5, 28], Morra has the potential to positively affect cognition and to improve cognitive and perceptual skills. Moreover, with its involvement of many perceptual, cognitive and motor processes, it is an ideal tool to test several cognitive processes, as well as their development and rehabilitation [16]. An artificial Morra player can sensibly expand the numerous educational, rehabilitation and research applications of Morra.

Several steps need to be taken in order to use Gavina at the best of its computational capability, including increasing accuracy rates of finger counting, the integration of speech recognition software within the system to automatically recognize spoken numbers. Also, a virtual reality rendition of Gavina would allow a less complicated implementation and easier reproducibility than the physical robotic agent.

References

1. Granic, I., Lobel, A., Engels, R.C.M.E.: The benefits of playing video games. Am. Psychol. **69**, 66–78 (2014). https://doi.org/10.1037/a0034857
2. Boyle, E.A., et al.: An update to the systematic literature review of empirical evidence of the impacts and outcomes of computer games and serious games. Comput. Educ. **94**, 178–192 (2016). https://doi.org/10.1016/j.compedu.2015.11.003
3. Pallavicini, F., Ferrari, A., Mantovani, F.: Video games for well-being: a systematic review on the application of computer games for cognitive and emotional training in the adult population. Front. Psychol. **9** (2018). https://doi.org/10.3389/fpsyg.2018.02127

4. Noda, S., Shirotsuki, K., Nakao, M.: The effectiveness of intervention with board games: a systematic review. BioPsychoSoc. Med. **13**, 22 (2019). https://doi.org/10.1186/s13030-019-0164-1
5. Gobet, F., Retschitzki, J., de Voogt, A.: Moves in Mind: The Psychology of Board Games. Psychology Press, London (2004). https://doi.org/10.4324/9780203503638
6. Bevilacqua, V., et al.: Design and development of a forearm rehabilitation system based on an augmented reality serious game. In: Rossi, F., Mavelli, F., Stano, P., Caivano, D. (eds.) WIVACE 2015. CCIS, vol. 587, pp. 127–136. Springer, Cham (2016). https://doi.org/10.1007/978-3-319-32695-5_12
7. Oh, Y., Yang, S.: Defining exergames & exergaming (2010)
8. Zaina, L., Castro, E., Martinelli, S., Sakata, T.: Educational games and the new forms of interactions. Smart Learn. Environ. **6**(1), 1–17 (2019). https://doi.org/10.1186/s40561-019-0099-9
9. Batchelor, K.E., Bintz, W.P.: Hand-clap songs across the curriculum. Read. Teach. **65**, 341–345 (2012). https://doi.org/10.1002/TRTR.01052
10. Fauziddin, M., Mufarizuddin, M.: Useful of clap hand games for optimize cognitive aspects in early childhood education. Jurnal Obsesi Jurnal Pendidikan Anak Usia Dini. **2**, 162–169 (2018). https://doi.org/10.31004/obsesi.v2i2.76
11. Han, Y.-S., et al.: Development and effect of a cognitive enhancement gymnastics program for elderly people with dementia. J. Exerc. Rehabil. **12**, 340–345 (2016). https://doi.org/10.12965/jer.1632624.312
12. Baek, K., et al.: Response randomization of one- and two-person Rock–Paper–Scissors games in individuals with schizophrenia. Psychiatry Res. **207**, 158–163 (2013). https://doi.org/10.1016/j.psychres.2012.09.003
13. Wang, Z., Xu, B., Zhou, H.-J.: Social cycling and conditional responses in the Rock-Paper-Scissors game. Sci. Rep. **4**, 5830 (2014). https://doi.org/10.1038/srep05830
14. Miyake, A., Friedman, N.P., Emerson, M.J., Witzki, A.H., Howerter, A., Wager, T.D.: The unity and diversity of executive functions and their contributions to complex "Frontal Lobe" tasks: a latent variable analysis. Cogn. Psychol. **41**, 49–100 (2000). https://doi.org/10.1006/cogp.1999.0734
15. Dehaene, S.: Varieties of numerical abilities. Cognition **44**, 1–42 (1992). https://doi.org/10.1016/0010-0277(92)90049-N
16. Delogu, F., Barnewold, M., Meloni, C., Toffalini, E., Zizi, A., Fanari, R.: The Morra game as a naturalistic test bed for investigating automatic and voluntary processes in random sequence generation. Front. Psychol. **11** (2020). https://doi.org/10.3389/fpsyg.2020.551126
17. Hasuda, Y., Ishibashi, S., Kozuka, H., Okano, H., Ishikawa, J.: A robot designed to play the game "Rock, Paper, Scissors". In: 2007 IEEE International Symposium on Industrial Electronics, pp. 2065–2070 (2007). https://doi.org/10.1109/ISIE.2007.4374926
18. Lungu, I.-A., Corradi, F., Delbrück, T.: Live demonstration: convolutional neural network driven by dynamic vision sensor playing RoShamBo. In: 2017 IEEE International Symposium on Circuits and Systems (ISCAS), p. 1 (2017). https://doi.org/10.1109/ISCAS.2017.8050403
19. Ahmadi, E., Pour, A.G., Siamy, A., Taheri, A., Meghdari, A.: Playing Rock-Paper-Scissors with RASA: a case study on intention prediction in human-robot interactive games. In: Salichs, M.A., et al. (eds.) ICSR 2019. LNCS (LNAI), vol. 11876, pp. 347–357. Springer, Cham (2019). https://doi.org/10.1007/978-3-030-35888-4_32
20. Brock, H., Ponce Chulani, J., Merino, L., Szapiro, D., Gomez, R.: Developing a lightweight Rock-Paper-Scissors framework for human-robot collaborative gaming. IEEE Access. **8**, 202958–202968 (2020). https://doi.org/10.1109/ACCESS.2020.3033550
21. Ahn, H.S., Sa, I.-K., Lee, D.-W., Choi, D.: A playmate robot system for playing the Rock-Paper-Scissors game with humans. Artif Life Robot. **16**, 142 (2011). https://doi.org/10.1007/s10015-011-0895-y

22. Zizi Antonello: Il progetto Gavin 1.0: un esperimento di scienze integrate. https://www.lulu.com/shop/antonello-zizi/shop/antonello-zizi/il-progetto-gavin-10-un-esperimento-di-scienze-integrate/paperback/product-159zvve8.html?page=1&pageSize=4
23. Towse, J.N., Neil, D.: Analyzing human random generation behavior: a review of methods used and a computer program for describing performance. Behav. Res. Methods Instrum. Comput. **30**, 583–591 (1998). https://doi.org/10.3758/BF03209475
24. Liu, J., Kavakli, M.: A survey of speech-hand gesture recognition for the development of multimodal interfaces in computer games. In: 2010 IEEE International Conference on Multimedia and Expo, pp. 1564–1569 (2010). https://doi.org/10.1109/ICME.2010.5583252
25. Ito, K., Sueishi, T., Yamakawa, Y., Ishikawa, M.: Tracking and recognition of a human hand in dynamic motion for Janken (Rock-Paper-Scissors) robot. In: 2016 IEEE International Conference on Automation Science and Engineering (CASE), pp. 891–896 (2016). https://doi.org/10.1109/COASE.2016.7743496
26. Ramsden, E.: Hall-Effect Sensors: Theory and Application. Elsevier (2011). https://doi.org/10.1016/B978-0-7506-7934-3.X5000-5
27. Zhang, F., et al.: MediaPipe hands: on-device real-time hand tracking. arXiv:2006.10214 [cs] (2020). https://doi.org/10.48550/arXiv.2006.10214
28. Kuo, C.-Y., Huang, Y.-M., Yeh, Y.-Y.: Let's play cards: multi-component cognitive training with social engagement enhances executive control in older adults. Front. Psychol. **9** (2018). https://doi.org/10.3389/fpsyg.2018.02482

Integration of Point Clouds from 360° Videos and Deep Learning Techniques for Rapid Documentation and Classification in Historical City Centers

Yuwei Cao, Mattia Previtali(✉), Luigi Barazzetti, and Marco Scaioni

Department of Architecture, Built Environment and Construction Engineering, Politecnico di Milano, via Ponzio 31, 20133 Milan, Italy
{yuwei.cao,mattia.previtali,luigi.barazzetti,
marco.scaioni}@polimi.it

Abstract. Digital metric documentation of historical city centers is challenging because of the complexity of the buildings and monuments, which feature different geometries, construction technologies, and materials. We propose a solution for rapid documentation and classification of such complex spaces using 360° video cameras, which can capture the entire scene and can be pointed in any direction, making data acquisition rapid and straightforward. The high framerate during image acquisition allows users to capture overlapping images that can be used for photogrammetric applications. This paper aims to quickly capture 360° videos with low-cost cameras and then generate dense point clouds using the photogrammetric/structure from motion pipeline for 3D modeling. Point cloud classification is the prerequisite for such applications. Numerous deep learning methods (DL) have been developed to classify point clouds due to the expansion of artificial intelligence (AI) capabilities. We aim to pave the way toward utilizing the convolutional neural network (CNN) to classify point clouds generated by 360° videos of historic cities. A preliminary case study in a historic city center demonstrates that our method achieves promising results in the generation and classification of point clouds, with an overall classification accuracy of 96% using the following categories: ground, buildings, poles, bollards, cars, and natural.

Keywords: 360° videos · Digital recording · Historic city center · Point cloud · Deep learning · Classification

1 Introduction

Historic city centers feature complex geometry, heterogeneous buildings with different materials and construction technologies, and other elements such as monuments and vegetation, among others. This paper aims to develop a rapid mapping method for digital documentation and classification. The developed solution for data acquisition relies on 360° videos acquired with low-cost cameras. 360° cameras can capture the entire scene

P. L. Mazzeo et al. (Eds.): ICIAP 2022 Workshops, LNCS 13374, pp. 254–265, 2022.
https://doi.org/10.1007/978-3-031-13324-4_22

using the equirectangular projection, in which longitude-latitude coordinates are mapped to pixel coordinates.

Low-cost 360° cameras are made up of two or more cameras and can be pointed in any direction, so video acquisition is rapid and straightforward. The relative orientation of the cameras is known and can be exploited to stitch the various images into a single equirectangular projection. Images feature a resolution of about 16–24 megapixels, whereas videos can be acquired at 5k resolution (5120 × 2880), notwithstanding that a few camera models already provide higher resolution.

The high framerate during image acquisition allows users to capture overlapping images that can be used for photogrammetric applications. Moreover, the large field of view (360° × 180°) ensures overlap between several consecutive frames. The equirectangular camera model (also called spherical) is already available in some packages (Agisoft Metashape, Pix4DMapper, OpenDroneMap) that can produce dense point clouds using the photogrammetric/structure from motion pipeline for 3D modeling.

After detecting and matching image tie points, bundle adjustment based on the spherical camera model allows the computation of exterior orientation parameters. Then, dense point clouds can be extracted using the workflow proposed in [1], in which approximated EO parameters measured with a mobile phone are used to initialize image matching and orientation.

The generated dense point clouds are employed in many historical heritage preservation applications, such as historic heritage recording [2]. Classification of point clouds is a necessary condition for such applications. The emergence of artificial intelligence (AI) technology sparked the development of numerous machine learning (ML) and deep learning (DL) methods for classifying point clouds. DL methods can fully automate the extraction of features and classification of point clouds end-to-end, allowing us to classify historical city point clouds. The goal of this paper is to pave the way toward adapting the convolutional neural network (CNN) architecture designed for scene interpretation to a point cloud generated by 360° videos.

In summary, our contributions are:

- this paper uses 360° videos captured by low-cost 360° cameras and a photogrammetric/structure from motion pipeline for 3D modeling to produce dense point clouds; and
- the deep learning approach is applied to the 3D point cloud generated by 360° videos to automatically produce semantic labels for each point of a large-scale historic city point cloud. Without the need for handcrafted features and labeled data, we achieve an overall accuracy of 96% on the point cloud generated by 360° videos using the following classes: ground, buildings, poles, bollards, cars, and natural.

2 Background

The commercial market offers different cameras at variable prices, usually under 1,000 USD. A few professional 360° cameras (such as the Insta Pro 2, with an 8k resolution) are also available. They offer better image resolution but at the cost of more than 5,000 USD. Such expensive cameras are not considered in this paper.

Low-cost 360° cameras can be manually carried using an extension pole or be installed on a mobile platform, such as cars or drones. Nowadays, most cameras also provide specific software for downloading the acquired frames and performing automatic video stitching. The procedure also works directly on mobile devices (such as mobile phones or tablets), but the resolution of the resulting video is usually reduced. Desktop-based applications instead allow users to exploit maximum resolution.

360° videos are mainly used for immersive visualization. Different sharing services (e.g., Facebook, Visbit, Youtube, etc.) allow users to upload their videos and generate interactive bubble visualizations. Videos can also be exploited in virtual reality (VR) headsets. However, this paper aims to use 360° videos for metric applications, especially for generating dense point clouds with photogrammetric techniques [1].

For the classification of generated dense point clouds, ML approaches [3, 4] begin by manually extracting various features (e.g., planarity, linearity, perpendicularity, etc.) from the input point cloud to describe its shape or structure. After extracting the features, they are incorporated into ML methods such as random forests [5], decision trees [6], etc. However, such methods necessitate expert knowledge and handcrafted geometric features for different datasets.

On the other hand, DL methods can fully automate the extraction of features and classification of point clouds in an end-to-end manner. Various DL methods have been proposed to process 3D point clouds, which can be divided into two different categories: indirect and direct methods. The former requires first projecting the point cloud onto a regular structure (e.g., multiple images [7] or voxels [8]), where convolution operations can be more easily defined. The latter directly processes raw point clouds using multilayer perceptron (MLP) [9] or convolution operations [10, 11] techniques. Existing 3D large-scale point cloud datasets (e.g., Semantic3D [12], Paris-Lille-3d [13]) provide training data for large-scale scene classification, allowing us to automatically classify our historical city point clouds without the need for manual annotation.

3 Method

3.1 Generation of Point Clouds from 360° Videos

The concept of spherical photogrammetry was introduced more than a decade ago [14, 15]. The main idea is to extend the collinearity equation typical of frame cameras to spherical geometry. The ray connecting the perspective center with the object points projects the image points on a sphere. The positions of the points on the arbitrary sphere, expressed as latitude (φ) and longitude (λ) are then mapped on the cartographic plane with the so-called equirectangular projection that is neither conform nor equivalent by using the following formula:

$$x = r\lambda \text{ } and \text{ } y = r0 \tag{1}$$

with the angles (φ, λ) expressed in radiant. The radius of the sphere r (in pixels) can be estimated as $r = h/\pi = w/(2\pi)$, where (w, h) are the image width and height in pixels. The radius corresponds to the focal length of the camera, due to the characteristics of the equirectangular projection $h = w/2$. Starting from those conditions, the collinearity equations for spherical cameras can be derived by extending the method for adjusting geodetic networks based on angular measurements [16].

Nowadays some commercial softwares are supporting the processing of spherical cameras (e.g., Agisoft Metashape, Pix4D Mapper). As an example, in Agisoft Metashape the camera model is the following one:

$$\begin{aligned} u &= w0.5 + \frac{w}{2\pi}\tan^{-1}\left(\frac{X}{Z}\right) \\ v &= h0.5 + \frac{w}{2\pi}\tan^{-1}\left(\frac{Y}{\sqrt{X^2+Z^2}}\right) \end{aligned} \tag{2}$$

where (u, v) are the projected point coordinates in the image coordinate system (in pixels), w and h are the image width and height in pixels, and X, Y, Z are the point coordinates in the local camera coordinate system.

Even if the background of spherical photogrammetry is well defined, its applications in some real-world situations may pose some practical issues. Nowadays, spherical cameras allow video acquisition at a resolution similar to that of static spherical images (5.7K). The advantage of video acquisition compared with static images is the speed of the survey. Indeed, the operator has only to walk in the area to be surveyed with the camera mounted on a selfie stick. Then starting from the collected videos, some frames can be extracted and processed using the spherical camera model. However, in the case of long duration videos and paths, hundreds or even thousands of images can be extracted, making the successive steps of the processing quite time-consuming or, in the worst situation, the high amount of data may prevent a correct alignment of the photos.

For this reason, some optimization strategies may be set up to speed up those stages. In this paper, for example, approximated exterior orientation parameters captured with a mobile phone associated with the low-cost 360° are added in the image adjustment to reduce the search space of the matching stage only to images that are within a predefined camera distance. This strategy enables us to reduce processing time while maintaining the high metric quality of the final results [1].

3.2 Deep Learning-Based Point Cloud Classification

Point Set Convolution. Images have a regular grid structure so that convolutional neural networks (CNNs) can be adequately constructed and achieve SOTA in many image analyzing tasks. The points are spatially localized using their coordinates, which share the same characteristic as an image in that the features are also localized using their index/coordinates in a grid [10]. This means that point clouds can be classified using the CNN-based deep learning method.

To establish the correlation between the convolution kernel and the input points, CNN-based point classification methods (e.g., KPConv [10], ConvPoint [17]) proposed defining the kernel as an explicit set of points associated with weights.

Kernel Point Positions. Kernel point positions are critical to the convolution operator. KPConv [10] determines the location of kernel points by aligning an attractive force to the sphere center point and a repulsive force of each point to the others. While ConvPoint [17] first randomly samples K locations in a unit sphere and then uses gradient descent to learn more appropriate positions to continuously optimize the location of the core points when training the network. Considering that hyper-parameters are needed in the KPConv method, we employ ConvPoint as our baseline.

Input Point Patch Selection. A subsampling strategy is essential to select the input point cloud patch. As RandLA-Net [18] has demonstrated, sampling strategies such as Farthest Point Sampling (FPS) are computationally inefficient and are therefore inappropriate for large-scale point clouds. While random sampling is the most computationally efficient method, it suffers from information loss. A random sampling method with the number of previously selected times as a constraint is used when selecting output points. As illustrated in Fig. 1, the selected points are next utilized to determine their k-nearest neighbors (KNN) in order to construct the point patches, where KNN is based on the pointwise Euclidean distances.

Subsampling the input point cloud not only gradually reduces the number of points, but also increases the receptive fields of the later layers. By stacking convolution layers, the structure of the CNN can become hierarchical as the later layers can see the points within the receptive fields of the prior layers.

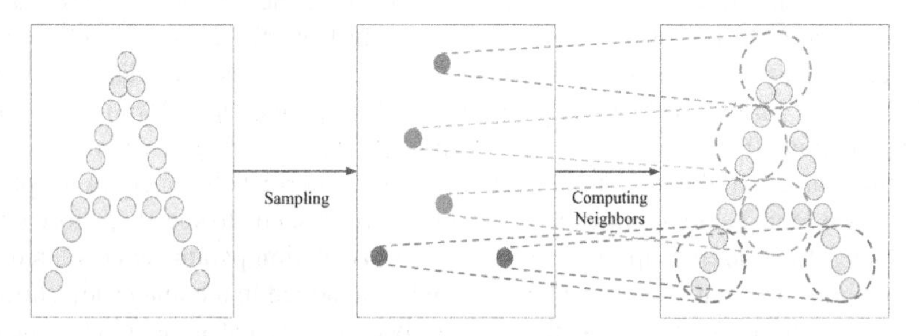

Fig. 1. Constructing local neighbors of random selected points.

Convolution Operation. A convolution layer is defined as a weighted sum of the input features. The correlation between the input point cloud patch and kernel points is calculated by a geometrical weighting function $g(\cdot)$, which takes a set of relative positions r_j of input point to kernel points as input and uses a multilayer perceptron (MLP) layer to approximate:

$$g(p_j, c) = MLP(p_j - c) \tag{3}$$

where $\{p_j | j < |X|\}$ is a point position in the input point cloud patch $X = \{(p, x)\}$, and c is the positions of kernel points $K = \{(\text{c}, \text{w})\}$.

Thus, the output feature map y of a point cloud patch X and a kernel K can be computed by:

$$y = \frac{1}{|X|} \sum\nolimits_{j}^{|X|} \sum\nolimits_{i}^{|K|} w_i x_j g(r_j) + b \tag{4}$$

where $g(\cdot)$ is the weighting function, x_j is the features of point p_j in the input point cloud patch X, and the associate weights of the i-th kernel point are denoted by w_i. In this function, each output feature is normalized by the input point size $|X|$ to guarantee the robustness of the network at different point cloud scales.

Architecture. As shown in Fig. 2, the DL-based classification network consists of an encoder and a decoder. A stack of convolutional layers that compresses and automatically extracts the features of the point cloud in the encoder and a symmetrical stack of deconvolutional layers as a decoder with skip connections. In the decoder, upsampling deconvolution layers are used in the decoder to obtain the same number of points as the corresponding layer in the encoder. Additionally, the features from the encoder and the decoder are concatenated to pass the features between the intermediate layers of the encoder and the decoder. Finally, a linear layer is used to make an output dimension that corresponds to the number of classes.

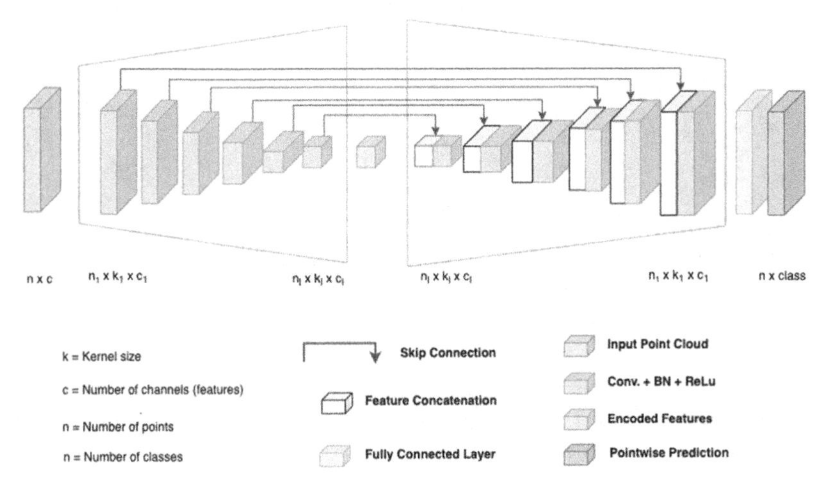

Fig. 2. The architecture of the ConvPoint network.

Evaluation Matrix. For the evaluation matrix of the semantic segmentation result, we use *Overall Accuracy* (OA), *Intersection-over-Union* (IoU), and *Mean Intersection over Union* (mIoU) to describe the performance of the used model. OA is used to describe the number of correct predictions over all points. IoU is used to describe the accuracy for each class. As shown in Eq. 5, where Intersection refers to the overlay between prediction and ground truth, whereas Union refers to the union of predicted and ground truth, and the mIoU represents the average IoU for each class.

$$IoU = \frac{Prediction \cap Ground\ Truth}{Prediction \cup Ground\ Truth} \tag{5}$$

4 Case Study

4.1 Dataset – Bassano Dataset

A dataset was acquired in the city center of Bassano del Grappa (VI, Italy) to test the proposed solution. The area selected for the test is in the northern part of the historical city center. In particular, the area covers (Fig. 3) the inner court of the Ezzelini's Castle and the exterior of the Cathedral of Santa Maria in Colle, the square in front of the Castle that is currently used as a car parking area, a couple of building blocks in the surroundings of the Castle characterized by narrow streets, and a part of the tree-lined avenue approaching the Castle. This area was selected since it presents a variety of architectural and natural objects typical of city centres (e.g., buildings – including situations typical of hystorical city centers like towers and belltowers, cars, trees, road signs, electric lines, etc.) and it allows a complete test of the proposed method for point cloud classification.

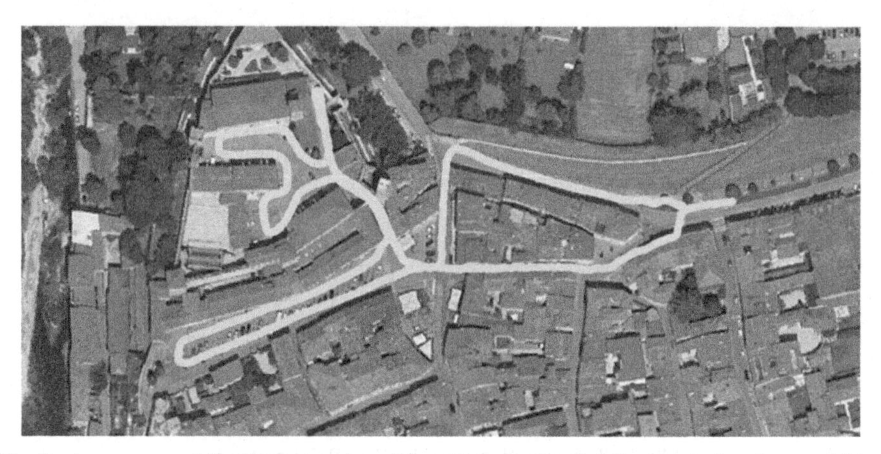

Fig. 3. Area surveyed for the generation of the dataset and path followed during the acquisition in yellow. (Color figure online)

4.2 Acquisition and Post-processing

For the acquisition of the dataset, a 5.7k camera was used. More specifically, an Insta360 One X2 cam was mounted on the top of a selfie stick, exposure was set to automatic mode, and a 5.7k 360° video at 30 frames per second was acquired for the entire area with the operator walking along the predefined path. The path was chosen to guarantee full coverage of the survey area by creating a closed-loop so that a further constraint is added during image alignment. The total distance covered by the acquisition is about 950 m, and the time duration of the acquired video is about 25 min. The trajectory provided by the GNSS receiver inside the mobile phone was synchronized with the image acquisition to have an initial rough positioning of the camera frames. Image stitching was carried out by using the software Insta360 Studio 2022 by using the dynamic stitching functionality available in the software.

Fig. 4. Image orientation and dense point cloud reconstruction results: oriented images (a); reconstructed dense point cloud with confidence filter set at 3 (b); and detail of the point cloud for the Cathedral of Santa Maria in Colle.

Starting from the collected video, a set of equirectangular frames has been extracted (1 frame per second), resulting in about 1500 frames. The position recorded by the GNSS receiver of the mobile phone was used as an initial estimate of the camera position (precision of this initial set of camera poses estimates was set to 10 m). Agisoft Metashape was used for image orientation and dense point cloud reconstruction (Fig. 4). For the processing, a pc with 10th Generation Intel® Core™ i9, NVIDIA® GeForce RTX™ 3090, and 128 GB of memory was used. Image orientation, using approximate frame position, took approximately 10 min. Dense reconstruction, "high" quality, was accomplished in 10 h on the whole dataset. The obtained point cloud was filtered to reduce noise in the data using only dense points with visibility higher or equal to 3. The final point cloud results in a total of 89 million points.

4.3 Classification

Dataset Settings. The Paris-Lille 3D dataset [13] was acquired by the Mobile Laser System (MLS) and it contains 38 million points taken from four scenes of two cities: Paris and Lille. In our training phase, we use a subset of the Paris-Lille 3D dataset (NPM3D) as our training data to reduce training time. Concerning the fact that the intensity feature of the point cloud generated by the 360° videos (testing data) and the data collected by the MLS (training data) are very different, testing the model trained with the NPM3D dataset on data collected by 360° videos would produce inaccurate predictions. Therefore, we train the network with a normalized intensity feature to reduce the differences. Let $P = \{p, x\}$ be the input point cloud, where p denotes the coordinates and x denotes the intensity feature. Due to the large size of both the training and test scenes, it is not possible for the network to feed the entire point cloud into the network at once. We follow the setting in ConvPoint and split point clouds into 8 m wide blocks in the horizontal direction, and sample 8,192 points in each block. In addition, the training data contains 10 coarse classes: unclassified, ground, buildings, poles, bollards, trash cans, barriers, pedestrians, cars, and natural. We reuse the classes in the NPM3D dataset, in which we only remove 4 classes that are irrelevant to our dataset: unclassified, trash can, barriers, and pedestrians.

Network Settings. The hyperparameters (e.g., number of output channels in each layer, size of output points that pass to the next layer, neighborhood size, and kernel size) of the network follow the original implementation of the fusion model in ConvNet. With a batch size of 4, a momentum of 0.98, and an initial learning rate of 10–2, we train the network with the NPM3D dataset on the Google Colab. At the test time, we used a batch size of 2 on the CPU.

Classification Results. Table 1 summarizes our results. Experiment results show that our deep learning method achieves promising point cloud classification results with an overall classification accuracy of 96%. The overall qualitative classification result of the Bassano historic city center is reported in Fig. 5. As shown, our classification model is capable of producing very smooth results on the test point cloud.

Furthermore, the per-class semantic segmentation results in Table 1 demonstrate that the model is quite efficient in identifying ground, buildings, bollards, and cars. As seen in Fig. 6, even when point clouds in the category of cars are extremely sparse, we can still reliably identify the category to which they belong.

However, since the point cloud classification model does not include a low vegetation category in the training dataset, the DL model is inefficient for low natural point clouds. As shown by the per-class result in Table 1 and the qualitative result (see Fig. 7), the low natural points are incorrectly identified as cars and buildings.

Table 1. Classification results (%) for the Bassano dataset in terms of overall accuracy (OA), mIoU (mean Intersection over Union), and per-class classification results Intersection over Union (IoU).

OA	mIoU	Ground	Buildings	Poles	Bollards	Cars	Natural
96.3	70.4	100.0	94.7	51.4	82.1	84.8	8.7

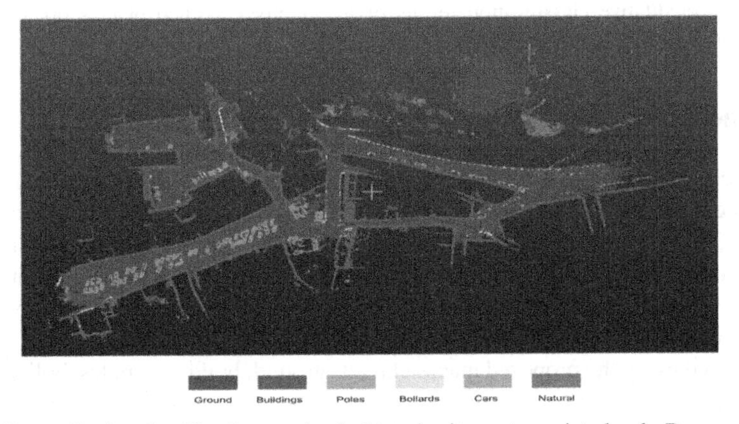

Fig. 5. The qualitative classification result of a historic city center point cloud - Bassano dataset. Different colors correspond to different categories. (Color figure online)

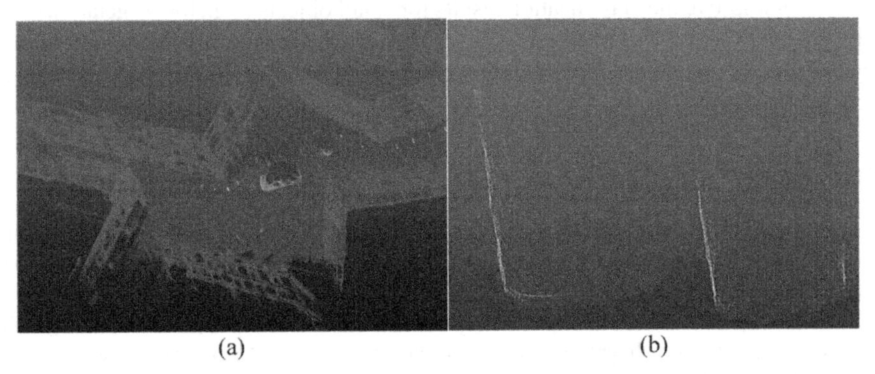

(a) (b)

Fig. 6. The qualitative classification results of correctly classified objects: (a) buildings and (b) cars.

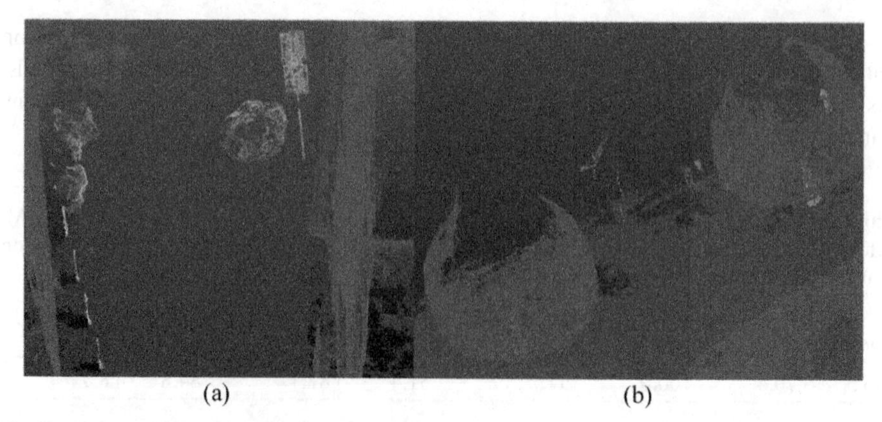

(a) (b)

Fig. 7. The qualitative classification results of incorrectly classified objects (a) low natural classified as cars and (b) classified as buildings.

5 Conclusion

This article demonstrates how to quickly generate and classify digital metric documentation of complicated historical city centers. We began by collecting 360° videos using low-cost 360° cameras. The dense point cloud was then generated using photogrammetry from captured high framerate and overlapping images. Finally, the generated point clouds were automatically classified, and each point was labeled using the deep learning method. Without the requirement of labels from the test scene, 96% overall accuracy was achieved using the proposed macro classes: ground, buildings, poles, bollards, cars, and natural.

In future work, we will explore the following aspects: 1) incorporating a subset of the semantic segmentation results from the test dataset to fine-tune the deep neural network and further improve the classification results for small objects (e.g., low vegetations); 2) validating the proposed method on additional historical heritage city datasets and more detailed classes (e.g., identifying different types of buildings) to further demonstrate the method's feasibility.

References

1. Barazzetti, L., Previtali, M., Roncoroni, F.: 3D modeling with 5K 360° videos. Int. Arch. Photogramm. Remote Sens. Spat. Inf. Sci. **XLVI-2/W1-2022**, 65–71. https://doi.org/10.5194/isprs-archives-XLVI-2-W1-2022-65-2022
2. Sánchez-Aparicio, L.J., Del Pozo, S., Ramos, L.F., Arce, A., Fernandes, F.: Heritage site preservation with combined radiometric and geometric analysis of TLS data. Autom. Constr. **85**, 24–39 (2018)
3. Weinmann, M., Jutzi, B., Mallet, C., Weinmann, M.: Geometric features and their relevance for 3D point cloud classification. ISPRS Ann. Photogramm. Remote Sens. Spat. Inf. Sci. **IV-1/W1**, 157–164 (2017)
4. Grilli, E., Farella, E.M., Torresani, A., Remondino, F.: Geometric features analysis for the classification of cultural heritage point clouds. Int. Arch. Photogramm. Remote Sens. Spat. Inf. Sci. **XLII-2/W15**, 541–548 (2019)

5. Bassier, M., Van Genechten, B., Vergauwen, M.: Classification of sensor independent point cloud data of building objects using random forests. J. Build. Eng. **21**, 468–477 (2019)
6. Babahajiani, P., Fan, L., Kamarainen, J., Gabbouj, M.: Automated super-voxel based features classification of urban environments by integrating 3D point cloud and image content. In: 2015 IEEE International Conference on Signal and Image Processing Applications (ICSIPA), pp. 372–377 (2015)
7. Fan, H., Su, H., Guibas, L.J.: A point set generation network for 3D object reconstruction from a single image. In: Proceedings of the IEEE Conference on Computer Vision and Pattern Recognition, pp. 605–613 (2017)
8. Riegler, G., Osman Ulusoy, A., Geiger, A.: OctNet: learning deep 3D representations at high resolutions. In: Proceedings of the IEEE Conference on Computer Vision and Pattern Recognition, pp. 3577–3586 (2017)
9. Qi, C.R., Su, H., Mo, K., Guibas, L.J.: PointNet: deep learning on point sets for 3D classification and segmentation. In: Proceedings of the IEEE Conference on Computer Vision and Pattern Recognition, pp. 652–660 (2017)
10. Thomas, H., Qi, C.R., Deschaud, J.E., Marcotegui, B., Goulette, F., Guibas, L.J.: KPConv: flexible and deformable convolution for point clouds. In: Proceedings of the IEEE/CVF International Conference on Computer Vision, pp. 6411–6420 (2019)
11. Cao, Y., Scaioni, M.: 3DLEB-Net: label-efficient deep learning-based semantic segmentation of building point clouds at LoD3 level. Appl. Sci. **11**(19), 8996 (2021)
12. Hackel, T., Savinov, N., Ladicky, L., Wegner, J.D., Schindler, K., Pollefeys, M.: Semantic3D.net:: a new large scale point cloud classification benchmark. ISPRS Ann. Photogramm. Remote Sens. Spat. Inf. Sci. **IV-1-W1**, pp. 91–98 (2017)
13. Roynard, X., Deschaud, J.E., Goulette, F.: Paris-Lille-3D: a large and high-quality ground-truth urban point cloud dataset for automatic segmentation and classification. Int. J. Robot. Res. **37**(6), 545–557 (2018)
14. Fangi, G.: Multiscale multiresolution spherical photogrammetry with long focal lenses for architectural surveys. Int. Arch. Photogramm. Remote Sens. Spat. Inf. Sci. **38**(Part 5), 1–6 (2010)
15. Fangi, G.: The Multi-image spherical panoramas as a tool for architectural survey. CIPA Herit. Doc. **21**, 311–316 (2011)
16. Fangi, G.: The Book of Spherical Photogrammetry: Theory and Experiences. Edizioni Accademiche Italiane (2017). 300 pages
17. Boulch, A.: ConvPoint: continuous convolutions for point cloud processing. Comput. Graph. **88**, 24–34 (2020)
18. Hu, Q., et al.: RandLA-Net: efficient semantic segmentation of large-scale point clouds. In: Proceedings of the IEEE/CVF Conference on Computer Vision and Pattern Recognition, pp. 11108–11117 (2020)

Towards the Creation of AI-powered Queries Using Transfer Learning on NLP Model - The THESPIAN-NER Experience

Alessandro Bombini[1(✉)], Lisa Castelli[1], Achille Felicetti[2], Franco Niccolucci[2], Anna Reccia[2], and Francesco Taccetti[1]

[1] INFN Florence Section, Via Bruno Rossi 1, Florence, Italy
bombinl@fi.infn.it
[2] PIN, Prato, Italy

Abstract. Tools for HEritage Science Processing, Integration, and ANalysis (THESPIAN) is a cloud system that offers multiple web services to the researchers of INFN-CHNet, from storing their raw data to reusing them by following the FAIR principles for establishing integration and interoperability among shared information.

The injection in the CHNet cloud database of data and metadata (the latter modelled on a CIDOC-based ontology called CRMhs [20]) is performed by using the cloud service THESPIAN-Mask.

THESPIAN-NER is a tool based on a deep neural network for Named Entity Recognition (NER), which will ease the data extraction from the database, enabling users to upload .pdf or .txt files and obtain named entities and keywords to be fetched in the metadata entries of the database.

The neural network, on which THESPIAN-NER relies, is based on a set of open-source NLP models; transfer learning was employed to customise the Named Entity Recognition output of the models to match the CRMhs ontology properties.

The service is now available in alpha version to researchers on the CHNet cloud.

Keywords: AI in natural language processing and cultural heritage applications · Named entity recognition · FAIR data management · Cloud services

1 Introduction

In the framework of the European projects ARIADNEplus and EOSC-Pillar initiative of the European Open Science Cloud (EOSC) framework, *Tools for HEritage Science Processing, Integration, and ANalysis* (THESPIAN) was developed [2]. THESPIAN is a cloud system offering multiple web services to the researchers of the Cultural Heritage Network (CHNet) of INFN (Istituto Nazionale di Fisica

P. L. Mazzeo et al. (Eds.): ICIAP 2022 Workshops, LNCS 13374, pp. 266–277, 2022.
https://doi.org/10.1007/978-3-031-13324-4_23

Nucleare). The mission of CHNet is to harmonise and enhance the expertise of the Institute in the field of Cultural Heritage, expertise distributed among many structures spread over the whole Italian territory. CHNet includes the INFN research groups whose activity is devoted to the development and application of technologies for the study and conservation of Cultural Heritage, but it is open also to other national partners with expertise complementary to that of the Institute and to international partners engaged in diagnostics in cultural heritage.

The purpose of the THESPIAN platform is to create a complete ecosystem for scientific data and metadata of physical analysis on cultural heritage, modelled according to the latest ontologies and standards, and to make them interoperable with data generated by other disciplines and accessible on other platforms according to the FAIR principles for data: findability, accessibility, interoperability, and reusability [21].

1.1 The Digital Infrastructure of INFN-CHNet

The Digital Heritage Laboratory (DHLab) of the INFN-CHNet consists of a set of software services developed targeting the needs of CHNet researchers, hosted in a cloud environment.

The main goal is to host a cloud service to store raw data and their metadata regarding scientific analysis on Cultural Heritage through a shared ontology implemented via a web service. After that, each researcher of the network may use and re-use their data, as well as the available data of all the other researcher of the network, and analyse such data using a set of web services hosted in the cloud.

This follows the step of the FAIR program in the European Science Cloud (EOSC) project; the goal is thus to have data which are *findable*, *accessible*, *interoperable* and *reusable* [21].

These principles are implemented in the web service THESPIAN-Mask [2,5] hosted in the CHNet cloud environment. It consists of a web platform for assisted metadata generation, and a service for persistent storage of scientific data and their metadata. The tool is tailored on a metadata model based on CRMhs, an extension of the CIDOC CRM ontology [1], designed for modelling the complex entities typical of heritage science, developed by INFN and VAST-LAB PIN [5].

The use of CRMhs also makes the information fully interoperable with other CIDOC-CRM-compatible data, and allows their integration in existing cloud environments and extended semantic graphs, such as the semantic data cloud developed by ARIADNE-plus for archaeological data.

The tool we are going to describe in the following, THESPIAN-NER, is a Natural Language Processing (NLP) tool for automatic Named Entity Recognition (NER) using two deep learning models based on different techniques, namely a Convolutional Neural Network (CNN)-based model and Transformer model[1]. It allows users to automatically annotate either archaeological documents or scientific reports written in Italian (either .txt or .pdf files), by identifying and labelling

[1] For some references about the usage of NER in Cultural Heritage applications, see [8,13,15,19] and references therein.

relevant semantic entities in the text, according to the CRMhs ontology used for describing the metadata. The entities can then be extracted and used for building custom queries to fetch related records available on the CHNet database, exploiting the NoSQL, JSON-based realisation of the database structure.

The service is based upon two ad-hoc trained neural network: *ArcheoNER*, for processing archaeological texts, and *hsNER*, for Scientific Reports of Analysis on Cultural Heritage. These models were built using the open-source python package *spaCy* [10–12,18] and trained using transfer learning to classify named entities with custom labels.

Users may upload their .txt or .pdf files, containing their archaeological or scientific reports, to THESPIAN-NER, and, employing automatic named entity recognition, automatically extract keywords and their labels, sorted by number of occurrences in the text, and use them to create custom queries to fetch the CHNet database for similar entries. THESPIAN-NER offers also a small tool for *extractive summarization* of the uploaded text.

The service is under current development at CHNet, and its alpha version is available for early use on the CHNet cloud.

In synthesis, THESPIAN-NER offers authorised users the possibility of generating query (semi)automatically on the CHNet database, starting from Italian written text files of either archaeological or heritage scientific topic, via an user-friendly web graphical interface; it relies on the two ad-hoc trained DNN models. The app is deployed as a service in an integrated cloud environment, compliant to FAIR rules.

2 The NER Model

As previously said in Sect. 1.1, the deep learning model employed has been chosen from the open-source library spaCy. The first version of the model, from spaCy v2, was a convolutional neural network (CNN)-based model [11]; the second version, from spaCy v3, was a transformer model [18]. We apply transfer learning to train and fine tune these models, and to have our custom named entity labels as outputs.

Initially, the spaCy v2 library was chosen because it offered three italian-based pre-trained models; we have employed as base model the largest one, IT_CORE_NEWS_LG (549 MB), originally trained on the *Italian Stanford Dependency Treebank annotated according to the UD annotation* [3,4], with four output NER labels: *Organisation* (ORG), *Miscellanea* (MISC), *Location* (LOC) and *Person* (PER).

The pipeline of the model is: TOK2VEC, MORPHOLOGIZER, TAGGER, PARSER, ATTRIBUTE_RULER, LEMMATIZER, NER; it has 500k keys, 500k unique vectors as word vectors, embedded in a 300-dimensional vector space.

During the development of the alpha-version of the web app, spaCy v3 was released, adding the Transformer model architecture. It was thus possible to have a different pipeline: TRANSFORMER, NER.

It was thus possible to train with transfer learning the two architectures, compare their scores, and employ the most suitable ones for our needs.

2.1 The Training Dataset: Archaeological Documents and Scientific Reports Annotated with INCEpTION

The dataset for the ArcheoNER training consists of 92 Italian-written archaeological documents containing 5230 entities, annotated using the open-source program INCEpTION [16]; the 9 entities to be identified within the text have been chosen for their relevance in the archaeological sector and according to their compatibility with top-level classes of CRMhs and the CIDOC CRM ecosystem. They are (Name, three-letter label, HEX code for HTML rendering[2]):

Artefact (ART, # 6196A4), *Site* (SIT, # D4AE60), *Person* (PER, # BF98D0), *Time span* (TSP, # 61A46A), *Activity* (ACT, # D12153), *Organisation* (ORG, # B86F9A), *Place* (LOC, # EDE89B), *Period* (PRD, # ABEF70), *Biological remains* (BIO, # 89CBCA)

The dataset for the hsNER training was formed by 43 italian-written scientific reports containing 5676 entities, annotated again with INCEpTION; in this case, the labels, chosen for relevance and their compatibility with the CRMhs ontology, were 15, extending the previous:

Artefact (ART, # 6196A4), *Person* (PER, # BF98D0), *Time span* (TSP, # 61A46A), *Activity* (ACT, # D12153), *Organisation* (ORG, # B86F9A), *Place* (LOC, # EDE89B), *Natural object* (BIO, # 89CBCA), *Sample* (SAM, # 8E44AD), *Analysis* (ANL, # 909497), *Material* (MAT, # F8C471), *Method* (MET, # A2D9CE), *Device* (DEV, # FADBD8), *Software* (SOF, # E67E22), *Result* (RES, # 58D68D).

The scientific reports employed refer to analysis carried out at the INFN-CHNet node of Florence, LABEC laboratory (*Laboratorio di tecniche nucleari applicate all'Ambiente e ai BEni Culturali*, laboratory of nuclear techniques for environment and cultural heritage) [7]; the analyses in the reports were either X-Ray Fluorescence (XRF) imaging of Pictorial Artworks or Radiocarbon dating of various samples, either *Natural objects* or *Artefacts*.

In future development, we plan to add two additional labels: *Site* (SIT, # D4AE60) and *Period* (PRD, # ABEF70), by enlarging the training dataset to account for reports with such entries.

The Challenges in the Data Harvest: One of the major issues faced in the creation of the training dataset, for both ArcheoNER and hsNER, is the harvest of appropriate data.

Indeed, the Italian written documents have to be chosen carefully, since the main goal of the project is the development of a web service for (semi)automatic query generation for the CHNet database, where the users (i.e. INFN-CHNet researchers and collaborators) may employ their written reports on scientific

[2] Using the open-source add-on library *displacy*, is possible to render the annotated text as HTML, where named entities are highlighted using a HEX colour. Such hex colour is employed in the web application as a visual aid to users, in order to visually discriminate entities among NER labels.

analysis on archaeological and other CH-related items to fetch the database for similar and/or related analysis entries.

This fact imposes a peculiar constraint on the nature of Italian written documents we have to employ for creating the dataset. Such constraint strongly limits the pace at which the dataset size can be increased, thus impacting the precision reachable by the two models.

Nevertheless, this fact do not constitute an issue. Indeed, as we will remark later, any sufficiently working model will allow for the creation of a working AI-powered query generator web service.

2.2 Training and Evaluation of the Model

Table 1. Comparison of the results of the two ArcheoNER models.

ArcheoNER	P	R	F
CNN	34.63	**35.78**	35.19
Transformer	**30.75**	38.94	**34.36**

Table 2. Comparison of the results of the two hsNER models.

hsNER	P	R	F
CNN	80.13	**63.02**	**70.55**
Transformer	**70.91**	76.82	73.75

The two models were trained with transfer learning [17], either on the archaeological dataset (i.e. ArcheoNER model) or the Scientific dataset (i.e. hsNER). Starting from the aforementioned spaCy models, we removed the output layer, inserting our ad-hoc output layer, and trained the model. In the end, we performed the training fine-tuning.

The training was performed using an Intel(R) Core(TM) i9-10900 K CPU 3.70 GHz, and 4×16 Gb DDR4 3600 MHz RAM.

The scores employed are Precision (P), Recall (R) and F-score (F)[3], and are reported in Table 1 for the two ArcheoNER models, and in Table 2 for the two hsNER models. In Tables 3, 4 we report the scores divided per entity label for the ArcheoNER models (CNN and Transformer, respectively), while in Tables 5, 6 we report the scores divided per entity label for the hsNER models (CNN and Transformer, respectively).

Discussion of the Results: Looking at Tables 1, 2, it is immediate to realise that the Transformer model performances are strikingly similar to the ones of the CNN model, furnishing little (for hsNER) to none (for ArcheoNER) improvement. Also, it may seem easy to state that hsNER achieves way better results than ArcheoNER, whose results are quite far to be good.

[3] We recall that

$$\mathrm{P} = \frac{TP}{TP + FP}, \quad \mathrm{R} = \frac{TP}{TP + FN}, \quad \mathrm{F} = 2\,\frac{\mathrm{P} \cdot \mathrm{R}}{\mathrm{P} + \mathrm{R}}, \tag{1}$$

where TP are the true positive counts, FP are false positive counts, and FN are false negative counts.

Table 3. Results for CNN-ArcheoNER by labels.

	P	R	F
ACT	41.21	35.19	37.96
ART	34.61	37.96	36.21
BIO	23.91	23.91	23.91
LOC	23.85	24.31	24.08
ORG	28.79	29.69	29.23
PER	32.73	12.95	18.56
PRD	54.09	65.24	59.14
SIT	24.68	17.12	20.21
TSP	36.23	50.78	42.29

Table 4. Results for Transformer-ArcheoNER by labels.

	P	R	F
ACT	29.61	22.75	25.73
ART	23.33	38.42	29.03
BIO	15.38	30.43	20.44
LOC	37.61	38.17	37.88
ORG	32.89	39.06	35.71
PER	50.24	37.05	42.65
PRD	64.04	62.66	63.34
SIT	20.27	13.51	16.22
TSP	43.04	52.11	47.14

Table 5. Results for CNN-hsNER by labels.

	P	R	F
ACT	31.58	24.00	27.27
ANL	94.12	72.73	82.05
ART	55.56	38.46	45.45
DEV	83.33	90.91	86.96
LOC	75.00	100.00	85.71
MAT	60.00	90.00	72.00
MET	60.00	27.27	37.50
NAT	70.97	64.71	67.69
ORG	85.71	60.00	70.59
PER	66.67	40.00	50.00
SAM	83.12	50.00	62.44
SOF	100.00	100.00	100.00
RES	90.62	87.88	89.23
TSP	94.83	87.30	90.91

Table 6. Results for Transformer-hsNER by labels.

	P	R	F
ACT	35.29	24.00	28.57
ANL	58.62	77.27	66.67
ART	62.50	38.46	47.62
DEV	100.00	90.91	95.24
LOC	100.00	33.33	50.00
MAT	69.23	90.00	78.26
MET	38.46	45.45	41.67
NAT	89.29	73.53	80.65
ORG	88.24	75.00	81.08
PER	9.26	100.00	16.95
SAM	79.37	78.12	78.74
SOF	100.00	100.00	100.00
RES	90.91	90.91	90.91
TSP	100.00	96.83	98.39

From the comparison of Tables 3, 4 with Tables 5, 6, it may appear that the performances of hsNER are better than the ones of ArcheoNER. Instead, the hsNER may be *overfitted* to the LABEC scientific reports. In fact, applying both models to Italian-written documents of different origin (i.e. not on Archaeological articles nor LABEC scientific reports), it is possible to notice that the hsNER performance is worse than the ArcheoNER one. ArcheoNER, instead, was trained on a more statistically sparse dataset, and thus we have more difficulty in obtaining a good learning scores, which suggest that an enlargement of the training dataset is at need, for both hsNER and ArcheoNER.

Nevertheless, the result is not daunting, as previously stated in Sect. 2.1; in fact, both the models can be employed for the main goal of the project, which was building a web service for (semi)automatic query generation on CHNet database.

Indeed, both hsNER and ArcheoNER are capable of correctly identifying the most relevant nominal entities present in the text, which usually are the most important ones (with their labels) to construct the {KEY: VALUE} query. A most performing NER would be able to correctly identify more statistically uncommon named entities, which are usually less important in the process of building the query.

It is clear that an improvement on the models must, and will, be planned in the future: either by increasing the dataset size and quality, changing and fine-tuning the models, and so on; this will constitute an improvement in the efficiency of the whole web service, which is now available on the CHNet cloud.

3 The Web Service: AI-powered Queries

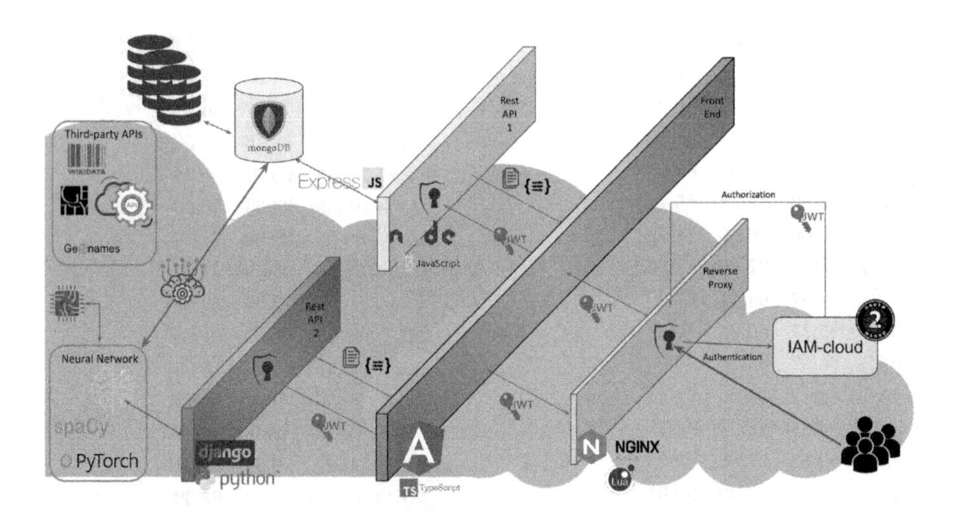

Fig. 1. Full-stack architecture of THESPIAN-NER service, and its interaction with the THESPIAN-Mask service.

The web service THESPIAN-NER comprises a Front-End and a Back-End service. The Front-End, client-side service, for uploading text items (in .txt or .pdf format), via a POST HTTP request to the THESPIAN-NER Back-End, server-side RESTful API, and to easily compose the (AI-powered) query via the Web User Interface, and thus, via a GET HTTP request to the CHNet database-connected Back-End RESTful API, fetch the database (for a pictorial representation of the service architecture and user-experience-workflow, see Fig. 1).

The Front-End was developed using TypeScript (TS), the JavaScript strict syntactical superset which adds optional static typing, and bundled using the open-source web application framework *Angular*.

Firstly, users have to upload an Italian-written document; it will be sent to the Back-End and processed, using the best trained model; after that, the Back-End replies by sending the annotated document as HTML, a summary of the text obtained via an extractive summarisation process, as well as the list of annotated entities as an array. The annotated text is thus rendered on the page and the named entities found in the text are displayed on the text, divided for labels and ordered by number of occurrences (see Fig. 2).

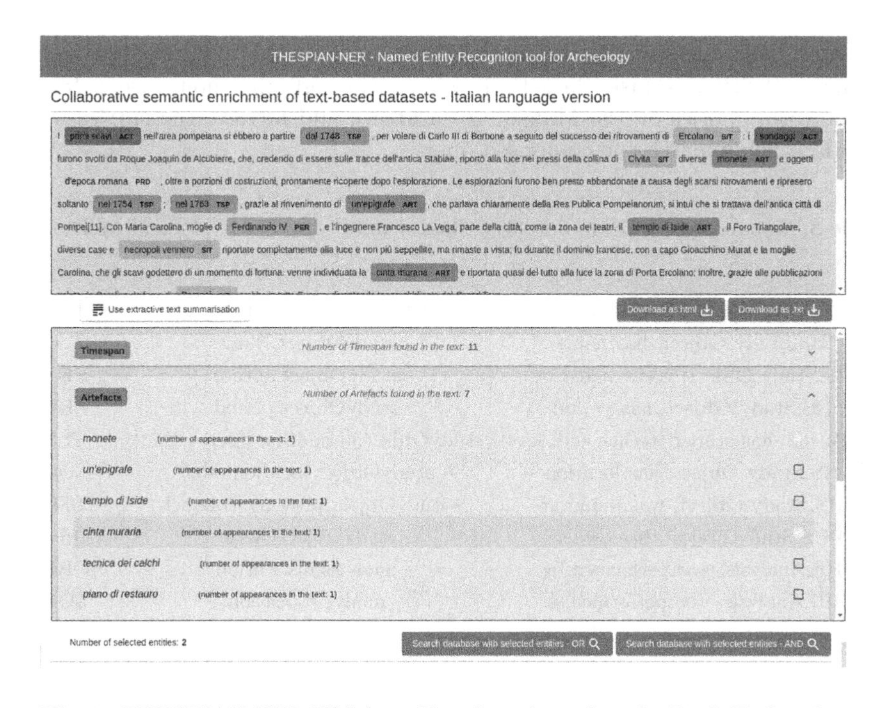

Fig. 2. THESPIAN-NER WebApp: User Interface after the Back-End reply.

At this point, users may select named entities to compose their queries by simple clicks on the User Interface (see Table 8 for a relation between database keys, their associated CRMhs ontology (Classes, Properties) [20], and the NER labels); they can either perform a logic OR or a logic AND on the labels. For example, if the user has selected from the *Person* (PER, #BF98D0) list the named entity 'LEONARDO DA VINCI', he/she may perform the query on the database with the corresponding {KEY: VALUE} of {STUDYOBJECT.AUTHOR: 'LEONARDO DA VINCI'}, as visually explained in Table 7.

Table 7. Example of a match between the annotated entity and the relative MongoDB query.

	label	entity
NER	PER	Leonardo da Vinci
MongoDB	studyObject.author	Leonardo da Vinci
	key	value

After their click on the appropriate button, a HTTP GET request is sent to the THESPIAN-Mask Back-End, which, through the mongoose.js library, can perform queries on the MongoDB database, and retrieve the entries whose metadata fields match the composed query.

Table 8. Map between CRMhs ontology Classes and Properties (C, P) and ArcheoNER labels.

CRMhs Entry (C, P)	THESPIAN-Mask JSON key	NER label
HS_Study_Object, has_name	studyObject.Name	ART, BIO
HS_Study_Object, was_created_by	studyObject.author	PER
HS_Study_Object, has_period	studyObject.period	PRD
HS_Period, has_start_date(has_end_date)	studyObject.periodStart(periodEnd)	TSP
HS_Study_Object, has_location	studyObject.locationLabel	LOC
HS_Study_Object, was_found_at	studyObject.provenanceLabel	SIT
HS_Study_Object, has_owner	studyObject.owner	ORG
HS_Analysis, was_performed_by	analysis.institution	ORG
HS_Analysis, was_performed_at	analysis.location	LOC
HS_Analysis, was_performed_during	analysis.startDate(endData)	TSP
HS_Sample_Preparation, used_method	sample.preparation.method	ACT

The THESPIAN-NER Back-End is developed using Python and the open-source web framework Django, and arranged as a RESTful API, capable of responding to POST requests (the ones described above). After the file is received, it is parsed for extracting the text as a string[4] and funrished to the stored model. After the analysis of the text, the results are sent back to the Front-End user.

The THESPIAN-Mask Back-End is developed using the open-source, back-end JavaScript runtime environment Node.js; it employs the modular web application framework package Express.js, Moongose.js for creating a connection between MongoDB and Express.js, and the node.js-middleware Multer for handling MULTIPARTFORM-DATA, which is used for handling uploaded files. The database employed is the NoSQL database MongoDB (see [2] for more details on THESPIAN-Mask).

[4] Using PyPDF2 for the .pdf parsing.

The platforms hosting the cloud have a modular architecture based on containers, i.e. each of the previous services were containerised using Docker.

In front of these an additional container hosts a reverse proxy that acts as an SSL/TLS terminator and enforces access control policies through a token-based authorisation mechanism relying on the IAM service [6] developed by the INDIGO project [9], and based on the OAuth2 protocol [14]. The container architecture allows easy deployment of the whole cloud service using an orchestrator, such as Kubernetes.

4 Conclusions and Outlook

THESPIAN-NER, a web service for (semi-)automatic query generation to fetch the INFN-CHNet database, is currently available in alpha-version on the CHNet cloud. It relies on the NoSQL, JSON-organised CHNet database and its connection with the ad-hoc developed CRMhs ontology. It relies also on the two customised, open-source NER models to annotate Italian-written archaeological text or reports of scientific analysis on cultural heritage. In order to employ the so-found named entity to build a query, it employs the match between NER labels, the NoSQL-database JSON metadata entries, and the ad-hoc defined CRMhs ontology.

The app is containerised and deployed on the CHNet cloud, which employs an OAuth2-based authorisation mechanism for enforcing connection cyber security.

The two deep learning models, ArcheoNER and hsNER, were customised from spaCy v3 models, and trained on two ad-hoc datasets, annotated via the open-source software INCEpTION, employing transfer learning, to recognise named entities with archaeological and/or scientific relevance.

At first sight, the scores of the two models may appear to be not completely satisfactory, but their performances are vastly sufficient for the development goal; of course, it implies that further studies are required for the NER models, both about varying model architectures, and/or by enlarging ad diversifying the training datasets.

Nevertheless, these two models allowed us to develop and deploy the web service THESPIAN-NER, which is currently available in the alpha version on the CHNet cloud, leaving for the future the goal of tuning the performances of the models, but offering right now to users easy-to-generate queries on the CHNet database, via a cloud-based web application.

Acknowledgments. The present research has been partially funded by the European Commission within the Framework Programme Horizon 2020 with the projects ARIADNEplus (GA no. H2020-INFRAIA-01-2018-2019-823914) and EOSC-Pillar (GA no. H2020-INFRAEOSC-05-2018-2019-857650).

References

1. Bekiari, C., Bruseker, G., Doerr, M., Oreand, C.E., Stead, S., Velios, A.: CIDOC CRM. International Committee for Documentation (CIDOC) of the International Council of Museums (ICOM). Version 7.1.1. https://doi.org/10.26225/FDZH-X261

2. Bombini, A., et al.: CHNet cloud: an EOSC-based cloud for physical technologies applied to cultural heritages. In: GARR (ed.) Conferenza GARR 2021 - Sostenibile/Digitale. Dati e tecnologie per il futuro, vol. selected papers. Associazione Consortium GARR, 10.26314/GARR-Conf21-proceedings-09 (2021). https://doi.org/10.26314/GARR-Conf21-proceedings-09
3. Bosco, C., Lenci, A., Montemagni, S., Simi, M.: Universal dependencies 2.9 - italian corpus, LINDAT/CLARIAH-CZ digital library at the Institute of Formal and Applied Linguistics (ÚFAL), Faculty of Mathematics and Physics, Charles University (2021). http://hdl.handle.net/11234/1-4611,
4. Bosco, C., Montemagni, S., Simi, M.: Converting Italian treebanks: Towards an Italian Stanford dependency treebank. In: Proceedings of the 7th Linguistic Annotation Workshop and Interoperability with Discourse, Sofia, Bulgaria, pp. 61–69. Association for Computational Linguistics, Aug 2013. https://aclanthology.org/W13-2308
5. Castelli, L., Felicetti, A., Proietti, F.: Heritage science and cultural heritage: standards and tools for establishing cross-domain data interoperability (2019). https://doi.org/10.1007/s00799-019-00275-2
6. Ceccanti, A., Vianello, E., Caberletti, M., Giacomini, F.: Beyond x.509: token-based authentication and authorization for hep. EPJ Web Conf. **214**, 09002 (2019). https://doi.org/10.1051/epjconf/201921409002
7. Chiari, M., et al.: LABEC, the INFN ion beam laboratory of nuclear techniques for environment and cultural heritage. Eur. Phys. J. Plus **136**(4), 472 (2021). https://doi.org/10.1140/epjp/s13360-021-01411-1
8. van Dalen-Oskam, K., et al.: Named entity recognition and resolution for literary studies. In: CLIN 2014 (2014)
9. DataCloud-Collaboration: INDIGO-DataCloud: A data and computing platform to facilitate seamless access to e-infrastructures
10. Honnibal, M., Johnson, M.: An improved non-monotonic transition system for dependency parsing. In: Proceedings of the 2015 Conference on Empirical Methods in Natural Language Processing, Lisbon, Portugal, pp. 1373–1378. Association for Computational Linguistics, Sept 2015. https://aclweb.org/anthology/D/D15/D15-1162
11. Honnibal, M., et al.: Explosion/spaCy: v2.1.7: Improved evaluation, better language factories and bug fixes, Aug 2019. https://doi.org/10.5281/zenodo.3358113
12. Honnibal, M., Montani, I., Van Landeghem, S., Boyd, A.: spaCy: industrial-strength Natural Language Processing in Python (2020). https://doi.org/10.5281/zenodo.1212303
13. van Hooland, S., De Wilde, M., Verborgh, R., Steiner, T., Van de Walle, R.: Exploring entity recognition and disambiguation for cultural heritage collections. Digital Sch. Humanit. **30**(2), 262–279 (2013). https://doi.org/10.1093/llc/fqt067
14. (IETF): The OAuth 2.0 Authorization Framework (2012). https://datatracker.ietf.org/doc/html/rfc6749
15. Jain, N., Krestel, R.: Who is mona L.? identifying mentions of artworks in historical archives. In: Doucet, A., Isaac, A., Golub, K., Aalberg, T., Jatowt, A. (eds.) TPDL 2019. LNCS, vol. 11799, pp. 115–122. Springer, Cham (2019). https://doi.org/10.1007/978-3-030-30760-8_10
16. Klie, J.C., Bugert, M., Boullosa, B., de Castilho, R.E., Gurevych, I.: The inception platform: Machine-assisted and knowledge-oriented interactive annotation. In: Proceedings of the 27th International Conference on Computational Linguistics: System Demonstrations, pp. 5–9. Association for Computational Linguistics, June 2018. http://tubiblio.ulb.tu-darmstadt.de/106270/

17. Mesnil, G., et al.: Unsupervised and transfer learning challenge: a deep learning approach. In: Guyon, I., Dror, G., Lemaire, V., Taylor, G., Silver, D. (eds.) Proceedings of ICML Workshop on Unsupervised and Transfer Learning. Proceedings of Machine Learning Research, Bellevue, Washington, USA, vol. 27, pp. 97–110. PMLR, 02 July 2012. https://proceedings.mlr.press/v27/mesnil12a.html
18. Montani, I., et al.: explosion/spaCy: v3.1.4: Python 3.10 wheels and support for AppleOps, Oct 2021. https://doi.org/10.5281/zenodo.5617894
19. Mosallam, Y., Abi-Haidar, A., Ganascia, J.-G.: Unsupervised named entity recognition and disambiguation: an application to old French journals. In: Perner, P. (ed.) ICDM 2014. LNCS (LNAI), vol. 8557, pp. 12–23. Springer, Cham (2014). https://doi.org/10.1007/978-3-319-08976-8_2
20. Niccolucci, F., Felicetti, A.: A cidoc CRM-based model for the documentation of heritage sciences. In: 2018 3rd Digital Heritage International Congress (DigitalHERITAGE) held jointly with 2018 24th International Conference on Virtual Systems Multimedia (VSMM 2018), pp. 1–6 (2018). https://doi.org/10.1109/DigitalHeritage.2018.8810109
21. Wilkinson, M.D., Dumontier, M., Aalbersberg, I.J., et al.: The fair guiding principles for scientific data management and stewardship. Sci. Data **3**, 160018 (2016). https://doi.org/10.1038/sdata.2016.1

Detecting Fake News in MANET Messaging Using an Ensemble Based Computational Social System

Amit Neil Ramkissoon(✉) and Wayne Goodridge

Department of Computing and Information Technology, The University of the West Indies at St Augustine, St Augustine, Trinidad & Tobago
amit.ramkissoon@my.uwi.edu, wayne.goodridge@sta.uwi.edu

Abstract. Mobile Adhoc Networks (MANETs) are utilised in a variety of mission-critical situations and as such, it is important to detect any fake news that exists in such networks. This research proposes an Ensemble Based Computational Social System for fake news detection in MANET messaging. As such this research combines the power of Veracity, a unique, computational social system with that of Legitimacy, a dedicated ensemble learning technique, to detect fake news in MANET messaging. Veracity uses five algorithms namely, VerifyNews, CompareText, PredictCred, CredScore and EyeTruth for the capture, computation and analysis of the credibility and content data features using computational social intelligence. To validate Veracity, a dataset of publisher credibility-based and message content-based features is generated to predict fake news. To analyse the data features, Legitimacy, a unique ensemble learning prediction model is used. Four analytical methodologies are used to analyse these experimental results. The analysis of the results reports a satisfactory performance of the Veracity architecture combined with the Legitimacy model for the task of fake news detection in MANET messaging.

Keywords: Computational social system · Content · Credibility · Ensemble learning · Fake news detection · MANET

1 Introduction

Given the mission-critical nature of Mobile Ad hoc Networks (MANETs), it is essential to predict fake news in its messages. According to [7] Mobile Ad Hoc Networks (MANETs) can be defined as a group of mobile devices that are wirelessly connected to each other to form a dynamic, independent, and self-configuring network.

MANETs are used in a variety of applications, especially in a social setting by the employment of message sharing amongst network members. As stated in [20] the social and commercial benefits of MANETs exceed the technical issues associated with them making any deployment of MANET technology for mobile applications a success. MANETs are deployed in situations where infrastructure networks are near-to or impossible e.g., war, post-disaster management and recovery and rural settings.

P. L. Mazzeo et al. (Eds.): ICIAP 2022 Workshops, LNCS 13374, pp. 278–289, 2022.
https://doi.org/10.1007/978-3-031-13324-4_24

An example of the application of MANETs can be seen in the current Ukraine-Russia War. In the current war, cyberattacks targeting Ukraine's Critical Infrastructure have been one of the main methods utilised by Russia. As stated by [21] Russia has systematically attacked certain critical infrastructure in Ukraine to ensure that the people of Ukraine have little to no access to timely, accurate and legitimate information about the war. As such with Ukraine's connectivity to the world compromised, a MANET can be used to allow the Ukrainian people the opportunity to communicate with each other and the outside world and have access to information about the situation in their country. This example will be used throughout this research to illustrate the importance of this work.

One of the major problems that MANETs face is the spread of fake news and the inability to detect it. As stated in [12] in MANETs, one class of attack is known as a fabrication attack where the attacker sends fake news to its neighbours in the network.

According to [18] fake news is defined as any news article that is intentionally and verifiably false. Predicting and detecting fake news is a challenge. As stated in [9] many consumers of online news, especially via social media, internet-enabled platforms and message sharing networks ingest and spread fake news without cross-referencing and verifying the news posted to such websites. Hence automated credibility validation of news is a necessity.

A major methodology used for the prediction of fake news is the analysis of the features of fake news to determine if any relationships exist amongst these features. According to [28] fake news detection is subdivided into four categories based upon the viewpoint of the detection strategy. These viewpoints are (i) knowledge-based (ii) style-based (iii) propagation-based and (iv) credibility-based.

As stated above credibility based fake news detection focuses on detection techniques based upon the established reputation of the news publisher. Credibility based detection, a part of context-based detection, pays particular attention to the trustworthiness of the news publisher as trust and fake news are intimately intertwined [3].

Alongside credibility based fake news detection, content-based fake news detection, a sub-genre of style-based detection is also utilised to identify fake news. As stated in [27] fake news is comprised of both the physical and non-physical content.

One major problem that computational social systems can be applied to is the credibility and content detection of the spread of fake news in MANET messaging. Computational Social Systems attempt to model and reproduce human reactions via algorithms and processes to recognise patterns in behaviour.

Hence, to accomplish the task of fake news detection, this work proposes the ensemble-learning based Veracity architecture. The Veracity architecture was introduced by [16]. Veracity investigates how does a publisher's social behaviour relates to the legitimacy of his content. Veracity is a combined credibility and content based multidimensional computational social system for fake news detection that focuses on the capture, computation, and analysis of news publisher credibility on MANETs.

For prediction, the Legitimacy ensemble learning model is used. Legitimacy is a unique ensemble learning model for the task of credibility based fake news detection. This model consists of two native classification models namely, a Two-Class Boosted Decision Tree and a Two-Class Neural Network. The Legitimacy model follows a pseudo-mixture-of-experts methodology of combining ML techniques. To accomplish the functionality of a gating model, Logistic Regression is implemented to combine the output of the two main models. The Legitimacy model was proposed by [15].

Given the relationship between social behaviour and machine learning, this work combines both of the above architectures and proposes the unique contribution of an Ensemble-Based Computational Social System and the five associated algorithms. This system employs computational social systems, using an ensemble model, in an attempt to detect fake news in MANET messaging.

The contributions of this study are as follows:

1. The design of an Ensemble-Based Computational Social System
2. The investigation of the effect of ensemble learning on the performance of crowdsourced feedback-based prediction of fake news
3. The application of the Ensemble-Based Computational Social System to that of Fake News Detection in MANET Messaging
4. The discussion of the architecture and evaluation results of the Ensembled Based Computational Social System

The remainder of this paper is structured as follows. Section 2 presents the related work in this field. Section 3 describes the Legitimacy ensemble model whilst Sect. 4 presents an understanding of the Veracity architecture and the VerifyNews, PredictCred, CredScore, CompareText and EyeTruth algorithms. Experiments are conducted in Sect. 5 using the algorithms and the results of these are discussed in Sect. 6. Section 7 concludes the paper.

2 Literature Review

Prior research in fake news detection for MANET message sharing has been very sparse. Most of the prior research efforts have focused on Fake News Detection in the MANET subcategory of Vehicular Ad Hoc Networks (VANETs).

2.1 Vehicular Ad Hoc Networks

According to [19], there is a need for trusted information sharing in future vehicular networks to provide a platform for road safety and news sharing. To address this problem their work provides the Three-Valued Subjective Logic (3VSL) as a solution. Their system utilises roadside units and trust calculations. It ignores the knowledge and credibility of the message which can help to further identify the fake news. Finally, their system allows the message to spread before it can conclude that the news is fake.

As stated in [23], road safety and traffic efficiency in the Internet of Vehicles (IoV) is severely impacted by the unabated spread of fake news. To this end, they proposed Quick Fake News Detection (QcFND) in their paper, which operates as a network computing framework. Their framework employs Software Defined Networking (SDN), Edge Computing, Bayesian Networks and Blockchain technologies. Their scheme requires a constant connection to the roadside units and infrastructure network for fake news detection as detection is done at a central location.

2.2 Computational Social Systems

An example of computational social intelligence at work is provided by [11] which states that presently enormous amounts of information are generated on the Internet in the field of emergency management research, especially on public social platforms and this information is not being fully utilised. Their solution raises the profile of the need for greater emphasis to be placed on the field of emergency decision making. Their solution does not take into consideration the credibility of the publisher which has a heavy impact on the quality of the news.

According to another similar work proposed by [11] when news is posted to a network in the area of automatic reading and decision support systems, one of the tasks executed is the automatic segmentation of words. Though the text classifier of their solution seems similar to that proposed in this work, their work does not make use of the crowdsourced feedback that helps to validate the decision made by the system.

2.3 ML-Based Fake News Detection

Prior research in fake news detection and machine learning has been conducted.

As stated in [2], ensemble learning has been used previously in Fake News Detection. According to [2] In the current fake news corpus, there have been multiple instances where both supervised and unsupervised learning algorithms are used to classify text. In their paper, they propose a solution to the fake news detection problem using the machine learning ensemble approach and validates the use of such a solution. Their work focuses only on textual features and does not make use of the credibility features of the news.

According to [17], most of the existing machine learning based fake news detection schemes are composed of classical supervised models. In their paper, they attempt to develop an ensemble-based architecture for fake news detection. The individual models are based on Convolutional Neural Networks (CNN) and Bi-directional Long Short-Term Memory (LSTM). The representations obtained from these two models are fed into a Multi-layer Perceptron (MLP) for multi-class classification, but this model is computationally expensive.

3 Legitimacy Ensemble Model

The Legitimacy ensemble model discussed in this research has been proposed by [15]. As described in [15], their research analyses the performance of an ensemble learning model for fake news detection based upon models proposed by Microsoft Azure Machine Learning Studio Classic (AzureML) and described in [8].

The ensemble model consists of the following classification models.

3.1 Two-Class Boosted Decision Tree (BDT)

The Two-Class Boosted Decision Tree model has been proposed by AzureML. This paper utilises the Two-Class Boosted Decision Tree based upon the results stated in [14]. According to [14], from the experiments performed and the results obtained it is noted that the Two-Class Boosted Decision Tree performed the best. Hence, it can be concluded that based on our selected dataset the Two-Class Boosted Decision Tree is the best method suited for predicting Credibility Based Fake News.

3.2 Two-Class Neural Network

According to [8], a neural network is a set of interconnected layers. The inputs are the first layer and are connected to an output layer by an acyclic graph comprised of weighted edges and nodes. Between the input and output layers, you can insert multiple hidden layers.

3.3 Mixture of Experts

According to [24], they describe the original ME regression and classification models. In the ME architecture, a set of experts and a gate cooperate with each other to solve a nonlinear supervised learning problem by dividing the input space into a nested set of regions used for classification. The gate makes a soft split of the whole input space, and the experts learn the simple parameterized surfaces in these partitions of the regions. The parameters of these surfaces in both the gate and the experts can be learned using the EM algorithm.

3.4 Two Class Logistic Regression

According to [8], logistic regression is a well-known statistical technique that is used for modelling many kinds of problems. This algorithm is a supervised learning method; therefore, you must provide a dataset that already contains the outcomes to train the model. Logistic regression is a well-known method in statistics that is used to predict the probability of an outcome and is especially popular for classification tasks. The algorithm predicts the probability of occurrence of an event by fitting data to a logistic function. According to [10], one of the most commonly utilized linear statistical models for discriminant analysis is logistic regression.

4 Ensemble Based Veracity Architecture

This research presents the Ensemble Based Veracity architecture for fake news detection in MANET messaging. The Veracity architecture, as detailed by [16], accomplishes the task of gathering the credibility data of the news publisher and the content-based data of the intended message of the publisher. This is done in an infrastructureless MANET network setting in a decentralized, social computing environment utilising computational social intelligence. Veracity attempts to model social behaviour and such as Ukrainian reactions to news of the war spread over a MANET. To improve performance, Veracity uses the Legitimacy model along with the data features to predict the validity of the news, producing this novel ensemble-based combination.

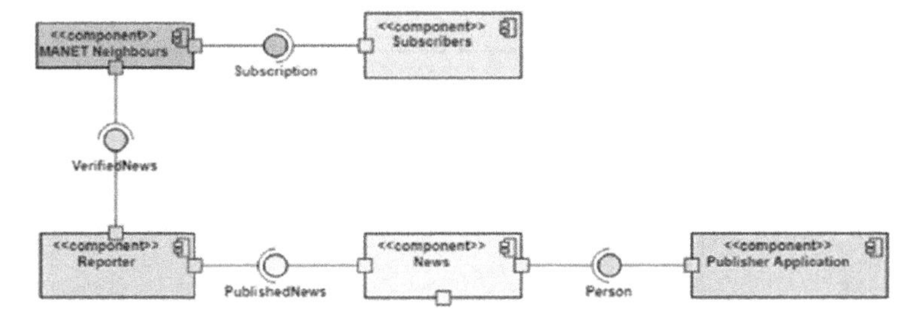

Fig. 1. Components of the veracity architecture

Veracity involves the detection of fake news from the publisher. The components of Veracity can be seen in Fig. 1 above. As seen in Fig. 1, Veracity is composed of the publisher application, the news, the monitoring agent Reporter, the various MANET neighbours, and the subscribers to the topic e.g., citizens of Ukraine. The time complexity of this unique ensemble based computational social system. is $O(n)$.

5 Experimental Design

The Veracity architecture is designed as a multidimensional computational social system. This paper however only explores one dimension of the Veracity architecture, i.e., its ability to detect and predict fake news generated in MANET messaging. As such the experimental design of this work only assesses the ability of the architecture to achieve its main purpose.

The following are the test environment parameters:

The environment as seen in Fig. 2 comprises the use of OMNET++ for the purpose of building a MANET simulation. This simulation involves the use of AODV as a MANET routing protocol and implementing the Veracity architecture alongside it.

5.1 ML Model

For the prediction model, the Microsoft Azure Machine Learning Studio (classic) environment is used to predict using the Legitimacy ensemble model as proposed by [15] and discussed above.

5.2 Dataset

The dataset is inspired by that proposed by [25] and contains 17,551 generated news records with 17,055 fake and 496 genuine messages. The dataset is a generated and synthetic one, that has been produced by the execution of the simulation environment. The experiments are validated on only this one dataset as there are no other that contains the features set required for the Veracity architecture. The features generated include: the text, eyewitness, label, source, date/time, language, listed count, location, statuses count, followers count, favourites count, time zone, user language, friends count, screen name, credibility score, text similarity and eyewitness score.

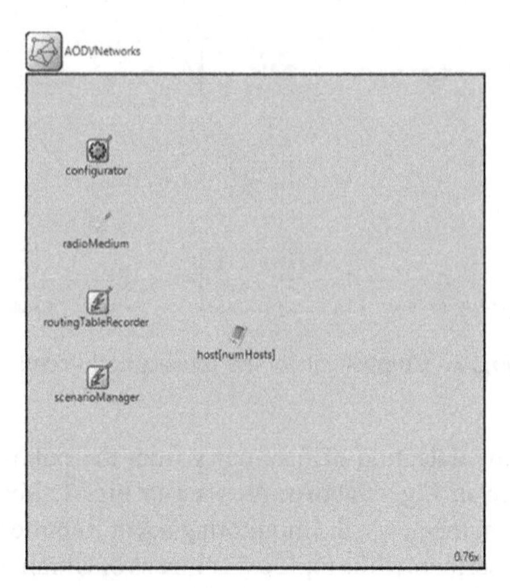

Fig. 2. OMNET++ environment.

5.3 Libraries

The FogNetSim++ library as proposed by [13] is utilised to provide the publish/subscribe functionality of the network. This library provides the functionality of publish/subscribe by providing network members such as message brokers, routers, and wireless devices.

The experiment is conducted using the above environmental setup. The MANET consists of 20 nodes all communicating with each other. Each node is configured as an AodvRouter with the Veracity functionality added to them, converting each into our unique Txc2Router. Each Txc2Router is simple in design, with all of the functionality and characteristics of the AodvRouter, inclusive of multicast forwarding.

Any missing data in the dataset is replaced with the mode value of that range. The data is then split using a separation threshold of 65% for training data and 35% for testing data. The selected ML model is then trained using the training data after which the model's performance is scored and evaluated against that of the testing dataset.

6 Results and Observations

For Veracity, the results are shown below.

6.1 Initial Results

The results are evaluated based on the Accuracy, Precision, Recall, F1-score, and AUC values and three types of graphs.

ROC Curve

As illustrated in Fig. 3, the results are first analysed based on the receiver operating characteristics (ROC) curve generated based on the performance of the model. A ROC curve is an ML evaluation method that visualizes, organises, and selects classifiers based on their performance at the task of classification [6].

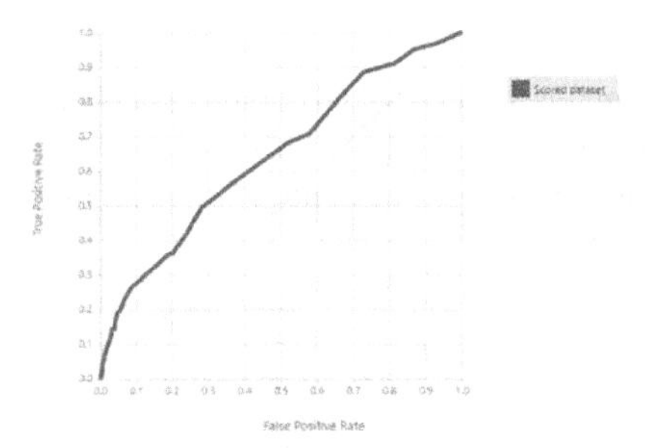

Fig. 3. ROC curve.

Precision/Recall Curve

The second method employed to analyse the performance of Veracity is the Precision/Recall Curve as seen in Fig. 4. Precision is the total value calculated when the true positive and false positive values are added together [4]. Recall is the total value calculated when the true positive and false negative values are added together.

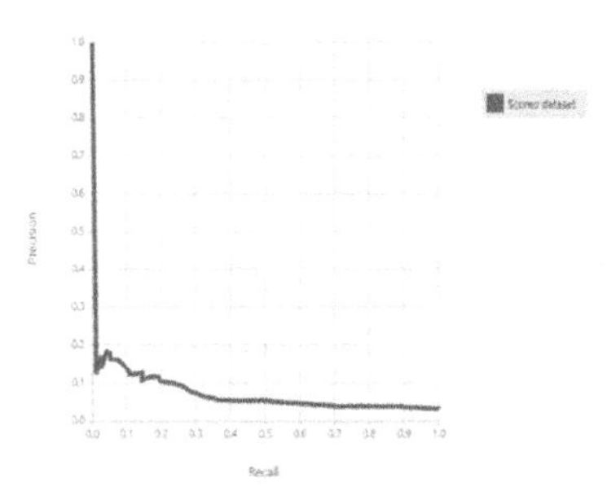

Fig. 4. Precision/Recall curve

Lift Curve

The third type of analysis is Lift Curve analysis. The Lift Curve as defined by [5] is the ratio of the result obtained with and without the classifier model applied to the prediction task known as the lift score.

The lift curve is presented in Fig. 5.

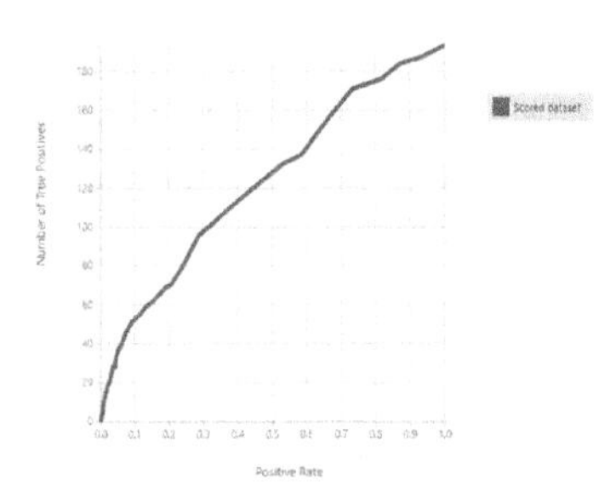

Fig. 5. Lift curve

Evaluation Metrics

The final type of results analysis used to analyse the experimental results are five evaluation metrics. For this type of analysis, the positive value is identified by 't,' the negative value is identified by 'f,' and the threshold value is 0.5 for all methods analysed. Accuracy or value of correctness is defined by [22] as:

$$Accuracy = \frac{TP + TN}{P + N} \tag{1}$$

where TP means the number of true positives predicted by the model, TN is the number of true negatives predicted by the classifier, P is the total of positives and N is the total of negatives.

Precision or value of trueness is defined by [22] as:

$$Precision = \frac{TP}{TP + FP} \tag{2}$$

where FP stands for the false positives predicted by the model.

Recall or true positive rate is defined by [22] as:

$$Recall = \frac{TP}{P} \tag{3}$$

The F1 Score or the harmonic mean of precision and recall is defined by [26] as:

$$F1\ Score = \frac{2 * precision * recall}{precision + recall} \tag{4}$$

The Area Under the Curve (AUC) value is defined by [1] as a comprehensive measure of the performance of the model by measuring the area of the shape formed by the curves produced in the experimental analysis. The AUC value ranges in value from 0 to 1.

The results show that Veracity performs well at the task of detection. The accuracy of the predictive algorithm was seen to be at 96.9%. The experiment had a 100% Precision and a 0% Recall. The F1-Score was seen to be at 0% and the AUC value was seen to be at 0.643. These results indicate that the predictive model was highly accurate and precise at the task. The same can be seen from the ROC, Precision-Recall and Lift curves in Figs. 3, 4 and 5 above. This can also be identified by the fact that 193 values were misclassified and 5950 were classified correctly leading to a high accuracy value. The model can also be seen to be highly precise by the fact that the precision of the predictions was seen to be 100%. The model also reported a 0% recall, which is a bit misleading given that the model reports zero true positives. Given the fact that integer division using zero results in zero, this recall value defaults to zero.

6.2 Comparative Analysis

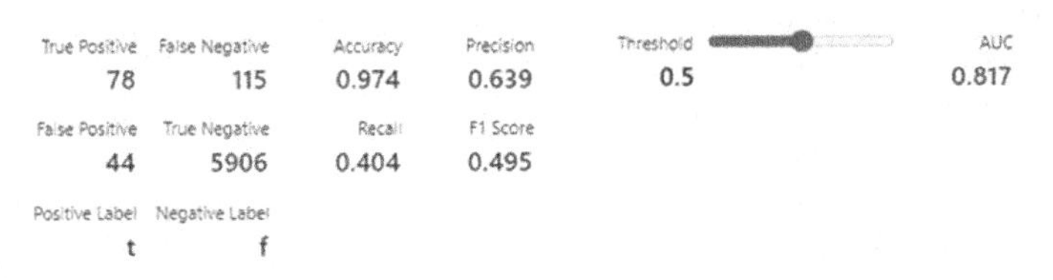

Fig. 6. Non-ensemble based veracity architecture evaluation metrics

Though the results of the non-ensemble-based Veracity architecture, as proposed by [16] and as seen in Fig. 6 above, appear better, the results obtained from the Ensemble Based Veracity Architecture show a significant improvement over them. By splitting the feature set the ensemble model allows for two subsets of the dataset to be assessed with the optimum number of features and provides a more targeted result.

7 Conclusion

This research introduced an ensemble-based computational social system to detect fake news in MANET messaging. This research attempted to answer the question of the effect of ensemble learning on the performance of crowdsourced feedback-based prediction of fake news. To this end, this work utilised Veracity, a known, multidimensional, fake news detection architecture for MANET messaging alongside Legitimacy an ensemble learning model for prediction. Veracity attempted to model social behaviour and human reactions by capturing, computing and analysing the credibility based and content-based features of messages posted to the MANET. To accomplish this task, Veracity, operating in a fully distributed and infrastructureless environment, employed five algorithms namely, VerifyNews, CompareText, PredictCred, CredScore and EyeTruth. Veracity utilised Legitimacy to predict, based upon these credibility and content features, the validity of the news. The prediction results were analysed using four machine learning methodologies. From the experiments conducted and the results obtained it is noted that Ensemble Based Veracity performed excellently and successfully modelled social reactions to the news on a MANET. It was also noted that the Ensemble prediction model enhanced the accuracy of the prediction and as such had a positive effect on the performance. Hence, based on preliminary results, it was concluded that the effect of ensemble learning on the performance of crowdsourced feedback-based prediction of fake news is positive. It was also noted that the Ensemble Based Veracity architecture is an appropriate method for detecting and predicting Fake News in MANET Messaging. Future work in this area involves investigating the energy efficiency considerations and performance of the Ensemble Based Veracity Architecture.

References

1. (2020). https://developers.google.com/machine-learning/crash-course/classification/roc-and-auc

2. Ahmad, I., Yousaf, M., Yousaf, S., Ahmad, M.O.: Fake news detection using machine learning ensemble methods. Complexity **2020** (2020)
3. Blackstock, O., Blackstock, U.: Opinion | we're not calling out Nicki Minaj. We're calling her in. The Washington Post (2021). https://www.washingtonpost.com/opinions/2021/09/17/nicki-minaj-vaccine-tweet-covid-infertility-misinformation/. Accessed 20 Apr 2022
4. Brownlee, J.: How to use ROC curves and precision-recall curves for classification in Python. Machine Learning Mastery (2019)
5. Choudhary, R., Gianey, H.K.: Comprehensive review on supervised machine learning algorithms. In: 2017 International Conference on Machine Learning and Data Science (MLDS), pp. 37–43 (2017)
6. Fawcett, T.: An introduction to ROC analysis. Pattern Recogn. Lett. **27**(8), 861–874 (2006)
7. Ibrahim Salim, M., Razak, T.A., Murugan, R.: A hybrid outlier detection approach with multi dimensional features to prevent black hole attack in MANET. Inf. Technol. Ind. **9**(1), 541–548 (2021)
8. Khan, J.M., Younus (2019). https://docs.microsoft.com/en-us/azure/machine-learning/studio
9. Khan, J.Y., Khondaker, M.T.I., Afroz, S., Uddin, G., Iqbal, A.: A benchmark study of machine learning models for online fake news detection. Mach. Learn. Appl. **4**, 100032 (2021)
10. Kirasich, K., Smith, T., Sadler, B.: Random forest vs logistic regression: binary classification for heterogeneous datasets. SMU Data Sci. Rev. **1**(3), 9 (2018)
11. Liang, X.: Introduction. In: Liang, X. (ed.) Social Computing with Artificial Intelligence, pp. 1–7. Springer, Singapore (2020). https://doi.org/10.1007/978-981-15-7760-4_1
12. Nazir, M.: A novel review on security and routing protocols in MANET. Commun. Netw. **8**(4), 205–218 (2016)
13. Qayyum, T.: FogNetSim++: a toolkit for modeling and simulation of distributed fog environment. IEEE Access **6**, 63570–63583 (2018)
14. Ramkissoon, A.N., Mohammed, S.: An experimental evaluation of data classification models for credibility based fake news detection. In: 2020 International Conference on Data Mining Workshops (ICDMW), pp. 93–100 (2020)
15. Ramkissoon, A.N., Goodridge, W.: Legitimacy: an ensemble learning model for credibility based fake news detection. In: 2021 International Conference on Data Mining Workshops (ICDMW), pp. 254–261. IEEE (2021)
16. Ramkissoon, A.N., Goodridge, W.: Veracity: a fake news detection architecture for MANET messaging. In: 2021 8th International Conference on Information, Cybernetics, and Computational Social Systems (ICCSS), pp. 402–407. IEEE (2021)
17. Roy, A., Basak, K., Ekbal, A., Bhattacharyya, P.: A deep ensemble framework for fake news detection and classification. arXiv preprint arXiv:1811.04670 (2018)
18. Shu, K.: Fake news detection on social media: a data mining perspective. ACM SIGKDD Explor. Newsl. **19**(1), 22–36 (2017)
19. Sohail, M.: Multi-hop interpersonal trust assessment in vehicular ad-hoc networks using three-valued subjective logic. IET Inf. Secur. **13**(3), 223–230 (2019)
20. Stieglitz, S., Fuchß, C.: Challenges of MANET for mobile social networks. Procedia Comput. Sci. **5**, 820–825 (2011)
21. Ukraine: A timeline of cyberattacks. CyberPeace Institute, 8 March 2022. https://cyberpeaceinstitute.org/ukraine-timeline-of-cyberattacks/. Accessed 11 Mar 2022
22. Vuk, M., Curk, T.: ROC curve, lift chart and calibration plot. Metodoloski zvezki **3**(1), 89 (2006)
23. Xiao, Y., Liu, Y., Li, T.: Edge computing and blockchain for quick fake news detection in IoV. Sensors **20**(16), 4360 (2020)
24. Yuksel, S.E., Wilson, J.N., Gader, P.D.: Twenty years of mixture of experts. IEEE Trans. Neural Netw. Learn. Syst. **23**, 1177–1193 (2012)

25. Zahra, K., Imran, M., Ostermann, F.O.: Automatic identification of eyewitness messages on Twitter during disasters. Inf. Process. Manage. **57**(1), 102107 (2020)
26. Zhang, D., Wang, J., Zhao, X.: Estimating the uncertainty of average F1 scores. In: Proceedings of the 2015 International Conference on the Theory of Information Retrieval, pp. 317–320 (2015)
27. Zhang, X., Ghorbani, A.A.: An overview of online fake news: characterization, detection, and discussion. Inf. Process. Manage. **57**(2), 102025 (2020)
28. Zhou, X., Zafarani, R.: Fake news: a survey of research, detection methods, and opportunities (2018)

PergaNet: A Deep Learning Framework for Automatic Appearance-Based Analysis of Ancient Parchment Collections

Marina Paolanti[1,2], Rocco Pietrini[1(✉)], Laura Della Sciucca[2], Emanuele Balloni[1], Benedetto Luigi Compagnoni[4], Antonella Cesarini[4], Luca Fois[4], Pierluigi Feliciati[3], and Emanuele Frontoni[1,2]

[1] VRAI Vision Robotics and Artificial Intelligence Lab, Università Politecnica delle Marche, Dipartimento di Ingegneria dell'Informazione (DII) , Via Brecce Bianche 12, 60131 Ancona, Italy
emanuele.balloni@unimc.it, r.pietrini@pm.univpm.it

[2] Department of Political Sciences, Communication and International Relations, University of Macerata, 62100 Macerata, Italy
{marina.paolanti,emanuele.frontoni}@unimc.it

[3] Department of Humanities, University of Macerata, 62100 Macerata, Italy
pierluigi.feliciati@unimc.it

[4] Archivio di Stato di Milano, Via Senato 10, 20121 Milano, Italy
as-mi@beniculturali.it

Abstract. Archival institutions and program worldwide work to ensure that the records of governments, organizations, communities and individuals be preserved for the next generations as cultural heritage, as sources of rights, and to hold the past accountable. The digitalization of ancient written documents made of parchment were an important communication mean to humankind and have an invaluable historical value to our culture heritage (CH). Automatic analysis of parchments has become an important research topic in fields of image and pattern recognition. Moreover, Artificial Intelligence (AI) and its subset Deep Learning (DL) have been receiving increasing attention in pattern representation. Interest in applying AI to ancient image data analysis is becoming mandatory, and scientists are increasingly using it as a powerful, complex, tool for statistical inference. In this paper it is proposed PergaNet a lightweight DL-based system for historical reconstructions of ancient parchments based on appearance-based approaches. The aim of PergaNet is the automatic analysis and processing of huge amount of scanned parchments. This problem has not been properly investigated by the computer vision community yet due to the parchment scanning technology novelty, and it is extremely important for effective data recovery from historical documents whose content is inaccessible due to the deterioration of the parchment. The proposed approach aims at reducing hand-operated analysis and at the same time at using manual annotation as a form of continuous learning. PergaNet comprises three important phases: classification of parchments recto/verso, the detection of text, then the

P. L. Mazzeo et al. (Eds.): ICIAP 2022 Workshops, LNCS 13374, pp. 290–301, 2022.
https://doi.org/10.1007/978-3-031-13324-4_25

detection and recognition of the "signum tabellionis". PergaNet concerns not only the recognition and classification of the objects present in the images, but also the location of each of them. The analysis is based on data from the ordinary use and does not involve altering or manipulating techniques in order to generate data.

1 Introduction

Ancient written documents made of parchment were an important communication mean to humankind and have, for that motive, an invaluable historical value to our culture heritage (CH). Before the middle ages, these documents were the primary writing equipment and they were the vehicle of culture, up to their replacement by paper [4]. Therefore, it becomes imperative that they be preserved and perpetuated in order to extend their life span in the interests of population and future generations [18].

Currently, digitization of historical parchments is extraordinarily convenient as it allows easy access to the documents from remote positions and removes the need for possible adverse physical management [6]. This arrangement is particularly suitable to archives and museums who retain such invaluable historical documents whose contents are unavailable and which cannot be fully restored by conventional tools, and are laborious to read directly due to high levels of damage and the delicate nature of the material. Such damaged parchments are notably prevalent in archives all over the world [11]. Digital representations of parchments allows to reduce this problem by giving to the operators the possibility to read their contents at any moment, from remote locations, and without necessitating harmful physical management of the document.

Thus, automatic analysis of parchments has become an important research topic in fields of image and pattern recognition. It has also been a considerable research issue for several years, gaining attention recently due to the potential value that can be unlocked from extracting the information stored in historical documents [7]. Moreover, Artificial Intelligence (AI) and its subset Deep Learning (DL) have been receiving increasing attention in pattern representation. Unlike simple artificial neural networks, DL algorithms are not only used for the mapping from representation to output but also to learn the representation itself [8]. Interest in applying AI to ancient image data analysis is becoming mandatory, and scientists are increasingly using it as a powerful, complex, tool for statistical inference. Computer-based image analysis provides an objective method of scoring visual content independent of subjective manual interpretation, while potentially being more sensitive, consistent and accurate. Learned representations often result in much better performance than can be reached with hand-designed representations. To date, parchment analysis has required user interaction, which is very time consuming for such data. Hence, the effective automatic feature extraction competence of Deep Neural Networks (DNNs) decrease the demand for a independent handcrafted feature extraction process.

Considering the above, in this paper it is proposed PergaNet a lightweight DL-based system for historical reconstructions of ancient parchments. The aim of PergaNet is the automatic analysis and processing of huge amount of scanned parchments. This problem has not been investigated by the computer vision community properly yet due to the parchment scanning technology novelty, and it is extremely important for effective data recovery from historical documents whose content is inaccessible due to the deterioration of the parchment. The proposed approach aims at reducing hand-operated analysis and at the same time at using manual annotation as a form of continuous learning. The whole system needs manual tagging of large training data. Up until now, large datasets have been necessary to boost the performance of DL models and all manually verified data will be used as continuous learning and will be maintained as training datasets.

The institute “Archivio di Stato di Milano”, the State Archives in Milan holds a wide asset of parchments belonging to 12th and 13th centuries. The Archive hosts a great collection of parchment documents that is identified by the “signum tabellionis” of notary. These signa are useful to establish a historical reconstruction of the documents since the notary who signed the document belonged to a specific historical period. PergaNet comprises three important phases: classification of parchments recto/verso, the detection of text, then the detection and recognition of the “signum tabellionis”. PergaNet concerns not only the recognition and classification of the objects present in the images, but also the location of each of them. The study expands the implementation of AI in archival science: a method that could be reproduced by many other Archives and for different types of documents. The analysis is based on data from the ordinary use and does not involve altering or manipulating techniques in order to generate data. This provides actionable insights that are helpful to identify text as style and not as reading. The preliminary results of the proposed approach show the effectiveness and the suitability of this method. It is a significant step towards finding the best path to automatically process and analyse a huge amount of scanned parchments data.

The paper is organized as follows. Section 2 provides a description of the approaches that were adopted for analysis of ancient documents. Section 3 describes the proposed DL-based pipeline. In Sect. 4, an evaluation of our approach is offered, as well as a detailed analysis of each component of our framework. Finally, in Sect. 5, conclusions and discussion about future directions for this field of research are drawn.

2 Related Works

In literature, in several works it has emerged that the main difficulty in automatic analysis of old parchment is the deterioration of the ink of the original text. This fact is due to the iron corrosion in the residual ink particles that were absorbed into the parchment. Thus, several researcher have proposed the application of image enhancement techniques to improve the quality of the digital document [13]. With the advancement of DNNs the features based approaches evolved in DL approaches.

The restoration of ancient document was the aim of the DL approach proposed in [1]. The authors proposed Pythia, an ancient text restoration model that recovers missing characters from a damaged text input using DNNs. Its architecture was designed to handle long-term context information, and deal efficiently with missing or corrupted character and word representations. For the training, they wrote a pipeline to convert PHI, the largest digital corpus of ancient Greek inscriptions, to machine actionable text, called PHI-ML.

In [15], the authors proposed a DL method for under-text separation. In particular, they have used a deep generative networks, by leveraging prior spatial information of the under-text script. To optimize the under-text, they emulated the process of palimpsest creation. This was performed by generating the under-text from a separately trained generative network to match it to the palimpsest image after mixing it with foreground text.

A Generative approach is also adopted in [16]. Tamrin et al., in fact, proposed a framework that followed a two-stage approach. The first one was devoted to data augmentation. A Generative Adversarial Network (GAN), was trained on degenerated documents. This network allowed the generation of synthesized new training document images. In the second stage, the document images generated before, was advanced with the choice of an inverse problem model with a DNN architecture.

The recent success of DL methods in image analysis related tasks has been also inspired the work proposed in [3]. In this paper, the authors adopted a fully convolutional neural network (FCNN) [9] for semantic labeling of pixels. Their idea was to use skip connections to combine the coarse features from deep layers with fine features from shallow layers to improve the final segmentation. The framework described allowed end-to-end training of an energy minimization function along with a semantic labeling network namely as Primal-Dual Net (PDNet).

A DL-based system was also proposed in [5] for Ethiopian ancient Geez character recognition system. The authors used a deep convolutional neural network for recognizing twenty-six base characters of this alphabet. Their system comprised the pre-processing phase of digitized ancient images, the segmentation stage to extract each character, and the feature extraction within the DNN.

Wiggers et al. [17] proposed two approaches for content-based image retrieval and pattern spotting in ancient document images by DL. Firstly, they pre-trained DNN model to deal with the inadequacy of training data. Then, they used a Siamese Convolution Neural Network trained on a previously collected subset of image pairs from the ImageNet dataset to determine the similarity-based feature maps.

In [2], it is proposed a system for recognizing ancient character for allowing users to organize the ancient Asian documents by DL algorithms. They applied the state-of-the-art DL-self-attention. The experiment were assessed on Oracle Bone Inscriptions and Kuzushi characters recognition, which also prove the effectiveness self-attention in ancient character recognition.

Nguyen et al. [10], adopted also a DL approach for facing with noisy conditions. The DNN followed an encoder-decoder structure by combining convolution/deconvolution layers with symmetrical skip connections and residual blocks for improving reconstructed image. They also described a global attention fusion to learn the significant regions in the image.

Considering the development of AI technologies, and in particular DL, in this paper it is proposed PergaNet a DL-based system for the automatic establishment of historical reconstruction of scanned ancient parchments. PergaNet could be adopted by archives and museums for reducing manual annotation by human operator and for preserving CH. The main contributions could be summarized as follows:

- the design of an efficient lightweight DL-based system to realize the automatic learning and analysis of ancient documents.
- the automatic design method of DNN combined with an object detector for improving the recognition accuracy.
- The study can also support scholars in the humanities in doing research concerning historical documents. In fact, it also contributes by enhancing easy access to the historical documents.

3 Materials and Methods

The overall framework of PergaNet for learning parchment features is depicted in Fig. 1. As stated in the Introduction, PergaNet framework consists of three stages: classification of parchments recto/verso, the detection of text, then the detection and recognition of the "signum tabellionis". Firstly, a network trained on a dataset of scanned parchments is needed to solve a classification task: recto/verso. This phase is performed by VGG16 Network [14]. After, inspired by the work in [19] the text in the image is detected. Then, YOLOv3 [12] was used, an architecture that learns to predict bounding box locations and classify these locations in one pass.

In the following subsections, we describe each part of our framework as well as the dataset used for evaluation.

3.1 Parchments Digitalization and Annotation: Dataset Collection

The dataset used for the evaluation comprises 2700 images. The dataset is collected by the parchments of the "Archivio di Stato di Milano" the State Archives in Milan holds a wide asset of parchments belonging to 12th and 13th centuries. These images were labelled in different classes belong to notarial family. In particular, for each class two kind of parchments is available: recto and verso of the document. The recto contains the signa tabellionis of the notary and in some cases in the parchment there is two different signa. Table 1 reports the classes of our dataset and the number of recto and signa present in the collection (Fig. 2).

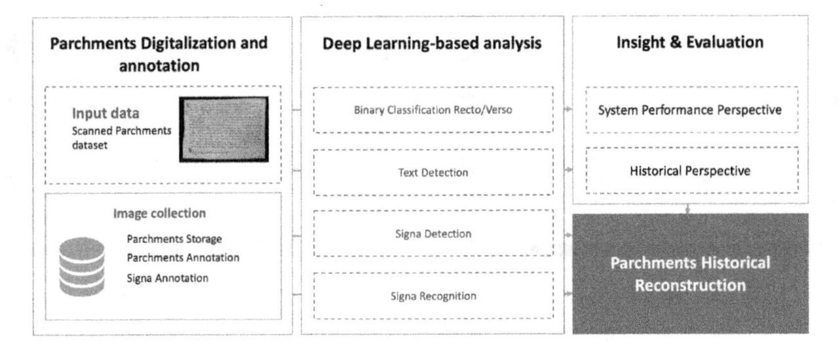

Fig. 1. PergaNet workflow. The scanned parchments are labelled based on signa tabellionis. This signum gives to images an historical period. Then a DL-bases system is able to detect text and signa in parchments as well recognizes the signum of the notary. Other data analytic layers are designed to provide insights from historical and system performance perspectives.

Table 1. Dataset of scanned parchments of the "Archivio di Stato di Milano"

Scanned Parchment	Recto	Verso
Notary 1	44	44
Notary 2	74	74
Notary 3	50	48
Notary 4	116	116
Notary 5	134	130
Notary 6	99	99
Notary 7	142	142
Notary 8	378	378
Notary 9	316	316

3.2 Deep Learning Pipeline

As stated, PergaNet is based on DL approaches devoted to different tasks. Firstly, it is important to classify the recto and the verso of the parchment. This step is fundamental, since the recognition and the historical reconstruction is based on the signum tabellionis, this object is present only in the recto side. For this reason, a binary classification is performed. We chose VGG16 Network [14] for its suitability and effectiveness in image classification task. After that inspired by the work of Zhou et al. [19], PergaNet detects the text in the image. This phase allows to exclude the text included in the parchment in the phase of recognition of the signa. The DNN model chosen is EAST for the word detection [19]. Finally, a Convolutional Neural Network has been employed for the signa detection. Our

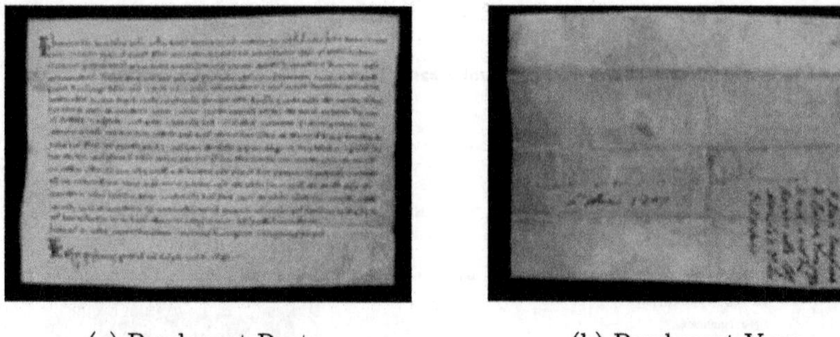

(a) Parchment Recto. (b) Parchment Verso.

Fig. 2. Examples of scanned parchments dataset images.

approach uses YOLOv3 [12] an algorithm that processes images in real time. We chose this algorithm because of its efficiency in computational terms and for its precision for detection and classification of objects. The network is pre-trained using COCO[1], a publicly available dataset, this choice aims to reduce the need of having a large amount of training data, with a high computational cost.

The DL pipeline is depicted in Fig. 3.

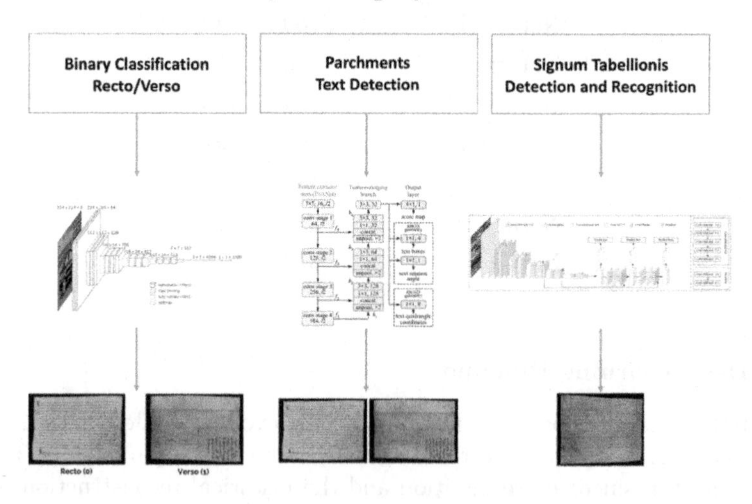

Fig. 3. PergaNet DL pipeline.

[1] https://cocodataset.org/#home.

4 Results and Discussions

In this section, the results of the experiments conducted on dataset is collected by the parchments of the “Archivio di Stato di Milano” are reported.

The first task is the binary classification of the parchment sides: recto and verso. As stated before we have chosen as DNN Model VGG16 Network. For the training phase the following setting has been used:

- Image reshape to 224 × 224;
- Data normalization: mean subtraction, std scaling;
- Augmentation with image rotation and flip;
- Batch size: 16;
- Adam optimizer;
- Learning Rate: 10e−4 (from 0th to 10th epoch) and 10e−5 (from 10th to 20th epoch);
- Cross-Entropy loss;
- Training from scratch;
- 20 epochs.

Table 2 reports and depicts the results of the binary classification. It is possible to note that the accuracy is high, thus confirming the suitability and the effectiveness of the proposed approach.

Table 2. Binary classification Recto/Verso of the parchments results.

Class	Precision	Recall	F1	Accuracy
Recto	1.00	0,70	0,83	0,85
Verso	0.77	1.00	0.87	0,85

Figure 4 represents the classification prediction compared with the ground truth. Label 0 refers to Recto side, Label 1 to the Verso one

The second experiments is devoted to the detection of the words in the text of the parchments. Figure 5 depicts the results of the text detection.

The last task is performed for the detection and the recognition of the signum tabellionis. YOLOv3 algorithm is trained and validated by using total 2700 images for signa detection of nine classes of notary. In the training phase, a group of pre-trained weights on COCO dataset is used as the initial parameters. Each model is trained by 200 epochs. After training process, the model is tested on 300 test data. The ground truth of the signa in each image of test is determined by the manual annotation. The performances of the proposed method are shown in Table 3. From Table 3, it can be quantitatively observed that the proposed method provide good results in the classification of nine notary classes. Moreover, the qualitative results are shown in Fig. 6.

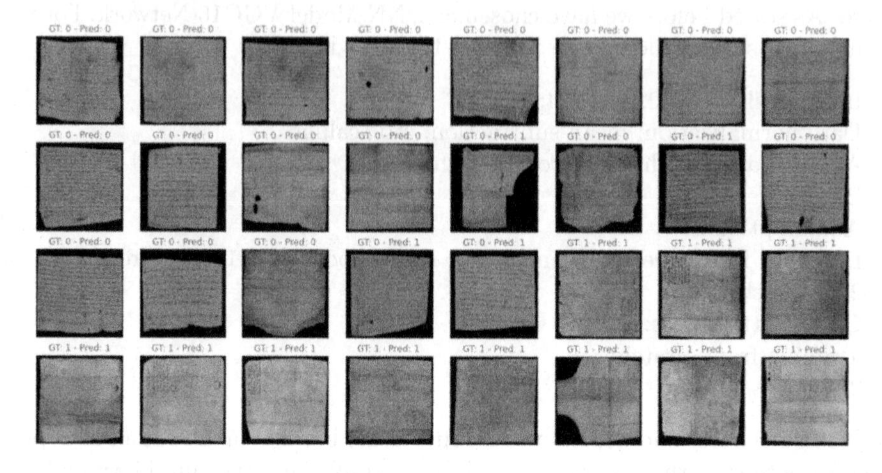

Fig. 4. Binary classification prediction compared with the ground truth. Label 0 refers to Recto side, Label 1 to the Verso one.

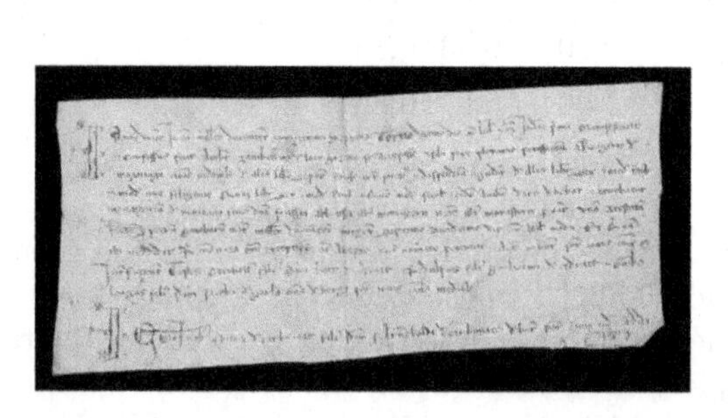

(a) Word Detection on Parchment Recto.

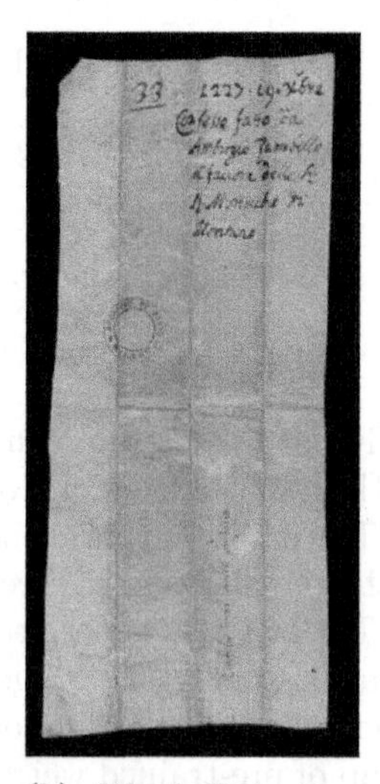

(b) Word Detection on Parchment Verso.

Fig. 5. Word detection results.

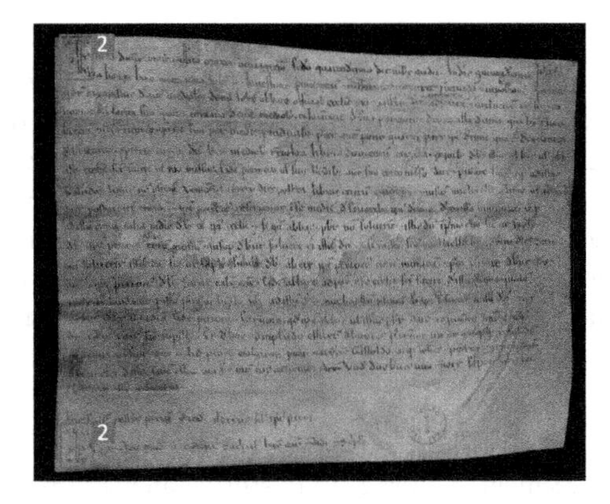

Fig. 6. Detection of signum tabellionis.

Table 3. Results obtained in the test phase on 300 images never seen in the training phase.

Class	Precision	Recall	mAP	F1 score
Notary 1	0.748	0.779	0.734	0.777
Notary 2	0.66	0.7	0.661	0.729
Notary 3	0.576	0.679	0.607	0.624
Notary 4	0.576	0.559	0.607	0.604
Notary 5	0.666	0.669	0.667	0.664
Notary 6	0.666	0.699	0.607	0.604
Notary 7	0.676	0.779	0.707	0.724
Notary 8	0.86	0.8	0.761	0.829
Notary 9	0.768	0.789	0.734	0.777

5 Conclusion and Future Works

Computer-based image analysis provides an objective method of scoring visual content independent of subjective manual interpretation, while potentially being more sensitive, consistent and accurate. In this paper, PergaNet a lightweight DL-based system is proposed for historical reconstructions of ancient parchments. The aim of PergaNet is the automatic analysis and processing of huge amount of scanned parchments of 12th and 13th centuries belonging to "Archivio di Stato di Milano". PergaNet is based on a DL pipeline that comprises three stages: classification of parchments recto/verso, the detection of text, then the detection and recognition of the "signum tabellionis". Firstly, a network trained on a dataset of scanned parchments is needed to solve a classification task: recto/verso. This phase is performed by VGG16 Network. After, the text in the image is detected. Then, YOLOv3 was used, an architecture that learns to predict bounding box locations and classify these locations in one pass. The study expands the implementation of AI in archival science: a method that could be reproduced by many other Archives and for different types of documents. The experiments yield high accuracy and demonstrate the effectiveness and suitability of our approach. Further investigation will be devoted to improve PergaNet by employing a larger dataset and extracting additional informative features for a better analysis and for deducing further insights. Moreover, we will extend the evaluation by comparing our DL pipeline with other existing systems.

References

1. Assael, Y., Sommerschield, T., Prag, J.: Restoring ancient text using deep learning: a case study on Greek epigraphy. arXiv preprint arXiv:1910.06262 (2019)
2. Atsumi, M., Kawano, S., Morioka, T., Lin, M.: Deep learning based ancient Asian character recognition. In: 2020 International Conference on Advanced Mechatronic Systems (ICAMechS), pp. 296–301. IEEE (2020)
3. Ayyalasomayajula, K.R., Malmberg, F., Brun, A.: PDNet: semantic segmentation integrated with a primal-dual network for document binarization. Pattern Recogn. Lett. **121**, 52–60 (2019)
4. Carvalho, H.P., et al.: Diversity of fungal species in ancient parchments collections of the archive of the University of Coimbra. Int. Biodeterior. Biodegrad. **108**, 57–66 (2016)
5. Demilew, F.A., Sekeroglu, B.: Ancient Geez script recognition using deep learning. SN Appl. Sci. **1**(11), 1–7 (2019)
6. Francomano, E.C., Bamford, H.: Whose digital middle ages? Accessibility in digital medieval manuscript culture. J. Mediev. Iber. Stud. **14**, 15–27 (2022)
7. Frinken, V., Fischer, A., Martínez-Hinarejos, C.D.: Handwriting recognition in historical documents using very large vocabularies. In: Proceedings of the 2nd International Workshop on Historical Document Imaging and Processing, pp. 67–72 (2013)
8. Granell, E., Chammas, E., Likforman-Sulem, L., Martínez-Hinarejos, C.D., Mokbel, C., Cîrstea, B.I.: Transcription of Spanish historical handwritten documents with deep neural networks. J. Imaging **4**(1), 15 (2018)
9. Long, J., Shelhamer, E., Darrell, T.: Fully convolutional networks for semantic segmentation. In: Proceedings of the IEEE Conference on Computer Vision and Pattern Recognition, pp. 3431–3440 (2015)
10. Nguyen, T.-N., Burie, J.-C., Le, T.-L., Schweyer, A.-V.: On the use of attention in deep learning based denoising method for ancient Cham inscription images. In: Lladós, J., Lopresti, D., Uchida, S. (eds.) ICDAR 2021. LNCS, vol. 12821, pp. 400–415. Springer, Cham (2021). https://doi.org/10.1007/978-3-030-86549-8_26
11. Pal, K., Terras, M., Weyrich, T.: 3D reconstruction for damaged documents: imaging of the great parchment book. In: Proceedings of the 2nd International Workshop on Historical Document Imaging and Processing, pp. 14–21 (2013)
12. Redmon, J., Farhadi, A.: YOLOv3: an incremental improvement. arXiv preprint arXiv:1804.02767 (2018)
13. Saxena, L.P.: Document image analysis and enhancement-a brief review on digital preservation. i-manager's J. Image Process. **8**(1), 36 (2021)
14. Simonyan, K., Zisserman, A.: Very deep convolutional networks for large-scale image recognition. arXiv preprint arXiv:1409.1556 (2014)

15. Starynska, A., Messinger, D., Kong, Y.: Revealing a history: palimpsest text separation with generative networks. IJDAR **24**(3), 181–195 (2021)
16. Tamrin, M.O., El-Amine Ech-Cherif, M., Cheriet, M.: A two-stage unsupervised deep learning framework for degradation removal in ancient documents. In: Del Bimbo, A., et al. (eds.) ICPR 2021. LNCS, vol. 12667, pp. 292–303. Springer, Cham (2021). https://doi.org/10.1007/978-3-030-68787-8_21
17. Wiggers, K.L., Junior, A.d.S.B., Koerich, A.L., Heutte, L., de Oliveira, L.E.S.: Deep learning approaches for image retrieval and pattern spotting in ancient documents. arXiv preprint arXiv:1907.09404 (2019)
18. Yahya, S.R., Abdullah, S.S., Omar, K., Zakaria, M.S., Liong, C.Y.: Review on image enhancement methods of old manuscript with the damaged background. In: 2009 International Conference on Electrical Engineering and Informatics, vol. 1, pp. 62–67. IEEE (2009)
19. Zhou, X., et al.: EAST: an efficient and accurate scene text detector. In: Proceedings of the IEEE conference on Computer Vision and Pattern Recognition, pp. 5551–5560 (2017)

Transformers with YOLO Network for Damage Detection in Limestone Wall Images

Koubouratou Idjaton[1(✉)], Xavier Desquesnes[1], Sylvie Treuillet[1], and Xavier Brunetaud[2]

[1] Université d'Orléans, INSA CVL, PRISME EA 4229, F45072 Orléans, France
koubouratou.idjaton@univ-orleans.fr
[2] Université d'Orléans, INSA CVL, LaMé, Orléans, France

Abstract. Cultural heritage buildings damage detection is of a great significance for planning restoration operations. However, the buildings analysis is generally performed by experts through on-site qualitative visual assessments. A highly time-consuming task, hardly possible at the scale of large historical buildings.

This paper proposes a new neural network architecture for automatic detection of spalling zones in limestone walls with color images. This architecture consists of the latest YOLO network, enhanced with layers of transformers encoder providing more comprehensive features. The performances of the proposed network improve significantly those of the YOLO core network on our dataset of over 1000 high resolution images from the Renaissance style Château de Chaumont in the Loire Valley (France).

1 Introduction

The preservation of historic buildings requires a careful examination of the status of their facades and structure in order to plan restoration operations. This expert analysis of heritage buildings is usually done by qualitative visual assessments on site. This is a time-consuming task that is difficult to carry out exhaustively on the scale of large historic buildings, on the entire facade of a castle, for example, due to the size of the buildings and the areas that remain unobservable by an expert from the ground. On the other hand, the increasing use of techniques of 3D acquisition and color camera mounted on a drone to scan the entire building gives experts access to more and more complete data. But the amount of data is such that it is necessary to develop algorithms to automatically detect the damaged areas on color images to facilitate their work.

In this paper, we propose to use state-of-the-art deep learning techniques for object detection to identify spalling areas on the facades of Loire Valley castles. Within the Loire Valley, a group of castles emblematic of Renaissance architecture present facades of blocks cut in tuffeau stone, a fine-grained limestone of white color, which allows the creation of magnificent ornaments sculpted

P. L. Mazzeo et al. (Eds.): ICIAP 2022 Workshops, LNCS 13374, pp. 302–313, 2022.
https://doi.org/10.1007/978-3-031-13324-4_26

in the stone. But this soft and very porous limestone is particularly sensitive to humidity variations and atmospheric pollution. The facades of the castles show a deterioration of the stones with time. The monitoring of these buildings requires an inventory of the different areas and types of deterioration, according to the illustrated ICOMOS ICIS glossary on stone deterioration patterns generally used as a reference [1]. One of the most damaging degradation in limestone masonry of Loire Valley castles is scaling in forms of spalling, a crack that develop in parallel to the stone surface, forming plates with a few centimeters depth that eventually fall, leaving a powdery surface [5].

We have built a database of more than thousand tagged images of limestone walls to train convolutional neural networks such as YOLO [6], a one-stage end-to-end model capable of detecting objects in real time, to test these capabilities on a castle scale (Fig. 1). We then propose an improved architecture including multi-head attention layers in the YOLO head that shows better performance (Fig. 2).

Fig. 1. Orthophoto of the east inner facade of the château de Chaumont-sur-Loire.

The rest of the paper is structured as follows. Section 2 reviews existing damage detection methods using deep learning. Section 3 describes the proposed network, its architecture, and its main strengths. Section 4 presents the dataset and technical implementation details, before presenting the results of the comparative study. Finally, Sect. 6 concludes this paper by considering the perspectives of this work.

2 Related Work

Different types of stone deterioration patterns: scratch, crack, detachment, flaking, blistering, discoloration, biological colonization (moss, lichen),etc., are listed by experts according to the illustrated glossary ICOMOS-ISCS [1]. Some of them represent losses of material and irregularity of the relief, while others are only chromatic alterations.

Early work on automatic stone damage detection in color images used traditional thresholding methods based on color histograms [10]. In [15], machine learning techniques has been introduced for the detection of stone-by-stone alterations on the walls of the Stirling Chapel in Scotland. The selected areas show either singular relief irregularities on the 3D model for losses or inhomogeneity in color images for chromatic alterations. The extracted features are used as inputs in a logistic regression classification algorithm.

Recently, work has adapted deep convolutional neural networks trained on large databases of generic images of objects and pets to detect stone deterioration. AlexNet [7] and GoogleLeNet [14] were used to classify about 2,000 images of the bricks cropped from the orthophotos of the wall of the Forbidden City into four categories: intact, spall, crack, and efflorescence [18]. A continuation of this work used on a more recent and powerful network like Faster R-CNN based on backbone ResNet101 for detecting spalling and efflorescence on two orthophotos (57,780 × 11,400 pixels) of the Palace Museum, China [19].

Authors in [8] also used Faster R-CNN, but with Inception as backbone, which is an older and less efficient model than ResNet101, because it does not use the residual information. They detect four types of damages: crack, loss, detachment, biological colonization. The images are from the regular report of nationally designated cultural properties in 2017 by the Cultural Heritage Administration of South Korea.

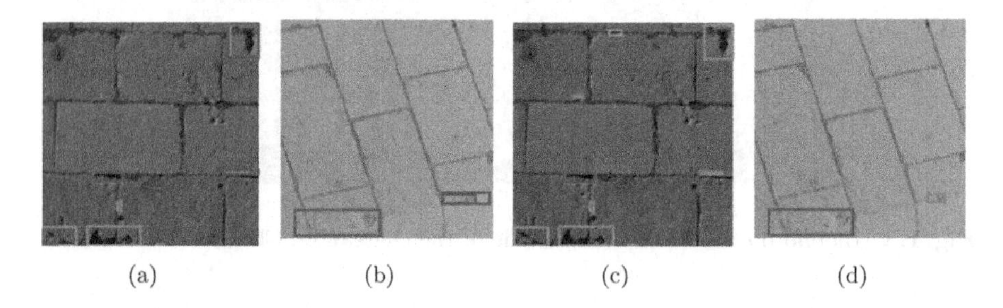

Fig. 2. The detection are in blue and the ground truth in orange. (a), (b) are example of detection performed with the proposed network and (c), (d) are example of detection performed with YOLOv5x network on the same images. (Color figure online)

Following the example of previous works, we have tested the YOLO network [6] for the detection of surface spalling on the exterior limestone walls of two Renaissance castles. To our knowledge, YOLO has never been tested in the context of detecting alterations on the walls of historical monuments, especially on walls made of fine limestone (tuffeau). This type of masonry, emblematic of the famous castles of the Loire Valley, is characterized by the search for a very homogeneous visual effect limiting the contrast of the joints with very similar colors of the stones in shades of cream to yellowish according to the layers of the quarry, which makes the task of segmentation more difficult.

This paper proposes a new neural network architecture for automatic detection of spalling zones in limestone walls with color images. This architecture consists of the latest YOLO network, enhanced with layers of transformers encoder providing more comprehensive features. The next section gives more details on the proposed architecture

3 Proposed Method

The proposed network architecture use convolution and self-attention and is mainly based on the YOLOv5x [6]. The input image is introduced into the YOLOv5x backbone, which is made of convolution 3×3 with stride 2, followed by a cross stage partial(CSP) bottleneck [17] with depth 3. The final layer of the backbone is a spatial pyramid pooling layer [3], sometimes its named the neck. It helps the network performed detection of objects at different size and scale. The output of the backbone is a feature map. This feature map from the backbone is additionaly refine in the Head. In this step, the classic YOLOv5x architecture performed more CSP Bottleneck. Our proposed network instead performed transformer self-attention modules with upsampling and concatenation with corresponding layers outputs from the backbone . Finally, the detect part is used to predict boxes, label and the confidence of the model.

The recently proposed Vision Transformers (ViT-YOLO) network [20] also achieves a possible way of combining YOLO architecture with transformer. However, that approach differs from our solution in methodology and in terms of desire output. ViT-YOLO receive as input a vector of the patches of size $16 \times 16 \times 3$ of a color image and predict as output the class of the image. It is an image classification network. Transformers requires a huge database to achieve state of the art performance. On the other hand, the combination with convolution layers only in the head, as in our proposed network allows to benefit from the convolution operations in the backbone and prevent the need of a huge dataset to achieve good performance (Fig. 3).

3.1 Transformer

Transformer architecture is a recent introduction in image processing technique. It has been introduced in 2017 and is a self-attention based architecture that become a standard in natural language processing [16]. Recent work has introduced self-attention in image processing and demonstrate its efficiency in rich features extraction for more accurate detection [2].

As shown in Fig. 4, a self-attention building block is made of the softmax of a scaled output of the matrix multiplication between a query (Q) and a key(K) vectors define in Formula (1), and the value(V) vector.

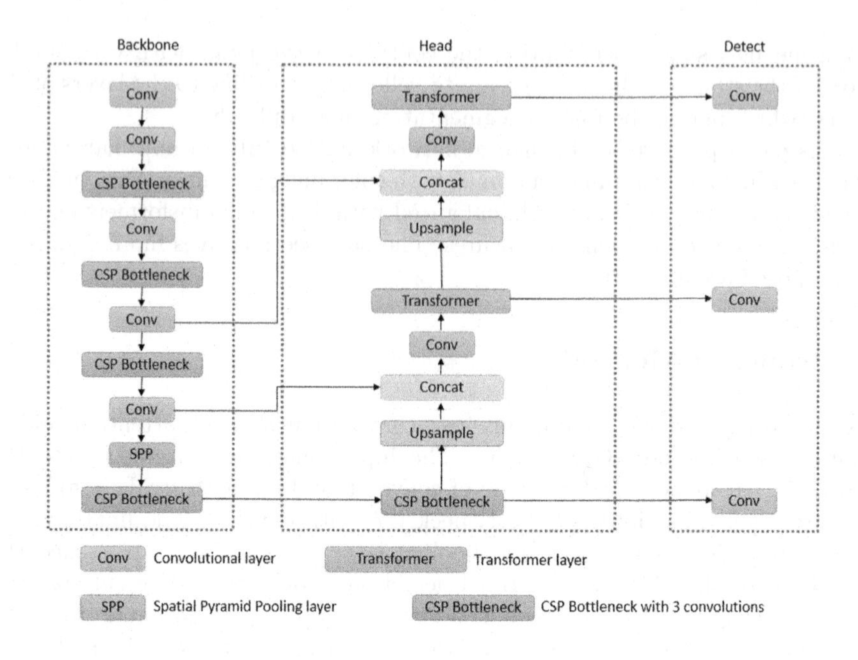

Fig. 3. Architecture of the proposed network.

$$Q, K, V = mW_{q,k,v}W_{q,k,v} \in R^{Dx3D} \tag{1}$$

with,
m the input matrice;
D the dimensions for V and K;
and W a learned weights respectively for q, k and v.

$$\begin{gathered} S = softmax(QK^T) \\ S \in R^{N \times N}, N = hw/P^2 \end{gathered} \tag{2}$$

with,
hw the image resolution;
and P the number of patches,

$$softmax(x) = \frac{e^{x_i}}{\sum e^{x_i}} \tag{3}$$

$$SA = SV \tag{4}$$

SA : self-attention output

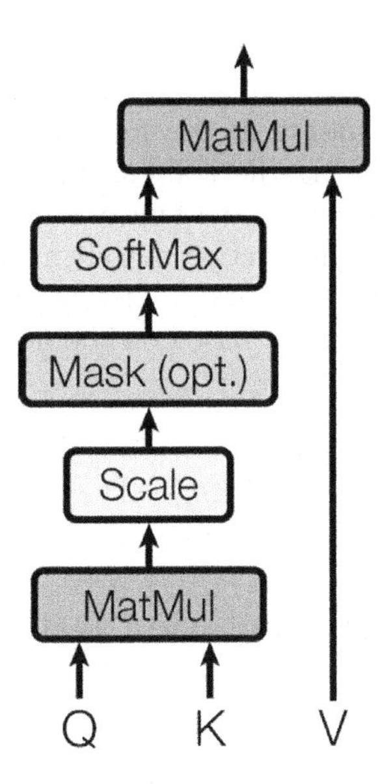

Fig. 4. Architecture of a self-attention building block [16].

3.2 Anchors

YOLO is an anchor based object detection model. The model generates thousands of anchors of multiple size (height and width) and at various ratios from which it maintains only those that have a high probability of containing an object. Then the selected anchors are refined progressively to remain few with the highest score of fitting an object.

The anchor generation process is typically configured by setting the anchor parameter that is a list of the desire default anchor size. Default anchor size are defined in the model architecture. More the anchor size adjusts to the variability of the possible bounding boxes in the dataset better the model learn and the greater is its precision in the detection. To obtain the most fitted anchors, we choose to learn the anchor parameter from a k-means algorithm, which find out by clustering the size of overall bounding boxes of the dataset.

3.3 Bounding Box Filtering

As in the process on detection several redundant bounding boxes are generated, non-maximum suppression(NMS) [4] is generally used to perform the selection of the relevant bounding box to keep forward. It considers the overlap area between the bounding boxes as the main criteria to decide which box to suppress.

In recent studies, different type of non-maximum suppression are introduced, GIoU-NMS [12], DIoU-NMS [21]. In our proposed network, we used DIoU-NMS which consider the overlap area and also the normalized distance between the bounding boxes central points. This approach helps the network to converge faster than with a typical NMS or a GIoU-NMS.

3.4 Loss Function

The loss function is an evaluation metric for the errors between the network prediction and the target during the training process. The proposed network use is a sigmoid activation followed by a binary cross-entropy loss. The sigmoid activation adjusts the network prediction output in the range $[0, 1]$. The binary cross-entropy loss compares the network prediction probability p ($p \in [0, 1]$) with the target y and calculate the error based on the negative average of the log of the correct p (see Formula 5).

$$Loss = -\frac{1}{N}\sum_{i=1}^{N} -(y_i log(p_i) + (1 - y_i) log(1 - p_i))) \quad (5)$$

4 Database and Implementation Details

4.1 Dataset

The dataset is made of 1012 sub-images with the size of 256×256 pixels extracted from the color images of the outer walls of the Château de Chaumont-sur-Loire (Fig. 1). The ground truth is provided by manual labeling conducted by experts. Experts usually label on orthomosaics from an on-site observation. Orthomosaics are created from a set of photos taken along the facade and assembled and rectified automatically from 3D modeling by photogrammetric software, MicMac [11]. This cartographic representation is very convenient for monitoring large historic buildings and provides access to actual surface measurements.

But for the detection of stone alterations it is better to work on the original color images which have much better quality and resolution than orthomosaics. The areas tagged by the experts on the orthomosaic were therefore projected onto the original color images based on geometric modeling and camera pose, as ground truth to train and test the networks.

Table 1 presents the distribution of the spalling areas in the dataset. A total of 3955 areas of spalling alterations was tagged on 1012 different images from the database. The training set represents 75% (759 images/2859 spalling areas) and the test set 25% (253 images/1096 spalling areas) of the overall database.

4.2 Implementation Details

The training is performed on a workstation running with the operating system OpenSUSE Leap 15.2. The Intel gpu-node used Intel Xeon Gold 6248 2.5 GHz processor with four(04) NVidia Tesla V100 graphics card of 32 GB memory each. The node contained 40 core with 192 GB of RAM. The source code environment is python 3.7, CUDA 11.2, cuDNN 8.1, Torch 1.9, Torchvision 0.10, along with others.

Table 1. Distribution of spalling areas in the image dataset

Data	Images	Spalling zones
Train set	759	2859
Test set	253	1096
Total	1012	3955

The testing a performed on a personal laptop with 16 GB memory, CPU Xeon E5-2620 v4 and a single NVIDIA Quadro P5000. This confirms the ability of the proposed network to be exploited by cultural heritage experts without the need of an extensive computational power.

The network training settings has been adjust to build in line with our workstation configuration and running environment. The network size is 256×256, aligned with the image size. One class is considered in the experiment, the spalling damage. The learning rate parameter is set to 0.01. We use SGD with a momentum of 0.937, a weight decay of 0.0005, a batch size of 12 and we use the one cycle learning rate policy [13].

Typically, the YOLOv5x network is pretrained on the Microsoft COCO database [9]. We refine the pretrained weights on our dataset by transfer learning. For our proposed network, a layer matching has been applied to input the pretrained weights from the COCO database into the corresponding layer in our network and learn the transformers layers we introduce from scratch with three (3) warm-up epoch at initial momentum 0.8 to help reduces instabilities in early training.

The results of the experiments performed are detailed in the following section.

5 Experiments

We present the experiments on the proposed network architecture and compare its performance to the classic YOLOv5x network performance.

As shown in Table 2 the proposed network obtain a mean average precision of 79% while the classic YOLOv5x is at 74% at IoU = 0.50. Same precision of 88% is realized by the two models, which means when they detect a damage bounding boxes much of them are real damages which is a good indicator for the cultural heritage experts.

However the proposed network does much better in recall than YOLOv5x, which means the ratio of the accurate damage zone detection performed by the proposed network to the actual relevant number of damage zone detection is higher. This offer better reliability to avoid excluding multiple zone of damage in the detection process.

The Fig. 5 below shows the detection of the two networks on various images from the test set. The test set contains several challenging limestone wall images. Both network relatively performed well in the particular case of images. However, the proposed network detects more damage zone in complex, crowded images while the classic YOLOv5x detect often less. Thus, it improved the efficiency of the cultural heritage experts in achieving better monitoring.

Table 2. Metrics comparison of the proposed network and YOLOv5x.

Methods	Precision	Recall	F1 score	mAP at IoU = 0.50
YOLOv5x	0.88	0.77	0.82	0.74
Ours	0.88	0.80	0.83	0.79

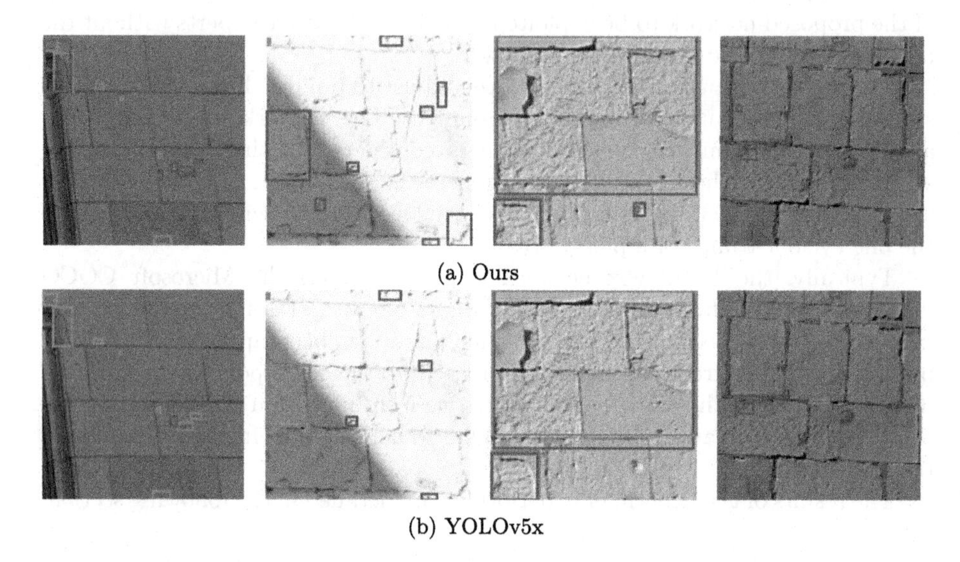

(a) Ours

(b) YOLOv5x

Fig. 5. Comparison of detection results: (a) proposed network, (b) YOLOv5x. In each case, the blue bounding boxes represent the detection and the orange the ground truth. (Color figure online)

Figure 6, present the evolution of the loss, mAP, precision and recall for the proposed network and YOLOv5x during the training process. The mean average precision of the proposed network progress better than the other. Same observation for the precision and the recall metrics value progression.

However the loss convergence is less than the YOLOv5x loss convergence during the training process. This may be explained by the fact that YOLOv5x network has been pretrained on a large dataset, Microsoft COCO, of approximately 328 000 images. While the proposed network has been mainly trained from scratch on the custom data details in Sect. 4.

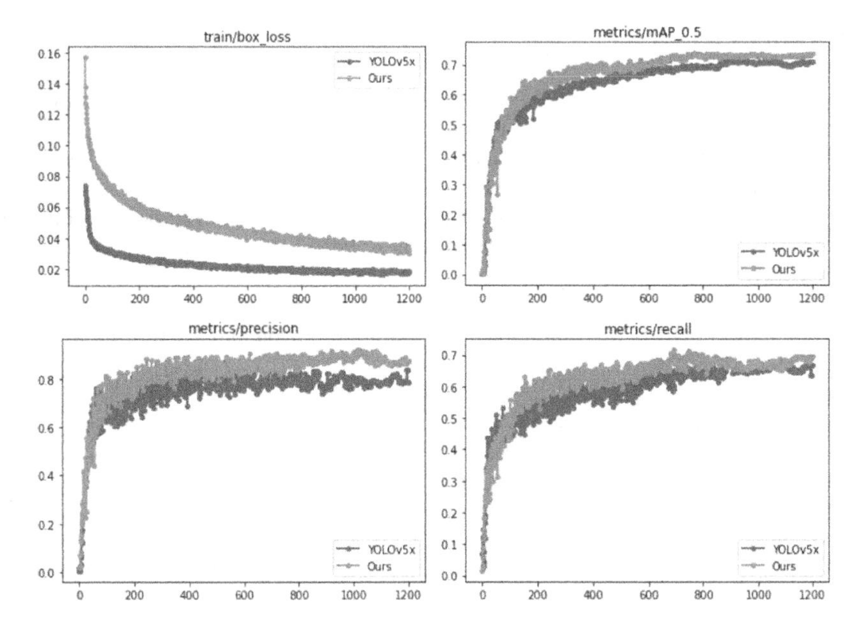

Fig. 6. Comparison of the evolution of the loss, mAP, precision and recall for the proposed network and YOLOv5x during the training.

Despite the small size of our custom dataset the convergence obtain might also be improved by the use of Distance IoU a box filtering approach known for its contribution to the rapid convergence of the trained network.

6 Conclusion

This paper proposed a novel network architecture for the detection of damage on limestone wall images. The proposed network combining the transformer building layer and YOLOv5 network. We enhance the network learning capabilities with new practice such as anchor learning with k-means and Distance-IoU Loss.

As shown in the experiments, this improved the detection of the spalling on limestone wall and outperformed the YOLOv5x network pretrained Microsoft COCO dataset and fine-tuned. This represents a significant improvement to facilitate the process of damage annotations by cultural heritage experts.

Further improvement would likely be to experiment the proposed network architecture on a large dataset containing images from various cultural heritage buildings in limestone, expecting to increase the performance on a larger dataset of limestone wall damage.

Acknowledgment. The authors benefited from the use of the cluster at the Centre de Calcul Scientifique en région Centre-Val de Loire.

References

1. ICOMOS-ISCS: Illustrated glossary on stone deterioration patterns = icomos-iscs: Glossaire illustré sur les formes d'altération de la pierre (2008)
2. Dosovitskiy, A., et al.: An image is worth 16x16 words: transformers for image recognition at scale. arXiv preprint arXiv:2010.11929 (2020)
3. He, K., Zhang, X., Ren, S., Sun, J.: Spatial pyramid pooling in deep convolutional networks for visual recognition. IEEE Trans. Pattern Anal. Mach. Intell. **37**(9), 1904–1916 (2015)
4. Hosang, J., Benenson, R., Schiele, B.: Learning non-maximum suppression. In: Proceedings of the IEEE Conference on Computer Vision and Pattern Recognition (CVPR), July 2017
5. Janvier-Badosa, S., Beck, K., Brunetaud, X., Guirimand-Dufour, A., Al-Mukhtar, M.: Gypsum and spalling decay mechanism of Tuffeau limestone. Environ. Earth Sci. **74**(3), 2209–2221 (2015). https://doi.org/10.1007/s12665-015-4212-2
6. Jocher, G., Nishimura, K., Mineeva, T., Vilariño, R.: Yolov5. Code repository (2020) https://github.com/ultralytics/yolov5
7. Krizhevsky, A., Sutskever, I., Hinton, G.E.: Imagenet classification with deep convolutional neural networks. Adv. Neural Inf. Process. Syst. **25**, 1097–1105 (2012)
8. Kwon, D., Yu, J.: Automatic damage detection of stone cultural property based on deep learning algorithm. Int. Arch. Photogram. Remote Sens. Spatial Inf. Sci. **42**(2/W15) (2019)
9. Lin, T.-Y., Maire, M., Belongie, S., Hays, J., Perona, P., Ramanan, D., Dollár, P., Zitnick, C.L.: Microsoft COCO: common objects in context. In: Fleet, D., Pajdla, T., Schiele, B., Tuytelaars, T. (eds.) ECCV 2014. LNCS, vol. 8693, pp. 740–755. Springer, Cham (2014). https://doi.org/10.1007/978-3-319-10602-1_48
10. Manferdini, A.M., Baroncini, V., Corsi, C.: An integrated and automated segmentation approach to deteriorated regions recognition on 3d reality-based models of cultural heritage artifacts. J. Cult. Herit. **13**(4), 371–378 (2012)
11. Pierrot, D.M.: Producing orthomosaic with a free open source software (micmac), application to the archeological survey of meremptah's tomb. In: Workshop Digital Specimen, Berlin, pp. 8–12, September 2014
12. Rezatofighi, H., Tsoi, N., Gwak, J., Sadeghian, A., Reid, I., Savarese, S.: Generalized intersection over union: A metric and a loss for bounding box regression. In: Proceedings of the IEEE/CVF Conference on Computer Vision and Pattern Recognition, pp. 658–666 (2019)
13. Smith, L.N.: A disciplined approach to neural network hyper-parameters: Part 1-learning rate, batch size, momentum, and weight decay. arXiv preprint arXiv:1803.09820 (2018)
14. Szegedy, C., et al.: Going deeper with convolutions. In: Proceedings of the IEEE Conference on Computer Vision And Pattern Recognition, pp. 1–9 (2015)
15. Valero, E., Forster, A., Bosché, F., Hyslop, E., Wilson, L., Turmel, A.: Automated defect detection and classification in ashlar masonry walls using machine learning. Autom. Construct. **106**, 102846 (2019)
16. Vaswani, A., et al.: Attention is all you need. In: Advances in Neural Information Processing Systems, pp. 5998–6008 (2017)
17. Wang, C.Y., Liao, H.Y.M., Wu, Y.H., Chen, P.Y., Hsieh, J.W., Yeh, I.H.: CspNet: a new backbone that can enhance learning capability of CNN. In: Proceedings of the IEEE/CVF Conference on Computer Vision and Pattern Recognition Workshops, pp. 390–391 (2020)

18. Wang, N., Zhao, Q., Li, S., Zhao, X., Zhao, P.: Damage classification for masonry historic structures using convolutional neural networks based on still images. Comput. Aid. Civil Infrastruct. Eng. **33**(12), 1073–1089, (2018)
19. Wang, N., Zhao, X., Zhao, P., Zhang, Y., Zou, Z., Ou, J.: Automatic damage detection of historic masonry buildings based on mobile deep learning. Autom. Construct. **103**, 53–66 (2019)
20. Zhang, Z., Lu, X., Cao, G., Yang, Y., Jiao, L., Liu, F.: ViT-YOLO: transformer-based yolo for object detection. In: Proceedings of the IEEE/CVF International Conference on Computer Vision, pp. 2799–2808 (2021)
21. Zheng, Z., Wang, P., Liu, W., Li, J., Ye, R., Ren, D.: Distance-IoU loss: faster and better learning for bounding box regression. In: Proceedings of the AAAI Conference on Artificial Intelligence, vol. 34, pp. 12993–13000 (2020)

Medical Transformers - MEDXF

On the Effectiveness of 3D Vision Transformers for the Prediction of Prostate Cancer Aggressiveness

Eva Pachetti[1,2](✉), Sara Colantonio[1], and Maria Antonietta Pascali[1]

[1] "Alessandro Faedo" Institute of Information Science and Technologies (ISTI), National Research Council of Italy (CNR), Pisa, Italy
{eva.pachetti,sara.colantonio,maria.antonietta.pascali}@isti.cnr.it
[2] Department of Information Engineering (DII), University of Pisa, Pisa, Italy

Abstract. Prostate cancer is the most frequent male neoplasm in European men. To date, the gold standard for determining the aggressiveness of this tumor is the biopsy, an invasive and uncomfortable procedure. Before the biopsy, physicians recommend an investigation by multiparametric magnetic resonance imaging, which may serve the radiologist to gather an initial assessment of the tumor. The study presented in this work aims to investigate the role of Vision Transformers in predicting prostate cancer aggressiveness based only on imaging data. We designed a 3D Vision Transformer able to process volumetric scans, and we optimized it on the ProstateX-2 challenge dataset by training it from scratch. As a term of comparison, we also designed a 3D Convolutional Neural Network, and we optimized it in a similar fashion. The results obtained by our preliminary investigations show that Vision Transformers, even without extensive optimization and customization, can ensure an improved performance with respect to Convolutional Neural Networks and might be comparable with other more fine-tuned solutions.

Keywords: Vision Transformers · Prostate cancer · ProstateX-2

1 Introduction

According to the World Health Organization, prostate cancer (PCa) is the most common tumor among European men [1]. For PCa patients, a biopsy followed by a microscopic examination of the collected specimen is, at the moment, the gold standard for diagnosis. Usually, before resorting to biopsy, the patient undergoes a multiparametric magnetic resonance imaging (mpMRI) examination. mpMRI investigations typically involve the acquisition of axial T2-weighted (T2w) images, used to investigate the anatomy, and diffusion-weighted images (DWI), from which the apparent diffusion coefficient (ADC) maps are derived. By comparing T2w images and ADC maps, radiologists make an early qualitative diagnosis according to the Prostate Imaging Reporting and Data System (PI-RADS) [2] guidelines. The PI-RADS score assigns a numerical value between

P. L. Mazzeo et al. (Eds.): ICIAP 2022 Workshops, LNCS 13374, pp. 317–328, 2022.
https://doi.org/10.1007/978-3-031-13324-4_27

1 and 5 to the suspected lesion, which is an index of probability that the lesion constitutes an aggressive prostate neoplasm. The higher the PI-RADS score, the greater the likelihood that the suspected nodule is malignant. Typically if PI-RADS $\geqslant 3$, the patient undergoes a biopsy. At this point, the tumor's aggressiveness is assessed by examining the biopsy specimen, and a grade known as the Gleason Score (GS) is associated with the lesion. If GS $\geqslant 3{+}4$, the tumor is considered clinically significant [3]. In particular, for patients with lesions having GS $> 3{+}4$, treatment is foreseen; in all other cases, the patient usually undergoes active surveillance [4].

However, this early diagnosis is affected by inter-operator variability since most depend on the radiologist's experience and the acquisition protocol used. For this reason, the patient may be over-diagnosed if the biopsy reveals a tumor that is not clinically significant [5]. Because of all these reasons, there is now an increasing need for an automated tool that can diagnose PCa in a non-invasive, robust, and reliable manner. Several studies to date are focusing on building machine learning models that exploit the potential of deep learning for the automatic classification of PCa lesions from mpMRI images. Most of the works attempt to classify clinically significant from non-significant PCa (i.e., GS $\leqslant 3{+}3$ vs. GS $\geqslant 3{+}4$) [6–9]. Only a few studies have addressed the issue of PCa aggressiveness, i.e., low-grade (LG) (GS $\leqslant 3{+}4$) vs. high-grade (HG) (GS $\geqslant 4{+}3$) lesions. In [10], the authors exploited the AlexNet model pre-trained on the ImageNet dataset, and they fine-tuned it on axial T2w images, sagittal T2w images, and ADC maps following a transfer learning approach. In particular, they leveraged a combined loss function that reduces feature variances between the same classes obtaining an AUROC of 0.869. In [11], the authors optimized several 2D CNNs with and without Attention Gates, training them with only T2w images, with only ADC maps, and with the combination of the two modalities, building a multimodal CNN. The CNN with Attention Gates trained on T2w images produced 0.875 AUROC.

Assessment of PCa aggressiveness is a challenging task for several reasons. First of all, the lesion occupies very few pixels within the image. In addition, it may occur in different areas of the prostate; therefore, the network must be able to identify it among other tissues before classifying it. For this reason, many works are now focusing on building an end-to-end model, which first detects the lesion and then classifies it [12–14].

Recently, Vision transformers (ViTs) have gained popularity in Computer Vision, exceeding the performance of CNNs in almost all tasks: classification [15], object detection [16] and segmentation [17]. They have seen an increase in their application also in medical imaging [18]. Classic ViTs require large amounts of data to be trained. Because of this, usually transfer learning approach is exploited. In this work, we wanted to verify ViTs' effectiveness in addressing a challenging task as the prediction of PCa aggressiveness without any pre-training steps but by training them from scratch on 3D acquisitions.

In the following sections, we describe our experiments with 3D ViTs, and basic 3D CNNs applied to a freely available dataset (i.e., ProstateX-2 [19]). Firstly, we introduce the dataset used and how this was prepared for training the deep learning models. Afterward, we give a description of the 3D ViT architecture used and of the training pipeline. We do the same for the 3D CNN models that we exploited to compare and evaluate the performance ensured by 3D ViTs. Therein, we report the results and compare our work to one belonging to the state-of-the-art addressing the same task. Finally, according to the results, we establish the potential effectiveness of 3D ViTs in determining PCa severity.

2 3D Vision Transformer and 3D CNN Development for Prostate Cancer Classification

The work aims to develop a 3D ViT model for assessing PCa aggressiveness based on axial volumetric T2w imaging data. Starting from the ViT model proposed in [15], we modified the architecture by reducing the number of parameters to train the model from scratch on the ProstateX-2 challenge dataset [19]. We also designed a 3D CNN and trained it from scratch on the same dataset as a reference model against which we compared our 3D ViT.

2.1 Dataset Composition

The dataset for the ProstateX-2 challenge [19] was acquired at the Radboud University Medical Centre (Radboudumc) in the Prostate MR Reference Center. The dataset contains 112 lesions from 99 patients. GS is provided for each lesion to be used as ground truth. Each study was performed through mpMRI, of which we exploited only axial T2w acquisitions since according to [11], they provide better results in the application of deep learning models for the assessment of PCa severity. In terms of aggressiveness, the dataset is composed as follows: 77 LG (69%) and 35 HG (31%). As for the location of the lesion, the dataset is organized as follows: 50 peripheral (PZ) (44%), 47 anterior fibromuscular stroma (AS) (43%), and 15 transition (TZ) (13%).

2.2 Data Preparation

To provide the model with only the most meaningful information, we selected only a subset of slices for each MRI scan, thus reducing the size of the 3D volume processed by the deep learning models. Based on the supplementary information provided with the dataset, we first selected the slice that contains the lesion. Hence, starting from that slice, we selected two slices above and below for a total of five slices per lesion. This approach allowed us to consider slices that contain the lesion or are strictly around it. Next, we harmonised the pixels dynamic from $[0 - 2^{16}]$ to $[0 - 2^{8}]$, and we converted each image type from *uint*16 to *uint*8. This operation did not affect the image quality since the *uint*16 range is barely exploited. Indeed, the maximum value assumed by the pixels in all acquisitions was 800. This procedure ensures that each image has the same range of pixel values.

Since not all the patients had equal image sizes, to make the procedure reproducible to further processing, we rescaled all the images to the most common and largest ones in the dataset (i.e., 384×384). This approach limited the number of patients that required resampling and avoided losing information due to down-sampling.

Assuming that the prostate is placed within the center of the image, we center-cropped each slice to let the model focus only on the prostate gland. The final size of each image was 128×128. Through a visual inspection, we verified that this size was appropriate to include the prostate glands of all sizes in the crop's field of view and yet, at the same time, remove most of the tissue that does not belong to the gland. Eventually, for each lesion, we obtained a volume of size $128 \times 128 \times 5$.

Since the dataset was unbalanced, we applied, to the training dataset only, three data augmentation techniques: vertical flip, horizontal flip, and rotation. Since the training set was composed of 54 LG and 27 HG volumes, we chose 9 HG volumes randomly with a fixed seed, and, for each one, we added three augmented versions to the set. In the end, the training set was composed of 54 LG and 54 HG volumes.

Eventually, we applied a mean normalization by calculating the mean value of the pixels across all the volumes within the training set only and subtracting it from all the slices in the training, validation, and test sets.

2.3 3D ViT Architectures

The ViT model used in this work stemmed from the one introduced in [15]. Since this model was designed to be trained on 2D images, we modified its structure so that it could work on 3D volumes by processing 3D patches. As a result, the only change necessary was in the embeddings' processing, as they could no longer be derived from 2D images but from 3D volumes. For this, we introduced an input variable z that represents the number of slices of the input volume and so of each patch, and we replaced the 2D convolutional layer used to compute the patch embeddings with a 3D convolutional layer.

All the three architectures described in the original work [15] were designed to be pre-trained on large datasets and then fine-tuned on smaller datasets. As we were working on 3D data, we avoided transfer learning and trained the 3D ViT from scratch. Considering the limited size of the ProstateX-2 dataset, we then rescaled the original architecture to significantly reduce the number of parameters to be set. We determined the most suitable architecture with a grid search on 18 different configurations (see Table 1), designed by varying the following parameters: Multi Layer Perceptron (MLP) size (d), hidden size (D), number of layers (L), and number of attention heads (k). In all configurations, we used a patch size (p) of 16 on the plane (i.e., the shape of the patch was $16 \times 16 \times 5$). This value seemed reasonable to allow the ViT to process enough information for each patch. In addition, some preliminary tests using $p = 8$

showed significantly worse results. We chose L and k values with the purpose of significantly reducing the number of parameters w.r.t the architecture proposed in [15]. After, we derived D value by exploiting the relation (1)

$$D = \frac{p^2 c}{k} \quad , \tag{1}$$

where c is the number of channels in the image. Finally, we calculated d value according to (2)

$$d = p^2 cn \quad , \tag{2}$$

where n is the number of patches. We also tested the d value used in the ViT-Base architecture described in [15], which is equal to 3072.

Table 1. The values considered in the grid-search.

Patch size	d	L	D	k	N configuration
16	2048	4	64	4	1
			32	8	2
			16	16	3
		6	64	4	4
			32	8	5
			16	16	6
		8	64	4	7
			32	8	8
			16	16	9
	3072	4	64	4	10
			32	8	11
			16	16	12
		6	64	4	13
			32	8	14
			16	16	15
		8	64	4	16
			32	8	17
			16	16	18

2.4 3D ViTs Training

Training, validation, and test of the models were coded in Python by employing the following modules: Pytorch (v. cuda-1.10.0) [20], Keras (v. 2.7.0) [21], Tensorflow (v. 2.7.0) [22], Numpy (v.1 .20.3) [23], Scikit-learn (v. 0.24.2) [24], Pydicom (v. 2.1.2) [25], Pillow (v. 9.0.1) [26] and Pandas [27].

Since the goal of this work was a preliminary investigation of the effectiveness of ViTs in PCa aggressiveness, we did not perform a comprehensive hyperparameter optimization; instead, we focused mainly on optimizing the architectural features of ViTs via the grid search described above. The hyperparameters' values used are: Learning rate = 1e–4, Weight decay = 1e–2, Number of steps = 1000, Batch size = 4, Warmup steps = 1000, Optimization algorithm = Adam, Loss function = Binary Cross Entropy. To make each training run reproducible, we exploited the reproducibility flags provided by Pytorch [20], Numpy [23], and Random [28] libraries, choosing a seed equal to 42.

We split the entire dataset into two: 90 lesions (80%) were used for the grid search and the final training of the best-performing architecture; 22 lesions (20%) were kept for the final test of the best-performing architecture. We ensured a strict patient separation by this split. This means that all the lesions of the same patient were contained only in one of the two splits to avoid any data leakage. In addition, we stratified w.r.t the aggressiveness label ($\frac{2}{3}$ LG and $\frac{1}{3}$ HG) and the lesion location ($\frac{2}{5}$ PZ, $\frac{2}{5}$ AS, and $\frac{1}{5}$ TZ).

We used the 90-lesion sub-set to carry out the grid search. This sub-set was further split into two sub-sets: 90% used for training and 10% used for validation. As a result, the validation set comprised 9 lesions (4 PZ [3 LG + 1 HG] + 4 AS [3 LG + 1 HG] + 1 TZ HG). For each ViT configuration, we evaluated the following metrics: specificity, sensitivity, accuracy, AUROC, and F2-score. The training was performed according to an ad-hoc early-stopping criterion defined as follows.

Early-Stopping Criterion. On the validation set, we measure both the specificity and the sensitivity at each epoch. If both metrics are greater than 0.6, we save the model at that epoch. In the subsequent epochs, if the specificity and sensitivity condition still occurs, as well as an increase in AUROC, the best model is updated. If this condition is never met, we save the model that has the higher AUROC. When possible, this criterion ensures that the model can distinguish between both classes more accurately.

At the end of the grid search, we chose the best configuration based on the performance on the validation set, and we re-trained it with a 5-fold cross-validation (CV) to obtain more statistically reliable results. Namely, the training set was divided into five equally distributed folds, of which, in turn, one was used as a validation set. This way, we minimized possible splitting bias. Moreover, also, in this case, we stratified w.r.t classes and lesion zones. The five models were finally evaluated on the same test set (i.e., the 22 lesions mentioned above). We reported performance as mean and standard deviation across each training run.

2.5 CNNs Architectures

As a comparison, we designed a 3D CNN and trained it by following the same approach used to train the 3D ViTs. The 3D CNN model consisted of three convolutional blocks (the composition of each block is described in Table 2) and four fully connected layers. We performed an architecture optimization of this model as done for ViT's architecture. A total of five configurations were considered. In each configuration, we varied the size of the Max Pooling kernel within the three convolutional blocks. As detailed in the Keras library [21] documentation, a kernel consists of $(dim1, dim2, dim3)$, where $dim1$ corresponds to the depth, $dim2$ to the height and $dim3$ to the width of the kernel. So, we investigated five different combinations of the placement and number of kernels acting only on the plane ((1,2,2)) and kernels acting also on the third dimension ((2,2,2)). We provide a complete description of the different configurations in the Table 3.

Table 2. The composition of the 3D CNN convolutional blocks. In the first block, k = 7, while k = 3 in the other two blocks.

Convolutional block
3D Convolutional layer (kernel $k \times k \times k$)
3D Max pooling layer
Batch normalization layer
3D Convolutional layer (kernel $1 \times 1 \times 1$)

Table 3. The composition of the five alternative configurations of the 3D CNN. MP: Max Pooling.

N configuration	MP kernel size		
1	(1,2,2)	(1,2,2)	(2,2,2)
2	(1,2,2)	(2,2,2)	(2,2,2)
3	(2,2,2)	(1,2,2)	(1,2,2)
4	(1,2,2)	(2,2,2)	(1,2,2)
5	(2,2,2)	(2,2,2)	(1,2,2)

To train each 3D CNN's configuration, we exploited the same dataset partitioning used for 3D ViTs. To make the results comparable, we again evaluated the performance of each configuration by training the model with the fixed splitting of the dataset. Regarding the early-stopping criterion, we established that if the validation loss did not decrease for more than five consecutive epochs, training was stopped. We then re-trained the best configuration by applying the 5-fold CV, and we evaluated all five models on the test set, reporting the mean and standard deviation results. The training hyperparameters were set as follows: Learning rate = 1e–4, Epochs = 20, Batch size = 4, Optimization algorithm = Adam, Loss Function = Cross Entropy.

3 Results

3.1 3D ViT Results

The following parameters led to the best-performing 3D ViT: $p = 16$, $d = 2048$, $L = 6$, $D = 32$ and $k = 8$. This corresponds to the configuration number five in Table 1. An overview of the model is depicted in Fig. 1. On the 5-fold CV training this model provided 0,775 AUROC and 0,523 F2-score. In particular, the best split w.r.t the AUROC metric yielded 0,927 AUROC and 0,735 F2-score. We reported complete results for all the five CV models in Table 4.

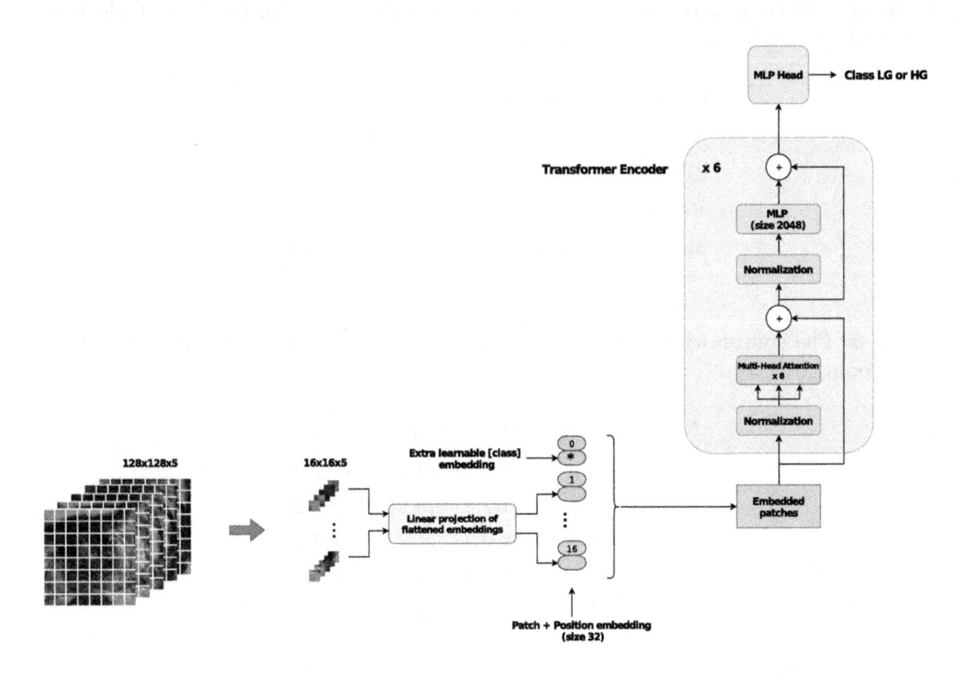

Fig. 1. Our best-performing ViT architecture.

3.2 CNN Results

The best 3D CNN configuration resulted as the number four of those shown in Table 3. By applying the 5-fold CV on the test set, this model yielded 0.585 mean AUROC and 0.215 mean F2-score. The best split w.r.t the AUROC metric provided 0.635 AUROC and 0.3125 F2-score. We reported all the results for the five CNN models in Table 5.

4 Discussion and Conclusions

This study aimed to evaluate the effectiveness of 3D ViTs in assessing the aggressiveness of PCa, as this deep learning model is emerging as a new gold standard in

Table 4. Results on the test set of the best-performing 3D ViT configuration for each CV splitting.

Cross-validation fold	Specificity	Sensitivity	Accuracy	AUROC	F2-score
1	0,688	0,667	0,682	0,74	0,606
2	0,875	0,167	0,682	0,698	0,185
3	0,75	0,333	0,636	0,708	0,333
4	0,75	0,833	0,773	0,802	0,758
5	0,688	0,833	0,727	0,927	0,735
Mean (SD)	0,750 (0,076)	0,567 (0,303)	0,700 (0,052)	0,775 (0,094)	0,523 (0,254)

Table 5. Results on the test set of the best-performing 3D CNN configuration for each CV splitting.

Cross-validation fold	Specificity	Sensitivity	Accuracy	AUROC	F2-score
1	1.0	0.167	0.773	0.583	0.2
2	0.813	0.167	0.636	0.604	0.179
3	0.938	0.167	0.727	0.552	0.192
4	0.625	0.333	0.545	0.635	0.313
5	0.9375	0.167	0.727	0.552	0.192
Mean (SD)	0,8625 (0,145)	0,2 (0,068)	0,682 (0,215)	0.585 (0,089)	0.215 (0,050)

several computer vision tasks. As a starting point, we exploited the architecture proposed in [15], and we modified it to preprocess 3D patches and significantly reduce the number of parameters. In this way, we could train it from scratch using a small amount of data, such as the ProstateX-2 challenge dataset [19]. With a grid search of the architectural features of the newly defined 3D ViT model, we selected the best-performing architecture and evaluated it via a CV approach. We designed and trained from scratch a 3D CNN model to have a basic reference model against which to compare our 3D ViT. It is worth noting that, to our knowledge, this is the first study in which a 3D CNN is trained on volumetric scans to predict the aggressiveness of PCa. Indeed, three-dimensional CNN models have been previously exploited only to distinguish clinically significant from non-significant lesions [6,7]. As a result of our comparison, we found that 3D ViT outperformed 3D CNN when trained with the same pipeline. Although both models exploited volumetric information, the 3D CNN likely suffered more from the lack of data. The best-performing 3D ViT instead, despite the limited amount of data and without any specific structural optimization, provided quite good results, reaching an AUROC of 0.927 on the test set in the best dataset partitioning strategy.

As a further means of comparison with state-of-the-art methods, we compared our results with those obtained in [10], which is the only work, to the best of our knowledge, that addressed our same clinical task on the ProstateX-2 challenge dataset. For the sake of clarity, we highlight the differences between our work and [10]. In [10], the authors fine-tuned the AlexNet model instead of training a CNN from scratch; in addition, they exploited 2D images cropped around the center of the lesion rather than 3D volumes cropped around the prostate. They also performed training with more data since, in addition to the ProstateX-2 challenge dataset, they used 132 additional lesions from a private dataset. Finally, they split the dataset randomly, while we ensured a stratified and complete separation among patients. We reported the results of the comparison in Table 6. We must emphasize that despite our model's lower performance, it was obtained with a smaller training set. Furthermore, unlike [10], we ensured a complete separation of patients between training and test sets, as well as a double stratification, w.r.t. class and the lesion's zone. This approach suppressed any bias in favor of the model's classification capabilities.

Table 6. Comparison between our 3D ViT and the 2D CNN from [10].

Model	Specificity	Sensitivity	Accuracy	AUROC	F2-score
Our ViT	0,750 (0,076)	0,567 (0,303)	0,700 (0,052)	0,775 (0,094)	0,523 (0,254)
CNN from [10]	-	0.794 (0.012)	0.738 (0.014)	0.809 (-)	-

Our study has been conceived as a preliminary investigation and, as such, it has some limitations. Indeed, we did not apply any image enhancement steps nor any architectural optimization of the original ViT model by, for instance, including anatomical priors or employing diverse loss functions. ProstateX-2 is a challenging dataset as it contains lesions in different areas of the prostate gland. We applied the 3D ViT only to T2w scans, as these appeared more informative according to our previous research in the field [11]. Nonetheless, the contribution of ADC maps in cancer lesions located in diverse gland zones might be informative, and they could enable a multimodal 3D ViT to predict lesions' aggressiveness more accurately. Overall, as a first exploratory step, our results are encouraging and suggest that 3D ViTs, trained from scratch, might be a viable strategy for assessing PCa aggressiveness. Further research is needed to verify this claim, especially on larger datasets and on datasets acquired with different protocols and from various institutions. This approach would be necessary to validate the robustness and generalization capabilities of the 3D ViT model. All these additional experiments will be the subject of our future works.

Acknowledgements. The research leading to these results has received funding from the European Union's Horizon 2020 research and innovation programme under grant agreement No 952159 (ProCAncer-I) and from the Regional Project PAR FAS Tuscany - NAVIGATOR. The funders had no role in the design of the study, collection, analysis and interpretation of data, or writing the manuscript.

References

1. World Health Organization, I.A.f.R.o.C: Fact sheet on cancer incidence in Europe (2020). https://gco.iarc.fr/today/data/factsheets/populations/908-europe-fact-sheets.pdf
2. Turkbey, B., et al.: Prostate imaging reporting and data system version 2.1: 2019 update of prostate imaging reporting and data system version 2. Eur. Urol. **76**(3), 340–351 (2019). https://doi.org/10.1016/j.eururo.2019.02.033
3. Barentsz, J.O., et al.: PI-RADS prostate imaging-reporting and data system: 2015, version 2. Eur. Urol. **69**, 16–40 (2016). https://doi.org/10.1016/j.eururo.2015.08.052
4. Mohler, J.L., et al.: Prostate cancer, version 2.2019, NCCN clinical practice guidelines in oncology. J. Natl Compreh. Cancer Netw. **17**(5), 479–505 (2019). https://doi.org/10.6004/jnccn.2019.0023
5. Vickers, A.J.: Effects of magnetic resonance imaging targeting on overdiagnosis and overtreatment of prostate cancer. Eur. Urol. **80**(5), 567–572 (2021). https://doi.org/10.1016/j.eururo.2021.06.026
6. Liu, S., Zheng, H., Feng, Y., Li, W.: Prostate cancer diagnosis using deep learning with 3d multiparametric MRI. Med. Imaging 2017: Comput. Aid. Diagn. **10134**, 581–584 (2017). SPIE https://doi.org/10.48550/arXiv.1703.04078
7. Mehrtash, A., et al.: Classification of clinical significance of MRI prostate findings using 3D convolutional neural networks. In: Medical Imaging 2017: Comput. Aid. Diagn. **10134**, 101342 (2017). International Society for Optics and Photonics https://doi.org/10.1117/12.2277123
8. Mehta, P., Antonelli, M., Ahmed, H.U., Emberton, M., Punwani, S., Ourselin, S.: Computer-aided diagnosis of prostate cancer using multiparametric mri and clinical features: A patient-level classification framework. Med. Image Anal. **73**, 102153 (2021). https://doi.org/10.1016/j.media.2021.102153
9. Song, Y., et al.: Computer-aided diagnosis of prostate cancer using a deep convolutional neural network from multiparametric MRI. J. Magn. Reson. Imaging **48**(6), 1570–1577 (2018). https://doi.org/10.1002/jmri.26047
10. Yuan, Y., et al.: Prostate cancer classification with multiparametric MRI transfer learning model. Med. Phys. **46**(2), 756–765 (2019). https://doi.org/10.1002/mp.13367
11. Bertelli, E., et al.: Machine and deep learning prediction of prostate cancer aggressiveness using multiparametric MRI. Front. Oncol. **11**, 802964–802964 (2021). https://doi.org/10.3389/fonc.2021.802964
12. Mehta, P., et al.: Autoprostate: towards automated reporting of prostate MRI for prostate cancer assessment using deep learning. Cancers **13**(23), 6138 (2021). https://doi.org/10.3390/cancers13236138
13. Wang, Z., Liu, C., Cheng, D., Wang, L., Yang, X., Cheng, K.-T.: Automated detection of clinically significant prostate cancer in MP-MRI images based on an end-to-end deep neural network. IEEE Trans Medi. Imag. **37**(5), 1127–1139 (2018). https://doi.org/10.1109/TMI.2017.2789181
14. Yoo, S., Gujrathi, I., Haider, M.A., Khalvati, F.: Prostate cancer detection using deep convolutional neural networks. Sci. Rep. **9**(1), 1–10 (2019). https://doi.org/10.1038/s41598-019-55972-4
15. Dosovitskiy, A., et al.: An image is worth 16x16 words: Transformers for image recognition at scale. arXiv preprint arXiv:2010.11929 (2020). https://doi.org/10.48550/arXiv.2010.11929

16. Carion, N., Massa, F., Synnaeve, G., Usunier, N., Kirillov, A., Zagoruyko, S.: End-to-end object detection with transformers. In: European Conference on Computer Vision, pp. 213–229 (2020). https://doi.org/10.48550/arXiv.2005.12872
17. Ranftl, R., Bochkovskiy, A., Koltun, V.: Vision transformers for dense prediction. In: Proceedings of the IEEE/CVF International Conference on Computer Vision, pp. 12179–12188 (2021). https://doi.org/10.48550/arXiv.2103.13413
18. Matsoukas, C., Haslum, J.F., Söderberg, M., Smith, K.: Is it time to replace CNNs with transformers for medical images. arXiv preprint arXiv:2108.09038 (2021). https://doi.org/10.48550/arXiv.2108.09038
19. Litjens, G., Debats, O., Barentsz, J., Karssemeijer, N., Huisman, H.: Prostatex challenge data. Cancer Imag. Arch. **10**, 9 (2017)
20. Paszke, A., Gross, S., Massa, F., Lerer, A., Bradbury, J., Chanan, G.: PyTorch: An Imperative Style, High-Performance Deep Learning Library. Curran Associates, Inc. (2019). https://doi.org/10.48550/arXiv.1912.01703
21. Chollet, F., et al.: Keras. GitHub (2015). https://github.com/fchollet/keras
22. Developers, T.: TensorFlow. Zenodo (2021). https://doi.org/10.5281/zenodo.5593257
23. Harris, C.R., et al.: Array programming with NumPy. Nature **585**(7825), 357–362 (2020). https://doi.org/10.1038/s41586-020-2649-2
24. Pedregosa, F.: Scikit-learn: Machine learning in Python. J. Mach. Learn. Res. **12**, 2825–2830 (2011). https://doi.org/10.5555/1953048.2078195
25. Mason, D., Scaramallion, Rhaxton, Mrbean-Bremen, Suever, J., Vanessasaurus: pydicom/pydicom: pydicom 2.1.2. Zenodo (2020). https://doi.org/10.5281/zenodo.4313150
26. Clark, A.: Pillow (PIL Fork) Documentation. readthedocs (2015)
27. Reback, J., McKinney, W., jbrockmendel, den Bossche, J.V., Augspurger, T., Cloud, P.: pandas-dev/pandas: Pandas 1.2.4. Zenodo (2021). https://doi.org/10.5281/zenodo.4681666
28. Van Rossum, G.: The Python Library Reference, release 3.8.2. Python Software Foundation (2020)

Exploring a Transformer Approach for Pigment Signs Segmentation in Fundus Images

Mara Sangiovanni[1(✉)], Maria Frucci[1], Daniel Riccio[1,3], Luigi Di Perna[2], Francesca Simonelli[2], and Nadia Brancati[1]

[1] Institute for High Performance Computing and Networking National Research Council of Italy (ICAR-CNR), Via P. Castellino 111, 80131 Naples, Italy
{mara.sangiovanni,maria.frucci,nadia.brancati}@cnr.it

[2] Eye Clinic, Multidisciplinary Department of Medical, Surgical and Dental Sciences, University of Campania "Luigi Vanvitelli", Via Pansini 5, 80131 Naples, Italy
francesca.simonelli@unicampania.it

[3] University of Naples Federico II, Via Claudio 21, 80125 Naples, Italy
daniel.riccio@unina.it

Abstract. Over the past couple of years, Transformers became increasingly popular within the deep learning community. Initially designed for Natural Language Processing tasks, Transformers were then tailored to fit to the Image Analysis field. The self-attention mechanism behind Transformers immediately appeared a promising, although computationally expensive, learning approach. However, Transformers do not adapt as well to tasks involving large images or small datasets. This propelled the exploration of hybrid CNN-Transformer models, which seemed to overcome those limitations, thus sparkling an increasing interest also in the field of medical imaging. Here, a hybrid approach is investigated for Pigment Signs (PS) segmentation in Fundus Images of patients suffering from Retinitis Pigmentosa, an eye disorder eventually leading to complete blindness. PS segmentation is a challenging task due to the high variability of their size, shape and colors and to the difficulty to distinguish between PS and blood vessels, which often overlap and display similar colors. To address those issues, we use the Group Transformer U-Net, a hybrid CNN-Transformer. We investigate the effects, on the learning process, of using different losses and choosing an appropriate parameter tuning. We compare the obtained performances with the classical U-Net architecture. Interestingly, although the results show margins for a consistent improvement, they do not suggest a clear superiority of the hybrid architecture. This evidence raises several questions, that we address here but also deserve to be further investigated, on how and when Transformers are really the best choice to address medical imaging tasks.

Keywords: Retinitis Pigmentosa · Segmentation · Transformers

P. L. Mazzeo et al. (Eds.): ICIAP 2022 Workshops, LNCS 13374, pp. 329–339, 2022.
https://doi.org/10.1007/978-3-031-13324-4_28

1 Introduction

Transformers are the hype of the moment within the deep learning community. The topic was propelled by the 2017 seminal paper of Vaswani [20], that proposed the Transformer architecture in the Natural Language Processing context. Rapidly, the Transformer was adapted to tackle various image processing tasks. Several architectures were proposed, ranging from hybrid approaches mixing traditional CNNs with self-attention based mechanisms, to pure Transformer-based ones, such as ViT [4] that showed the possibility to perform image classification with an architecture directly inspired to the NLP one. A Transformer is based on the self-attention mechanism, whose ability to learn both long- and short-range dependencies, if provided with enough input, has been claimed in several works including the cited ViT. However, the core of the self-attention calculation is based on a dot product among all the input pairs, thus making it computationally unfeasible when working on big-sized images, while requiring large datasets to ensure performances. Indeed the first successful applications of Transformers usually involved classification tasks on large dataset of small images. Hybrid CNN-Transformer approaches tried to overcome these problems, with the aim to exploit self-attention also for smaller datasets, bigger images and more complicated tasks such as segmentation or detection [5,6]. The adaptation of Transformers to address those issues sparkled a great interest in the field of medical imaging, with an always increasing number of proposed new architectures addressing several topics [6], including medical images segmentation. Here, we address the task of segmenting Pigment Signs (PSs) in ocular Fundus Images (FI). PSs are the hallmarks of the Retinitis Pigmentosa (RP), a disease encompassing a group of genetically heterogeneous eye disorders, whose effects involve a progressive visual field loss with night visual impairment that will eventually lead to complete blindness. No cure is available for RP at the moment, but therapies can be used to delay the degeneration effects if early administered. The typical signs of the disease (that also include attenuated retinal vessels and optical disc pallor) are characteristic PSs slowly accumulating in the retina. Segmenting PSs in FI is challenging: FI are very variable from patient to patient, hence resulting in a wide range of different colors and intensities; RP is not a common disease, and the publicly available dataset are few and include a limited number of images (our private dataset contains 100 images from 10 patients); FI are large (in the order of thousands of pixels both for height and width); they contain information at a very small scale, thus preventing any successful resizing strategy; furthermore PSs are usually very close to blood vessels with which they share very similar coloration. Due to these features, and the complexity of the segmentation task, we opted for a hybrid architecture, to exploit the advantages coming from the spatial inductive bias of a classical CNN alongside the Transformer long range detection ability. Among the existing architectures we chose the Group Transformer U-Net (GT U-Net) described in [7] because: i) it couples the bottleneck approach described in [18] with the sub-patches extraction process inspired by the successful swin transformer [8]. Hence it is able to capture both long-range and small-sized dependencies while being less demanding than a pure

Transformer; ii) it has been used by the authors to segment retinal blood vessel in FI - a somehow similar problem - with very interesting performances. It is known that deep learning architectures, when working from the scratch on a new task, need a very careful parameter tuning process. Moreover, it appears that Transformers, although generating smoother loss functions (as a result of their architecture), very often show non-convex loss landscapes when the number of training data is reduced [12]. The main aim of this work is to explore the impact of the training choices on the network performances. We worked on the number and size of the input patches, the batch size, the loss function, and the optimizer and scheduler used for the learning process. The preliminary results, that we would like to share with the community, is that not all that glitters is gold: with a careful set of choices, the GT-UNet displays interesting performances, but not at all better than a classical U-Net on which a similar tuning strategy is applied. This finding leaves us with some unanswered questions: is it a Transformer architecture really the best choice for a medical segmentation problem, where usually the number of available images is not high? is it a hybrid Transformer the best choice for the task of PSs segmentation? Are really Transformers worth all the computational resources they require? We try to get some intuitions about these topics, that deserve to be further investigated in the future.

2 Methods

The problem of segmenting PSs in FI with a deep learning approach has been already investigated in [3], where the authors use a U-Net architecture on the public RIPS dataset [14]. Here we use GT U-Net, a hybrid Transformer model that tries to combine and exploit the benefits of both the U-Net structure and the bottleneck multihead self-attention mechanism proposed by Srinivas et al. in [18]. In detail, the first convolutional block of the U-Net architecture is replaced with a bottleneck block, both in the encoding and in the decoding paths. Each bottleneck block implements a multihead self-attention (MHSA) mechanism in which locality, inside the input image or feature map, is obtained including a relative distance-aware positional encoding. Furthermore the GT U-Net introduces a grouping strategy (called Group Transformer), inspired by the famous Swin Transformer [8]: the input is split in smaller patches, on which the before described MHSA is applied. In order to learn long-range relationships among distant features, the visual field becomes bigger as the network goes deeper. The GT U-Net combines the ability to learn local and invariant features (typical of the convolutional networks) with the detection of long range relationships, although at the cost of an increased computational complexity.

Here we investigate the performances of the GT U-Net, taking into account several settings including the input size, both in terms of number of patches and of patch dimension, the batch size, the loss function, the chosen optimizer and its learning rate (i.e. initial learning rate and associated scheduling strategy). See Table 1 for a listing. We adopted a commonly used approach in the field: since input images are too big to fit in memory we extracted patches in a random

Table 1. List of the investigated parameters

Parameter name	Values (unit)
Patch size (H = W)	64/128 (pixels)
Batch size	64/128
Number of patches	6000/20000/40000
Optimizer	Adam/SGD
Scheduler	Cosine/Cyclic/One Cycle
Loss	Cross Entropy/Lovasz/Focal Tversky/Dicetopk
Initial learning rate	0.25–0.00005 (see text for details)

fashion from them. Following the hypothesis that a larger patch could enhance the long-range dependencies captured by Transformers, we increased the size of the considered patch with respect to the vanilla GT U-Net.

Coupling a well-chosen LR with a good scheduling strategy is an important choice, that, in practice, also depends from the used optimizer. We selected adaptive optimizers supporting both the 'warm restart' strategy and a scheduling approach in which the LR is changed over each batch. We used: i) an implementation of the Adam optimizer [1] that includes the warm restarts, the per-batch scheduling approach (based on an inner cosine-annealing strategy) and a decoupled weight decay technique to reinforce regularization [10]); ii) we used the classical pytorch Stochastic Gradient Descent (SGD) coupled with two cyclical schedulers, namely the Cyclic and the One Cycle Scheduler [16].

To select the initial LR we followed the advice and indications given in [17]: to have some hints on how the loss is dependent from the LR, we calculated the values of the loss corresponding to different learning rates. The LR was changed in a predefined range spanning different orders of magnitudes. The calculation was performed along several batches of training data. In principle, the LR corresponding to the steepest gradient could be automatically extracted as the best, but in practice it often happens that it corresponds to local minima or singularities of the loss function. To avoid choosing a wrong LR, we visually inspected the loss-vs-LR plot to select the initial value. We adapted the code provided in [13] to perform both the calculation and visualization. We repeated the process for each combination of chosen loss/optimizer/batch size.

Since there is a close interplay among the batch size and the chosen LR when using SGD, with a larger batch size enforcing stability but lowering a bit the performances, we initially kept the number of extracted patches the same of our previous work and investigated the variability induced on both the Cyclical and One Cycle schedulers when doubling the batch size. We also assessed which of the two scheduler displayed a more reliable behaviour. After, we raised the number of extracted patches, under the hypothesis that more input would be beneficial for a (hybrid) Transformer architecture.

We also investigated the use of different loss functions, to try to understand their impact on the network performances. Inspired by the work of Ma et al. [11], we decided to use: i) a Distribution-based loss, namely the Cross-Entropy loss; ii) two region-based ones, namely the Lovasz loss and the Focal Tversky loss; iii) a compound loss, namely the Dice TopK loss. Except for the Cross Entropy and Lovasz losses (for which we used the pytorch standard implementation, and the code proposed by [2], respectively) the others were taken from the code associated to the cited Ma et al. paper [11].

Lastly, we picked the models with the best performances and run them with a even higher number of patches to push further the investigation on how the size of the input could influence the obtained results.

3 Data

A private dataset of Fundus Images containing PS of RP, acquired at the Eye Clinic of University of Campania "Luigi Vanvitelli", has been adopted for the experiments. The dataset is composed by 100 FI acquired from ten patients. Five FI per eye, overlying different parts of FI were acquired for each patient. Indeed, 96 images contain PS while in 4 images PS are absent. The images were acquired using the digital retinal camera Canon CR4-45NM (Canon UK, Reigate, UK). The images have a resolution of 1440 $\times$ 2160 pixels. Also for the same patient, the images display a wide variability in terms of contrast, color balancing and focus/sharpness. An expert in the field of ophthalmology was asked to manually segment each image, marking all the pixels, assumed to belong to a PS, with a high degree of confidence. So, for each image was generated a binary mask containing PS and which represents the Ground Truth of our dataset. In addition, binary masks delineating the Fields of View were generated. This private dataset is different from a benchmark dataset namely RIPS [14], mainly for two aspects: i) the number of patients is higher, but for each patient only one session was acquired, and ii) the PS are located on the pericentral region closer to the blood vessels with which they share a similar color. These aspects make the present dataset more appealing as a benchmark.

4 Experiments and Results

To compare the performances obtained by the various combination of models, losses, optimizers, schedulers, learning rates, batch sizes, and input size we adopt the standard metrics, i.e. Accuracy, Sensitivity, Specificity, and F-Measure. They are based on the definitions of: True Positives (TP), i.e. the number of pixels correctly classified as PSs; True Negatives, the number of pixels correctly classified as normal fundus; False Positives (FP), the number of pixels wrongly classified as PS; False negatives (FN), the number of pixels wrongly assigned to normal fundus. The standard metrics are then defined as follows: Sensitivity = TP/(TP + FN); Specificity = TN/(TN + FP); Accuracy = (TP + TN)/(TP + TN + FP + FN); F-measure = 2TP/(2TP + FP + FN).

The input dataset is split in a training set containing 8 patients, and a validation and test sets containing a single patient each, for a total of $train_n = 80$,

$valid_n = 10$ and $test_n = 10$ images, respectively. We consider the best model as the one having higher F-measure on the validation set. The metrics for each experiment are averaged on all the test images.

Code was run for both the models on a single NVIDIA V100 GPU used in exclusive mode. In Fig. 1 the computing time is shown for both the architectures and on the three different input sizes used. The difference, as expected, is striking, with the U-Net running - for the largest input size - in less than two days, and the GT U-Net requiring more than a month.

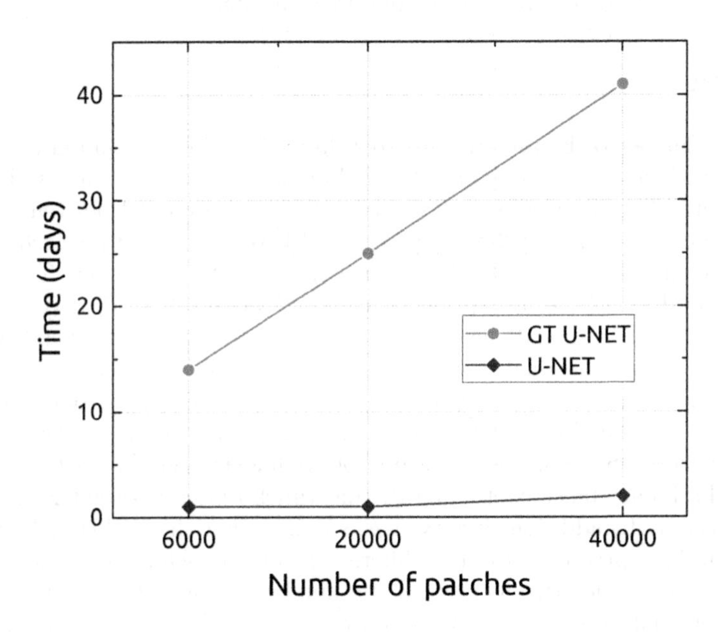

Fig. 1. Elapsed times (in days) for the GT U-Net/U-Net models with respect to the input patches number.

We started our experiments by investigating the effect of the input size, batch size and their interplay with the SGD optimizer. We started from a number of patches $n_{patches} = 6000$, and selected the initial LR by visual inspection of the Loss-vs-LR plot (LR = 0.05). Pach size p_{size} is set to 64 (as in the original paper) or 128. Also the batch size b_{size} is set to 64 or 128. The loss function is the standard Cross-Entropy. We also repeated the experiments with both the Cyclic (CY) and One Cycle (OC) Scheduler, to have some indication on what was the most reliable between them. On the combination $p_{size} = 64$, $b_{size} = 64$ the OC scheduler seems to show one of the superconvergence behaviour claimed in [17] with the F1 on the test being 0.72 (i.e. better and faster than all the other tests in the same conditions). On all the other experiments the CY scheduler has the higher F1. Hence we kept it as the default scheduler for SGD in all the subsequent experiments. Concerning the batch size, when raising it from 64 to 128 we noticed a slight improvement (F1 = 0.65) that is more evident when the

p_{size} is set to 128. In this case we have an interesting F1 = 0.72, thus reinforcing the intuition that using a larger patch size is beneficial for the learning. Since the batch size seems not to be so relevant for the results, we decided to keep it low (i.e. b_{size} = 64) whereas we set p_{size} = 128 throughout all the other tests.

Table 2. Performance measures of the GT U-Net and U-Net models on the different approaches, i.e. using Adam or SGD as an optimizer, and Cross-Entropy (CE), Lovasz (LO), Dicetopk (DT) or Focal Tversky (FT) as loss. The number of input patch is 20000. For each approach the best values are in bold, whereas absolute best values are underlined.

Architecture	Optimizer	Loss	F1	Accuracy	Precision	Sensitivity	Specificity
GT U-Net	SGD	CE	0.643	0.995	0.610	0.679	0.997
		DT	**0.708**	**0.997**	**0.854**	0.605	**0.999**
		LO	0.535	0.992	0.421	**0.734**	0.994
		FT	0.676	0.996	0.711	0.645	0.998
	ADAM	CE	<u>**0.788**</u>	**0.997**	**0.826**	0.754	**0.999**
		DT	0.758	**0.997**	0.799	0.720	**0.999**
		LO	0.723	0.996	0.692	**0.758**	0.998
		FT	0.676	0.996	0.642	0.714	0.997
U-Net	SGD	CE	0.741	**0.997**	**0.860**	0.651	**0.999**
		DT	**0.776**	**0.997**	0.806	**0.749**	**0.999**
		LO	0.228	0.972	0.138	0.656	0.974
		FT	0.517	0.991	0.396	0.745	0.993
	ADAM	CE	<u>**0.788**</u>	**0.997**	0.833	0.748	0.999
		DT	0.764	**0.997**	**0.901**	0.663	**1.000**
		LO	0.777	**0.997**	0.804	0.752	0.999
		FT	0.657	0.995	0.555	**0.804**	0.996

We then decided to increment the number of patches to $n_{patches}$ = 20000, and to investigate the role of two optimizers (i.e. AdamW and SGD) in combination with several loss functions. It is well known in literature that segmentation tasks may benefit from using a loss function that is closer to the measure used to assess the test results. Cross-Entropy is considered the standard, but is a per-pixel measure. Hence we used a wide range of functions which calculate the loss in a more sophisticated way with the aim to better guide the whole learning process. We calculated the loss-vs-LR plot to visualize the loss landscapes and identify the initial LRs for all the combinations of optimizers and losses. The losses parameters were left at the default values.

As a reference for the reader, we briefly recall the results presented in a preliminary work [15]. Models were run on the same machine above described, with the pytorch standard Adam implementation as optimizer, and on the same training, validation and test sets. The other parameters were: $n_{patches}$ = 6000, p_{size} = 64, b_{size} = 64. The best results were obtained by the U-Net with the

Lovasz Loss (F1 score = 0.70) and by the GT U-Net with the Cross-Entropy Loss (F1 score = 0.675). At the end of all the parameter tuning and loss exploration we obtained the results shown in Table 2.

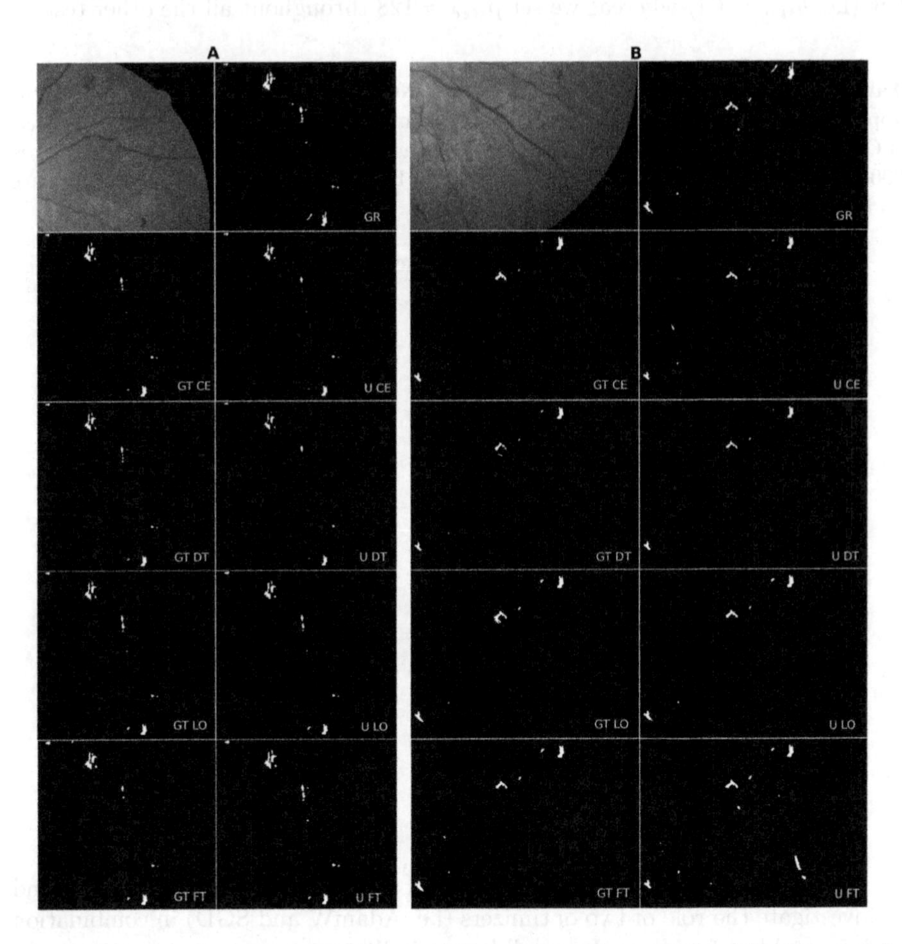

Fig. 2. Two sample image portions, and the segmentations obtained by the models with the Adam optimizer for all the losses. Legend: GR = ground truth, GT = GT U-Net; U = U-Net; CE = Cross-Entropy Loss; DT = Dice Topk loss; LO = Lovasz Loss; FT = Focal Tversky Loss

Strikingly, the best F1 is obtained, for both the models, using the classical combination of Adam with the CE. Even more striking is that the two models achieve a very similar result. It should be noted that the combined loss DiceTopK shows very interesting results for all the approaches, thus suggesting that - with a better tuning of its hyper-parameters - maybe even better results might be obtained. It seems to be the most reliable loss, since it gives very interesting results always, whereas all the others display oscillating behaviours, in particular when the SGD optimizer is used. The Lovasz loss confirms its promising results when combined with ADAM on the U-Net model.

The quality of the obtained segmentation could be visually inspected by looking at Fig. 2 where we show the results obtained for two image portions (A and B). Only a reduced part of the whole retina image is shown, to better visualize small PS. Segmentations are shown for all the used losses.

Finally, to get some intuition on how relevant is the amount of input patches, but also to try to further understand to which extent the Transformer contributes to the results, we set $n_{patches} = 40000$, following the suggestion that increasing the input would help the Transformer in having better performances. Nonetheless, also in this case the results do not show a superiority of the GT U-Net over its classical counterpart. However, this might be also due to a reduced amount of initial images: although extracting a lot of patches, this might not really extend the amount of information provided, since the variability is still restricted to 10 patients.

5 Discussion and Conclusions

Although being a preliminary result, it is striking to see that the GT U-Net performances are comparable but not better than its classical U-Net counterpart. This observation raises several questions that remain open and deserve to be further investigated. It is a Transformer always a better choice for medical image segmentation? Are really Transformers worth all the effort needed, in terms of both computational and time resources? Although generally Transformers seem to perform very well, there are for sure some drawbacks that should be taken into account. Apparently the MHSA mechanism induces smoother loss landscapes [12]. However, this seems to apply only when large input datasets are considered, with non-convex landscapes being generated when the number of input images is too low. This behaviour is not fully explored yet, but it seems to corroborate the well-known Transformer need of having very large datasets in input. When this necessity is not met, a very careful choice of the training parameters is mandatory to obtain interesting performances. This tuning process might be very time consuming: in our experiments the GT U-net training times were measured in weeks, whereas the U-Net completes in days. Another question to take into account regards the specific problem of PS segmentation: is it a Transformer approach the most suitable for our problem? We postulated the need for learning the long-range dependencies to discriminate among pigments and blood vessels. The latter have a well-defined spacial continuity on a larger scale than pigments, thus the blood vessels ideally would be better caught using a Transformer. Maybe this is again a problem related to the paucity of the training image dataset: this is a common problem in the field of medical images, where data collection is complicated and often limited. On the other side, getting more images would require having more computational resources, thus strongly limiting the research groups that could afford similar approaches. Bringing all together: given our problem, input dataset, chosen architecture and available computational resources it seems more efficient using a simple U-Net rather than the GT U-net. Interestingly, this could be a very reasonable choice as stated in the very recent paper of [9]. The authors note that the power of the hybrid CNN-Transformer architectures is usually attributed to the Transformer part. However is not really understood to what extent the Transformer contributes to the hybrid architectures performances, and authors show that - if carefully tuned - a classical CNN could perform as well as a hybrid architecture, but with the plus of having the well established CNN's simplicity and economy. Hence, we plan to further study the problem of PS segmentation exploring several directions: i) implementing a pure CNN with all the suggestions proposed in [9] to see if and how performances would improve; using a pure Transformer such as [19], both starting from the scratch or using a pre-trained model, to try to assess how the Transformer's MHSA mechanism is relevant for our problem. Concluding, we would like to share with the community of medical imaging a preliminary yet interesting finding: maybe CNN are perceived as old if compared with the way more fashionable Transformers, but they are still cheap, efficient and affordable. We think that further studies are still needed to fully understand which are the conditions and the tasks on which Transformers could be the best choice, especially in the field of medical imaging.

References

1. Adamw optimizer and cosine learning rate annealing with restarts. https://github.com/mpyrozhok/adamwr
2. Berman, M., Triki, A.R., Blaschko, M.B.: The Lovasz-Softmax loss: a tractable surrogate for the optimization of the intersection-over-union measure in neural networks. In: 2018 IEEE/CVF Conference on Computer Vision and Pattern Recognition, Salt Lake City, UT, pp. 4413–4421. IEEE, June 2018
3. Brancati, N., Frucci, M., Riccio, D., Di Perna, L., Simonelli, F.: Segmentation of pigment signs in fundus images for retinitis pigmentosa analysis by using deep learning. In: Ricci, E., Rota Bulò, S., Snoek, C., Lanz, O., Messelodi, S., Sebe, N. (eds.) ICIAP 2019. LNCS, vol. 11752, pp. 437–445. Springer, Cham (2019). https://doi.org/10.1007/978-3-030-30645-8_40
4. Dosovitskiy, A., et al.: An image is worth 16 × 16 words: transformers for image recognition at scale. arXiv:2010.11929 [cs], June 2021
5. Han, K., et al.: A survey on vision transformer. arXiv:2012.12556 [cs], August 2021
6. Khan, S., Naseer, M., Hayat, M., Zamir, S.W., Khan, F.S., Shah, M.: Transformers in vision: a survey. ACM Comput. Surv. (CSUR) (2021)
7. Li, Y., et al.: GT U-Net: a U-Net like group transformer network for tooth root segmentation. In: Lian, C., Cao, X., Rekik, I., Xu, X., Yan, P. (eds.) MLMI 2021. LNCS, vol. 12966, pp. 386–395. Springer, Cham (2021). https://doi.org/10.1007/978-3-030-87589-3_40
8. Liu, Z., et al.: Swin transformer: hierarchical vision transformer using shifted windows. In: Proceedings of the IEEE/CVF International Conference on Computer Vision, pp. 10012–10022 (2021)
9. Liu, Z., Mao, H., Wu, C.Y., Feichtenhofer, C., Darrell, T., Xie, S.: A ConvNet for the 2020s. arXiv:2201.03545 [cs], March 2022

10. Loshchilov, I., Hutter, F.: Decoupled weight decay regularization. arXiv:1711.05101 [cs, math], January 2019
11. Ma, J., et al.: Loss odyssey in medical image segmentation. Med. Image Anal. **71**, 102035 (2021)
12. Park, N., Kim, S.: How do vision transformers work? arXiv:2202.06709 [cs], February 2022
13. PyTorch learning rate finder. https://github.com/davidtvs/pytorch-lr-finder
14. The RIPS dataset. https://www.icar.cnr.it/sites-rips-datasetrips/
15. Sangiovanni, M., Brancati, N., Frucci, M., Di Perna, L., Simonelli, F., Riccio, D.: Segmentation of pigment signs in fundus images with a hybrid approach: a case study. Pattern Recogn. Image Anal. **32**(2), 312–321 (2022)
16. Smith, L.N.: Cyclical learning rates for training neural networks. In: 2017 IEEE Winter Conference on Applications of Computer Vision (WACV), pp. 464–472. IEEE (2017)
17. Smith, L.N.: A disciplined approach to neural network hyper-parameters: part 1 - learning rate, batch size, momentum, and weight decay. arXiv:1803.09820 [cs, stat], April 2018
18. Srinivas, A., Lin, T.Y., Parmar, N., Shlens, J., Abbeel, P., Vaswani, A.: Bottleneck transformers for visual recognition. In: 2021 IEEE/CVF Conference on Computer Vision and Pattern Recognition (CVPR), Nashville, TN, USA, pp. 16514–16524. IEEE, June 2021
19. Strudel, R., Garcia, R., Laptev, I., Schmid, C.: Segmenter: transformer for semantic segmentation. In: Proceedings of the IEEE/CVF International Conference on Computer Vision, pp. 7262–7272 (2021)
20. Vaswani, A., et al.: Attention is all you need. In: Advances in Neural Information Processing Systems, pp. 5998–6008 (2017)

Transformer Based Generative Adversarial Network for Liver Segmentation

Ugur Demir, Zheyuan Zhang, Bin Wang, Matthew Antalek, Elif Keles, Debesh Jha, Amir Borhani, Daniela Ladner, and Ulas Bagci(✉)

Northwestern University, Chicago, IL 60201, USA
ulas.bagci@northwestern.edu

Abstract. Automated liver segmentation from radiology scans (CT, MRI) can improve surgery and therapy planning and follow-up assessment in addition to conventional use for diagnosis and prognosis. Although convolution neural networks (CNNs) have became the standard image segmentation tasks, more recently this has started to change towards Transformers based architectures because Transformers are taking advantage of capturing long range dependence modeling capability in signals, so called attention mechanism. In this study, we propose a new segmentation algorithm using a hybrid approach combining the Transformer(s) with the Generative Adversarial Network (GAN) approach. The premise behind this choice is that the self-attention mechanism of the Transformers allows the network to aggregate the high dimensional feature and provide global information modeling. This mechanism provides better segmentation performance compared with traditional methods. Furthermore, we encode this generator into the GAN based architecture so that the discriminator network in the GAN can classify the credibility of the generated segmentation masks compared with the real masks coming from human (expert) annotations. This allows us to extract the high dimensional topology information in the mask for biomedical image segmentation and provide more reliable segmentation results. Our model achieved a high dice coefficient of 0.9433, recall of 0.9515, and precision of 0.9376 and outperformed other Transformer based approaches.

Keywords: Liver segmentation · Transformer · Generative adversarial network

1 Introduction

Liver cancer is among the leading causes of cancer-related deaths, accounting for 8.3% of cancer mortality [14]. The high variability in shape, size, appearance, and local orientations makes liver (and liver diseases such as tumors, fibrosis)

U. Demir and Z. Zhang—Contribute equally to this paper.

P. L. Mazzeo et al. (Eds.): ICIAP 2022 Workshops, LNCS 13374, pp. 340–347, 2022.
https://doi.org/10.1007/978-3-031-13324-4_29

challenging to analyze from radiology scans for which the image segmentation is often necessary [3]. An accurate organ and lesion segmentation could facilitate reliable diagnosis and therapy planning including prognosis [5].

As a solution to biomedical image segmentation, the literature is vast and rich. The self-attention mechanism is nowadays widely used in the biomedical image segmentation field where long-range dependencies and context dependent features are essential. By capturing such information, transformer based segmentation architectures (for example, SwinUNet [2]) have achieved promising performance on many vision tasks including biomedical image segmentation [7,15].

In parallel to the all advances in Transformers, generative methods have achieved remarkable progresses in almost all fields of computer vision too [4]. For example, Generative Adversarial Networks (GAN) [6] is a widely used tool for generating one target image from one source image. GAN has been applied to the image segmentation framework to distinguish the credibility of the generated masks like previous studies [9,11]. The high dimensional topology information is an important feature for pixel level classification, thus segmentation. For example, the segmented mask should recognize the object location, orientation, and scale prior to delineation procedure, but most current segmentation engines are likely to provide false positives outside the target region or conversely false negatives within the target region due to an inappropriate recognition of the target regions. By introducing the discriminator architecture (as a part of GAN) to distinguish whether the segmentation mask is high quality or not, we could proactively screen poor predictions from the segmentation model. Furthermore, this strategy can also allow us to take advantage of many unpaired segmentation masks which can be easily acquired or even simulated in the segmentation targets. To this end, in this paper, we propose a Transformer based GAN architecture as well as a Transformer based CycleGAN architecture for automatic liver segmentation, a very important clinical precursor for liver diseases. By combining two strong algorithms, we aim to achieve both good recognition (localization) of the target region and high quality delineations.

2 Proposed Method

We first investigated the transformer architecture to solve the liver segmentation problem from radiology scans, CT in particular due to its widespread use and being the first choice in most liver disease quantification. The self-attention mechanism of the Transformers has been demonstrated to be very effective approach when finding long range dependencies as stated before. This can be quite beneficial for the liver segmentation problem especially because the object of interest (liver) is large and pixels constituting the same object are far from each other. We also utilized an adversarial training approach to boost the segmentation model performance. For this, we have devised a conditional image generator

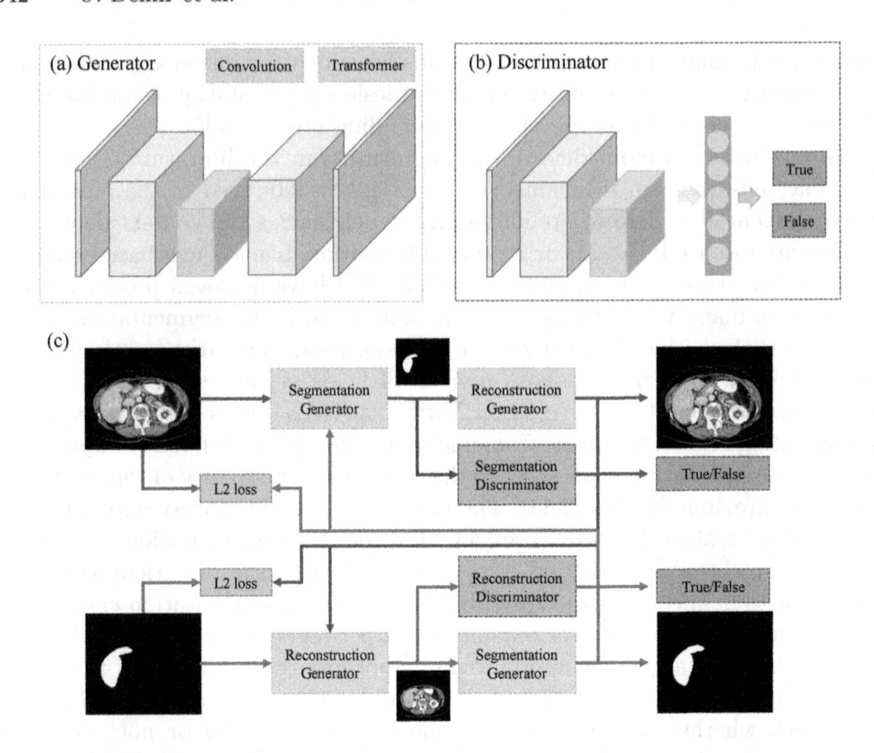

Fig. 1. Block diagram of the Transformer GAN. Generator (a) and discriminator components (b) include Transformer blocks (red). (c) CycleGAN with two generators and two discriminators is illustrated. (Color figure online)

in a vanilla-GAN that learns a mapping between the CT slices and the segmentation maps (i.e., surrogate of the truths or reference standard). The adversarial training forces the generator model to predict more realistic segmentation outcomes. In addition to vanilla-GAN, we have also utilized the CycleGAN [13,17] approach to investigate the effect of cycle consistency constraint on the segmentation task. Figure 1 demonstrates the general overview of the proposed method.

2.1 Transformer Based GAN

Like other GAN architectures [10], Transformer based GAN architecture is composed of two related sub-architectures: the generator and the discriminator. The generator part could generate the segmentation mask from the raw image (i.e., segmentation task itself), while the discriminator can provide us with the confidence of such generated mask compared with the expert-human annotated masks. GAN provides a better way to distinguish the high-dimensional morphology information. The discriminator can provide the similarity between the predicted masks and the ground truth (i.e., surrogate truth) masks. Vanilla GAN considers the whole segmentation to decide whether it is fake or not.

2.2 Transformer Based CycleGAN

One alternative extension to the standard GAN approach is to use transformer based segmentation model within the CycleGAN setup. Unlike a standard GAN, CycleGAN consists of two generators and two discriminator networks. While the first discriminator accepts the raw images as input and predicts the segmentation masks, the second generator takes the predicted segmentation maps as input and maps them back to the input image. The first discriminator classifies the segmentation masks as either real or fake, and the second discriminator distinguishes the real and the reconstructed image. Figure 1 illustrates this procedure with liver segmentation from CT scans. The CycleGAN does not require paired examples to conduct the classification and predictions, and the training can be performed in an unsupervised manner which is a real strength compared to vanilla GANs.

To embed transformers within the CycleGAN, we utilized the encoder-decoder style convolution transformer model [13]. The premise behind this idea was that the encoder module takes the input image and decreases the spatial dimensions while extracting features with convolution layers. This allowed processing of large-scale images. The core transformer module consisted of several stacked linear layers and self-attention blocks. The decoder part increased the spatial dimension of the intermediate features and makes the final prediction. For the discriminator network, we tried three convolution architectures. The vanilla-GAN discriminator evaluates the input image as a whole. Alternatively, we have adopted PatchGAN discriminator architecture [8] to focus on small mask patches to decide the realness of each region. It splits the input masks into NxN regions and asses their quality individually. When we set the patch size to a pixel, PatchGAN can be considered as pixel level discriminator. W have observed that the pixel level discriminator tends to surpass other architecture for segmentation. Figure 1 demonstrates the network overview. In all of the experiments, the segmentation model uses the same convolution transformer and pixel level discriminator architectures.

3 Experimental Setup

We have used Liver Tumor Segmentation Challenge (LiTS) dataset. LiTS consists of 131 CT scans. This dataset is publicly available under segmentation challenge website and approved IRB by the challenge organizers. More information about the dataset and challenge can be found here[1].

[1] https://competitions.codalab.org/competitions/17094#learn_the_details.

All our models were trained on NVIDIA RTX A6000 GPU after implemented using the PyTorch [12] framework. We have used 95 samples for training and 36 samples for testing. All models are trained on the same hyperparameters configuration with a learning rate of $2e^{-4}$, and Adam optimizer with beta1 being 0.5 and beta2 being 0.999. All of the discriminators use the pixel level discriminator in both GAN and CycleGAN experiments. We have used recall, precision, and dice coefficient for quantitative evaluations of the segmentation. Further, segmentation results were qualitatively evaluated by the participating physicians. Our algorithms are available for public use.

Table 1. Performance of Transformer based methods on the LITS dataset [1].

Method	Dice coefficient	Precision	Recall
Transformer [13,16]	0.9432	0.9464	0.9425
Transformer - CycleGAN (ours)	0.9359	**0.9539**	0.9205
Transformer - GAN (ours)	**0.9433**	0.9376	**0.9515**

4 Results

We presented the evaluation results in Table 1. Our best performing method was Transformer based GAN architecture, achieved a highest dice coefficient of 0.9433 and recall rate of 0.9515. Similarly, our transformer based CycleGAN architecture has the highest precision, 0.9539. With Transformer based GAN, we achieved 0.9% improvement in recall and 0.01% improvement in dice coefficient with respect to the vanilla Transformers. It is to be noted that we have used also post-processing technique which boosts the performance for "all" the baselines to avoid biases one from each other.

Figure 2 shows our qualitative results for the liver segmentation. We have examined all the liver segmentation results one-by-one and no failure were identified by the participating physicians. Hence, visual results agreed with the quantitative results as described in Table 1.

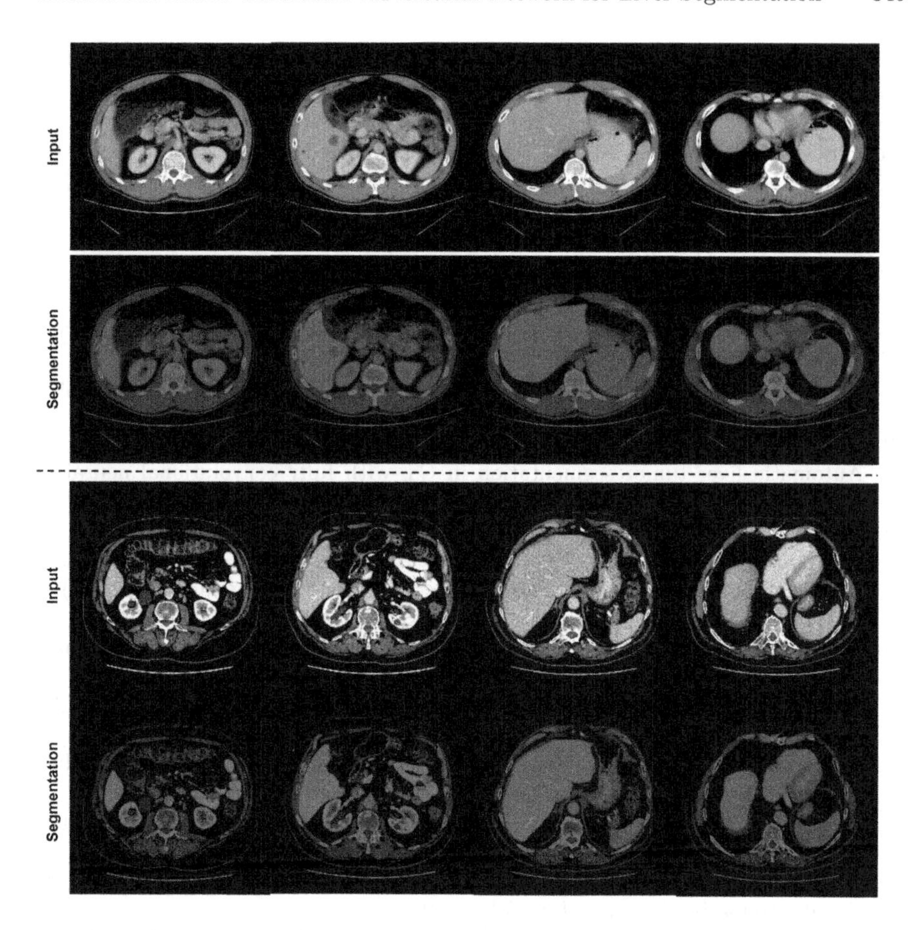

Fig. 2. Transformer based GAN liver segmentation results. Green: True positive, Red: False Positive, Blue: False Negative (Color figure online).

5 Conclusion

In this study, we explored the use of transformer-based GAN architectures for medical image segmentation. Specifically, we used a self-attention mechanism and designed a discriminator for classifying the credibility of generated segmentation masks. Our experimental result showed that the proposed new segmentation architectures could provide accurate and reliable segmentation performance as compared to the baseline Transformers. Although we have shown our results in an important clinical problem for liver diseases where image-based quantification is vital, the proposed hybrid architecture (i.e., combination of GAN and Transformers) can potentially be applied to various medical image segmentation tasks beyond liver CTs as the algorithms are generic, reproducible, and carries similarities with the other segmentation tasks in biomedical imaging field. We anticipate that our architecture can also be applied to medical scans within the semi-supervised learning, planned as a future work.

Acknowledgement. This study is partially supported by NIH NCI grants R01-CA246704 and R01-CA240639.

References

1. Bilic, P., et al.: The liver tumor segmentation benchmark (LiTS). arXiv preprint arXiv:1901.04056 (2019)
2. Cao, H., et al.: Swin-Unet: Unet-like pure transformer for medical image segmentation. arXiv preprint arXiv:2105.05537 (2021)
3. Chlebus, G., Schenk, A., Moltz, J.H., van Ginneken, B., Hahn, H.K., Meine, H.: Automatic liver tumor segmentation in CT with fully convolutional neural networks and object-based postprocessing. Sci. Rep. **8**(1), 1–7 (2018)
4. Chuquicusma, M.J., Hussein, S., Burt, J., Bagci, U.: How to fool radiologists with generative adversarial networks? A visual turing test for lung cancer diagnosis. In: 2018 IEEE 15th International Symposium on Biomedical Imaging (ISBI 2018), pp. 240–244. IEEE (2018)
5. Cornelis, F.H., et al.: Precision of manual two-dimensional segmentations of lung and liver metastases and its impact on tumour response assessment using RECIST 1.1. Eur. Radiol. Exp. **1**(1), 1–7 (2017). https://doi.org/10.1186/s41747-017-0015-4
6. Goodfellow, I., et al.: Generative adversarial nets. In: Advances in Neural Information Processing Systems, vol. 27 (2014)
7. Huang, X., Deng, Z., Li, D., Yuan, X.: Missformer: an effective medical image segmentation transformer. arXiv preprint arXiv:2109.07162 (2021)
8. Isola, P., Zhu, J.Y., Zhou, T., Efros, A.A.: Image-to-image translation with conditional adversarial networks. In: CVPR (2017)
9. Khosravan, N., Mortazi, A., Wallace, M., Bagci, U.: PAN: projective adversarial network for medical image segmentation. In: Shen, D., et al. (eds.) MICCAI 2019. LNCS, vol. 11769, pp. 68–76. Springer, Cham (2019). https://doi.org/10.1007/978-3-030-32226-7_8
10. Liu, Y., et al.: Cross-modality knowledge transfer for prostate segmentation from CT scans. In: Wang, Q., et al. (eds.) DART/MIL3ID -2019. LNCS, vol. 11795, pp. 63–71. Springer, Cham (2019). https://doi.org/10.1007/978-3-030-33391-1_8
11. Luc, P., Couprie, C., Chintala, S., Verbeek, J.: Semantic segmentation using adversarial networks. In: NIPS Workshop on Adversarial Training, Barcelona, Spain, December 2016. https://hal.inria.fr/hal-01398049
12. Paszke, A., et al.: Pytorch: an imperative style, high-performance deep learning library. In: Advances in Neural Information Processing Systems, vol. 32 (2019)
13. Ristea, N.C., et al.: CyTran: cycle-consistent transformers for non-contrast to contrast CT translation. arXiv preprint arXiv:2110.06400 (2021)
14. Sung, H., et al.: Global cancer statistics 2020: GLOBOCAN estimates of incidence and mortality worldwide for 36 cancers in 185 countries. CA Cancer J. Clin. **71**(3), 209–249 (2021)
15. Vaswani, A., et al.: Attention is all you need. In: Advances in Neural Information Processing Systems, vol. 30 (2017)

16. Wu, H., et al.: CvT: introducing convolutions to vision transformers. In: Proceedings of the IEEE/CVF International Conference on Computer Vision (ICCV), pp. 22–31, October 2021
17. Zhu, J.Y., Park, T., Isola, P., Efros, A.A.: Unpaired image-to-image translation using cycle-consistent adversarial networks. In: Proceedings of the IEEE International Conference on Computer Vision, pp. 2223–2232 (2017)

Learning in Precision Livestock Farming - LPLF

Suggestions for the Environmental Sustainability from Precision Livestock Farming and Replacement in Dairy Cows

Lovarelli Daniela[1](✉), Berckmans Daniel[2], Bacenetti Jacopo[1], and Guarino Marcella[1]

[1] Department of Environmental Science and Policy, Università degli Studi di Milano, Via Celoria, 10, 20133 Milan, Italy
daniela.lovarelli@unimi.it

[2] Department of Biosystems, Division M3-BIORES: Measure, Model and Manage Bioresponses, Catholic University of Leuven, Kasteelpark Arenberg 30, 3001 Heverlee, Belgium

Abstract. The livestock sector, like other sectors, has a high environmental impact and we must find solutions to reduce it to accomplish the requirements for a more sustainable production system in line with the European Green Deal requirements.

The aim of this paper is to show a case study in which it is evaluated the effect of PLF technology on the environmental impact of dairy cattle farming by using simulations of Life Cycle Assessment (LCA). This case study involves the use of pedometers for an improved detection of oestrus events in order to make more efficient the livestock activities and the related environmental impact. The results show that the application of LCA can work as a feasible approach to get insight in the significance of the environmental benefit of applying PLF tools on farms.

Keywords: Environmental sustainability · Dairy cows · Monitoring · Efficient management

1 Introduction

It is widely known and recognized that the livestock sector has both positive and negative impacts on the environment. These latter are related mostly to the emissions caused by animals such as methane from enteric fermentation of dairy cows and to the emissions from the storage, treatment, and field distribution practices of slurry [1]. Considering the big growth of the sector of the last decades and that it is continuing to expand to respond to the global increased demand for animal products, the livestock sector needs a critical reflection [2–4]. A possible trade-off between the positive and negative effects of livestock production on the environmental impact can be identified with an enhanced holistic efficiency and performance, partially made possible by technology and improved farm management [1, 3]. Measures to reduce emissions have been widely studied and proposed in the recent past, and often mitigation strategies introduced on farms [5].

P. L. Mazzeo et al. (Eds.): ICIAP 2022 Workshops, LNCS 13374, pp. 351–360, 2022.
https://doi.org/10.1007/978-3-031-13324-4_30

However, carrying out accurate measurements by experimental work is very expensive in equipment, monitoring techniques, time required for several seasons and manpower. Therefore, for evaluating the environmental impact of different processes, models and secondary data are commonly used, which have the advantage of being more widely adoptable, although less accurate. When it is required to compare different scenarios in which technologies are present or not, the main issue is to collect accurate data about the case in which no technologies are available, because less data are available. This makes comparisons more difficult. Moreover, for some cases like emissions from naturally ventilated buildings there are even no accurate monitoring techniques, therefore, the error might be even bigger than the positive effect of technology. In particular, ammonia concentrations could be measured with expensive techniques but measuring methane or monitoring the ventilation rate is much more challenging.

In the context of evaluating the environmental impact of processes in which different methods or technologies are adopted and need to be compared, the Life Cycle Assessment (LCA) approach is helpful because it allows to have a holistic view on the system and evaluate possible side effects on the environmental perspective of different mitigation strategies [6, 7]. However, also the effect of ecosystem services and territory maintenance need to be mentioned [6, 8] and is not properly included in LCA studies. Evaluations on the effect of global warming are widely increasing, due to its important role in current and future policies [1, 3], especially in view of the European Green Deal and Farm to Fork strategy that aim to abate greenhouse gases (GHG) emissions by 2030.

To achieve the primary goal of a farmer (i.e. production of milk, meat, eggs, fiber, etc.) high productivity in an economically sustainable way is fundamental. Growing healthy animals, with good performances, welfare, and a balanced use of inputs (e.g., feed) is the key point [9]. These aspects have also an environmentally sustainable façade, since a balanced use of inputs in respect to the outputs, good performances, efficiency, high yields, and high welfare and health indicators are positive aspects for an environmentally sustainable livestock system [10]. In addition, considering that farms are reducing in number while increasing the number of farmed animals, monitoring all individuals within the herd is challenging for a farmer, and automatic systems, sensors and technology can be of help [11, 12]. Technologies able to support farmers in monitoring big herds and single animals and in the decision-making process are spreading widely. They also bring benefits to the monitoring of variables that have become impossible to be continuously monitored by humans, such as the identification of night-time or silent oestrus events [13, 14], variations in behaviour [15–17], but also the monitoring of other variables that help improve welfare, such as the microclimate in the barn (i.e. temperature, relative humidity, wind speed) [18] and the air quality (i.e. pollutants concentration in air) [19]. Furthermore, until now farmers, researchers and policy makers have focused their attention on animals in their production stages, but it is important to provide enhancements to the non-productive phases as well, since they are the future productive herd: animals growing in healthy conditions will be more robust and resistant to illnesses or stresses during their productive stages. Paying attention to young animals also influences the environmental perspective. In fact, heifers not adequately farmed (i.e. fed and monitored) will postpone their first calving, thus prolonging their unproductive age. In this context it is much interesting to evaluate the environmental impact of this difference. However, as mentioned, measuring the potential environmental advantage of new PLF technology in comparison with the absence of technology is quite complex because setting up experimental studies that compare this condition (with PLF) with the one prior to the installation of PLF tools entails the difficulty of not having specific data in that previous period. Collecting accurate data during several seasons is expensive, takes a lot of time and manpower and is often not accurate enough to come up with reliable results.

The example that we use aims to show the potential of LCA to evaluate environmental impacts on dairy cows, thus comparing the environmental sustainability of a farm equipped or not with PLF, in particular with pedometers. In this condition, both the age of the first calving and the efficiency of the heat detection are evaluated. A simulation is carried out for a traditional dairy cattle farm, modelling the effect of PLF installation. The benefits of PLF in this example can serve as insights in the general use of PLF and show the interest in having the possibility to quantify the beneficial effect of PLF on farm.

2 Materials and Methods

2.1 Farm Description

To evaluate the environmental performance of a dairy cattle farm in which pedometers or other similar technological support is introduced to detect heat events, a dairy cattle farm of average dimensions located in Northern Italy was identified.

This farm has no technological support, since it is a traditional farm where the farmer is still evaluating the potential benefit of introducing sensors/tools (Traditional Scenario, TS). In Italy, in fact only about 30% of farms have PLF tools for heat detection [20], which is by far one of the most widespread tools. This farm has the characteristics reported in Table 1 regarding herd dimension, average dry matter intake (kg DMI d^{-1}) and milk production (kg milk d^{-1}) in the different phases of the lactation.

Considering the lack of technological support in the herd monitoring, the performances of these cows are not optimal; cows farmed in this farm have an average age at first calving equal to 28 months, and the pregnancy rate and fertilization success are quite unsatisfactory, on average with 3 months of failed inseminations. This is quite common in Italian dairy cattle farms of this type [21]. In addition, the lactation lasts longer than the theoretical 305 DIM, and it reaches 395 DIM before drying, which creates unnecessary environmental impacts and needs therefore to be avoided.

Table 1. Farm characteristics in the traditional scenario.

Variable	Unit	TS
Dairy cows	n	180
Dry cows	n	28
Heifers	n	82
Calves	n	70
Dairy cows	kg DM d^{-1}	23.0
Dry cows	kg DM d^{-1}	12.0
Heifers	kg DM d^{-1}	12.0
Calves	kg DM d^{-1}	8.0
Total lactations	n	3
Milk prod. at 0–90 DIM	kg d^{-1}	45
Milk prod. at 90–210 DIM	kg d^{-1}	37
Milk prod. at 210–305 DIM	kg d^{-1}	26
Milk prod. at 305–365 DIM	kg d^{-1}	22
Milk prod. at 365–395 DIM	kg d^{-1}	21

2.2 Modelled Scenarios

Beside these traditional farm characteristics, two improved situations are modelled and tested on their environmental impact consequences:

- "Optimal Scenario" (OS): in this case, the farmer adopts the best technologies that can support the heat events detection, such as the measurement of the progesterone level in milk. In this case, the farmer properly grows heifers and promptly identifies the oestrus window, even when it occurs at night-time, or it is silent. Thank to this improved management practice, the first calving takes place at 23 months, which is a proper timing for not encountering parturition problems [22]. The subsequent calving-conception interval (CCI) is minimized since monitoring progesterone in milk can allow identifying at best the oestrus events and defining when to inseminate the cow. This implies that the lactation proceeds in its optimal theoretical duration and the cow is dried off after 305 DIM;
- "Intermediate Scenario" (IS): in this case, the farmer installs common technology solutions like accelerometer sensors/pedometers on the cows. This allows the detection of heat events with a better accuracy than humans but with possible detection errors. In this case, the heifers are properly grown, but insemination and the subsequent calving take place later than in OS (calving at 25 months). This condition is quite common in Italian livestock farms of Northern Italy, where 25 months represents the average age of the first calving [23]. Due to not identifying all of the estrus events, some failures in cows' fertilization are considered, and therefore they are dried off after 365 DIM, which is also a common practice.

Table 2 reports the differences in the age at first calving and the duration of lactation of TS (traditional scenario), IS (intermediate scenario) and OS (optimal scenario).

Table 2. Age at first calving and duration of the lactation in the three scenarios (traditional, intermediate and optimal, respectively for TS, IS and OS).

Variable	Unit	TS	IS	OS
Age at first calving	months	28	25	23
Duration of lactation	days	395	365	305

2.3 LCA and Climate Change

After a literature survey and previous experience on the assessment of environmental sustainability of dairy farms in Northern Italy, an equation that links dairy efficiency with climate change was defined [7, 24, 25]. From this equation, the environmental impact for the category of Climate Change (CC; kg CO_2 eq.) was quantified and used to predict the environmental impact of the modelled scenarios.

3 Results and Discussion

Table 3 reports the average dry matter intake (DMI; kg DM) per scenario of farmed animals during the early growing stages and, for dairy and dry cows, for each lactation.

Table 3. Dry matter intake of the entire herd per lactation per scenario.

Variable	Unit	TS	IS	OS
Dairy cows	t DM	1635.3	1511.1	1262.7
Dry cows	t DM	13.4	13.4	13.4
Heifers	t DM	275.5	246.0	226.3
Calves	t DM	67.2	67.2	67.2
Total ingestion	t DM lact^{-1}	1991.5	1837.7	1569.7

Table 4, instead, shows the milk production of the 180 dairy cows in the 3 scenarios, depending on the length of their lactation period. Here, IS and OS show a lower total ingestion for dairy cows because of the shorter duration of the lactation compared to TS, and because of the shorter duration of the diet as a heifer (i.e. different age at first calving). Similarly, for milk production it is observed that IS and OS produce less milk than TS, because of the shorter duration of the lactation period. In this period, however, the lactation curve is decreasing, therefore milk production is lower and milk quality reduces as well. Both aspects of feed intake and milk production contribute to the calculation of the Dairy Efficiency (DE; kg milk kg DMI-1), which is expressed as the amount of milk produced per amount of feed ingested. DE results higher in OS and lower in IS and TS.

Table 4. Milk production of the farmed dairy cows (n. 180) per lactation per scenario, and the calculated average dairy efficiency (DE; kg milk kg DMI^{-1}).

Variable	Unit	TS	IS	OS
Milk prod. 0–90	$t\,d^{-1}$	729	729	729
Milk prod. 90–210	$t\,d^{-1}$	799.2	799.2	799.2
Milk prod. 210–305	$t\,d^{-1}$	444.6	444.6	444.6
Milk prod. 305–365	$t\,d^{-1}$	237.6	237.6	
Milk prod. 365–395	$t\,d^{-1}$	113.4		
Total milk production	t $lactation^{-1}$	2323.8	2210.4	1972.8
Dairy efficiency	$kg_{milk}\ kg_{DMI}{}^{-1}$	1.17	1.20	1.26

A literature search was carried out searching for studies in which both the Dairy Efficiency (DE) of livestock farms and the calculated Climate Change (CC) data were reported. From the analysis of these data have resulted the values shown in Table 5. The linear regression among these values is reported in Eq. (1) and shows quite satisfactory results, with $R^2 = 0.69$:

$$CC = -1.6763 * DE + 3.6122 \tag{1}$$

Table 5. Mean and standard deviation of DE and CC from literature for calculating Eq. (1).

n. farms	DE	CC	Authors
33	1.35 (0.26)	1.38 (0.32)	Lovarelli et al. (2019)
102	1.19 (0.25)	1.51 (0.53)	Zucali et al. (2017)

The application of Eq. (1) to the studied scenarios, results in CC ($kg_{CO2eq}\ kg_{milk}{}^{-1}$) values for the 3 scenarios as: 1.65 (TS), 1.60 (IS) and 1.50 (OS), respectively. This is a simplification in the quantification of CC, because it assumed that the composition of the animals' diet and milk quantity and quality are not affected by any difference in the 3 scenarios. Instead, the different CC values are caused by the duration of one diet instead of the other and of one lactation duration instead of the other. For a more detailed assessment, additional considerations could be done to include the possible effects of different productivity, milk quality and udder health due to the different management opportunities and technological equipment. In this preliminary study, these differences were excluded to avoid additional variability.

Some further considerations can be made by making equal lifetimes for the cows. Both a longer and a shorter lifetime have been tested as follows: (a) if cows in OS live longer, reaching equal levels of IS and TS (i.e. 66 or 72 months) or (b) if cows live shorter (i.e. also IS and TS have lifetime equal to 58 months as in OS). Of course, this is an assumption based on the need to compare all scenarios based on the common productive duration, set equal to 3 lactations. When changing the lifetime to make it comparable in the 3 scenarios, the number of lactations becomes the variable. In this condition, the cows in IS and TS need 8 and 14 months more, respectively, than cows in OS to conclude the 3rd lactation. Therefore, if cows in OS had a prolonged productive period, they would eat (with a diet for dry and dairy) and produce more, as reported in Table 6 (they would add one dry period and one lactation – partial in BS-66 or complete in BS-72).

Table 6. Results for prolonged lifetime of OS.

Variable	Unit	OS-66 months	OS-72 months
Lifetime	months	66	72
Additional milk prod.	t	1328.4	1972.8
Total milk prod.[a]	t	7247	7891
Additional feed	t	765.4	1323.7
Total feed[a]	t	5474	6033
Dairy efficiency	$kg_{milk}\ kg_{DMI}^{-1}$	1.32	1.31
Climate change	$kg_{CO2eq}\ kg_{milk}^{-1}$	1.39	1.42

[a] referred to lifetime

Conversely, if the cows in TS and IS had a shorter productive period to compare them with cows in OS, then the results would be as reported in Table 7, with a common lifetime equal to 58 months, thus not being able to start (TS) or conclude (IS) the 2nd lactation.

The results of CC reported in Table 6 and 7 were calculated based on the different DE values that result from the modelled assumptions. Both DE and CC result better in OS in all the modelled options (i.e. OS, OS-66 months and OS-72 months), followed by IS.

Table 7. Results for shortened lifetime of TS (-14 months) and IS (-8 months).

Variable	Unit	TS-58 months	IS-58 months
Lifetime	months	58	58
Lowered milk prod.	t	−2324	−1238
Total milk prod.[a]	t	4647.6	5392.8
Lowered feed	t	−9.1	−5.6
Total feed[a]	t	5965.3	5507.6
Dairy efficiency	$kg_{milk}\ kg_{DMI}{}^{-1}$	0.78	0.98
Climate change	$kg_{CO2eq}\ kg_{milk}{}^{-1}$	2.31	1.97

[a] referred to lifetime

4 Conclusions

The example used in this case study shows that using a PLF technology that permits a more accurate identification of oestrus events and that avoids missed inseminations in dairy cows, finally leads to relevant reductions in the Climate Change impact category. Therefore, it is important to direct farmers towards improved management practices supported by the high potentialities of technologies and of artificial intelligence predictive models. LCA can work as a feasible approach to understand the significance of an environmental benefit when applying a certain PLF technology on farms. Certainly, a more comprehensive LCA study can be done by considering not only the impact category of Climate Change but also the other environmental impact categories listed in an LCA study, in order to understand if additional benefits can be achieved in the system.

Acknowledgements. This research was funded by the Italian Ministry of Education, University and Research in Research projects of relevant national interest with grant number 20178AN8NC - "Smart Dairy Farming – Innovative solutions to improve herd productivity".

References

1. Opio, C., Gerber, P., Steinfeld, H.: Livestock and the environment: addressing the consequences of livestock sector growth. Adv. Anim. Biosci. **2**(3), 601–607 (2011). https://doi.org/10.1017/s204047001100286x
2. Pelletier, N., Tyedmers, P.: Forecasting potential global environmental costs of livestock production 2000–2050. Proc. Natl. Acad. Sci. U.S.A. **107**(43), 18371–18374 (2010). https://doi.org/10.1073/pnas.1004659107
3. Steinfeld, H., Gerber, P.: Livestock production and the global environment: consume less or produce better? Proc. Natl. Acad. Sci. U.S.A. **107**(43), 18237–18238 (2010). https://doi.org/10.1073/pnas.1012541107
4. Bellarby, J., Tirado, R., Leip, A., Lesschen, J.P., Smith, P.: Livestock greenhouse gas emissions and mitigation potential in Europe. Glob. Chang. Biol. **19**(1), 3–18 (2013). https://doi.org/10.1111/j.1365-2486.2012.02786.x

5. Herrero, M., et al.: Greenhouse gas mitigation potentials in the livestock sector. Nat. Clim. Chang. **6**(5), 452–461 (2016). https://doi.org/10.1038/NCLIMATE2925
6. Kiefer, R.L., Menzel, F., Bahrs, E.: Integration of ecosystem services into the carbon footprint of milk of South German dairy farms. J. Environ. Manage. **152**, 11–18 (2015). https://doi.org/10.1016/j.jenvman.2015.01.017
7. Pirlo, G., Lolli, S.: Environmental impact of milk production from samples of organic and conventional farms in Lombardy (Italy). J. Clean. Prod. **211**, 962–971 (2019). https://doi.org/10.1016/j.jclepro.2018.11.070
8. Chatterton, J., Graves, A., Audsley, E., Morris, J., Williams, A.: Using systems-based life cycle assessment to investigate the environmental and economic impacts and benefits of the livestock sector in the UK. J. Clean. Prod. **86**, 1–8 (2015). https://doi.org/10.1016/j.jclepro.2014.05.103
9. Brito, L.F., et al.: Large-scale phenotyping of livestock welfare in commercial production systems: a new frontier in animal breeding. Front. Genet. **11**, 1–32 (2020). https://doi.org/10.3389/fgene.2020.00793
10. Lovarelli, D., Bacenetti, J., Guarino, M.: A review on dairy cattle farming: is precision livestock farming the compromise for an environmental, economic and social sustainable production? J. Clean. Prod. **262**, 121409 (2020). https://doi.org/10.1016/j.jclepro.2020.121409
11. Berckmans, D., Guarino, M.: From the editors: precision livestock farming for the global livestock sector. Anim. Front. **7**(1), 4–5 (2017). https://doi.org/10.2527/af.2017.0101
12. Pezzuolo, A., Cillis, D., Marinello, F., Sartori, L.: Estimating efficiency in automatic milking systems. Eng. Rural Dev. **16**, 736–741 (2017). https://doi.org/10.22616/ERDev2017.16.N148
13. Arcidiacono, C., Mancino, M., Porto, S.M.C.: Moving mean-based algorithm for dairy cow's oestrus detection from uniaxial-accelerometer data acquired in a free-stall barn. Comput. Electron. Agric. **175**, 105498 (2020). https://doi.org/10.1016/j.compag.2020.105498
14. Zebari, H.M., Rutter, S.M., Bleach, E.C.L.: Characterizing changes in activity and feeding behaviour of lactating dairy cows during behavioural and silent oestrus. Appl. Anim. Behav. Sci. **206**, 12–17 (2018). https://doi.org/10.1016/j.applanim.2018.06.002
15. Cairo, F.C., et al.: Applying machine learning techniques on feeding behavior data for early estrus detection in dairy heifers. Comput. Electron. Agric. **179**, 105855 (2020). https://doi.org/10.1016/j.compag.2020.105855
16. Leliveld, L.M.C., Riva, E., Mattachini, G., Finzi, A., Lovarelli, D., Provolo, G.: Dairy cow behavior is affected by period, time of day and housing. Animals **2022**(12), 512 (2022). https://doi.org/10.3390/ani12040512
17. Riaboff, L., Shalloo, L., Smeaton, A.F., Couvreur, S., Madouasse, A., Keane, M.T.: Predicting livestock behaviour using accelerometers: a systematic review of processing techniques for ruminant behaviour prediction from raw accelerometer data. Comput. Electron. Agric. **192**, 106610 (2022). https://doi.org/10.1016/j.compag.2021.106610
18. Lovarelli, D., Finzi, A., Mattachini, G., Riva, E.: A survey of dairy cattle behavior in different barns in Northern Italy. Animals **10**, 713 (2020)
19. Tassinari, P., et al.: A computer vision approach based on deep learning for the detection of dairy cows in free stall barn. Comput. Electron. Agric. **182**, 106030 (2021). https://doi.org/10.1016/j.compag.2021.106030
20. Abeni, F., Petrera, F., Galli, A.: A survey of Italian dairy farmers' propensity for precision livestock farming tools. Animals **9**(5), 1–13 (2019). https://doi.org/10.3390/ani9050202
21. Holtz, W., Niggemeyer, H.: Reliable identification of pregnant dairy cows by double milk progesterone analysis. Livest. Sci. **228**, 38–41 (2019). https://doi.org/10.1016/j.livsci.2019.07.014

22. Pirlo, G., Miglior, F., Speroni, M.: Effect of age at first calving on production traits and on difference between milk yield returns and rearing costs in Italian Holsteins. J. Dairy Sci. **83**(3), 603–608 (2000). https://doi.org/10.3168/jds.S0022-0302(00)74919-8
23. CLAL: Italian Dairy Economic Consulting (2022). https://www.clal.it/en/?section=costi_latte. Accessed 7 Feb 2022
24. Lovarelli, D., et al.: Improvements to dairy farms for environmental sustainability in Grana Padano and Parmigiano Reggiano production systems. Ital. J. Anim. Sci. **18**(1), 1035–1048 (2019). https://doi.org/10.1080/1828051X.2019.1611389
25. Zucali, M., Tamburini, A., Sandrucci, A., Bava, L.: Global warming and mitigation potential of milk and meat production in Lombardy (Italy). J. Clean. Prod. **153**, 474–482 (2017). https://doi.org/10.1016/j.jclepro.2016.11.037

Intelligent Video Surveillance for Animal Behavior Monitoring

Souhaieb Aouayeb[1](✉), Xavier Desquesnes[1], Bruno Emile[1], Baptiste Mulot[2], and Sylvie Treuillet[1]

[1] PRISME Laboratory, Orléans University, 45100 Orléans, France
ao.souhaieb@gmail.com,
{xavier.desquesnes,bruno.emile,sylvie.treuillet}@univ-orleans.fr
[2] ZooParc de Beauval & Beauval Nature, 41110 Saint-Aignan, France
baptiste.mulot@zoobeauval.com

Abstract. The behavior of animals reflects their internal state. Changes in behavior, such as a lack of sleep, can be detected as early warning signs of health issues. Zoologists are often required to use video recordings to study animal activity. These videos are generally not sufficiently indexed, so the process is long and laborious, and the observation results may vary between the observers. This study looks at the difficulty of measuring elephant sleep stages from surveillance videos of the elephant bran at night. To assist zoologists, we propose using deep learning techniques to automatically locate elephants in each camera surveillance, then mapping the elephants detected onto the barn plan. Instead of watching all of the videos, zoologists will examine the mapping history, allowing them to measure elephant sleeping stages faster. Overall, our approach monitors elephants in their barn with a high degree of accuracy.

Keywords: Animal behavior · Object detection · Mapping

1 Introduction

In an internationally renowned modern zoo, improving animal management, farming, and welfare standards is an essential and ongoing scientific process. As a result, developing novel tools for monitoring and analyzing animal behavior is becoming a critical issue for animal park management. Especially, continuous and direct observation of animals by humans 24 hours a day, seven days a week is impractical, both financially and in terms of animal behavior.

The video surveillance systems for monitoring animal activity have several advantages over animal-based systems, including lower installation and maintenance costs, a modular and scalable system, and, most importantly, a non-invasive technique for animals. However, this approach faces some technical and scientific challenges, including poor image acquisition conditions at night, the presence of dust, and a large number of cameras, all of which make the duty of notation difficult for breeders and carers.

P. L. Mazzeo et al. (Eds.): ICIAP 2022 Workshops, LNCS 13374, pp. 361–371, 2022.
https://doi.org/10.1007/978-3-031-13324-4_31

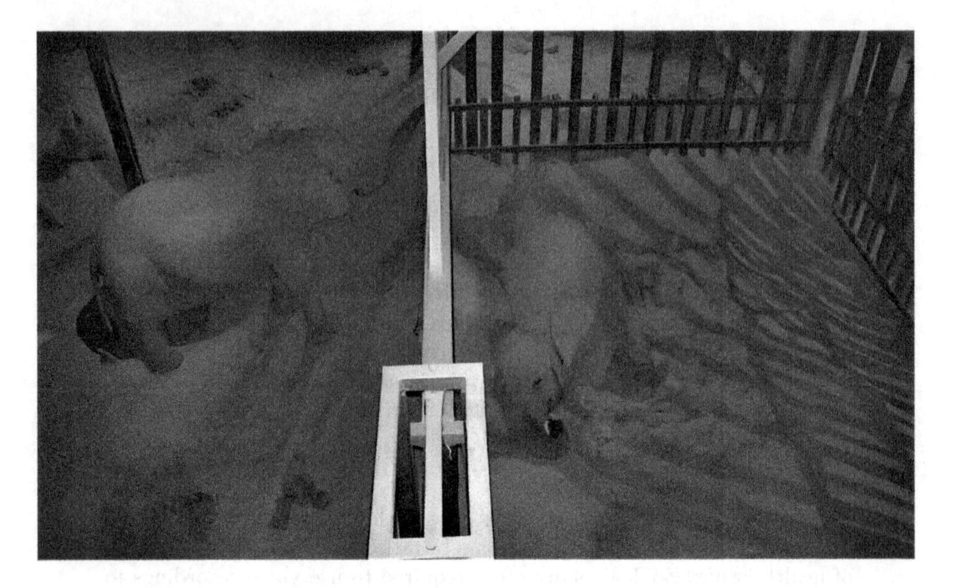

Fig. 1. An image was captured using camera 5 of box C. The image shows two elephants, one awake and the other lying down.

Fortunately, the deep learning (DL) techniques make this approach possible to address these challenges, as evidenced by the proliferation of research in recent years on the analysis of human behavior from videos [8]. As a result, it is possible to develop a methodology for analyzing animal behavior. An example of this is to use a video surveillance system to detect dairy estrus at night [36]. We can also refer to an automated tracking method for measuring behavioral changes in pigs [18]. Recently, a panda face recognition algorithm based on deep neural networks was developed using multi-camera and demonstrated pandas in their daily activities [4].

This framework will investigate elephants' behaviors, particularly at night. The central aspect is concerned with health management to identify sleep-deprived elephants. Another factor is to observe herd interactions at night. Do nocturnal herd movements, for example, bothersome elephants more than others? In this context, the idea of our work is to use a DL approach to automatically process nighttime images captured by surveillance cameras, which focuses on automated object detection of elephants. In addition, projections of each elephant's location on the enclosure plan to track the herd's nighttime movement. This study's findings will benefit animal parks by reducing zoologists' workload through intelligent video surveillance and help them to understand the nature of elephant sleep better. This study may also serve for future applied research on automatic animal monitoring in intensive farms. For example, based on the elephants' color and size, this work can easily be re-targeted to the cattle. The rest of this paper is structured as follows. Section 2 reviews the latest object detection approaches. Section 3 provides an overview of the framework and describes the proposed method for detecting elephant groups during sleep in the zoo enclosure. Section 4 discusses the experimental results of databases used in this work. Finally, Sect. 5 concludes the study with general debates and works for the future.

2 Object Detection

This section investigates approaches for object detection to identify the behavior of elephants in video clips. Not only should these approaches detect object location, but they should also help to classify objects to aid in semantic interpretation. Traditional solutions solve detection as a classification of candidate border boxes by using handcrafted features like HoG [5] and SIFT [20]. Recent DL is the basis of new solutions [15]. Typically, two components form a modern detector, a backbone for transforming images to feature maps and a head for predicting classes and boundary object boxes. VGG [25], ResNet [9], DenseNet [11], ResNeXt [35], and MobileNet [10] are the most common backbone models. There are two major types in the head part: two-stage and one-stage models.

The two-stage models first propose multiple object candidates, known as regions of interest (RoI). Then classify the region candidates and refine their location. Region-based Convolutional Neural Networks (R-CNN) is a popular two-stage model [7]. Over the years, many modifications have taken place in the networks of this family. To reduce redundant CNN computation in R-CNN for acceleration, Faster R-CNN [23] contains a Regional Proposal Network (RPN) instead of the Selective Search algorithm [29]. Due to overfitting at training and a quality mismatch at inference, Cascade R-CNN [2] concatenates several detectors trained with increased intersection-over-union (IoU) thresholds. Later, Libra R-CNN [21] integrates samples of IoU-balanced, balanced feature pyramid, and balanced L1 loss to address training imbalances.

Alternatively, the one-stage models directly predict classes and box locations without generating a sparse RoI set. YOLO [22] is one of the first efforts to propose a unified architecture that divides the image into a regular grid and makes two bounding boxes predict each cell. In the following years, researchers used anchor boxes, a better backbone, and several further tweaks to develop multiple YOLO variants [6,30]. In addition, SSD [16] predicts class scores and bounding boxes using multi-scale features. RetinaNet [13] employs a focal loss function to deal with class imbalance during training. In the case of EfficientDet [27], as its name suggests, EfficientNet [26] serves as a backbone network with Bidirectional Feature Pyramid Network (BiFPN) as the feature network and a shared prediction class/box network.

Besides these two models, the success of Transformers in many computer vision fields recently led to a new paradigm for object detection [12]. The crucial part of the Transformer is the multi-head attention, which can increase model capacities significantly. The first proposal was the TRansformer DEtection (DETR) network [3]. It uses a simple architecture by combining CNN and Transformer encoder-decoder. When compared to other detectors, DETR was the first end-to-end object detector that did not require handcrafted components such as rules-based training target assignment or non-maximum suppression post-processing. Later, a DETR variant [37], inspired by deformable convolution, developed multi-scale deformable attention modules.

3 Proposed Approach

This section discusses the proposed methodology for detecting sleeping zoo elephants and mapping the herd's nighttime movement. These elephants are brought to an elephant barn every evening (around 6 pm) until the following morning. They will often take short naps and choose to lie down to sleep. Figure 1 shows some sleeping zoo elephants lying down. During their slumber, which lasts between 4 and 6 h, the elephants awaken and wander around the barn. The barn has nine boxes, one of which is a central box that connects the four boxes on the left with the four boxes on the right. A total of ten surveillance cameras have been installed in the barn to detect sleeping zoo elephants. Figure 2 shows the cameras' positions on the enclosure plan. The amount of sleep each elephant is calculated by zoologists using video recordings from surveillance cameras. The process is lengthy and laborious due to the number of cameras complicating re-identification and the non-optimal image acquisition conditions resulting in missing detection.

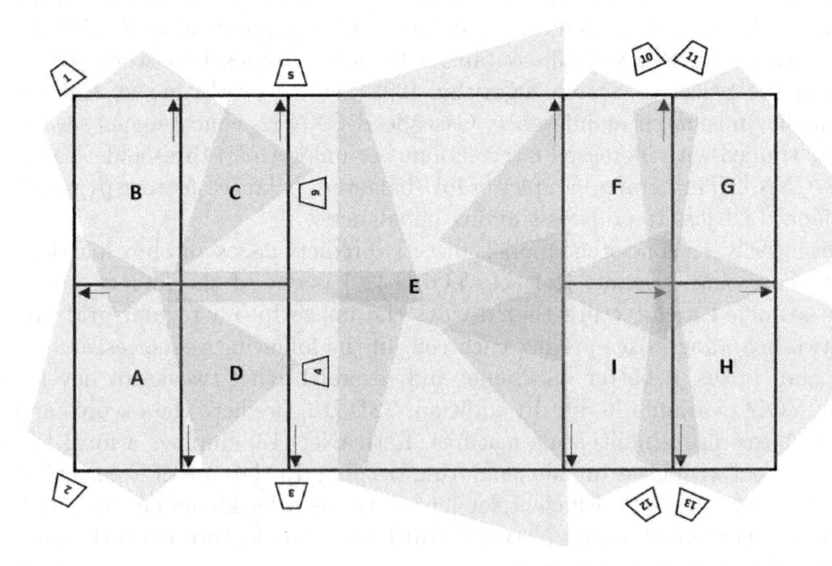

Fig. 2. Plan of the enclosure: boxes correspond to capital letters, small trapezoids with numbers indicate cameras' position within the building, large trapezoids indicate cameras' field of view, and arrows correspond to box doors.

The proposed solution included two steps. The first step is to use each camera surveillance to locate the elephants and determine which ones are awake or asleep. An object detection model detects elephants on the different surveillance cameras and then classifies them in a standing or lying down class. The second step is to project the position of each elephant detected to its actual location onto the plan of the enclosure. Consequently, zoologists will be able to track elephants without watching all of the videos. The homography method [1] is a well-known method for converting a monitored scene to a plane. Firstly, four or more corresponding points between the video and the enclosure's plan are selected. Then, based on these corresponding points, the conversion of the homography matrix is calculated. For instance, if points p are the video's feature points and the corresponding points in the enclosure plan are points P. The relationship between p and P is represented by the equation below. H (h11, h12, h13, h21, h22, h23, h31, h32, h33) is a 3×3 homography matrix.

$$\begin{bmatrix} Px \\ Py \\ 1 \end{bmatrix} = \begin{bmatrix} h11 \; h21 \; h31 \\ h21 \; h22 \; h23 \\ h31 \; h32 \; h33 \end{bmatrix} \begin{bmatrix} px \\ py \\ 1 \end{bmatrix} \tag{1}$$

Since there are several cameras, each camera has its homography matrix. Next, the position of the detected elephant is projected to its actual location onto the enclosure's plan using the homography matrix, as shown in Fig. 3.

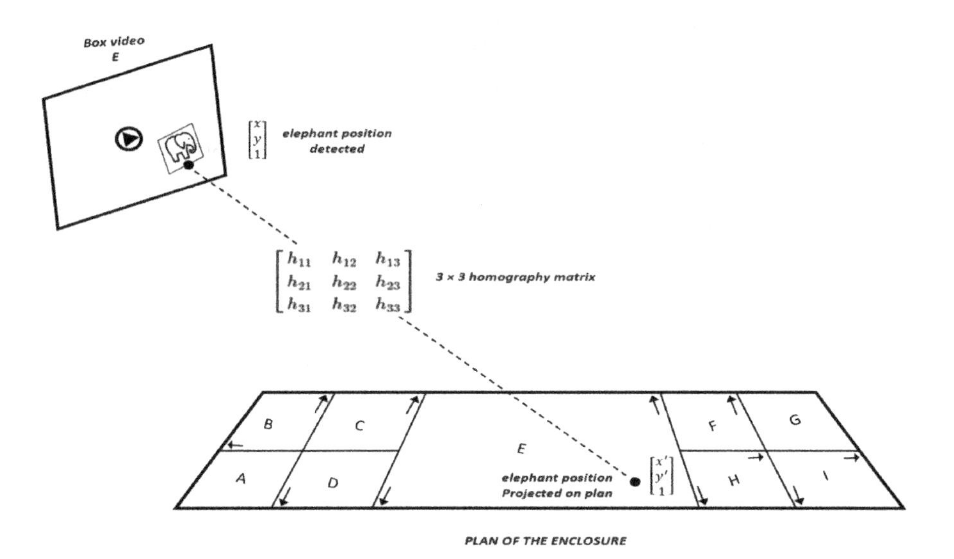

Fig. 3. Illustration of homography induced by a surveillance camera's image and the enclosure's plan. A 3×3 homography matrix connects the two points (x, y, 1) and (x', y', 1).

4 Experimental Evaluation

This section presents the results obtained using our approach. We first describe the database used. We compare in the second step the performance of three well-known DL approaches to object detection-based neural networks (namely two-stage, one-stage, and Transformer) using standard metrics from the literature such as mean average precision, precision, and recall. Then, we present some experimental mapping results. Finally, we show some improvements and some advantages of our approach.

4.1 Data

The videos used to create the samples were all obtained from the monitoring system at the Beauval Zoo in France. In the particular case of this project, it would require hours of watching surveillance videos to select and extract the desired videos, then manually create ground truth boxes for the elephants in each frame of the video using image annotation software such as LabelImg [28]. We assembled a set of 2160 nocturnal images and looked at two classes: awake elephant and asleep elephant. The images from all cameras are used to train the object detection models. It is worth noting that an elephant can be filmed by two cameras simultaneously in some situations. In this case, we label the clearest elephant due to the poor conditions for acquiring images at night and the fact that the cages of the other box hide this elephant. The training and test data contain representations of all possible scenarios. As shown in Table 1, the database was divided roughly 70% for training and 30% for testing.

Table 1. Configuration of the database

Total number of images	2167
Training images	1517
Test images	650
Number of classes	2

4.2 Detector Evaluation

Due to the growing popularity of DL, many open-source software libraries implement SoTA object detection algorithms. The results provided for Faster R-CNN and Cascade R-CNN are relied on the Detectron2 framework [34]. Similarly, we relied on the PyTorch implementation for Scaled YoloV4 [33], EfficientDet [32] and DETR [19]. In the mentioned implementations in Table 2, we used the same default model training settings as in the COCO dataset [14]. In all experiments, we used pre-trained backbones. It's worth noting that we choose smaller typical backbones for this comparison to set a standard benchmark for different sorts of detectors. The two- and one-stage detectors were trained using a stochastic gradient descent (SGD) optimizer [24], whereas the Transformer detector used an AdamW optimizer [17]. The initial learning rate values range from 10^{-3} to 10^{-5} for each model. We also employed common augmentation strategies for the selected models, such as flipping, cropping, and scaling the image. To evaluate the performance of object detectors, multiple criteria, such as precision and recall based on true/false positives (TP and FP) and false negatives (FN), are used. A low precision detector is prone to false alarms, whereas a low recall detector misses targets. Furthermore, we used the F-score, which is especially helpful in determining the capacity to limit both FP and FN. Their formulas are defined respectively as the following:

$$P = \frac{TP}{TP + FP} \tag{2}$$

$$R = \frac{TP}{TP + FN} \tag{3}$$

$$F - score = 2 \times \frac{P \times R}{P + R} \tag{4}$$

However, the final performance metric is mean Average Precision (mAP), which computes the average precision value for recall values ranging from 0 to 1. As for the experimental procedure, an Intel (R) Core (TM) i9-9900K 3.60 GHz processor, 2 Nvidia Titan RTX GPUs, and 64 GB DDR3 RAM were used for training. Table 2 lists the detection results of all the tested methods regarding the mean average precision, recall, precision and F-score. The findings reveal that the Scaled-YoloV4 architecture achieves higher performance for all computed measures. When compared to other detectors, the Scaled-YoloV4 uses and combines new features (most notably, Cross-StagePartial-Connections (CSP), a new backbone capable of improving CNN's learning capability [31]) that make its design suitable for efficient training and detection. On the other hand, DETR has the worst performance, which could be because the input resolution of the feature maps is limited in the DETR as a feature encoder, as the complexity of the attention module increases quadratically with increasing input resolution. As in our case, small datasets can present additional challenges for training Transformer models. For the rest of the work, we will apply the homography mapping based on the bottom center of the box as determined by the Yolov4-P5 detection.

4.3 Homography Evaluation

To assess the performance of homography estimation, we calculate the average distance between points transformed by annotated homographies and points transformed by estimated homographies. We examine a total of 100 annotated homography frames. The average error of homography is 0.7 cm on 1920 $\times$ 1080 images, which is small. Figure 4 depicts some of our method's detection and mapping results from the elephant barn surveillance cameras.

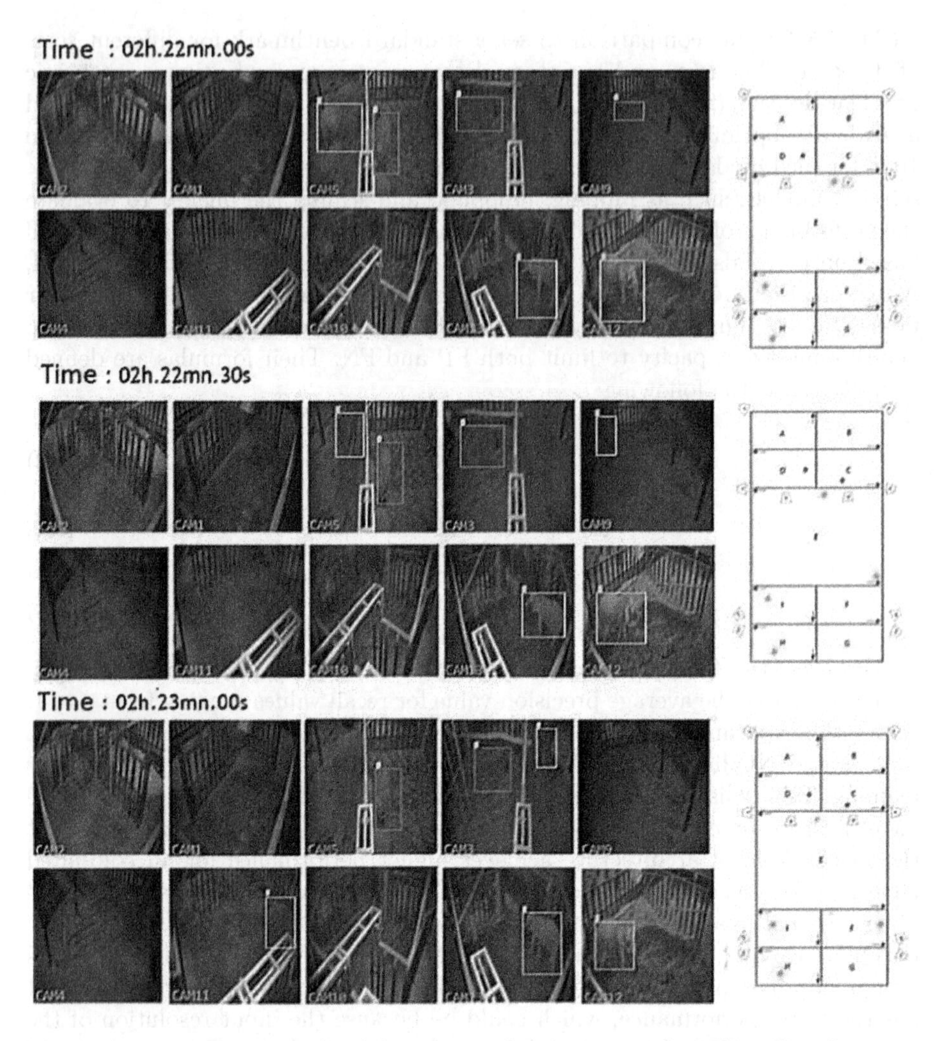

Fig. 4. A one-minute example of detection and mapping results. The yellow dots represent awake elephants, while the pink dots represent sleeping elephants. (Color figure online)

Table 2. Benchmarks for tested models of elephants dataset.

Type	Model	Backbone	mAP	P	R	F-score
Two-stage	Faster R-CNN	ResNet-50	0.593	0.616	0.689	0.650
	Cascade R-CNN	ResNet-50	0.612	0.649	0.662	0.655
One-stage	EfficientDet	EfficientNet-B2	0.683	0.663	0.713	0.687
	Scaled-YoloV4	CSP-P5	**0.711**	**0.891**	**0.942**	**0.915**
Transformer	DETR	ResNet-50	0.497	0.416	0.395	0.452

The proposed approach generates bounding boxes that confine the elephants, with their locations projected on an enclosure plan (as shown in Fig. 4). We can reliably recognize elephants with high accuracy and robustness, indicating that the approach is well-suited to supporting zoologists in their investigations and decreasing their effort. Using this software, the zoologist can jump directly to points of interest, which means they won't have to perform as much annotation work and follow a herd of elephants without sifting through all the surveillance cameras. We reach the approach's limits due to the difficulty of detecting elephants hiding behind other elephants. We conclude that this use case requires the integration of additional constraints related to adding a spatio-temporal module or setting the cameras to have a better perspective to ensure that we have at least a camera without an occlusion view.

5 Conclusions and Future Work

This work contributes a reliable method for detecting and mapping zoo elephants to track the herd's movement at night and identify sleeping elephants. By comparing SoTA object detection models, we demonstrated that our suggested approach achieves the best performances with Scaled-YOLOV4. The bounding boxes generated by the object detection model allow us to determine elephants' current locations. We then proceeded to map via homography the location of the elephants on the enclosure plan. The Breeders and the carers will be relieved of long stares in front of computers thanks to the findings of this study, which will enable them better comprehend herd dynamics at night. Future work for this project will include the development of automated ID identification of each elephant to reduce the labor of zoologists in elephant identification.

Acknowledgements. This work is part of ANIMOV project, supported by the Region Centre-Val de Loire (France). The authors would like to acknowledge the Conseil Régional of Centre-Val de Loire for its support as well as the ZooParc de Beauval and its zookeepers for the data.

References

1. Andrew, A.M.: Multiple view geometry in computer vision. Kybernetes (2001)
2. Cai, Z., Vasconcelos, N.: Cascade R-CNN: delving into high quality object detection. In: 2018 IEEE/CVF Conference on Computer Vision and Pattern Recognition, pp. 6154–6162 (2018)
3. Carion, N., Massa, F., Synnaeve, G., Usunier, N., Kirillov, A., Zagoruyko, S.: End-to-end object detection with transformers. In: Vedaldi, A., Bischof, H., Brox, T., Frahm, J.-M. (eds.) ECCV 2020. LNCS, vol. 12346, pp. 213–229. Springer, Cham (2020). https://doi.org/10.1007/978-3-030-58452-8_13
4. Chen, P., et al.: A study on giant panda recognition based on images of a large proportion of captive pandas. Ecol. Evol. **10**, 3561–3573 (2020)
5. Dalal, N., Triggs, B.: Histograms of oriented gradients for human detection. In: 2005 IEEE Computer Society Conference on Computer Vision and Pattern Recognition (CVPR 2005), vol. 1, pp. 886–893 (2005)

6. Farhadi, A., Redmon, J.: Yolov3: an incremental improvement. In: Computer Vision and Pattern Recognition, pp. 1804–2767. Springer, Heidelberg (2018)
7. Girshick, R.B., Donahue, J., Darrell, T., Malik, J.: Rich feature hierarchies for accurate object detection and semantic segmentation. In: 2014 IEEE Conference on Computer Vision and Pattern Recognition, pp. 580–587 (2014)
8. Gowsikhaa, D., Abirami, S., Baskaran, R.: Automated human behavior analysis from surveillance videos: a survey. Artif. Intell. Rev. **42**(4), 747–765 (2012). https://doi.org/10.1007/s10462-012-9341-3
9. He, K., Zhang, X., Ren, S., Sun, J.: Deep residual learning for image recognition. In: 2016 IEEE Conference on Computer Vision and Pattern Recognition (CVPR), pp. 770–778 (2016)
10. Howard, A.G., et al.: Mobilenets: efficient convolutional neural networks for mobile vision applications. arXiv abs/1704.04861 (2017)
11. Huang, G., Liu, Z., Weinberger, K.Q.: Densely connected convolutional networks. In: 2017 IEEE Conference on Computer Vision and Pattern Recognition (CVPR), pp. 2261–2269 (2017)
12. Khan, S.H., Naseer, M., Hayat, M., Zamir, S.W., Khan, F.S., Shah, M.: Transformers in vision: a survey. arXiv abs/2101.01169 (2021)
13. Lin, T.Y., Goyal, P., Girshick, R.B., He, K., Dollár, P.: Focal loss for dense object detection. In: 2017 IEEE International Conference on Computer Vision (ICCV), pp. 2999–3007 (2017)
14. Lin, T.-Y., et al.: Microsoft COCO: common objects in context. In: Fleet, D., Pajdla, T., Schiele, B., Tuytelaars, T. (eds.) ECCV 2014. LNCS, vol. 8693, pp. 740–755. Springer, Cham (2014). https://doi.org/10.1007/978-3-319-10602-1_48
15. Liu, L., et al.: Deep learning for generic object detection: a survey. Int. J. Comput. Vision **128**, 261–318 (2019)
16. Liu, W., et al.: SSD: single shot MultiBox detector. In: Leibe, B., Matas, J., Sebe, N., Welling, M. (eds.) ECCV 2016. LNCS, vol. 9905, pp. 21–37. Springer, Cham (2016). https://doi.org/10.1007/978-3-319-46448-0_2
17. Loshchilov, I., Hutter, F.: Decoupled weight decay regularization. In: ICLR (2019)
18. Matthews, S.G., Miller, A.L., Plötz, T., Kyriazakis, I.: Automated tracking to measure behavioural changes in pigs for health and welfare monitoring. Sci. Rep. **7**, 1–12 (2017)
19. Carion, N., Massa, F., Synnaeve, G., Usunier, N., Kirillov, A., Zagoruyko, S.: DETR (2020). https://github.com/facebookresearch/detr
20. Osuna, E., Freund, R.M., Girosi, F.: Training support vector machines: an application to face detection. In: Proceedings of IEEE Computer Society Conference on Computer Vision and Pattern Recognition, pp. 130–136 (1997)
21. Pang, J., Chen, K., Shi, J., Feng, H., Ouyang, W., Lin, D.: Libra R-CNN: towards balanced learning for object detection. In: 2019 IEEE/CVF Conference on Computer Vision and Pattern Recognition (CVPR), pp. 821–830 (2019)
22. Redmon, J., Divvala, S.K., Girshick, R.B., Farhadi, A.: You only look once: unified, real-time object detection. In: 2016 IEEE Conference on Computer Vision and Pattern Recognition (CVPR), pp. 779–788 (2016)
23. Ren, S., He, K., Girshick, R.B., Sun, J.: Faster R-CNN: towards real-time object detection with region proposal networks. IEEE Trans. Pattern Anal. Mach. Intell. **39**, 1137–1149 (2015)
24. Ruder, S.: An overview of gradient descent optimization algorithms. arXiv preprint arXiv:1609.04747 (2016)
25. Simonyan, K., Zisserman, A.: Very deep convolutional networks for large-scale image recognition. CoRR abs/1409.1556 (2015)

26. Tan, M., Le, Q.: Efficientnet: rethinking model scaling for convolutional neural networks. In: International Conference on Machine Learning, pp. 6105–6114. PMLR (2019)
27. Tan, M., Pang, R., Le, Q.V.: Efficientdet: scalable and efficient object detection. In: 2020 IEEE/CVF Conference on Computer Vision and Pattern Recognition (CVPR), pp. 10778–10787 (2020)
28. Tzutalin, D.: Labelimg (2015). https://github.com/tzutalin/labelImg
29. Uijlings, J.R.R., van de Sande, K.E.A., Gevers, T., Smeulders, A.W.M.: Selective search for object recognition. Int. J. Comput. Vision **104**, 154–171 (2013)
30. Wang, C.Y., Bochkovskiy, A., Liao, H.Y.M.: Scaled-yolov4: scaling cross stage partial network. In: 2021 IEEE/CVF Conference on Computer Vision and Pattern Recognition (CVPR), pp. 13024–13033 (2021)
31. Wang, C.Y., Liao, H.Y.M., Wu, Y.H., Chen, P.Y., Hsieh, J.W., Yeh, I.H.: CSPNet: a new backbone that can enhance learning capability of CNN. In: Proceedings of the IEEE/CVF Conference on Computer Vision and Pattern Recognition Workshops, pp. 390–391 (2020)
32. Wightman, R.: EfficientDet-PyTorch (2020). https://github.com/rwightman/efficientdet-pytorch
33. Wong, K.Y.: Scaledyolov4 (2021). https://github.com/WongKinYiu/ScaledYOLOv4
34. Wu, Y., Kirillov, A., Massa, F., Lo, W.Y., Girshick, R.: Detectron2 (2019). https://github.com/facebookresearch/detectron2
35. Xie, S., Girshick, R.B., Dollár, P., Tu, Z., He, K.: Aggregated residual transformations for deep neural networks. In: 2017 IEEE Conference on Computer Vision and Pattern Recognition (CVPR), pp. 5987–5995 (2017)
36. Yang, C.J., Lin, Y.H., Peng, S.: Develop a video monitoring system for dairy estrus detection at night. In: 2017 International Conference on Applied System Innovation (ICASI), pp. 1900–1903 (2017)
37. Zhu, X., Su, W., Lu, L., Li, B., Wang, X., Dai, J.: Deformable DETR: deformable transformers for end-to-end object detection. arXiv abs/2010.04159 (2021)

Quick Quality Analysis on Cereals, Pulses and Grains Using Artificial Intelligence

Bendadi Prayuktha[1](✉), Mankina Vishali[1], Distante Alessandro[1], and Guzzi Rodolfo[2]

[1] Instituto Scientifico Biomedico Euro Mediterraneo, Via Reali di Bulgaria, 72023 Mesagne, BR, Italy
prayukthabendadi@gmail.com, distante@isbem.it

[2] Emeritus of OPTICA (formerly OSA), Washington, DC 20036, USA

Abstract. Our purpose is to design a system for Quick Quality analysis of cereals, pulses, and grains using Artificial Intelligence implementing hardware and software technologies, which could swiftly analyze the type and quality of bulk grains without human intervention. Our aim is detecting and image processing to assess the grain's quality, which is also used as food for livestock, that is an important problem since high food quality has a great impact on animal health, overcoming the limitations of prior work in this field. In our methodology, the complete grain is analyzed to determine its quality by placing it in a controlled environment with a 13 Megapixel, 4K Camera Module to perform the initial step for image processing procedures, before implementing a Neural network.

Surpassing the hurdles of samples to estimate the quality of the complete grains, we demonstrated a full-grain scan technique, resulting in a unique hardware system that can more efficiently estimate the quality of the entire grains. Furthermore, our technology yields faster results since it captures the moving grains on the conveyor belt using a precise, fast camera module, which is analyzed by expeditious NVidia Jetson Nano. We applied automation in every step of this process by using vacuum tubes to collect the grains, a filter to align them, conveyor belt with a variable notch for optimum grain disposal. In summary, our study describes innovative developments stemming from a system that provides proper analysis of bulk grains, cereals, and pulses as well as automation of the system.

Keywords: Image processing · Neural network · NVidia Jetson Nano

1 Introduction

Grains are the most important commodity for peasants in our society to boost their agricultural and livestock farming revenue. The amount of automation used to assess grain quality is modest, and most of the job is done by hand. The task is so large that requires a lot of testing experience. It makes expensive and time-consuming as import and export trading have grown in popularity. This inconsistency is becoming increasingly apparent. One of the elements whose examination is more complex and complicated

P. L. Mazzeo et al. (Eds.): ICIAP 2022 Workshops, LNCS 13374, pp. 372–383, 2022.
https://doi.org/10.1007/978-3-031-13324-4_32

than that of other factors is varietal purity (Gowda and Alagasundaram 2016). Visual inspection is used to quickly check grain type and quality in the current grain handling system. This evaluation procedure, on the other hand, is arduous and time-consuming. Physical factors such as exhaustion and vision, mental state induced by prejudices and job pressure, and working environments such as incorrect lighting, temperature, and so on, can all have a significant impact on a grain inspector's decision-making ability. The quality of grain is determined entirely by the grain type and its purpose. It consists of a variety of physical (moisture content, kernel size), sanitary (fungi and mycotoxin count), and intrinsic (fat content, protein content, hardness, and starch) quality criteria. Color, composition, bulk density, odor, fragrance, size, and shape are all elements that influence grain quality. Grains and pulses will be degraded if there is one discolored kernel, according to US standards. Insects, heat, molds, weathering, sprouting, frost, illnesses, non-uniform maturity, and lack of/partial grain filling can all cause damage which reduces the whole quality of grains for animal food (Lasztity and Salgo 2002).

1.1 Problem Definition and Solution

The main issue of quality analysis is when the quality of the entire grain is judged upon analyzing a certain sample of grains. This problem can be addressed by building a dedicated hardware setup and software system for carrying out the process in a smooth quick and accurate manner. Quality analysis using image processing and neural networking technique is a popular research domain. It is time-efficient, which allows researchers to focus on the results rather than the analysis process. We created a technology that can instantly determine the kind and grade of any bulk grain without the need for significant human assistance. Automation is where our technique fits in, the system uses Artificial intelligence to detect the type of grains, and image processing involves taking measurements of objects within an image and assigning them to groups or classes. Depending on the type of grain or pulses and its purpose, grain processing will have various goals for different individuals. The minor axis length and major axis length of grains are used to evaluate grain form, and therefore the relationship between period and dimension of grains is so important. Detailed grain form size, period, and breadth are adopted by the library database.

2 Literature Review

Inspection of cereal and grain quality is crucial for both domestic and worldwide markets. For classification and quality analysis, the sample image is compared to the database image. When it comes to food grain quality, every aspect of supply and marketing profit is taken into account. In the existing grain-handling system, visual examination is employed to quickly determine grain kind and quality. A model for quality grade assessment and identification is constructed based on morphological and color parameters (R, G, B). Principal component analysis was a method of image processing that used the extended maxima operator to detect the chalky area in grain (Sidnal et al. 2013).

Using image processing techniques, a system for grading and analyzing rice grains based on grain size and form has been developed. The quality of grains is influence by both physical and chemical variables (Bergman et al. 2012). The edge detection approach is employed in this study to estimate the region of each grain's borders. To recognize the edges of grains, a sophisticated edge detector is used. Based on their dimensional characteristics, grains were appropriately identified as whole or fractured kernels (Group 2015). When it comes to analyzing the visual quality of kernels, the image processing method was shown to have a lot of potential. ICC, AACC, ISO, and CODEX-approved brand-making quality determination methodologies for wheat and rye are employed (Suismono 2013). The quality assurance system is the result of years of hard work. They were, however, attempting to analyze the sample grains using neural networks and various image processing approaches. As a result, there was no requirement for them to create a specific hardware configuration for carrying out their research method. However, the primary issue is that, we cannot determine the quality of complete grains only by analyzing their samples. This issue is handled in our technique, for which we had to create a dedicated hardware configuration in order to complete the procedure in a smooth, speedy, and precise manner with disciplined software system.

3 Methodology

3.1 Approach Technology

This arrangement works under the industrial internet of things (Fig. 1).

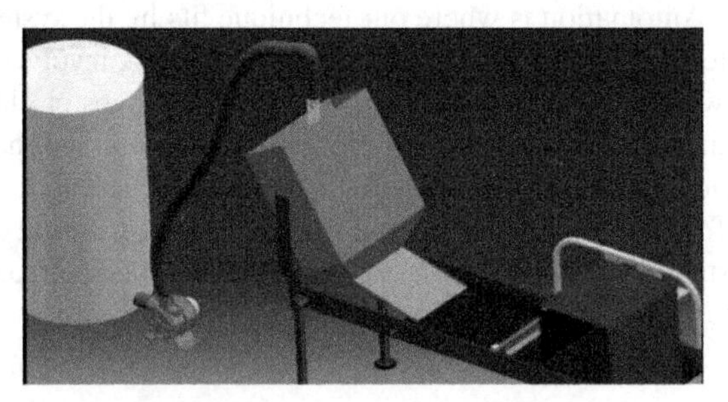

Fig. 1. Hardware setup of the system

Development of the system requires a Hardware setup to compensate for the software processing. The automation of collecting and disposal of grains for efficient estimation of quality. The Hardware Machinery is structured as follows (Fig. 2).

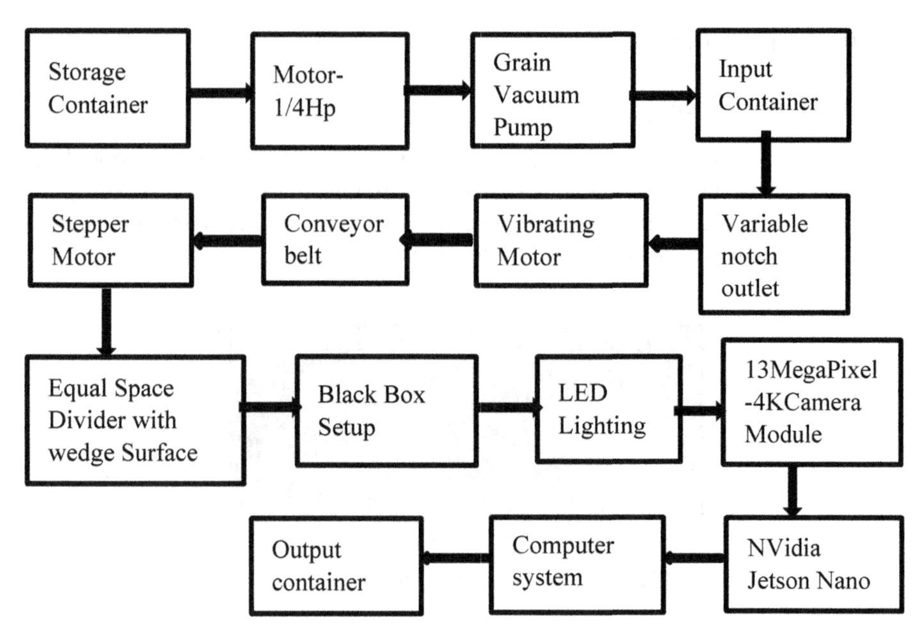

Fig. 2. Flowchart of the hardware setup

The automation of the system is handled with NVidia jetson Nano where the image capturing and processing technique is involved, with the Neural networking for a precise process that involves input layers, hidden layers, and output layers in the initial phase that gives the activation value. Based on the activation value we collect data for the knowledge base table (Fig. 3).

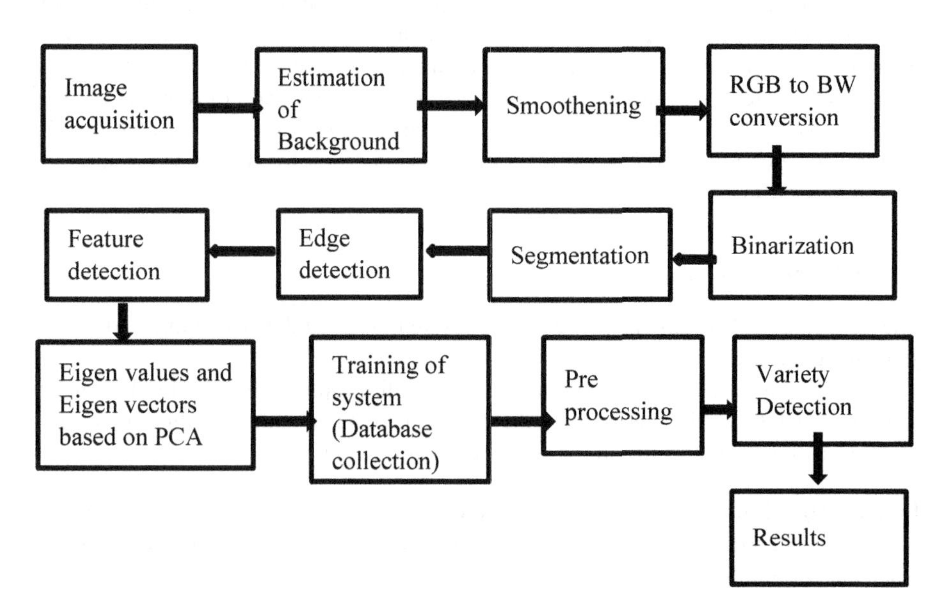

Fig. 3. Flowchart of software system

4 Implementation

4.1 Hardware System

The machinery of the system is organized in a way such that a sample of stored rice grains from a scalable container with a capacity of 10 tons starts the process. The grains are subsequently sucked from the storage container to the input container by the grain vacuum tube, which is powered by a 1/4-hp engine (Fig. 4).

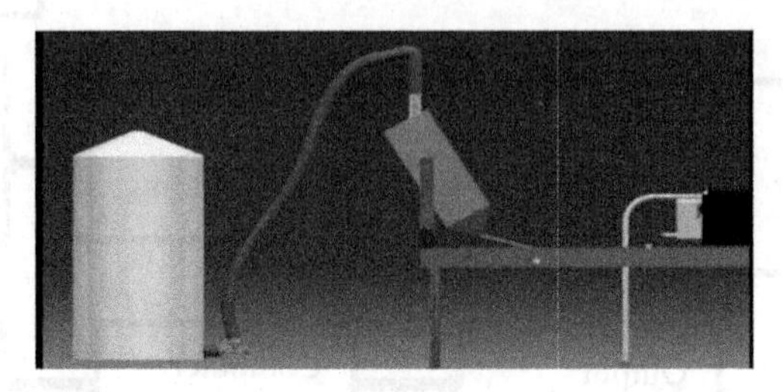

Fig. 4. Storage container with vacuum tube

The input container has a vibrating motor and a variable notch outlet on the conveyor belt, which is specially manufactured at high speed for grain analysis and filtration at high capacity. The unstructured grains are released onto a conveyor belt that is operated by a component stepper motor, and a filter with an equal space divider with a wedge surface assists the grains in keeping equal spacing in an ordered way. Make certain that the mass flow rate is managed and that the grains coming out do not overlap.

Consequently, these grains are allowed to enter the controlled environment of the black box setup, which is equipped with a broadband lighting system for superior illumination and a 13 megapixel, 4K camera module that is reliable in detecting defects and foreign materials in the grains, and is ideal for regular and conventional sorting applications, with a fixed focus, auto white balance, and auto-exposure control. The captured images are then sent to the NVidia Jetson Nano for image processing and Neural networking process.

4.2 Software System

The software technology involved in the system is neural networking and image processing techniques.

Prior to this result about the quality of grains, the captured images by the camera from the black box, which are sent to the NVidia Jetson Nano must go through a series of modifications. we perform in the sequence of following.

Image Acquisition - Initially, acquiring the image is the first stage in every image processing technique. We captured images of a sample of rice grains with an approximate black background.

Image Preprocessing - The obtained images may have an uneven background and unwanted distortions. Image preprocessing is a critical step in removing these undesired sources such as irregular background noise and blur. The acquired images are stored on the hard disc of the PC.

Background Estimation - The obtained images may have an uneven background and unwanted distortion. Gaussian method is used to reduce the undesired noise and distortion in the background, the estimated background is subtracted from the original image.

RGB to Binary Conversion - 0, which indicates black and 1, which indicates White are the two levels in the Binary image. After adjusting the contrast level, a binary version of the image is formed.

Edge Detection - For edge detection in our proposed methodology, a **Canny Edge Detector** is used (Asif et al. 2018). The detection of the edge is a major challenge since the quality of the grains is strongly related to the edges. These edges can be used to calculate the parameters of various objects in the image. The segmented images are then fed into a canny edge detector which detects the edge of the grains. It lowers the error rate and volume of data processed.

Segmentation - Binarized images are segmented. Segmentation breaks one image into several segments that may be easily analyzed. Segmentation is the important phase in calculating morphological characteristics.

Identifying the Objects - cv2.connectedcomponents() function in open CV that identifies all related objects in an image. It will display the image size, pixels, connectivity, and number of objects in an image.

Feature Detection - cv2.connectedcomponents(label, columns) function in python is used to measure the morphological features of the grains such as minor axis length, Major axis length, Minor axis length, and Aspect ratio, Hue, Saturation, Intensity of each kernel (Sanjaysdev 2021) (Fig. 5).

$$AspectRatio = \frac{Major\ Axis\ Length}{Minor\ Axis\ Length} \tag{1}$$

$$I = \frac{1}{3}(R + G + B) \tag{2}$$

$$H = Cos^{-1}\left\{\frac{\frac{1}{2}[(R-G)+(R-B)]}{[(R-G)^2+(R-B)(G-B)]^{1/2}}\right\} \tag{3}$$

$$S = 1 - \frac{3}{(R+G+B)}[min(R, G, B)] \tag{4}$$

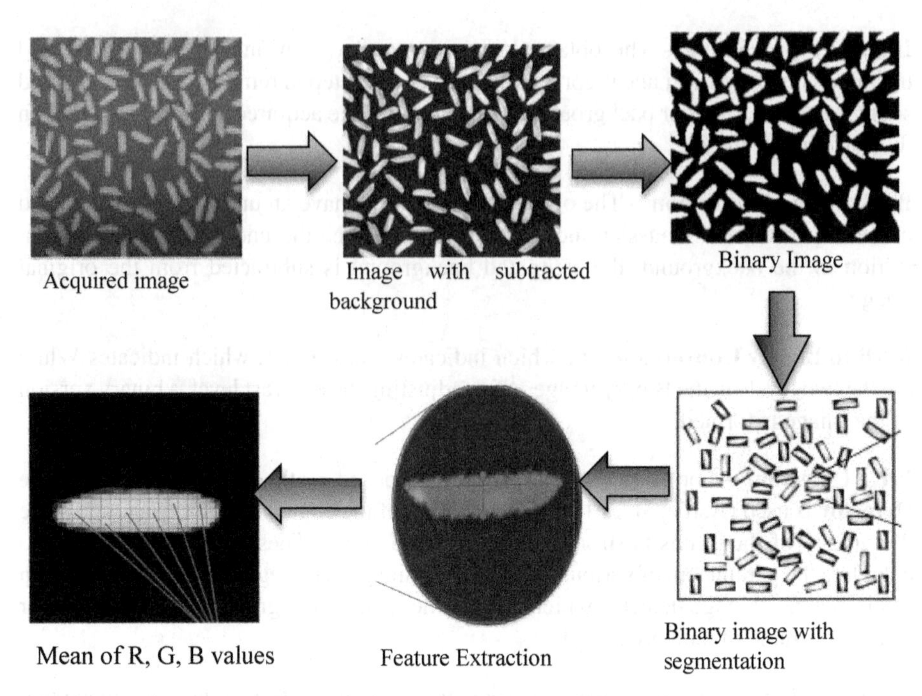

Fig. 5. Process chart of image processing

We then store these features of each kernel in a feature table (Table 1).

Table 1. Feature table for one kernel

Major axis length	Minor axis length	Length	Width	Area	Hue	Saturation	Aspect ratio
150	115	20	7	105	0.031	0.3130	1.26

Variety Detection Using PCA (Principal Component Analysis) - Using morphological features, mean and average values, Eigen values and vectors of the image are calculated. The following are the stages of putting PCA into action:

- First of all, standardize the data in the matrix form.
- Calculate the mean of data in the form of Column matrix.
- Subtract the calculated mean from each point of database.
- Calculate the variance and co-variance matrix.
- Calculate the Eigen value and Eigen vectors of the covariance matrix.

Images are then compared with images fed in database, if the samples match the database images, the grain varieties are verified and the image is subsequently entered into database for training.

The features extracted for each kernel are from an image of sample grains which are initially released onto the belt, we input these features to all the neural network models trained on a specific grain type. Otherwise, we simply extract the features and store them in a feature table (Asif et al. 2018).

In this example, only the neural network model which was specially trained on rice grain samples would produce high mean activation value.

4.3 How Does Our Neural Network Work?

The neural networking model is composed of several nodes and layers of nodes. A node is a computing unit that accepts input signals from the nodes of the layer above it. The input signal levels are multiplied by weighing factors and added together (Mohan and Raj 2020). The signal is then summed and fed through an activation function, which functions as a filter and modulates it (Fig. 6).

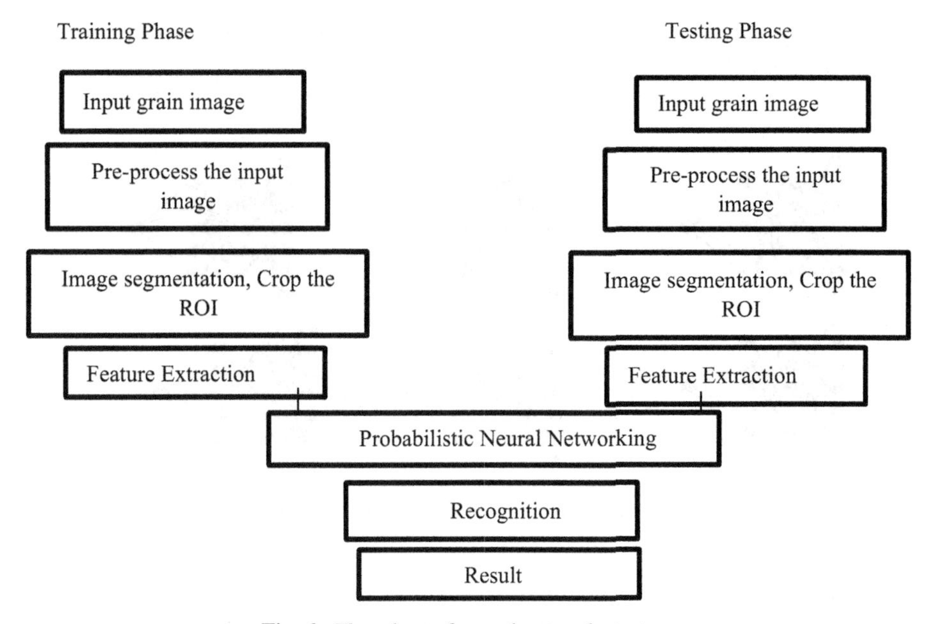

Fig. 6. Flowchart of neural network strategy

There are numerous activation functions that differ from one another based on filter functions, threshold, and shape. We adopted the multiclass classification using the SoftMax activation function for the study. Softmax is an activation function that outputs the probability for each class (Fig. 7).

The single modified signal from the activation function is then sent to every node in the following layer. Using an Optimizer, the weighting parameters for each node are tuned throughout the training operation to reduce the prediction error. The optimizer validates our function and performs gradient descent for efficiency. Utilizing the process

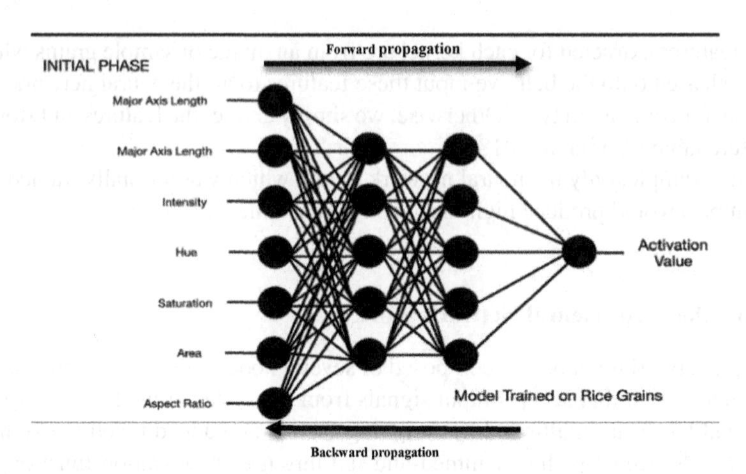

Fig. 7. Neural networking

of forwarding and backpropagation of each unit results in the efficient hypothesis of the output layer (Fig. 8).

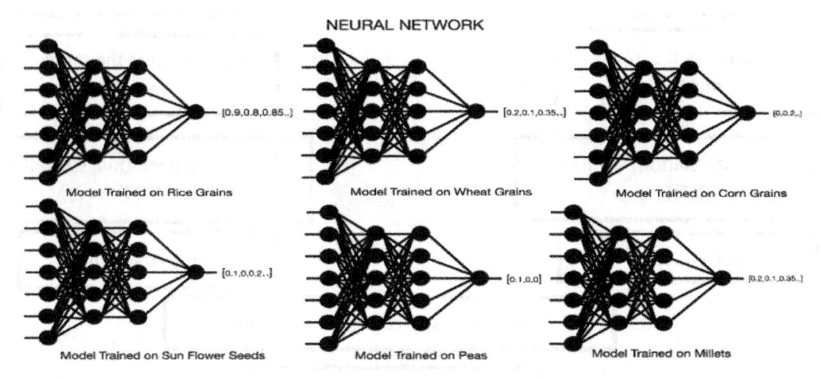

Fig. 8. Neural networking with activation values of different grains

Based on the activation values of different grains, the values are classified and stored in the knowledge base table which is the feature collection of purest samples. So, we collect the purest samples of all the grains and split them into training samples for training our neural network and testing samples for testing the efficiency of our model, where the input sample is 80% into training and 20% into testing (Wijaya 2019).

We then capture the pictures of all of these training samples of each grain type.

The knowledge base table is the collection of pure samples from the library database (Shetty 2021; Jafari et al. 2018; Karababa and Coskuner 2007) (Table 2).

Secondly, in the testing phase, the process is carried out by the same methodology and after the feature extraction and processing, the values are compared to the purest samples from the library database table to determine the quality check.

Table 2. Knowledge base table of pure samples from the library data base

Variants	Major axis length	Minor axis length	Length	Width	Area	Hue	Saturation	Aspect ratio
Rice (long grain)	150	115	20	7	105	0.031	0.3130	1.26
Rice (medium grain)	120	95	5.3	2.5	101.5	0.031	0.3130	0.015
Sun flower	110.8	53	10.7	5.2	127.6	0.34	0.3124	2.05
Corn	103.09	93	12.6	8.3	166.3	0.75	0.863	1.51
Wheat	35	16.5	5.9	2.8	110	0.39	0.26	2.12

For better understanding, we consider training our model with rice grains and we capture the images of pure rice grains, pre-process them and then extract the features of each kernel, and finally store the mean of all these features along with the standard deviation from each image into our knowledge base table. We then train our neural network with these pure features by backpropagation and adjusting the weights and biases of the model. Proper Hyperparameter tuning in this phase will definitely result in efficient models. We compare all these testing sample features with the features of the purest rice samples stored in our knowledge base table. Then, we transform these comparisons into a percentage, determining the mean quality of all the grain kernels. We perform the same analysis and find the quality of grains in all the images captured and finally, we display the mean of all these percentage values as the quality of rice grains.

5 Results

We compare and analyze the difference between the feature collection of pure sample grains from the knowledge base table and the mean of all the feature tables in one image that gives the output of quality in percentage. We then test our model with those testing samples and measure the accuracy of the model.

If the accuracy is more than 95%, we deploy them.

$$Model\ accuracy = \frac{Number\ of\ images\ correctly\ predicted}{Total\ number\ of\ images\ passed} \tag{5}$$

In our Testing and Training process of neural networking, we considered analyzing a sample of rice grains and predicted the values shown below. The mean of all the individual quality values gives us the total quality of the grain being analyzed.

In this case, the total quality of analyzed rice grains is 89% (Fig. 9).

Note: This Technology isn't just for evaluating rice grains; it can also be used to analyze wheat, barley, peas, corn, and a variety of other grains for feeding animals by adopting

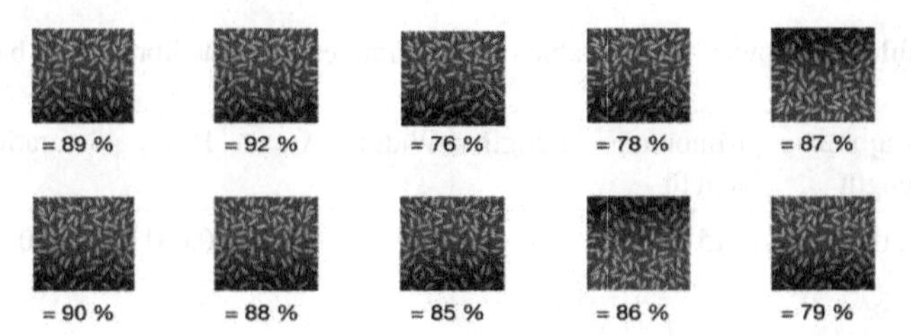

Fig. 9. Individual analysis of sample grains.

the same procedure. It can also be optimized to cover a wide range of types of grains within the same group.

6 Conclusion

In conclusion, this technological system is a simple, fast hardware machinery setup with a scalable capacity for analyzing the quality and filtering large amounts of grains. We utilized Neural networking with pattern recognition to achieve accurate and precise results by pre-processing, training, and testing the grains by considering the library database in the context of recovering knowledge on pure samples of wheat, rice, maize, corn, and many other cereals ranging from ancient to modern varieties. Grains are detected and analyzed using computer vision, which has 90% accuracy rate. The method described in this paper is for analyzing food quality for animals, and it introduces a strong grain analysis methodology. In addition, by offering a more accurate quality evaluation by means of automating processes and artificial intelligence in scanning grains, the outcomes of this approach can assist food quality of animals, farmers, industrialists and quality assessment experts.

7 Future Work

In Future, the varietal characteristics of each grain are considered, and working on the strategies to implement and analyze the quality of different varieties of the same grains. Involved in analyzing different pure samples from the library database and improving pattern recognition for optimal resolution of grains condition.

Acknowledgment. We hereby acknowledge the contribution of information providers to give us proper information and also enable us to write this paper. We would also like to thank the eminent personalities who spared their precious time during our research to respond in this regard. We once again thank all and everyone who contributed to this research paper by providing us with worthy information. Moreover, we are highly thankful to "Prof. Cosimo DISTANTE", General chair, Institute of Applied sciences and intelligent systems, CNR, Italy", who motivated us with their valuable knowledge and experience. Without the motivation of such dignitaries, it is not only difficult but even impossible to do such creative work.

References

Asif, M.J., Shahbaz, T., Rizvi, S.T.H., Iqbal, S.: RiceGrain identification and quality analysis using image processing based on prinicipal component analysis. IEEE, Turin (2018)

Bergman, C., Chen, M.-h., Delgado, J., Gipson, N.: Rice grain quality. USDA-ARS-Rice Research Unit Rice Quality Program, China (2012)

Mohan, D., Raj, M.G.: Quality analysis of rice grains using ANN and SVM. J. Crit. Rev. **7**(1), 395–402 (2020)

Gowda, N., Alagasundaram, K.: Use of thermal imaging to improve the food quality durimg storage. Indian Institute of Crop Processing Technology, Thanjavur (2016)

Group, I.R.: Quality control standards forcereal value chains. United States Agency for International Development, Washington (2015)

Jafari, S.S., Khazaei, J., Arabhosseini, A., Khoshtaghaza, M.H.: Study on mechnaical properties of sunflower seeds. ResearchGate, p. 08, 03 June 2018

Karababa, E., Coskuner, Y.: Moisture dependent physical properties of dry cornkernels. ResearchGate, p. 7, 17 July 2007

Lasztity, R., Salgo, A.: Quality assurance of cereals-past, present, future. Periodica Polytechnica Chemical Engineering, p. 9, 13 May 2002

Sanjaysdev: geeksforgeeks (2021). www.geeksforgeeks.org. Accessed 28 Nov 2021

Shetty, C.K.: Length and area of the rice grains of different types as identified by the image processing techniques. ResearchGate **4**(34), 9 (2021)

Sidnal, N., Patil, U.V., Patil, P.: Grading and quality testing of food grains using neural network. Int. J. Res. Eng. Technol. **2**, 545–549 (2013)

Suismono: Analysis of paddy grain and rice quality. Papriz-Japan International Cooperation agency (JICA), Madagascar (2013)

Wijaya, C.Y.: towardsdatascience (2019). http://towardsdatascience.com. Accessed 08 Apr 2019

Label a Herd in Minutes: Individual Holstein-Friesian Cattle Identification

Jing Gao, Tilo Burghardt, and Neill W. Campbell(✉)

Department of Computer Science, University of Bristol, Bristol, UK
{jing.gao,neill.campbell}@bristol.ac.uk, tilo@cs.bris.ac.uk

Abstract. We describe a practically evaluated approach for training visual cattle ID systems for a whole farm requiring only ten minutes of labelling effort. In particular, for the task of automatic identification of individual Holstein-Friesians in real-world farm CCTV, we show that self-supervision, metric learning, cluster analysis, and active learning can complement each other to significantly reduce the annotation requirements usually needed to train cattle identification frameworks. Evaluating the approach on the test portion of the publicly available Cows2021 dataset, for training we use 23,350 frames across 435 single individual tracklets generated by automated oriented cattle detection and tracking in operational farm footage. Self-supervised metric learning is first employed to initialise a candidate identity space where each tracklet is considered a distinct entity. Grouping entities into equivalence classes representing cattle identities is then performed by automated merging via cluster analysis and active learning. Critically, we identify the inflection point at which automated choices cannot replicate improvements based on human intervention to reduce annotation to a minimum. Experimental results show that cluster analysis and a few minutes of labelling after automated self-supervision can improve the test identification accuracy of 153 identities to 92.44% (ARI = 0.93) from the 74.9% (ARI = 0.754) obtained by self-supervision only. These promising results indicate that a tailored combination of human and machine reasoning in visual cattle ID pipelines can be highly effective whilst requiring only minimal labelling effort. We provide all key source code and network weights with this paper for easy result reproduction.

Keywords: Precision farming · Self-supervision · Active and metric learning

1 Introduction

Background. Individual animal identification is mandatory [21] in dairy farming and critical for managing aspects such as disease outbreaks and animal welfare. To date, invasive identification methods [5] such as ear tags, tattoos, radio-frequency tags or branding are most often deployed. However, ethical considerations aside, these techniques cannot provide continuous location and ID information which still requires specialist tracking systems. For Holstein-Friesians, which constitute the most

J. Gao—Was supported by the China Scholarship Council. This work also benefitted from ground work carried out under the EPSRC grant EP/N510129/1 and the John Oldacre Foundation. Many thanks to Andrew Dowsey and Will Andrew for their contributions.

P. L. Mazzeo et al. (Eds.): ICIAP 2022 Workshops, LNCS 13374, pp. 384–396, 2022.
https://doi.org/10.1007/978-3-031-13324-4_33

numerous and also highest milk-yielding [31] cattle breed, contactless visual biometric (re)identification methods [16,28] using their characteristic black-and-white skin markings [1–4,12,17] have become viable due to advances in deep learning. These approaches produce continuous coverage as long as cameras cover the whole farming area of interest. Similarly, face [36], muzzle [5], retina [5], or rear [23] biometrics may also be used in specific settings. However, modern biometric deep learning approaches that underpin systems for larger herds require significant amounts of identity-annotated visual data, demanding weeks of human annotation efforts [2].

Conceptual Approach. To address this problem and reduce labelling requirements, the literature has recently fielded self-supervision methods [9], which learn by exploiting the internal structure of data. However, although superior to traditional unsupervised approaches [37] the accuracy achieved with such systems still lags significantly behind benchmarks using supervised deep learning [2]. In response, here we advocate combining self-supervision, the analysis of the constructed identity space, and minimal active learning [29] to improve performance whilst limiting annotation requirements. Noting that research into visual cattle ID systems with reduced labelling is in its infancy [2,9,32], we propose a hybrid training approach [26,33–35] for learning cattle IDs from RGB videos. To isolate the impact of different sources of training information, we follow a three-phase strategy:

- PHASE #1: Exploit data-internal structure (via self-supervised metric learning).
- PHASE #2: Exploit latent identity space structure (via cluster analysis).
- PHASE #3: Utilise limited and targeted user input (via active learning).

We will show that each of these phases can contribute to improving learning iteratively and together lead to benchmarks closer to fully supervised learning performance – yet at a fraction of the labelling effort. Figure 1 illustrates the proposed approach visually.

Paper Contribution. Overall, this paper makes three key contributions to the field of visual learning in precision farming:

1. A principled and practical three-phase hybrid deep learning framework with reduced labelling requirements for training coat-based biometric cattle ID systems.
2. Integration of deep learning with a fast labelling approach (one-click same or different ID selection at 30 queries per min) avoiding specific identity annotation, cold start issues [14], or seed labelling requirements.
3. A detailed system evaluation and comparative analysis on real-world farming videos across the public Cows2021 dataset including the identification of inflection points in the learning where human intervention becomes vital.

We proceed by providing details on the dataset, implementation, experiments, and results. Finally, we discuss how these methods and insights can be utilised for more rapid ID system roll out in precision farming.

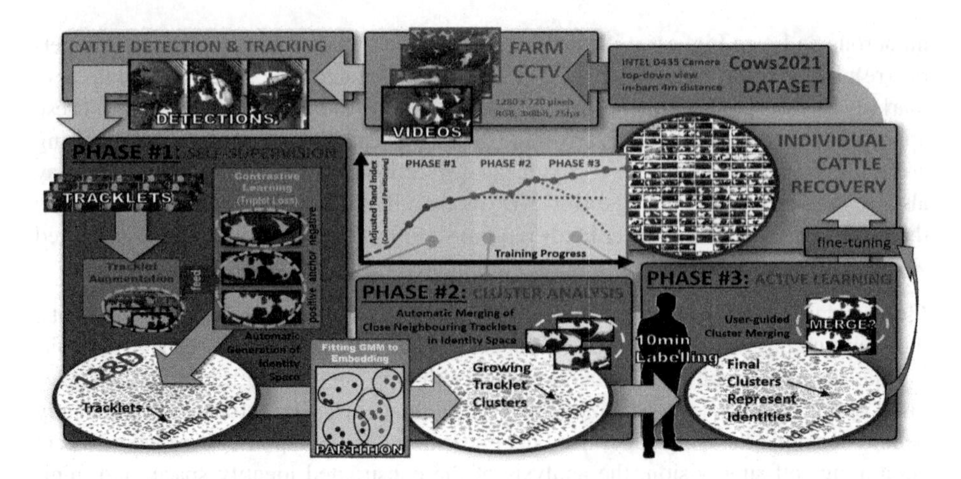

Fig. 1. Conceptual Overview. Using the public Cows2021 dataset *(purple)* we put forward a 3-phase identity learning process *(blue-green-orange)* using Holstein-Friesian detection sequences *(yellow)* in RGB videos from an operating farm. Our approach combines self-supervised metric learning *(blue)*, cluster analysis *(green)*, and active learning *(orange)* to iteratively improve performance *(middle plot)* under a minimal labelling regime. First, single animal tracklets are formed via automated oriented object recognition and tracking *(yellow)*. All frames from the tracklets are then mapped into a 128-dimensional identity space via metric learning and self-supervision *(blue)*. On this space we perform cluster analysis and merge tracklets to form growing clusters of individual cattle *(green)* used for fine-tuning. We identify the inflection point at which machine merging cannot compete with user input and thus determine when active labelling *(orange)* is needed. We track the evolving identity space via t-SNE and show significant improvements beyond self-supervision, reducing annotation requirements to around ten minutes. (Color figure online)

2 Dataset

The *Cows2021* dataset [6] contains 720 p HD digital RGB video data from a working dairy farm in the United Kingdom. It was captured by a single-view top-down camera placed 4 m above the ground between milking parlour and holding pens and operating 25 Hz (see Fig. 2 *(top)*). Excluding three particular animal identities (IDs: 155, 169, 182) with too little data for effective verification options or poor quality images, we utilise the remaining herd covering 179 individual cattle (see Fig. 2 *(bottom)*) in our study. The dataset is first automatically processed into 435 tracklets from 301 videos produced by a deep object detection (see Fig. 2 *(top)*) and tracking framework for cattle detailed in [9]. This ID-agnostic extraction performs rotational normalisation resulting in tracklets (see Fig. 2 *(middle)*) that contain exactly *one* individual with an average number of 1.45 tracklets per video. Further statistics and data splits are given in Fig. 2. Note that four totally black cows (IDs: 54, 69, 73, 173) were treated as one individual during training and were excluded in the validation and test sets to avoid systematic errors arising by mixing this particular physical anomaly with other aspects of performance. Finally, for open-set testing we withheld 24 individual cattle from training and regular testing altogether. Since each tracklet contains data of only a single individual, self-supervision can be used to associate images to identity classes.

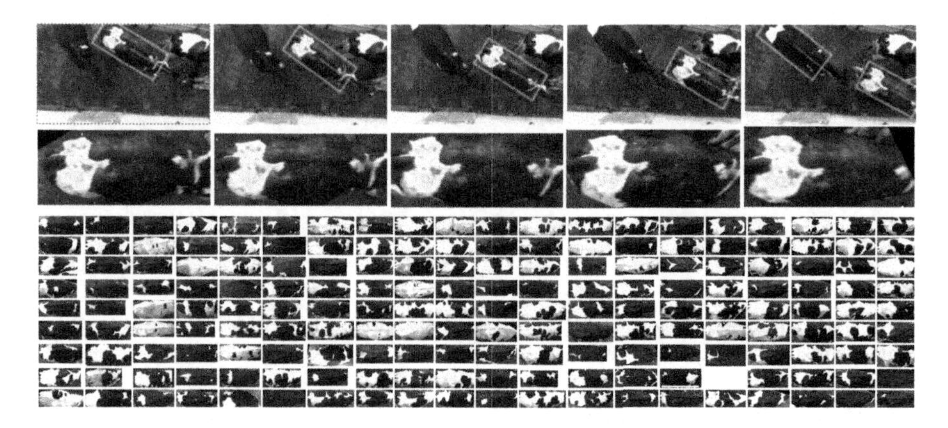

Fig. 2. Dataset. Our training data contains 23,350 frames (4670 frames and their four augmentations) from 301 RGB videos resolved at 1280 × 720 pixels. Automated oriented object detection and tracking [9] of these frames ***(top)*** yields 435 normalised tracklets ***(middle)***. ***(bottom)*** The tracklets cover 155 individuals, all of which feature in the training set. 153 individuals are used in the test set (5344 extra images), and 149 in the validation set (2480 extra images). The four all-black cattle are represented here by a single image. Red boundaries denote individuals not included in the test set, both red and blue boundaries are not included in the validation set, and the last 24 individuals (654 images) after the blank area are unseen cattle during training used for open-set evaluation only. (Color figure online)

3 Implementation

3.1 PHASE #1: Self-supervision

Metric Deep Learning and Identity Space Construction. As our base architecture we use a ResNet50 [10] backbone pre-trained on ImageNet [7] modified to have a fully-connected final layer mapping to a latent 128-dimensional vector from triplet image inputs detailed in [9]. To initialise this space for identity information without any supervision we treat each tracklet as a unique class representing the same, unknown individual forming a set of 'positive' image samples. One video may contain one or more tracklets. We pair these sets against 'negative' samples from cattle shown in other tracklets from different videos, i.e. we use a video-aware sampling strategy. Statistically, the fact that the same 'positive' individual may – with a small chance – appear in some 'negative' sample in a different video is accepted as training noise during this initialisation stage. For self-supervised learning we use reciprocal triplet loss ($\mathbb{L}_{RTL}$) [20] leading to an initial version of the identity space that optimises:

$$\mathbb{L}_{RTL} = d(x_a, x_p) + d(x_a, x_n)^{-1} \tag{1}$$

where d denotes the Euclidean distance, x_a and x_p are sampled from the 'positive' set and x_n is a 'negative' sample. We utilise online batch hard mining [11] expanded to a search that exploits both anchor and negative samples (see code base for implementation details). Training of this stage took approx. 7 h on an RTX2080 node using

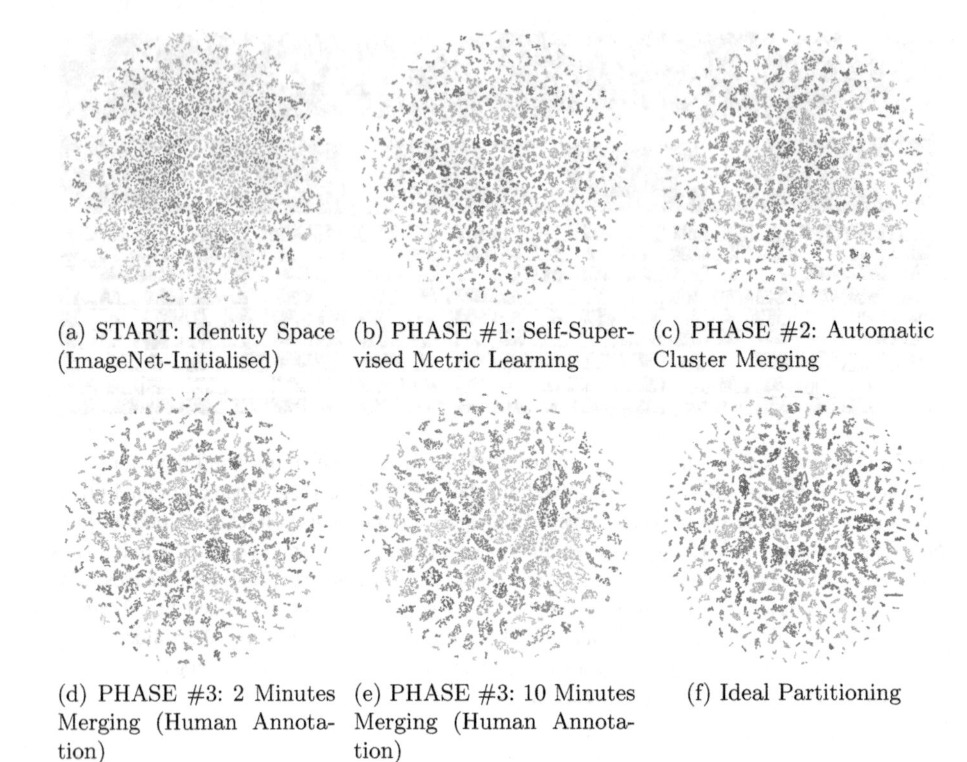

(a) START: Identity Space (ImageNet-Initialised)
(b) PHASE #1: Self-Supervised Metric Learning
(c) PHASE #2: Automatic Cluster Merging
(d) PHASE #3: 2 Minutes Merging (Human Annotation)
(e) PHASE #3: 10 Minutes Merging (Human Annotation)
(f) Ideal Partitioning

Fig. 3. Identity Space Evolution. 2D t-SNE visualisations of the training data projected into the generated 128D latent identity space through the different phases of learning where colours relate to ground truth identities. Note that after ***(a)*** ImageNet initialisation, ***(b)*** self-supervision yields a tracklet-structured space where around three times more clusters are present than individual identities. ***(c)-(e)*** Merging of these groupings and subsequent fine-tuning based on this leads to the step-wise discovery of actual identities and fewer and fewer overall clusters, first conducted via automatic cluster analysis in Phase #2 and finally via active learning in Phase #3.

SGD [27] over 50 epochs with batch size 16, learning rate 1×10^{-3}, margin $\alpha = 2$, and weight decay 1×10^{-4}. The pocket algorithm [30] against the validation set was used to address network overfitting according to the Adjusted Rand Index (ARI). Using t-distributed Stochastic Neighbour Embedding (t-SNE) [19] for visualisation, Fig. 3*(a)*) depicts the initial identity space after ImageNet pre-training and before self-supervision. As described above, the self-supervised metric learning then initialises this identity space as given in Fig. 3*(b)* leading to distinct local groupings of data points that relate to tracklets (i.e. image groups) of single cattle identities.

3.2 PHASE #2: Cluster Analysis

Clustering the Identity Space. Our aim is that the identity space should be partitioned such that all tracklets are grouped to relate to individual cattle identities.

Algorithm 1: Find candidates of tracklet pairs to merge based on GMM fit

Input: Data Points T in Identity Space;
Data Points $T_\theta \subset T$ with tracklet ID θ;
Output: Candidate List R of tracklet ID pairs;

```
1  T^m ← GMM_{g=450}(T), T is partitioned into clusters T^m with labels m;
2  T^m_θ = T_θ ∩ T^m, Data Points with tracklet ID θ in cluster T^m;
3  for k ← 1 to 4 do
4  |   if k = 1 then
5  |   |   τ_m ← Mode (T^m_θ);
6  |   else
7  |   |   l^m_k ← Select (k, GMM, T^m, T);
8  |   |   τ_{l^m_k} ← Mode (T^{l^m_k}_θ);
9  |   |   R ← list.append (τ_m, τ_{l^m_k});
10 |   end
11 end
```

However, we start by discovering clusters as given by the data in the space so far. We achieve such a partitioning by fitting a Gaussian Mixture Model (GMM) [22,25] with g = 450 components (this parameter is non-critical, provided that is is set to a number a little over the total number of tracklets) over 150 iterations to the tracklet-initialised identity space (see Fig. 4*(left)*). Our approach is to utilise the discovered clusters to merge tracklets automatically into groups that represent the same individuals.

Associating GMM Clusters with Tracklets. We hypothesise that different GMM clusters which cover tracklets of the same individual should be near to each other in identity space and that the similarity between tracklets can be evaluated based on Euclidean tracklet-to-tracklet distances. Therefore, we avoid calculating the distance of all the combinations of tracklet pairs and instead work with the top-k matches to each tracklet utilising the GMM for calculation. The pseudocode is presented in Algorithm 1. First, all data points $t \in T$ in the 128-dimensional embedding space are partitioned via the fitted GMM yielding $g = 450$ data point clusters (i.e. GMM components) $T^m \subset T$ where $m \in [1, g]$ denotes the cluster label. Next, the dominant tracklet ID marked as τ_m is assigned to each cluster label m. Let $T_\theta \subset T$ be the set of datapoints associated with the input tracklet ID θ. The intersection set $T_\theta \bigcap T^m$ denoted as T^m_θ is then used to find the most frequent tracklet ID θ therein denoted τ_m and derived via a function 'Mode' (line 8 of the Algorithm).

Selecting Candidates for Merging. Next, we find candidate tracklet pairs for merging based on the associated GMM information. We use the fact that the GMM can return probabilities that an input t belongs to *any* particular cluster m. Across all data points in any GMM cluster T^m we collect the $k^{th}(k \in [2, 4])$ most likely cluster labels, that is for each cluster T^m and k we find the most frequent cluster label l^m_k via a function 'Select', which effectively identifies the kth nearest cluster to T^m. Each cluster identified by l^m_k has itself a dominant tracklet ID $\tau_{l^m_k}$. With this information in hand, we finally compile the list R of candidate tracklet ID pairs by pairing the dominant tracklet ID τ_m with $\tau_{l^m_k}$.

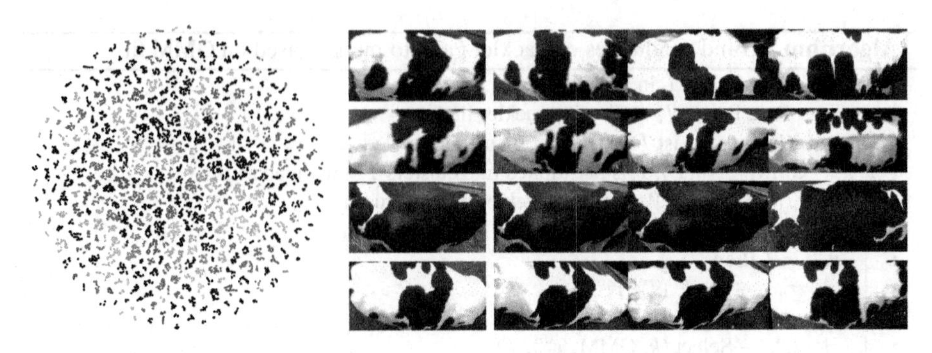

Fig. 4. GMM Clustering and Tracklet Ranking. ***(left)*** Training data of identities *(colours)* assigned to the ground truth correctly and *(black)* mismatched projected into the identity space after self-supervision with GMM. ***(right)*** Four examples *(rows)* of sample images from *(leftmost)* anchor tracklets (with ID τ_m) and the top three similar tracklets with IDs $\tau_{l_2^m}$, $\tau_{l_3^m}$ and $\tau_{l_4^m}$ forming candidate tracklet ID pairs $(\tau_m, \tau_{l_k^m}) \in R$ to be merged as described in Algorithm 1. (Color figure online)

Candidate Ranking and Cluster Merging. We now rank the generated candidate tracklet pairs (T_θ, T_η) associated to the list of tracklet ID pairs in R based on the tracklet-to-tracklet distance between each two tracklets as suggested in previous work [34]. The distance between two tracklets is calculated as the mean Euclidean distance between all point pairs that exist between the tracklets. Candidates ranked towards the top are considered closest in identity space and have indeed a high chance of representing the same individual. The dotted brown curve in Fig. 5 *(left)* confirms this hypothesis against the ground truth by tracking the rate of correct merging. The graph shows a clear downwards trend of this rate when moving lower down the ranking. The rate first declines slowly, but then drops rapidly.

Identity Space Update via Fine-Tuning. After merging, we fine-tune the network for a further 5h of training as described in Sect. 3.1 in order to incorporate the new information derived from cluster merging. This yields an updated network which defines a new embedding function for the identity space. By measuring the validation performance of this network against the number of tracklet mergers (or queries) we identify the point at which automatic merging cannot improve network performance (measured via ARI) anymore. This occurs after 173 queries as shown as the peak of the green curve in Fig. 5 *(left)*. Automatic merging after this point is thus too noisy to further improve performance. Consequently, user interaction via active learning should start at this point if further performance improvements are required.

3.3 PHASE #3: Active Learning

User-guided Fine-Tuning. Active learning [24,29] aims to interactively annotate the most informative samples from the training model, followed by a model update. A domain-relevant sample selection is absolutely critical [8,18,33]. For our case, having constructed a domain-informed ranked list R of potential tracklet ID mergers already, we will further utilise this information to guide active learning. Essentially, we will continue the tracklet merging strategy as before, but exploit user input to verify the merging instead of blindly accepting it. Thus, the final phase refines the identity space with a query sequence of one-click user inputs and utilises the new candidate ranking after another round of self-supervision retrained with the automatically merged tracklets. For manual annotation, a human is shown the next highest ranking potential merger tracklet pair from R presented as a single image from each of two tracklets. By confirming the identity to be the same or different only (one click), the human annotator throughput with this approach is extremely high at approximately 30 answers per min. The approach avoids specific identity annotations (i.e. selection from a catalogue) and it has no cold start issues [14]. Practically, whenever two tracklets sharing the same identity are identified by an annotator, they are merged into a single tracklet and the number of tracklets is reduced as before during automatic merging. Note that it is possible that – across various mergers – more than two tracklets may merge into a single tracklet due to the transitive nature of the process. As before: after merging, fine-tuning for 5 h, is used to incorporate new information derived from cluster merging into the deep network. The resulting networks are benchmarked as orange curves (marked 'Active') in Fig. 5.

4 Experimental Results

Closed Set Testing. The proposed pipeline is first evaluated on the 5,344 unseen test images across 153 individuals. This is closed-set testing since all these individuals have

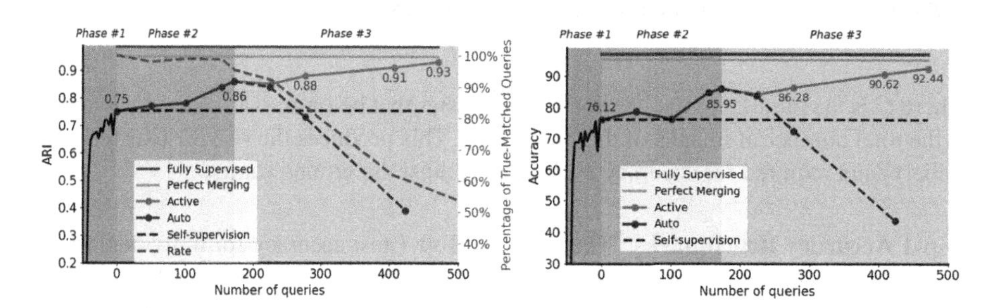

Fig. 5. Individual Identification Performance. System performance (closed-set) across the three learning phases shaded as blue, green, and orange. Benchmarks provide ***(left)*** ARI measurements and ***(right)*** Top-1 ID accuracy along evolving system training. Evaluation is conducted on 5k+ unseen test images. Phase #1 in blue depicts results of self-supervised metric learning over the last 50 epochs leading up to the 0 mark of the abscissa indicating the start of tracklet merging. Phases #2 and #3 show results of fine-tuning after automated and manual merging, respectively. The abscissa tracks the number of mergers (or queries). Note, that at 173 queries, automatic merging (green curve) peaks and further mergers are too noisy to improve performance without manual correctness checking via active learning (orange curve). Note further that performance after 10 min of labelling at accuracy 92.44% is close to the performance of perfectly merging *all* clusters exhaustively (95.17%) based on ground truth. The dotted brown curve on the left quantifies the proportion of correct merger suggestions). (Color figure online)

been seen (in different videos) during training. A projection of the closed testset into the learned identity space is visualised via t-SNE in Fig. 7*(a)*). In order to evaluate the clustering performance in identity space, we use two measures: the Top-N ID prediction accuracy and also the Adjusted Rand Index (ARI) [13] for assessing structural clustering similarity.

Structural Clustering Benchmarks. We evaluate the structural similarity of the testset clustering produced by the trained network against the ground truth. We quantify the similarity via the ARI measure which operates on two data partitions and does not require any knowledge of ID labels. In particular, the testset clustering is derived by fitting a GMM (with the component cardinality of the testset) to the test data in the fine-tuned identity space. Figure 5 *(left)* shows the evolution of ARI across the three learning phases. Self-supervision of Phase #1 as described in Sect. 3.1 leads to an initial ARI value of 0.75[1] (see blue curve). In Phase #2, automatic tracklet merging and subsequent fine-tuning as described in Sect. 3.2 steadily improves the ARI further to a peak value of 0.86 at query 173 (see green curve). At this point, we find that any further automatic merging is not beneficial as depicted by the dotted green curve. In Phase #3, user input is now utilised. The ARI can be increased further to 0.88 after 3.5 min of human labelling, 0.91 after 8 min, and 0.93 after 10 min. Perfectly merging all clusters exhaustively only leads to a small increase of ARI at 0.95.

Accuracy Stipulation. Next, we evaluate identification accuracy against ground truth ID labels of the testset. In order to obtain IDs for our network output, each fitted GMM cluster used for structural benchmarking is assigned to the one individual ID having the highest overlap ratio with the ground truth, defined as:

$$\mathbb{O}_l = C/L \tag{2}$$

where C is the number of images in a GMM cluster that belongs to an individual, and L is the total number of images of that individual. This produces the (GMM Cluster)-(ID Label) pairs required for accuracy evaluation against the ground truth.

Top-1 Accuracy Benchmarks. The resulting Top-1 test accuracy for individual identification over the three training phases is given in Fig. 5 *(right)*. Generally, the plot structure is consistent with our ARI analysis in Fig. 5 *(left)* showing the effectiveness of all three phases of our approach across the two different performance measures. Phase #1 leads to an accuracy of 76.12% already on the test set, emphasising the strength of a tracklet-driven self-supervision signal during metric learning. In Phase #2, automatic tracklet merging can further improve accuracy to a peak value of 85.95% (see green curve). Note that automatic merging is imperfect and accuracy is not optimised directly by fine-tuning, thus a temporary decrease of accuracy is possible. In Phase #3, human labelling allows a further increase in accuracy to 86.28% after 3.5 min of human labelling, 90.62% after 8 min, and finally 92.44% after 10 min.

[1] Note that this performance improves on the self-supervision state-of-the-art [9] in the domain by using the same network yet with our extended hard mining regime. Testing our Phase #1 method on their testset improves ARI from their published ARI of 0.53 to 0.65.

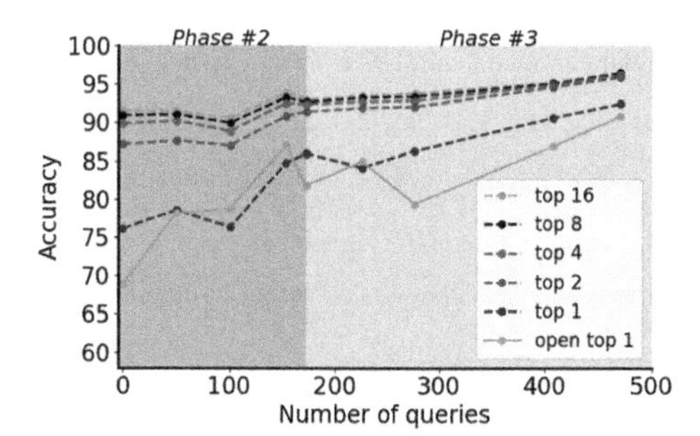

Query	N=1	N=2	N=4	N=8	N=16	open
0	76.12	87.13	89.86	90.92	91.56	68.83
50	78.56	87.59	90.21	91.04	91.47	78.16
101	76.29	86.99	88.87	89.95	90.53	78.77
154	84.69	90.85	92.50	93.26	93.90	87.05
173	85.95	91.39	92.35	92.66	93.00	81.78
226	84.11	91.90	92.68	93.23	93.51	84.94
277	86.28	92.05	92.85	93.39	93.81	79.37
408	90.62	94.59	94.91	95.08	95.23	86.90
471	92.44	95.92	96.20	96.41	96.59	90.81

Fig. 6. Top-N and Open-Set Performance. Depicted is the Top-N identification accuracy of 153 individuals across the final two learning phases shaded green and orange, respectively. Benchmarks provide accuracy curves for Top-N closed-set (24 individuals) accuracy (detailed also as values on the ***(right)***) and Top-1 open-set accuracy in yellow. Note that open-set performance benefits strongly from the proposed Phase #2 and #3, leading to near closed-set performance. (Color figure online)

Top-N Accuracy Benchmarks. For each GMM cluster assigned to an ID as per Eq. 2, one can alternatively rank all identities according to $\mathbb{O}_l$where identities that have a $\mathbb{O}_l =$ 0 form the tail of the ranking sequence with randomly assigned, remaining IDs. This process allows the creation of a Top-N [15] accuracy benchmark as shown in Fig. 6. In this statistic, a prediction is considered correct if and only if the ground truth ID is found amongst the top N ranked IDs. As expected, for $N = 1$ this equates to the traditional definition of accuracy. Figure 6 emphasises that across Phases #2 and #3 improve accuracy across all N. This setting is particularly interesting if semi-automatic identification is used to present a system user with a set of N candidate identities for a query. For $N = 16$ this setting leads to a near-perfect accuracy benchmark of 96.59% using only 10 min of labelling.

Open-Set Accuracy Benchmarks. The proposed pipeline is secondly evaluated on the 654 unseen test images of 24 never seen individuals. A projection of this open testset into the learned identity space is visualised via t-SNE in Fig. 7 *(c)*. This 'Open-Set' testing is critical for stipulating how far performance can translate across herds or farms in a zero-shot identification paradigm. Figure 6 depicts Top-1 open-set test performance across the final two phases of training. It can be seen that open-set performance generally lags behind closed-set performance. However, the margin (gap between blue and yellow curves) is only 1.53% after 10 min of manual labelling. This is promising, showing that the identity space created generalises well across the domain of Holstein-Friesians and not just the training herd.

(a) CLOSED TESTSET, 153 animals. (10min Annotation) (b) CLOSED TESTSET, 153 animals. (Fully Supervised) (c) OPEN TESTSET, 24 animals. (10min Annotation)

Fig. 7. t-SNE Testset Embeddings. 2D t-SNE visualisation of the closed and open test sets when projected into the identity space. ***(a)*** shows the result for the proposed method at accuracy 92.44% after all three stages of training and 10 min of manual annotation. ***(b)*** For comparison, we show the same testset projected into an identity space built from fully manual annotation at the Top-1 accuracy 96.05% using weeks of frame-by-frame identity labelling. ***(c)*** When projecting unseen images of unseen animals into the space used in ***(a)*** identities still cluster well (accuracy 90.81%) confirming generalisation to a domain wider than the herd trained on.

5 Conclusion

We have put forward a practical, three-phase deep learning approach for training an ID system for Holstein-Friesians on a farm, requiring only ten minutes of labelling. We showed that automatic identification of individual cattle in real-world farm CCTV can be achieved effectively by combining self-supervision, metric learning, cluster analysis, and active learning. We provided detailed explanations and key source code[2] for full result reproduction and evaluated the approach using the publicly available Cows2021 dataset. Self-supervised metric learning was first leveraged to initialise an identity space where tracklets are considered a distinct entity. Grouping entities is then performed by automated merging via cluster analysis and active learning feeding into fine-tuning. Experimental results showed that cluster analysis and a few minutes of labelling after automated self-supervision can indeed improve the identification accuracy of closed and open test sets compared to self-supervision only. Despite superior performance of fully supervised systems which required weeks of frame-by-frame labelling in the past, our 10 min labelling approach shows promising results indicating that human and machine reasoning in tandem can be integrated into visual cattle ID pipelines in a highly effective fashion requiring only minimal labelling effort.

References

1. Andrew, W.: Visual biometric processes for collective identification of individual Friesian cattle. Ph.D. thesis, University of Bristol (2019)

[2] https://github.com/Wormgit/LabelaHerdinMinutes.

2. Andrew, W., Gao, J., Mullan, S., Campbell, N., Dowsey, A.W., Burghardt, T.: Visual identification of individual holstein-friesian cattle via deep metric learning. Comput. Electron. Agric. **185**, 106133 (2021)
3. Andrew, W., Greatwood, C., Burghardt, T.: Visual localisation and individual identification of holstein friesian cattle via deep learning. In: Proceedings of the IEEE International Conference on Computer Vision, pp. 2850–2859 (2017)
4. Andrew, W., Hannuna, S., Campbell, N., Burghardt, T.: Automatic individual holstein friesian cattle identification via selective local coat pattern matching in rgb-d imagery. In: 2016 IEEE International Conference on Image Processing (ICIP), pp. 484–488. IEEE (2016)
5. Awad, A.I.: From classical methods to animal biometrics: a review on cattle identification and tracking. Comput. Electron. Agric. **123**, 423–435 (2016)
6. Campbell, N., Burghardt, T., Gao, J., Andrew, W.: The cows2021 dataset (2021). https://data.bris.ac.uk/data/dataset/4vnrca7qw1642qlwxjadp87h7
7. Deng, J., Dong, W., Socher, R., Li, L.J., Li, K., Fei-Fei, L.: Imagenet: a large-scale hierarchical image database. In: 2009 IEEE conference on computer vision and pattern recognition, pp. 248–255. Ieee (2009)
8. Freytag, A., Rodner, E., Denzler, J.: Selecting influential examples: active learning with expected model output changes. In: Fleet, D., Pajdla, T., Schiele, B., Tuytelaars, T. (eds.) ECCV 2014. LNCS, vol. 8692, pp. 562–577. Springer, Cham (2014). https://doi.org/10.1007/978-3-319-10593-2_37
9. Gao, J., Burghardt, T., Andrew, W., Dowsey, A.W., Campbell, N.W.: Towards self-supervision for video identification of individual holstein-friesian cattle: The cows2021 dataset (2021). arXiv preprint arXiv:2105.01938
10. He, K., Zhang, X., Ren, S., Sun, J.: Deep residual learning for image recognition. In: Proceedings of the IEEE Conference on Computer Vision and Pattern Recognition, pp. 770–778 (2016)
11. Hermans, A., Beyer, L., Leibe, B.: In defense of the triplet loss for person re-identification (2017). arXiv preprint arXiv:1703.07737
12. Hu, H., et al.: Cow identification based on fusion of deep parts features. Biosys. Eng. **192**, 245–256 (2020)
13. Hubert, L., Arabie, P.: Comparing partitions. J. Classi. **2**(1), 193–218 (1985)
14. Konyushkova, K., Sznitman, R., Fua, P.: Learning active learning from data. In: NIPS. pp. 4228–4238 (2017). http://papers.nips.cc/paper/7010-learning-active-learning-from-data
15. Krizhevsky, A., Sutskever, I., Hinton, G.E.: Imagenet classification with deep convolutional neural networks. In: Pereira, F., Burges, C.J.C., Bottou, L., Weinberger, K.Q. (eds.) Advances in Neural Information Processing Systems, vol. 25, pp. 1097–1105. Curran Associates, Inc. (2012). http://papers.nips.cc/paper/4824-imagenet-classification-with-deep-convolutional-neural-networks.pdf
16. Kühl, H.S., Burghardt, T.: Animal biometrics: quantifying and detecting phenotypic appearance. Trends Ecol. Evol. **28**(7), 432–441 (2013)
17. Li, W., Ji, Z., Wang, L., Sun, C., Yang, X.: Automatic individual identification of holstein dairy cows using tailhead images. Comput. Electron. Agric. **142**, 622–631 (2017)
18. Lindenbaum, M., Markovitch, S., Rusakov, D.: Selective sampling for nearest neighbor classifiers. Mach. Learn. **54**(2), 125–152 (2004)
19. van der Maaten, L.J., Hinton, G.E.: Visualizing high-dimensional data using t-sne. J. Mach. Learn. Res. **9**, 2579–2605 (2008)
20. Masullo, A., Burghardt, T., Damen, D., Perrett, T., Mirmehdi, M.: Who goes there? exploiting silhouettes and wearable signals for subject identification in multi-person environments. In: Proceedings of the IEEE International Conference on Computer Vision Workshops (2019)

21. Parliament, E., Council: Establishing a system for the identification and registration of bovine animals and regarding the labelling of beef and beef products and repealing council regulation (ec) no 820/97 (1997). http://eur-lex.europa.eu/legal-content/EN/TXT/?uri=celex:32000R1760 (Accessed 29-January 2016)
22. Pedregosa, F., et al.: Scikit-learn: machine learning in Python. J. Mach. Learn. Res. **12**, 2825–2830 (2011)
23. Qiao, Y., Su, D., Kong, H., Sukkarieh, S., Lomax, S., Clark, C.: Individual cattle identification using a deep learning based framework. IFAC-PapersOnLine **52**(30), 318–323 (2019)
24. Ren, P., et al.: A survey of deep active learning. ACM Comput. Surv. (CSUR) **54**(9), 1–40 (2021)
25. Reynolds, D.A.: Gaussian mixture models. Encycl. Biometrics **741**, 659–663 (2009)
26. Rizve, M.N., Duarte, K., Rawat, Y.S., Shah, M.: In defense of pseudo-labeling: An uncertainty-aware pseudo-label selection framework for semi-supervised learning. In: International Conference on Learning Representations (2020)
27. Robbins, H., Monro, S.: A stochastic approximation method. Ann. Math. Statist. **22**(3), 400–407 (1951)
28. Schneider, S., Taylor, G.W., Kremer, S.C.: Similarity learning networks for animal individual re-identification-beyond the capabilities of a human observer. In: Proceedings of the IEEE/CVF Winter Conference on Applications of Computer Vision Workshops, pp. 44–52 (2020)
29. Settles, B.: Active learning literature survey. Computer Sciences Technical Report 1648, University of Wisconsin-Madison (2009)
30. Stephen, I.: Perceptron-based learning algorithms. IEEE Trans. Neural Netw. **50**(2), 179 (1990)
31. Tadesse, M., Dessie, T.: Milk production performance of zebu, holstein friesian and their crosses in ethiopia. Livest. Res. Rural. Dev. **15**(3), 1–9 (2003)
32. Vidal, M., Wolf, N., Rosenberg, B., Harris, B.P., Mathis, A.: Perspectives on individual animal identification from biology and computer vision. Integr. Comp. Biol. **61**(3), 900–916 (2021)
33. Wang, K., Zhang, D., Li, Y., Zhang, R., Lin, L.: Cost-effective active learning for deep image classification. IEEE Trans. Circuits Syst. Video Technol. **27**(12), 2591–2600 (2016)
34. Wang, M., Lai, B., Jin, Z., Gong, X., Huang, J., Hua, X.: Deep active learning for video-based person re-identification (2018). arXiv preprint arXiv:1812.05785
35. Wu, Y., Lin, Y., Dong, X., Yan, Y., Ouyang, W., Yang, Y.: Exploit the unknown gradually: One-shot video-based person re-identification by stepwise learning. In: Proceedings of the IEEE Conference on Computer Vision and Pattern Recognition, pp. 5177–5186 (2018)
36. Yao, L., Hu, Z., Liu, C., Liu, H., Kuang, Y., Gao, Y.: Cow face detection and recognition based on automatic feature extraction algorithm. In: Proceedings of the ACM Turing Celebration Conference-China, pp. 1–5 (2019)
37. Yu, H.X., Wu, A., Zheng, W.S.: Cross-view asymmetric metric learning for unsupervised person re-identification. In: Proceedings of the IEEE International Conference on Computer Vision, pp. 994–1002 (2017)

Workshop on Small-Drone Surveillance, Detection and Counteraction Techniques - WOSDETC

DroBoost: An Intelligent Score and Model Boosting Method for Drone Detection

Ogulcan Eryuksel(✉), Kamil Anil Ozfuttu, Fatih Cagatay Akyon, Kadir Sahin, Efe Buyukborekci, Devrim Cavusoglu, and Sinan Altinuc

OBSS AI, OBSS Technology, Universiteler Mahallesi,
1606. Cad. No:4/1/307 Cyberpark Cyberplaza C blok 3.kat, Cankaya, Ankara, Turkey
{ogulcan.eryuksel,anil.ozfuttu,fatih.akyon,kadir.sahin,
efe.buyukborekci,devrim.cavusoglu,sinan.altinuc}@obss.com.tr
https://www.obss.com.tr

Abstract. Drone detection is a challenging object detection task where visibility conditions and quality of the images may be unfavorable, and detections might become difficult due to complex backgrounds, small visible objects, and hard to distinguish objects. Both provide high confidence for drone detections, and eliminating false detections requires efficient algorithms and approaches. Our previous work, which uses YOLOv5, uses both real and synthetic data and a Kalman-based tracker to track the detections and increase their confidence using temporal information. Our current work improves on the previous approach by combining several improvements. We used a more diverse dataset combining multiple sources and combined with synthetic samples chosen from a large synthetic dataset based on the error analysis of the base model. Also, to obtain more resilient confidence scores for objects, we introduced a classification component that discriminates whether the object is a drone or not. Finally, we developed a more advanced scoring algorithm for object tracking that we use to adjust localization confidence. Furthermore, the proposed technique won 1st Place in the Drone vs. Bird Challenge (Workshop on Small-Drone Surveillance, Detection and Counteraction Techniques at ICIAP 2021).

Keywords: Drone detection · Deep learning · Object tracking · Object detection · Synthetic data

1 Introduction

Unmanned Aerial Vehicles (UAVs) have been used for some time, and the recent technological and industrial developments have reduced the costs, making them more accessible and abundant in the commercial markets. After a few successful commercial use cases, the demand for drones has boomed. The use cases

Supported by OBSS Technology.

P. L. Mazzeo et al. (Eds.): ICIAP 2022 Workshops, LNCS 13374, pp. 399–409, 2022.
https://doi.org/10.1007/978-3-031-13324-4_34

include surveillance & security [15], photography, delivery [5], warehouse operations [23], environmental monitoring [9], etc. With the increase in market capacity, drones can be easily purchased on the Internet at low prices in the present era. Another perspective of drone technology is that they got smaller and better at the assigned tasks as time passed. Consequently, high reachability brings plenty of opportunities for commercial industry and individuals along with the defense market. On the other hand, as a side-effect of the ease of reachability raises issues about safety, privacy, and security. Thus, misuse of UAVs for illegal activities, invasion of privacy, and violation of regulations need to be addressed.

To address the problems that emerged due to the misuse of drones and to gather the potential solutions under a roof, Drone-vs-Bird Challenge was organized starting from 2017 at the International Workshop on Small-Drone Surveillance, Detection and Counteraction Techniques (WOSDETC) [7]. The challenge aims to assess discrimination between a drone and other (flying) objects at far distances, including similar objects (e.g., birds) from a video dataset. Another challenge apart from detecting drones is that it is not straightforward to make additions to the provided dataset as flying drones often require permission and are restricted in several areas.

Drone detection, a sub-field of object detection, has been studied especially within the past few years with the challenge's launch. The solutions proposed for drone detection and tracking were various in terms of methodology but mostly based on deep learning in the present era. Some works also include approaches that are not based on deep learning, such as SVMs [21], AdaBoost [10].

In this paper, we propose an approach composed of drone detection and object tracking components. We have taken the methodology of our previous work [3] as a baseline and further improved both the detection and tracking components. Firstly, we use a more diverse and balanced dataset that combines several sources. Secondly, we use false negative predictions from synthetically generated drone image data to improve detection performance. We also included various background images with no drones as negative samples to training data where our model created false predictions. We combined real data with these samples to improve cases where base model fails. Thirdly, we used a binary classification algorithm trained on bounding box crops of drone images from training datasets and our YOLOv5 models' false positive predictions on the training data. Model confidence is calculated by the geometric average of YOLOv5 confidence and binary classification confidence. Lastly, we used a specialized scoring algorithm to determine the confidence for the track and adjust the bounding box prediction confidences using this information.

2 Related Work

Deep learning-based detection approaches have produced good results for various applications in recent years, including drone detection.

A two-staged detection strategy has been proposed in [20]. First, the authors examined the suitability of different conventional image processing-based object detection techniques, i.e., frame differencing and background subtraction techniques, locally adaptive change detection, and object proposal techniques [16], to extract region candidates in video data from static and moving cameras. In the second stage, a shallow CNN classification network is used to classify each candidate region into drone and clutter categories.

In [6], Gagné and Mercier (referred to as the Alexis team) proposed a drone detection approach based on YOLOv3 [19] and taking a single RGB frame as input. By integrating an image tiling strategy, this approach can successfully detect small drones in high-resolution images. Alexis Team leveraged the public PyTorch implementation of YOLOv3 with Spatial Pyramid Pooling (YOLOv3-SPP) made available by Ultralytics [13]. In Spatial Pyramid Pooling [11], the input features are processed by pooling layers of varying sizes in parallel and then concatenated to yield fixed-length feature vectors.

Moreover, in [6] EagleDrone Team proposed a YOLOv5 [14] based drone detection modality with a linear sampling-based data sub-sampling method. They propose using computed loss per image to select the sampling probability. Furthermore, they detect small and low-resolution drones using an ESRGAN-based super-resolution approach.

Recently, authors of [3] and CARG-UOTTAWA team [6] proposed YOLOv5 [14] based drone detection modalities utilizing extra training data. Different from the CARG-UOTTAWA team, [3] proposed a novel track boosting technique to update the confidence scores of the predictions based on a Kalman-based tracker [4]. Moreover, [3] proposes a synthetic data augmentation technique to increase the detection performance further.

3 Proposed Technique

The proposed technique is based on our previous work in [3] which utilizes a YOLOv5 [14] detector and Kalman-based object tracker. Further modifications have been done by improving the synthetic data generation and track boosting stages. Moreover, an extra classification stage is added on top of the detector, and a new scoring algorithm is proposed to increase the tracker performance.

3.1 Detection Model

Like in our previous work [3], we use YOLOv5 [14] as our object detection algorithm. It has been shown that more computationally intensive detection methods can yield success and, in many cases, better than YOLOv5 [6]. YOLOv5 is preferred over other algorithms because it yields a good performance as a detector and computational efficiency makes it a better choice for our solution considering the use in real world applications allowing us to achieve real-time performance and advantages in rapid experimentation allowing us to perform more trainings and experiments in any given time (Fig. 1).

Fig. 1. Generated synthetic data similar to examples where the real model performs poorly. Synthetic data similar to FP predictions above (a), synthetic data similar to FN predictions below (b).

3.2 Synthetic Data

The use of synthetic data in machine learning training is an increasingly common method for problems where data is difficult to access. The problem of drone detection is one of the areas where it is difficult to obtain data with a large number and variety. In our previous work, we developed a method that produces synthetic data by placing 3D objects on 2D backgrounds and performed some experiments using synthetic and real data. As a result of the experiments, it was seen that the synthetic data alone showed an average performance, and when used with real data, it provided an increase in performance. In this study, various improvements were made to our previous method to overcome the difference between real and synthetic domains, also known as the "domain gap" [22].

The use of real-time rendering in synthetic data generation is causing various artifacts around the rendered object. It was observed that these artifacts increased as the object got smaller. Especially when rendered 3D objects on the backgrounds consisting of 2D real images are placed, these artifacts become more evident and negatively affect model performance [17]. For this reason, we developed a method to generate synthetic data using offline rendering techniques.

To increase the performance of the real model, we first identified the cases where the model gave false positive (FP) and false negative (FN) predictions on the test dataset. Then we found similar examples in environments that create FP predictions (such as city and nature environments) and placed 3D models of objects that could be confused with drone objects (such as bicycles and street lamps) on these backgrounds. As a second method, 3D drone objects were placed on backgrounds similar to backgrounds where drones could not be detected, and data similar to FN samples were generated. Also, post-processing effects such as motion blur, filmic noise, and depth of field were applied to generate synthetic data for blending 3D objects with background images.

Synthetic data that we generate according to the cases where the real model makes wrong predictions improves model performance and helps to eliminate FP and FN predictions successfully.

3.3 Binary Classifier Boosting

YOLOv5 detectors output category confidence scores in a coupled fashion, together with box regression prediction, from the detection head. Since box regression and classification outputs are trained in a coupled way, it occasionally results in incorrect classification bias, resulting in false positives (FP) having large confidence scores and true positives (TP) having low confidence scores. To overcome this issue, we add a second classification step after YOLOv5 object detector and train a separate image classifier for this purpose. We train this binary image classifier with the cropped TP and FP box predictions from the training set.

For each image in the training set, we perform prediction using the fine-tuned YOLOv5 object detector and acquire t^{th} box prediction B_t and corresponding confidence score C_t. Then we calculate intersection over union (IOU) over each predicted box B_t and ground truth box B_t^g with a match threshold of 0.3 and assign each prediction as TP or FP. After labeling each box prediction B_t, we crop these boxes with a constant margin and create a binary image classification dataset. Ultimately we fine-tune a Vision Transformer (ViT) [8] on this dataset to have a model that predicts whether a given image contains a *drone* or not. At inference time, we crop the predicted box B_t coming from the object detector and feed into the fine-tuned image classifier to get a classification score Cl_t and calculate the updated confidence score C_t^u as:

$$C_t^u = \sqrt{C_t \times Cl_t} \tag{1}$$

Regularizing the detector confidence score with an additional image classification stage decreases the number of FPs and increases the TP scores.

3.4 Improved Track Boosting and Scoring

In our previous work [3], we had implemented a Kalman-based tracking algorithm to put the images related to a certain drone on the same track. This allowed us to use the temporal information found in images. We had developed a track-boosting algorithm that increased the prediction scores for tracks that are consistent over time. We propose an advanced and more nuanced version of the track-boosting algorithm based on a scoring mechanism. We use the scoring mechanism along with other temporal information to put drones in certainty categories.

The scoring algorithm simply keeps a record of a score for each track. With each new image in the track, the score of the track is modified. As described in

Eq. 2, we update the score (S_i) with an amount equal to the difference between a confidence threshold (T_{conf}) and confidence of the current prediction (C_i). We used 0.3 as the threshold in our experiments.

$$S_i = S_{i-1} + T_{conf} - C_i \tag{2}$$

This scoring allows us to keep the negative effect from a low confidence prediction but still allows the system to recover from it if it can provide high confidence predictions in other frames. This performs better than just removing predictions below a certain threshold because low confidence scores do not necessarily mean incorrect predictions, difficult/rare examples may also result in lower confidence scores which can possibly be recovered later with the scoring approach.

As consecutive frames are accumulated in the system, the score usually converges to a positive or a negative value. The longer the track is, the more information is used in scoring, thus allowing better convergence. This allows us to define different categories of certainty for each track. This further improves our previous strategy of averaging the confidence of each frame with maximum confidence in the track by increasing or decreasing confidence values in a more controlled manner.

The calculation of scores for each confidence category is given below. Here $S'_{i,j}$ is the confidence for prediction on i'th track and j'th position in the track and s_i is the score vector for track i.

- For tracks that were able to collect higher than a certain score (>25 in our experiments) the tracks are considered highly likely to be a drone. The confidence is $S'_{i,j} = S_{i,j} * 0.3 + \max(s_i) * 0.7$
- For tracks with a negative score, we considered them unlikely to be drones. Instead of removing the predictions, we leave them as is since confidence is considerably low: $S'_{i,j} = S_{i,j}$
- For tracks that have a relatively low score (<5 in our experiments), if the object has a median velocity below a threshold (0.3 in our experiments), they are considered as stationary objects. The velocity is calculated by the tracker and smoothed out in a window. Confidence for stationary objects are penalized so confidence becomes $S'_{i,j} = S_{i,j} * 0.3$
- Remaining tracks are considered possible drones and confidences are boosted by averaging with the max confidence. $S'_{i,j} = S_{i,j} * 0.5 + \max(s_i) * 0.5$

This algorithm can improve its decisions as more information is collected, and it also makes it possible to make decisions on-the-fly without waiting for the end of the track if need be. In this case, the first predictions would be closer to the confidence levels of the detection algorithm but would improve over time. The main advantage of the explained algorithm is that it can make more accurate insights into the predictions by using temporal information. Not only it is helpful to increase the mAP score overall it can also provide some explainability better than using simple threshold values. Expanding the algorithm makes it possible to inject insight gathered from human experience to improve the results without requiring any training.

4 Experiments

4.1 Datasets

Different from our previous work [3], for training, we have used three real-world datasets in addition to the challenge dataset. Additional real-world datasets are:

- Real World Object Detection Dataset For Quadcopter Unmanned Aerial Vehicle [18] (named as Real World UAV Dataset for the rest of the paper)
- Det-Fly Dataset [24]
- Multirotor Aerial Vehicle VID (MAV-VID) Dataset [12]

The Real World UAV Dataset is created from public UAV videos gathered from popular video services. This dataset consists of 51446 images with different image resolutions, ranging from 640×480 to 4k. The dataset contains drone objects from different types, sizes, positions, environments, and lighting conditions.

The Det-Fly Dataset consists of approximately 13271 images. Image resolutions in the dataset range from 1080p to 4k. The dataset has four environmental backgrounds: sky, urban, field, and mountain. Also, these backgrounds are distributed equally across the dataset. Moreover, drone objects in the images are tiny and challenging.

The MAV-VID Dataset contains 29500 images for training and 10732 images for validation. Videos in this dataset are captured from other drones, surveillance cameras, and mobile devices.

Table 1. Details of datasets used in the experiments. * The value within brackets '()' represents the number of video samples.

Dataset	Type	Split	Total samples	Used samples
Drone-vs-Bird	Video*	Train	76818 (63)	38409
		Val	28182 (13)	1875
Real-world UAV	Image	Train	46299	46299
		Val	5145	1875
Det-Fly	Image	Train	13271	11280
		Val	1991	1875
MAV-VID	Image	Train	29500	9834
		Val	10732	1875
Synthetic & Negative	Image	Train	100000	5000
		Val	–	–

A large amount of data is produced as a result of the synthetic data generation infrastructure mentioned in Sect. 3.2. In order to find examples that will improve the real model from these data, 3616 false negative samples were extracted from 100k synthetic data. In addition, 1392 false positive samples were extracted from the negative sample pool generated in the same way. Thus, a set of nearly 5000 data consisting of synthetic and negative samples was created.

By combining four real-world datasets, we have increased diversity across drone types, drone sizes, background environments, and lighting conditions. In addition, synthetic and negative image samples are composed to build a synthetic dataset. With the combination of real-world and synthetic datasets, we further increased the diversity of data. The details of the datasets are shown in Table 1.

4.2 Training Details

In our experiments, we have used a popular object detection model, YOLOv5. There are lots of available architectures in YOLOv5. We have used YOLOv5m6 model architecture in experiments. We have conducted experiments using only real-world datasets and a combination of real-world datasets and synthetic data. In total four real-world datasets were used for training (Drone vs. Bird, Real World UAV, Det-Fly and MAV-VID). Moreover, we have created a validation dataset that contains 7500 images. The validation dataset is created by composing four real-world datasets.

In all experiments, the YOLOv5m6 model is fine-tuned for 10 epochs, with 4 batch size and 1920 image size. COCO pre-trained model weights are used in the fine-tuning stage. Also, the best model is chosen using the mAP score computed on the validation dataset.

4.3 Results

To evaluate our proposed methods, we randomly chose 13 videos from Drone vs. Bird dataset. For data sampling, inference and evaluation, our open-source vision framework SAHI [1,2] is utilized.

YOLOv5 models used in experiments:

- **Real Data Model:** YOLOv5 model trained on real-world datasets.
- **Real+Synthetic Data Model:** YOLOv5 model trained on combination of synthetic dataset and real-world datasets.

As seen in Table 2, Real+Synthetic Data Model outperforms Real Data Model, always increasing performance by up to 1.2 mAP. Moreover, the classifier boosting technique improves mAP scores by 1.8 and 1.0 for Real Data Model and the Real+Synthetic Data Model, respectively. Finally, applying the track boosting method with classifier boosting achieves 86.3 mAP for Real+Synthetic Data Model. Thus, we marginally increased mAP by 3.3 by combining all of our proposed techniques.

Table 2. Fine-tuning results for synthetic data augmentation, classifier boosting, and track boosting technique. In the 'Technique' column, 'YOLOv5' means YOLOv5m6 model is used as a detector, 'CB' means model detection confidence scores boosted by using a drone classifier, and 'TB' means the proposed track boosting algorithm is applied. 'mAP' corresponds to mean average precision at 0.50 intersection over union threshold.

Model	Technique	mAP
Real+Synthetic Data Model	YOLOv5 + CB + TB	**86.6**
Real Data Model	YOLOv5 + CB + TB	86.3
Real+Synthetic Data Model	YOLOv5 + CB	85.5
Real Data Model	YOLOv5 + CB	85.4
Real+Synthetic Data Model	YOLOv5	84.5
Real Data Model	YOLOv5	83.2

5 Conclusion and Discussion

Here, we have presented three main approaches that result in improvement of object detection performance:

- Adding synthetic drone image samples and negative background images
- Additional binary classification model
- Boosting confidence values using temporal information with a scoring algorithm

We observed that using synthetically generated data does not automatically increase detection performance by itself. Adding synthetic data to the model blindly would often perform worse than not including at all. This might result from a domain gap between real data and synthetic data. On the other hand, real-world drone datasets by themselves might fail to provide the necessary generalizability due to the lack in variety of backgrounds and objects. To decrease false negative rate of the model, it was essential to generate synthetic data with complex backgrounds that normally the model performs poorly on. Also, there are various complex objects and backgrounds that can be false positive predictions. Adding various complex objects and backgrounds also provides more generalizability to the model and decrease false positive rates. Using synthetic data with moderate amounts and based on the error analysis of the models outweighs the negative effects of the domain gap.

Objects detection models find objects anywhere on an image, and this requires operating in a very large variety of images and large sample space. However, the predictions of the detection model (whether TP or FP) is a more confined space of images that are drones or at least have drone like features. A binary classification model is shown to be more effective at operating in this confined space, thus separating false predictions from real drone predictions a little better, improving our results.

Using temporal information can improve drone detection performance by boosting the predictions found in the same track [3]. We showed that it is possible to go further and discriminate confident predictions from non-confident ones based on a scoring algorithm that accumulates confidence scores over time. This can both be used to increase the overall mAP score and introduce more explainability to prediction based on tracks.

References

1. Akyon, F.C., Altinuc, S.O., Temizel, A.: Slicing aided hyper inference and fine-tuning for small object detection. arXiv preprint arXiv:2202.06934 (2022)
2. Akyon, F.C., Cengiz, C., Altinuc, S.O., Cavusoglu, D., Sahin, K., Eryuksel, O.: SAHI: a lightweight vision library for performing large scale object detection and instance segmentation (November 2021). https://doi.org/10.5281/zenodo.5718950. https://doi.org/10.5281/zenodo.5718950
3. Akyon, F.C., Eryuksel, O., Ozfuttu, K.A., Altinuc, S.O.: Track boosting and synthetic data aided drone detection. In: 2021 17th IEEE International Conference on Advanced Video and Signal Based Surveillance (AVSS), pp. 1–5. IEEE (2021)
4. Alori, J., Descoins, A., Ríos, B., Castro, A.: tryolabs/norfair: v0.3.1 (2021). https://doi.org/10.5281/zenodo.5146254. https://github.com/tryolabs/norfair
5. Benarbia, T., Kyamakya, K.: A literature review of drone-based package delivery logistics systems and their implementation feasibility. Sustainability **14**(1), 360 (2021)
6. Coluccia, A., et al.: Drone vs. bird detection: deep learning algorithms and results from a grand challenge. Sensors **21**(8), 2824 (2021)
7. Coluccia, A., et al.: Drone-vs-bird detection challenge at IEEE AVSS2017. In: 2017 14th IEEE International Conference on Advanced Video and Signal Based Surveillance (AVSS), pp. 1–6 (2017). https://doi.org/10.1109/AVSS.2017.8078464
8. Dosovitskiy, A., et al.: An image is worth 16x16 words: transformers for image recognition at scale. arXiv preprint arXiv:2010.11929 (2020)
9. Fascista, A.: Toward integrated large-scale environmental monitoring using WSN/UAV/crowdsensing: a review of applications, signal processing, and future perspectives. Sensors **22**(5), 1824 (2022)
10. Gökçe, F., Üçoluk, G., Şahin, E., Kalkan, S.: Vision-based detection and distance estimation of micro unmanned aerial vehicles. Sensors **15**(9), 23805–23846 (2015). https://doi.org/10.3390/s150923805. https://www.mdpi.com/1424-8220/15/9/23805
11. He, K., Zhang, X., Ren, S., Sun, J.: Spatial pyramid pooling in deep convolutional networks for visual recognition. IEEE Trans. Pattern Anal. Mach. Intell. **37**(9), 1904–1916 (2015)
12. Isaac-Medina, B.K., Poyser, M., Organisciak, D., Willcocks, C.G., Breckon, T.P., Shum, H.P.: Unmanned aerial vehicle visual detection and tracking using deep neural networks: a performance benchmark. In: Proceedings of the IEEE/CVF International Conference on Computer Vision, pp. 1223–1232 (2021)
13. Jocher, G., et al.: ultralytics/yolov3: 43.1mAP@0.5:0.95 on COCO2014 (2020). https://doi.org/10.5281/zenodo.3785397. https://github.com/ultralytics/yolov3
14. Jocher, G., et al.: ultralytics/yolov5: v3.1 - bug fixes and performance improvements (October 2020). https://doi.org/10.5281/zenodo.4154370. https://github.com/ultralytics/yolov5

15. Li, X., Savkin, A.V.: Networked unmanned aerial vehicles for surveillance and monitoring: a survey. Future Internet **13**(7), 174 (2021)
16. Müller, T.: Robust drone detection for day/night counter-UAV with static VIS and SWIR cameras. In: Ground/Air Multisensor Interoperability, Integration, and Networking for Persistent ISR VIII, vol. 10190, p. 1019018. International Society for Optics and Photonics (2017)
17. Özfuttu, K.A.: Generating synthetic data with game engines for deep learning applications. Master's thesis, Hacettepe Üniversitesi (2022)
18. Pawełczyk, M., Wojtyra, M.: Real world object detection dataset for quadcopter unmanned aerial vehicle detection. IEEE Access **8**, 174394–174409 (2020). https://doi.org/10.1109/ACCESS.2020.3026192
19. Redmon, J., Farhadi, A.: YOLOv3: an incremental improvement. arXiv preprint arXiv:1804.02767 (2018)
20. Sommer, L., Schumann, A., Müller, T., Schuchert, T., Beyerer, J.: Flying object detection for automatic UAV recognition. In: 2017 14th IEEE International Conference on Advanced Video and Signal Based Surveillance (AVSS), pp. 1–6. IEEE (2017)
21. Srigrarom, S., Hoe Chew, K., Meng Da Lee, D., Ratsamee, P.: Drone versus bird flights: classification by trajectories characterization (2020). https://doi.org/10.23919/SICE48898.2020.9240313
22. Tobin, J., Fong, R., Ray, A., Schneider, J., Zaremba, W., Abbeel, P.: Domain randomization for transferring deep neural networks from simulation to the real world. CoRR abs/1703.06907 (2017). http://arxiv.org/abs/1703.06907
23. Wawrla, L., Maghazei, O., Netland, T.: Applications of drones in warehouse operations. Whitepaper. ETH Zurich, D-MTEC (2019)
24. Zheng, Y., Chen, Z., Lv, D., Li, Z., Lan, Z., Zhao, S.: Air-to-air visual detection of micro-UAVs: an experimental evaluation of deep learning. IEEE Robot. Autom. Lett. **6**(2), 1020–1027 (2021)

Drone-vs-Bird Detection Challenge at ICIAP 2021

Angelo Coluccia[1(✉)], Alessio Fascista[1], Arne Schumann[2], Lars Sommer[2], Anastasios Dimou[3], Dimitrios Zarpalas[3], Nabin Sharma[4], Mrunalini Nalamati[4], Ogulcan Eryuksel[5], Kamil Anil Ozfuttu[5], Fatih Cagatay Akyon[5], Kadir Sahin[5], Efe Buyukborekci[5], Devrim Cavusoglu[5], Sinan Altinuc[5], Daitao Xing[6], Halil Utku Unlu[6], Nikolaos Evangeliou[7], Anthony Tzes[7], Abhijeet Nayak[8,9], Mondher Bouazizi[11,12], Tasweer Ahmad[8,10], Artur Gonçalves[8], Bastien Rigault[8], Raghvendra Jain[12], Yutaka Matsuo[13], Helmut Prendinger[8], Edmond Jajaga[14], Veton Rushiti[14], Blerant Ramadani[14], and Daniel Pavleski[14]

[1] University of Salento, Lecce, Italy
angelo.coluccia@unisalento.it
[2] Fraunhofer IOSB, Karlsruhe, Germany
[3] Information Technologies Institute, Centre for Research and Technology Hellas (CERTH), Thessaloniki, Greece
[4] School of Computer Science, Sidney University of Technology, Sidney, Australia
[5] OBSS Technology, Universiteler Mahallesi, Cankaya, Ankara, Turkey
[6] New York University, New York, USA
[7] New York University, Abu Dhabi, UAE
[8] National Institute of Informatics, Tokyo, Japan
[9] University of Freiburg, Freiburg im Breisgau, Germany
[10] COMSATS University, Islamabad, Pakistan
[11] Keio University, Yokohama, Japan
[12] Optimays Inc., Tokyo, Japan
[13] The University of Tokyo, Tokyo, Japan
[14] Mother Teresa University, Skopje, North Macedonia

Abstract. This paper reports the results of the 5th edition of the "Drone-vs-Bird" detection challenge, organized within the 21st International Conference on Image Analysis and Processing (ICIAP). By taking as input video samples recorded by common cameras, the aim of the challenge is to devise advanced approaches aimed at spotlighting the presence of drones flying in the monitored area, while limiting the number of wrong alarms raised when similar flying entities such as birds suddenly appear in the scene. To this end, a number of important issues such as the dynamic variations in the scene and the background/foreground motion effects should be carefully considered, so as to allow the proposed solutions to correctly identify drones only when they are actually present. The paper summarizes the novel algorithms proposed by the four participating teams that succeeded in providing satisfactory detection performance on the 2022 challenge dataset.

Keywords: Drone detection · Deep learning · Image and video signal processing

P. L. Mazzeo et al. (Eds.): ICIAP 2022 Workshops, LNCS 13374, pp. 410–421, 2022.
https://doi.org/10.1007/978-3-031-13324-4_35

1 Introduction

Unmanned aerial vehicles (UAVs) are fast becoming more capable featuring improved autonomy, sensor payload, and processing capabilities. On the other hand, they are increasingly accessible to the public either off-the-shelf or custom-made from their parts. This combination of increased capability and availability opens a new world of opportunities [12], but it also unveils considerable risks for misuse of UAVs for illegal activities. The use of UAVs for smuggling goods, industrial espionage, unlawful surveillance, privacy violations, interference with aircraft operations and terrorist attacks are just some of the events that have been already identified.

While currently there are numerous unimodal [10,20,23] and multi-modal [11,19, 21,24] UAV detection systems under investigation in literature, detection based on video analysis presents a good trade-off between cost and performance, offering significant detection range. However, there are still challenges related to the visual similarity of UAVs and other small flying objects such as birds that can have a similar appearance, especially at long distances or environments with reduced visibility. The agility and speed of a UAV that can surpass 100 km/h [5] can further stress a vision-based detection system.

Aiming at promoting research on UAV detection based on video-analytics, in 2017 the first edition of the *International Workshop on Small-Drone Surveillance, Detection and Counteraction Techniques* (WOSDETC) [9] was organized as part of *IEEE International Conference on Advanced Video and Signal based Surveillance* (AVSS), held in Lecce, Italy. In conjunction with the workshop, a grand challenge called *drone-vs-bird detection challenge* was launched, supported by the SafeShore project[1]. In 2019, a second edition of the challenge was organized, again as part of WOSDETC and co-located with the 16th edition of AVSS held in Taipei, Taiwan [8]. A third edition of the Drone-vs-Bird challenge was organized in 2020, initially planned as part of the 17th edition of AVSS in Washington DC, USA, but then run as virtual event due to the COVID-19 pandemic [7]. The fourth edition of the challenge was organized in conjunction with the 17th AVSS in 2021 as a virtual event [6].

Since its first edition, the challenge has attracted interest from hundreds of research groups, spread all over the world. The main goal is to correctly identify the presence of drones suddenly appearing in a video sequence, without being confused by birds or additional disturbing objects in the scene. The 2022 challenge dataset comprises different video sequences covering both maritime and land scenarios, acquired under different cameras and background conditions. All the participants submitted a set of score files containing the detection results, together with a companion paper describing the proposed methodology. More than 50 different research groups requested access to the dataset for participation in this edition of the challenge.

2 Challenge Dataset and Evaluation Protocol

The Drone-vs-Bird Detection Challenge dataset consists of a pool of 77 different video sequences, released to all the participating teams as additional training data. The videos

[1] The project "SafeShore" has received funding from the European Union's Horizon 2020 research and innovation programme under grant agreement No. 700643.

have been recorded using MPEG4-coded cameras, in part during the experimental campaigns conducted within the ALADDIN[2] and SafeShore projects, and in part provided under courtesy of the Fraunhofer IOSB research institute. More specifically, the dataset contains sequences acquired with both static and moving cameras, with different resolutions ranging between 720 × 576 and 3840 × 2160 pixels.

The average number of frames in each sequence is about 1,384, with each frame containing on average 1.12 annotated drones. As shown in Fig. 1, the dataset includes 8 different types of commercial drones, namely Parrot Disco, DJI Inspire, DJI Phantom, DJI Mavic, DJI Matrice, 3DR Solo Robotics, and two additional custom fixed-wing drones. Overall, 3 types have fixed wings while the other 5 have rotary wings. Each video sequence is associated with a separated annotation file, which reports the number of frames in which drones enter the scenes, together with their exact location expressed in terms of bounding boxes surrounding the drones (specified as [top_x top_y width height]).

Fig. 1. Examples of drone types present in the training set, i.e., Parrot Disco, 2 custom fixed-wing drones, DJI Inspire, DJI Phantom, DJI Mavic, DJI Matrice and 3DR Solo Robotics.

The dataset has been conceived with the idea of providing a rather large variability of scenes, resulting in different levels of difficulty for the detection algorithms. In particular, the sequences include scenes with rather different backgrounds, from clear sky to dense vegetation, as well as different weather conditions ranging from sunny to cloudy. To further challenge the detection task, some sequences also include direct sun glares and possible variations in the camera characteristics, as reported in Fig. 2. In addition, drones may appear at different distances from the camera, leading to strong size and appearance variations among the collected frames. A large number of scenes include drones that are significantly distant from cameras, making the detection task extremely challenging [7]. Differently from drones, birds are not annotated and suddenly appear at some point in several scenes as disturbing objects.

The dataset released to test the proposed approaches consists of an additional set of 20 video sequences, provided without annotations. About fourteen video sequences

[2] The project "ALADDIN" has received funding from the European Union's Horizon 2020 research and innovation programme under grant agreement No. 740859.

Fig. 2. Sample frames extracted from the training videos exhibit the large variability of the dataset.

share similar characteristics compared to the videos included in the training set, being them collected during the same acquisition campaigns, at almost the same physical locations, though with some notable differences. In the past 2021 edition of the challenge, the test set was extended with seven additional video sequences featuring completely new scenes, backgrounds, and including two new types of rotating drones. In addition, they also include novel disturbing objects such as airplanes. The test set consists of five different sequences captured with a moving camera. In order to make the performance evaluation fair, all videos were shortened to a total duration of 1 min. The dataset is freely available for download upon signing a Data Usage Agreement (DUA), related to the usage of the dataset for research purposes only, while the corresponding annotations are available at https://github.com/wosdetc/challenge.

The rules of the Drone-vs-Bird Detection Challenge 2022 require that each submission from participating teams include a set of result files, one for each video sequence, with each file explicitly stating the frame numbers in which drones were detected, along with the corresponding predicted position given in terms bounding boxes ([top_x top_y width height]). Moreover, result files should also contain the confidence scores for each frame. If a frame is not reported in the file, it is assumed that no detection has been raised for that frame. The Average Precision (AP) is adopted as main performance metric to rank the proposed approaches. For this purpose, the rules of the challenge provide that a detection is counted as a true positive detection only if the Intersection over Union (IoU) criterion with a ground truth annotation exceeds 0.5. Conversely, detections for which the IoU is less than 0.5 are counted as false positive detections, while ground truth annotations without an assigned detection are counted as false negative detections. Test sequences were made publicly available to participants one week before the submission deadline.

3 Participation and Best Proposed Algorithms

We briefly summarize the algorithms adopted by the four best performing approaches.

OBSS AI team proposed a drone detection framework called DroBoost, that utilizes drone detection and object tracking approaches. They extended the previous work [3] which utilizes synthetic data and tracking methods to use temporal information by improving both detection and tracking components. As in the previous approach, they have used the YOLOv5 [14] object detection model. To increase generalization of the drone detection model, they have combined four different real-world drone detection datasets. Moreover, they generated synthetic data to overcome poor drone detection performance on complex backgrounds. Also, they have integrated synthetically generated negative images containing negative samples with commonly confused objects to training to reduce false positive predictions. They used a binary classifier trained on bounding box crops of the detection model predictions to improve accuracy further. A confidence value of a predicted bounding box is computed as the geometric mean of binary classifier's confidence score and object detector's confidence score. Furthermore, they proposed an intelligent scoring algorithm to determine the confidence of a track. With the scoring method, they can eliminate false positive tracks and also increase confidence values for trusted tracks prediction. Lastly, they open-sourced some of the operations used in their proposed method [1,2].

Daitao et al. from RISC-NYUAD lab proposed a long-term drone surveillance system, named DroneNet, by fusing object detection, tracking and classification methods. The system takes RGB images as input and feed them into YOLOv5 [14] based detector to find drone candidates. A high efficient object tracker is integrated to update status for those objects from detector. In this way, the system improves the computational efficiency by avoiding running complex detection module on each frame. Moreover, a drone classification model is applied on the outputs of the detection and tracking modules to further distinguish drones from other background distractors (airplanes, birds). By leveraging inference optimization with TensorRT and ONNX, the system achieves extremely high efficiency on NVIDIA GPUs. The proposed method is validated on both Drone-vs-Bird competition and real-world field tests. The results demonstrates its effectiveness and robustness in drone detection and tracking applications.

Nayak et al. from the NII team investigated the Fully Convolutional One-Stage (FCOS) object detector [22] and selected it for drone detection because it claims better performance for small-scale object detection. FCOS is an anchor-box free single-stage object detector with high inference speed and good performance at the same time. The FCOS architecture includes input, backbone network, feature pyramid network and classification-regression network for detection. The backbone network is responsible for creating different feature maps, and the feature pyramid network generates different scales of feature maps. Finally, a classification-regression network computes the classification loss, centerness loss and regression loss for final detection. By observing early experimentation, the authors carefully noted the samples that generated false-positives, and augmented their training dataset with such images including flying objects, leaves and objects with sharp edges.

A YOLOv4 approach for optical-based recognition is proposed by the UMT team. The solution is part of a data fusion system, which also includes radar-based detection, recognition and tracking of drones. In particular, a generic dataset refine process for YOLO-based approaches is proposed. As the authors argue, a dataset improvement life-cycle should follow a well-defined methodology in order to avoid image and annotation errors, and consequently improve model predictions. Thus, each dataset sequence should pass through the stages of extracting, reducing and selecting images. The generated dataset integrates various open source datasets classifying drones (multi-rotor and fixed-wing types) and birds. Furthermore, the authors indicate that the fusion of radar-based decisions with optical ones, will improve the system performance, especially on the sequences showing a drone appearing on complex backgrounds or with small size.

4 Results

The results for each team in the Drone-vs-Bird Detection Challenge 2022 are reported in Table 1. In addition to the AP computed over all sequences, which is used as criterion for the final ranking, the AP is given for every sequence of the test set. The results are further compared to the winning entry of the Drone-vs-Bird Detection Challenge 2021 [3] and two baseline detection methods, *i.e.* Faster R-CNN [18] with Feature Pyramid Network (FPN) [15] and RetinaNet [16]. The winner of the Drone-vs-Bird Detection Challenge 2021 applied YOLOv5 [14] in combination with Kalman-based object tracking. To train the YOLOv5 model, real data and synthetically generated drone images were used. The applied tracker was used to remove short-lived predictions and boost the confidence of a detection in a particular track. To train the two baselines, we employed the object detection toolbox MMDetection [4]. ResNet-50 [13] was used as backbone and weights pre-trained on MS COCO [17] for initialization. All input images are scaled to 1920×1080 pixels during training and evaluation.

The best AP computed over all sequences is obtained by DroBoost, ranking first in the Drone-vs-Bird Detection Challenge 2022. DroBoost clearly outperforms the other proposed methods as well as the baselines in case of fixed-wing drones (*i.e.* parrot_disco_long_session_2_1m) and on sequenceswith drones in front of structured background

Table 1. Detailed comparison for each team in the Drone-vs-Bird Detection Challenge 2022, the winning entry of the Drone-vs-Bird Detection Challenge 2021 and two baselines. The AP is given for every sequence of the test set. While the first block gives the results for sequences originating from the previous installment of the challenge, the second block reports the results for the novel sequences. The overall averaged result is given at the bottom.

Sequence	AP						
	DroBoost	DroneNet	NII	UMT	Winner 2021	FRCNN	RetinaNet
GOPR5843_004	0.982	0.640	0.890	0.128	0.953	**1.000**	0.989
GOPR5847_001	0.676	0.551	0.676	0.226	0.663	0.693	**0.736**
GOPR5853_002	**0.883**	0.778	0.488	0.200	0.618	0.606	0.470
GOPR5868_001	0.999	**1.000**	0.818	0.788	0.823	0.989	0.919
dji_mavick_mountain_cross	0.890	0.743	0.651	0.439	0.877	0.892	**0.903**
dji_phantom_mountain	0.764	0.762	**0.769**	0.261	0.666	0.640	0.731
parrot_disco_long_session.	**0.738**	0.222	0.215	0.131	0.546	0.133	0.086
2019_08_19_C0001_27_46.	0.494	0.806	0.306	0.555	0.810	**0.855**	0.737
2019_08_19_C0001_57_00.	0.731	0.891	0.830	0.751	0.873	**0.898**	0.887
2019_10_16_C0003_52_30.	0.792	**0.796**	0.433	0.726	0.519	0.697	0.699
2019_11_14_C0001_11_23.	0.529	0.073	0.591	0.418	0.100	0.779	**0.788**
4k_2020-06-22_C0006_split.	0.932	**0.933**	0.000	0.000	0.000	0.014	0.080
4k_2020-07-29_C0020_01	**1.000**	0.872	0.009	0.009	0.541	0.522	0.812
VID_20210606_141851_01	**0.907**	0.733	0.002	0.013	0.117	0.054	0.039
VID_20210606_143947_04	**0.473**	0.438	0.002	0.001	0.212	0.168	0.000
GOPR5856_001	**0.993**	0.980	0.982	0.575	–	0.990	0.986
GOPR5862_001	0.998	0.996	0.785	0.777	–	**1.000**	0.981
VID_20210417_143930_02	**0.899**	0.858	0.355	0.460	–	0.606	0.630
VID_20211010_143610_01	**1.000**	**1.000**	0.979	0.951	–	0.964	0.994
VID_20211012_175158_02	0.987	**1.000**	0.118	0.159	–	0.905	0.680
Overall	**0.796**	0.761	0.459	0.378	–	0.663	0.622

(*e.g.* VID_20210606_141851_01). Compared to the results of the winning entry of the previous edition, DroBoost achieves clearly higher AP values on most sequences. While the winning entry of the previous edition exhibits poor detection results on sequences recorded at twilight (*e.g.* 2019_11_14_C0001_11_23_inspire_1m) and on sequences with complex background (*e.g.* 4k_2020-06-22_C0006_split_01_01 that contains a drone in front of a streetlamp) due to a high number of missed detections, DroBoost is more robust in such scenarios. DroBoost and DroneNet outperform the other teams in case of disturbing objects, *e.g.* airplanes in sequences GOPR5862_001 and GOPR5868_001, as less false positive detections with high confidence scores are caused.

The recall rates for each sequence and the overall recall are given in Table 2. For this, all submitted predictions are considered. DroBoost exhibits the best recall rates on almost all sequences and the best overall recall. Even for sequences with drones in front of complex backgrounds (*e.g.* VID_20210417_143930_02), DroBoost achieves recall values close to 1. Reason for this is the high number of submitted predictions, *i.e.* more than 200k, whereas the other submissions comprise clearly less submissions. In particular, the submitted results of UMT only comprises about 12k predictions, which yields poor recall rates on several sequences.

Table 2. Resulting recall for each team in the Drone-vs-Bird Detection Challenge 2022, the winning entry of the Drone-vs-Bird Detection Challenge 2021 and two baselines. The recall is given for every sequence of the test set and the overall recall is given at the bottom.

	DroBoost	DroneNet	NII	UMT	Winner 2021	FRCNN	RetinaNet
GOPR5843_004	0.989	0.721	0.902	0.155	0.968	**1.000**	0.995
GOPR5847_001	0.686	0.585	0.682	0.226	0.667	0.695	**0.748**
GOPR5853_002	**0.913**	0.804	0.529	0.200	0.677	0.612	0.557
GOPR5868_001	**1.000**	**1.000**	**1.000**	0.992	0.998	**1.000**	**1.000**
dji_mavick_mountain_cross	**0.943**	0.843	0.772	0.537	0.896	0.914	0.931
dji_phantom_mountain	**0.895**	0.640	0.830	0.278	0.725	0.669	0.779
parrot_disco_long_session.	**0.910**	0.368	0.383	0.307	0.897	0.242	0.266
2019_08_19_C0001_27_46.	0.616	0.836	0.624	0.650	0.843	**0.889**	0.873
2019_08_19_C0001_57_00.	0.826	**0.905**	0.854	0.787	0.884	0.902	0.896
2019_10_16_C0003_52_30.	**0.866**	0.855	0.546	0.799	0.645	0.783	0.813
2019_11_14_C0001_11_23.	0.672	0.313	0.697	0.484	0.111	0.779	**0.792**
4k_2020-06-22_C0006_split.	**0.995**	0.965	0.000	0.017	0.000	0.248	0.531
4k_2020-07-29_C0020_01	**1.000**	0.901	0.089	0.067	0.733	0.874	0.914
VID_20210606_141851_01	**0.933**	0.760	0.037	0.095	0.188	0.181	0.256
VID_20210606_143947_04	**0.967**	0.445	0.094	0.011	0.673	0.394	0.029
GOPR5856_001	**0.998**	0.985	0.997	0.607	–	0.992	0.995
GOPR5862_001	**1.000**	0.997	**1.000**	0.825	–	**1.000**	**1.000**
VID_20210417_143930_02	**0.972**	0.858	0.444	0.472	–	0.658	0.736
VID_20211010_143610_01	**1.000**	**1.000**	**1.000**	0.952	–	0.992	**1.000**
VID_20211012_175158_02	0.995	**1.000**	0.375	0.276	–	0.933	0.821
Overall	**0.910**	0.821	0.611	0.451	–	0.757	0.777

To analyze the detection results with respect to the IoU criterion used to accept detections as true positive detections or not, we vary the applied IoU threshold value in the range between 0.1 and 0.6. The corresponding AP values are given in Table 3. Decreasing the IoU threshold value improves the overall AP for all methods, as more predictions are considered as true positives. The largest gain in AP is observed for NII, which indicates that the predictions are less accurately located.

Qualitative detection results for all teams and the corresponding ground truth are given in Figs. 3 and 4. Note that only detections with a confidence score above 0.5 are considered. All teams achieve good results for different drone types in case of backgrounds without complex structures (see Fig. 3). However, drones in front of structured

Table 3. Average Precision for various IoU threshold values used to accept detections as true positives.

IoU	AP			
	DroBoost	DroneNet	NII	UMT
0.1	0.900	0.864	0.652	0.519
0.2	0.897	0.864	0.649	0.518
0.3	0.891	0.852	0.619	0.492
0.4	0.863	0.833	0.542	0.449
0.5	0.796	0.761	0.459	0.378
0.6	0.652	0.454	0.330	0.263

background such as buildings may yield missed detection (see Fig. 4 top row). While DroBoost and DroneNet detect several drones in such scenarios, NII and UMT exhibit more missed detections. False positive detections due to disturbing objects such as airplanes often cause false positive detections in case of NII and UMT (see Fig. 4 bottom left). Further challenges are distant drones, as small drones are often inaccurately localized (see Fig. 4 bottom right).

Fig. 3. Image crops showing qualitative detection results for DroBoost (cyan), DroneNet (red), NII (blue) and UMT (yellow) as well as the corresponding GT (green). All methods achieve good detection results for different drone types in case of backgrounds without complex structures. (Color figure online)

Fig. 4. Image crops showing qualitative detection results for DroBoost (cyan), DroneNet (red), NII (blue) and UMT (yellow) as well as the corresponding GT (green). Drones in front of structured background such as buildings may yield missed detections, while disturbing objects such as airplanes often cause false positive detections in case of NII and UMT. Further challenges are distant drones, as small drones are often inaccurately localized. (Color figure online)

5 Conclusions

This paper presented the results of the 5th edition of the Drone-vs-Bird Detection Challenge 2022, launched in conjunction with the 21st International Conference on Image Analysis and Processing (ICIAP). The challenge stimulated a rich participation from several teams worldwide. The four succeeding solutions achieved interesting detection performance, and have been compared also with the winner of the past 2021 edition. The common characteristics of the proposed methods are the use of convolutional neural networks coupled with additional advanced processing steps used in moving object detection tasks. Some of the approaches have been also empowered with augmented training sets obtained via both synthetic generation or addition from existing open repositories.

References

1. Akyon, F.C., Altinuc, S.O., Temizel, A.: Slicing aided hyper inference and fine-tuning for small object detection. arXiv preprint arXiv:2202.06934 (2022)
2. Akyon, F.C., Cengiz, C., Altinuc, S.O., Cavusoglu, D., Sahin, K., Eryuksel, O.: obss/sahi: v0.9.2 (2021). https://doi.org/10.5281/zenodo.5718950. https://github.com/obss/sahi
3. Akyon, F.C., Eryuksel, O., Ozfuttu, K.A., Altinuc, S.O.: Track boosting and synthetic data aided drone detection. In: 2021 17th IEEE International Conference on Advanced Video and Signal Based Surveillance (AVSS), pp. 1–5. IEEE (2021)

4. Chen, K., et al.: MMDetection: open MMLab detection toolbox and benchmark. arXiv preprint arXiv:1906.07155 (2019)
5. Coluccia, A., Fascista, A., Ricci, G.: Online estimation and smoothing of a target trajectory in mixed stationary/moving conditions. In: IEEE International Conference on Acoustics, Speech and Signal Processing (ICASSP), pp. 4445–4449 (2019). https://doi.org/10.1109/ICASSP.2019.8683255
6. Coluccia, A., et al.: Drone-vs-bird detection challenge at IEEE AVSS2021. In: 2021 17th IEEE International Conference on Advanced Video and Signal Based Surveillance (AVSS), pp. 1–8 (2021). https://doi.org/10.1109/AVSS52988.2021.9663844
7. Coluccia, A., et al.: Drone vs. bird detection: deep learning algorithms and results from a grand challenge. Sensors **21**(8), 2824 (2021)
8. Coluccia, A., et al.: Drone-vs-bird detection challenge at IEEE AVSS2019. In: 2019 16th IEEE International Conference on Advanced Video and Signal Based Surveillance (AVSS), pp. 1–7 (2019). https://doi.org/10.1109/AVSS.2019.8909876
9. Coluccia, A., et al.: Drone-vs-bird detection challenge at IEEE AVSS2017. In: 14th IEEE International Conference on Advanced Video and Signal Based Surveillance (AVSS), pp. 1–6. IEEE (2017)
10. Coluccia, A., Parisi, G., Fascista, A.: Detection and classification of multirotor drones in radar sensor networks: a review. Sensors **20**(15) (2020). https://doi.org/10.3390/s20154172. https://www.mdpi.com/1424-8220/20/15/4172
11. Cubber, G.D., et al.: The SafeShore system for the detection of threat agents in a maritime border environment. In: IARP Workshop on Risky Interventions and Environmental Surveillance, Les Bons Villers, Belgium, May 2017 (2017)
12. Fascista, A.: Toward integrated large-scale environmental monitoring using WSN/UAV/Crowdsensing: a review of applications, signal processing, and future perspectives. Sensors **22**(5) (2022). https://doi.org/10.3390/s22051824
13. He, K., Zhang, X., Ren, S., Sun, J.: Deep residual learning for image recognition. In: Proceedings of the IEEE Conference on Computer Vision and Pattern Recognition (CVPR) (June 2016)
14. Jocher, G., et al.: ultralytics/yolov5: v3.1 - bug fixes and performance improvements (October 2020). https://doi.org/10.5281/zenodo.4154370. https://github.com/ultralytics/yolov5
15. Lin, T.Y., Dollár, P., Girshick, R., He, K., Hariharan, B., Belongie, S.: Feature pyramid networks for object detection. In: Proceedings of the IEEE Conference on Computer Vision and Pattern Recognition, pp. 2117–2125 (2017)
16. Lin, T.Y., Goyal, P., Girshick, R., He, K., Dollár, P.: Focal loss for dense object detection. In: Proceedings of the IEEE International Conference on Computer Vision, pp. 2980–2988 (2017)
17. Lin, T.-Y.: Microsoft COCO: common objects in context. In: Fleet, D., Pajdla, T., Schiele, B., Tuytelaars, T. (eds.) ECCV 2014. LNCS, vol. 8693, pp. 740–755. Springer, Cham (2014). https://doi.org/10.1007/978-3-319-10602-1_48
18. Ren, S., He, K., Girshick, R., Sun, J.: Faster R-CNN: towards real-time object detection with region proposal networks. In: Advances in Neural Information Processing Systems, vol. 28, pp. 91–99 (2015)
19. Samaras, S., et al.: Deep learning on multi sensor data for counter UAV applications-a systematic review. Sensors **19**(22) (2019). https://doi.org/10.3390/s19224837. https://www.mdpi.com/1424-8220/19/22/4837
20. Schumann, A., Sommer, L., Müller, T., Voth, S.: An image processing pipeline for long range UAV detection. In: Emerging Imaging and Sensing Technologies for Security and Defence III; and Unmanned Sensors, Systems, and Countermeasures, vol. 10799, p. 107990T. International Society for Optics and Photonics (2018)

21. Svanström, F., Englund, C., Alonso-Fernandez, F.: Real-time drone detection and tracking with visible, thermal and acoustic sensors. In: 2020 25th International Conference on Pattern Recognition (ICPR), pp. 7265–7272 (2021). https://doi.org/10.1109/ICPR48806.2021.9413241
22. Tian, Z., Shen, C., Chen, H., He, T.: FCOS: fully convolutional one-stage object detection. In: Proceedings of the IEEE/CVF International Conference on Computer Vision, pp. 9627–9636 (2019)
23. Wang, H., Peng, Y., Liu, L., Liang, J.: Study on target detection and tracking method of UAV based on LiDAR. In: 2021 Global Reliability and Prognostics and Health Management (PHM-Nanjing), pp. 1–6 (2021). https://doi.org/10.1109/PHM-Nanjing52125.2021.9612936
24. Xing, D., Tsoukalas, A., Giakoumidis, N., Tzes, A.: Computationally efficient RGB-T UAV detection and tracking system. In: 2021 International Conference on Unmanned Aircraft Systems (ICUAS), pp. 1410–1415 (2021). https://doi.org/10.1109/ICUAS51884.2021.9476750

An Image-Based Classification Module for Data Fusion Anti-drone System

Edmond Jajaga[1], Veton Rushiti[1](✉), Blerant Ramadani[1], Daniel Pavleski[1], Alessandro Cantelli-Forti[2], Biljana Stojkovska[3], and Olivera Petrovska[1]

[1] Mother Teresa University, Mirçe Acev nr. 4, 1000 Skopje, North Macedonia
{edmond.jajaga,veton.rushiti,blerant.ramadani,daniel.pavleski, olivera.petrovska}@unt.edu.mk

[2] Lab RaSS CNIT, 56124 Pisa, Italy
alessandro.cantelli.forti@cnit.it

[3] Faculty of Computer Science and Engineering, Intelligent Systems Department, Ss. Cyril and Methodius University, Rugjer Boshkovikj 16, 1000 Skopje, North Macedonia
biljana.stojkoska@finki.ukim.mk

Abstract. Means of air attack are pervasive in all modern armed conflict or terrorist action. We present the results of a NATO-SPS project that aims to fuse data from a network of optical sensors and low-probability-of-intercept mini radars. The requirements of the image-based module aim to differentiate between birds and drones, then between different kind of drones: copters, fixed wings, and finally the presence or not of payload. In this paper, we outline the experimental results of the deep learning model for differentiating drones from birds. Based on the trade-off between speed and accuracy, the YOLO v4 was chosen. A dataset refine process for YOLO-based approaches is proposed. The experimental results verify that such an approach provide a reliable source for situational awareness in a data fusion platform. However, the analysis indicates the necessity of enriching the dataset with more images with complex backgrounds as well as different target sizes.

Keywords: Anti-drone system · Deep learning · YOLO · Data fusion

1 Introduction

The human's multisensory system has been extensively studied in order to provide more accurate and more efficient machine decisions. This process includes integration of multi-source data and is called data fusion. In processing perspective, data fusion represents an area which includes a combination of batch and stream processing features. Namely, in data fusion systems, data is collected over time from continuous data streams and follows with continuous processing of a bunch of data. Thus, the system requires fast and lengthy performance. In situational awareness perspective, a data fusion system achieves refined position, identifies estimates and complete and timely assessments of situations, threats and their significance [1]. The final goal of using data fusion in multisensory

P. L. Mazzeo et al. (Eds.): ICIAP 2022 Workshops, LNCS 13374, pp. 422–433, 2022.
https://doi.org/10.1007/978-3-031-13324-4_36

environments is to obtain a lower detection error probability and higher reliability by using data from multiple distributed sources [2]. The same goal applies also for the deep learning (DL) approaches, which utilize convolutional neural networks (CNNs) for object detection and recognition. CNNs tend to look for meaningful features that can help to classify the images or, in the case of object detection, to draw the boundary boxes enclosing the target of interest [3].

Data fusion systems are especially important for the domain of Means of Air Attack (MoAA). One of the most developing MoAA category are the Unmanned Aerial Vehicles (UAVs) i.e., "drones". Killer drones represent a real threat to people's life and health. For example, just recently a drone of unknown origin crashed near Zagreb (Croatia) by flying undetected on a number of states. Fortunately no-one was injured. In order to facilitate the neutralization of killer-drones and minimize the risk for people and assets, a NATO SPS Anti-Drones project[1] has been focalized on the development of a new concept of an anti-drone system able to detect, recognize and track killer-drones. The project scope is to progress the state of the art exploiting mini-radar technology and signal processing, as well as data processing and fusion subsystem, for improving the performance and eliminating the environmental impact (e.g., ECM pollution) in an urban environment.

The system infrastructure includes a network of LPI (Low-probability-of-intercept) mini-radar with FMCW or noise-like waveform, and on-demand, fully digital, optical camera-integrated imaging capability, capable of working in all weather conditions, to be deployed and appropriately placed on the ground in the area of the asset to be protected. Although being less conventional compared to visual-based approaches, the adoption of FMCW radars enables complementary analyses in the joint time-frequency domain using, e.g., spectrograms and related signal processing tools [4]. The optical part is essential to support correct classification and therefore identification of the threat and thus to eliminate false alarms. To the best of our knowledge, this is the first attempt that proved data fusion by integrating image data with radar ones for differentiating drones from birds.

This paper covers the optical subsystem and automatic recognition, in particular the ability to distinguish drones from birds and is organized as follows. Section 2 gives an overview of the system design. Section 3 describe details for the dataset generation methodology of the proposed approach. Section 4 examines the experimental results. In Sect. 5, relevant related work from the literature is presented. Finally, the paper is concluded in Sect. 6, with directions for future work.

2 System Design

One of the main challenges of our system is establishing an efficient data fusion algorithm. Data fusion takes action in different levels of our system. In a higher-level perspective, as depicted in Fig. 1, the system should fuse together radar and camera data. We follow a similar approach to Liu et al. [5]. However, instead of integrating camera and acoustic data, our solution will combine Support Vector Machine (SVM) radar data (direction of arrival, range, angular coordinates, elevation and radar cross section) with

[1] https://antidrones-project.org/, last access 24.03.2022.

You Only Look Once (YOLO) camera-based images. In lower levels, the data fusion takes place only within the modules of the corresponding data source. As per the optical part, the images provided by the camera are processed by DL methods to support the data fusion algorithm with additional confidence score for radar-detected targets.

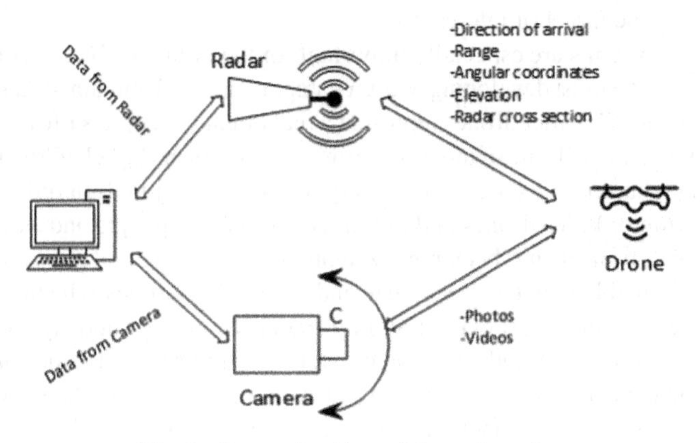

Fig. 1. Processing flow of drone detection

During the phase of literature review for object recognition approaches, a number of different DL approaches were considered. Namely, state-of-the-art ML frameworks, including: YOLO, TensorFlow and PyTorch, were examined. YOLO framework was chosen based on the project objectives, high accuracy and ability to detect objects in real-time by processing 67 FPS [6]. Moreover, it's power efficient compared to other DL detectors [7], open source, flexible network architecture, low hardware requirements i.e., minimum 4 GB memory and is able to detect relatively small objects. The YOLO architecture model is mainly based on Darknet [8], which typically consists of 19 convolutional layers and 5 pooling layers.

3 Dataset Generation Methodology

Although measurement campaigns to verify the quality of the mini-radar have provided some static and dynamic optical images of drones also equipped with synthetic payloads, unfortunately, to date there is a lack of existing drones' dataset [2, 9]. It's even harder to have sufficient number of images of drones with payload. Furthermore, a very sensitive issue represents the quality of the images in terms of drone or bird size and positions, as well as the background characteristics. For this purpose, the researchers have considered different drone dataset generation techniques, e.g., the randomization method described in [10] or combining background-subtracted real images as described in [9]. Our focus was rather on building a methodology for more qualitative dataset.

The number of classes to be recognized by the model should also be considered, because it reflects on the performance of the model. Thus, in line with radar-based recognition fusion, we plan to consider five classes on the optical side, including: drone, bird, fixed-wing, copter and drone with payload. Following the lack of images and the purposes of the challenge, we have decided to firstly enrich the dataset for the first two classes (drone and bird), then to follow up with the next two classes (fixed-wing and copter) and finally detect and classify drones with payload.

To ensure better recognition results, the following criteria were considered during dataset selection:

- different types of drones: copters and fixed wing ones, as well as different models including: DJI Phantom, Inspire, Mavic, RTK 300, and Matrice,
- multiple drones shown in different positions and distances,
- different backgrounds and sizes of drones, and
- a number of fixed wing drones, which are currently classified as drones.

As per our dataset a number of open-source datasets were considered, including DroneNet[2], Drone vs Bird [3], Skagen and Klim [15] and other free web images. From DroneNet and Skagen and Klim datasets, 2395 and 1709 images are used, respectively. The annotations on these datasets are already in YOLO format. Around 200 of images found on the web were manually annotated using LabelImg[3], a graphical image annotation tool. The largest dataset that was used for our approach is Drone vs Bird [3] one.

In summary, the dataset contains a total of 14 549 images, consisting of 12 370 images with drones only (including 3 261 with only fixed-wings), 1 857 only birds and 322 images with annotated drones and birds. Regarding the size of the targets, they mainly fall between the sizes of 16^2 and 48^2, and over 96^2 (see Fig. 2).

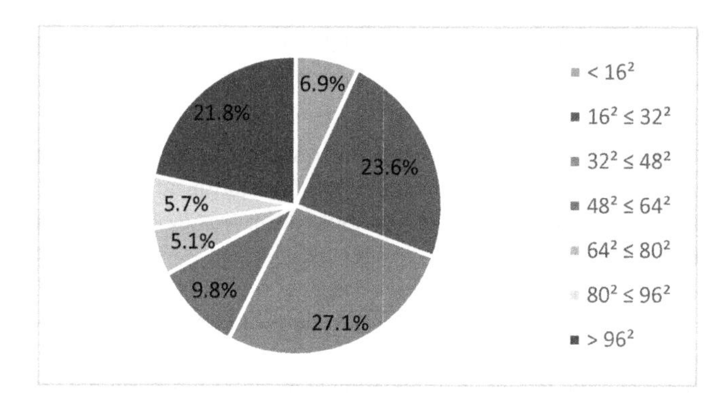

Fig. 2. Distribution of target sizes based on the annotations in the train and test data

As previously mentioned, a well-defined process of building the dataset was iteratively performed by the team. Namely, in order to match the YOLO format, a number of pre-processing steps were iteratively performed on each dataset, as depicted in Fig. 3. Each step is described in detail in the following subsections.

[2] https://github.com/chuanenlin/drone-net, last access 19.03.2022.

[3] https://github.com/tzutalin/labelImg, last access 19.03.2022.

Image Extraction and Selection. If the dataset contained video, then the step of image extraction per frame was performed. For this purpose, *Free Video to JPG Converter* application[4] was utilized. The tool supports customized extraction of images per frame and per second. The images were extracted frame-by-frame. The image filename was constructed in the following format "`<video_filename> <frame_number>`". The image selection was done by human intervention manually i.e., by removing unimportant images, which were selected based on the following criteria:

- images without drone or bird, or
- the target being so small that it causes confusion to the prediction model, or
- the body of the target object being mostly behind another visible object.

The image selection step was also performed for image-based datasets.

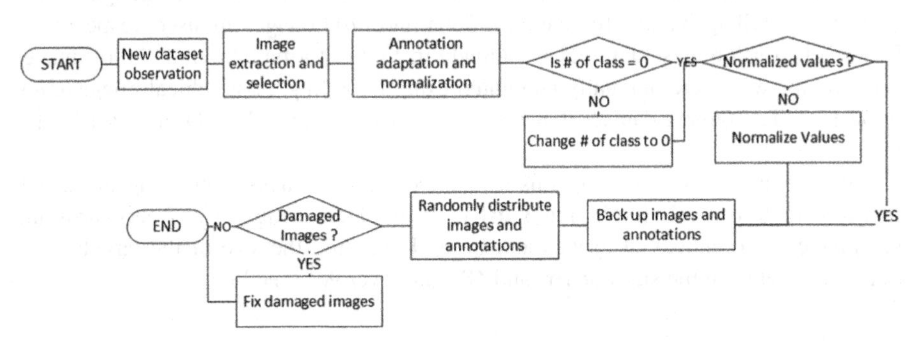

Fig. 3. The dataset refine process for using on YOLO-based models

Annotation Adaptation and Normalization. As YOLO framework requires, for each image, a single annotation file, in the next step the annotation adaptation task took place. Specifically, for this task a simple desktop application was developed. The application supports the following steps in order:

1. Load the folder images. Browse for the folder containing images (extracted from videos in the previous step).
2. Set annotated text file. Browse and open the annotation file containing the annotation list for every frame of the video. The general format and an example annotation of the Drone vs Bird dataset format consists like in Table 1.

[4] https://www.dvdvideosoft.com/products/dvd/Free-Video-to-JPG-Converter.htm, last access 19.03.2022.

Table 1. Input annotation's format.

Format	`framenum num_objs_in_frame obj1_x obj1_y obj1_w obj1_h obj1_class`
Example	34 1 1 241 55 43 drone

The annotation adaptation task outputs a text file for each image with a filename the same as the picture with the format described in Table 2. For multiple objects present in a single frame multiple rows were appended to the text file.

Table 2. YOLO annotation format.

Format	`obj1_classnumber obj1_x obj1_y obj1_w obj1_h`
Example	0 1 241 55 43

Since we are dealing with a huge number of files a validation function is needed to check for valid pairs of image files with corresponding annotation files. For this purpose, the output folder files generated from the last step, consisting of a list of couples (photo and text files), are firstly loaded and then get checked for invalid couples.

Furthermore, YOLO expects object annotations to be in the normalized format. For this purpose, a simple conversion tool was also developed to check and normalize the annotations.

Image Reduction. Since the images on successive frames are very similar and as such add little information to the model, we decided to reduce the dataset by removing every third image of each video file. This task was performed manually by using Windows Explorer feature to arrange files three per each row followed up by selecting the third column and removing files.

Train and Test Images Distribution. Finally, as per our solution the dataset should be organized into train and test folder, based on the specified 75–25% ratio. A Power Shell (PS) script was utilized to support this feature.

Fix Object Class, Annotations and Images. In different versions of our dataset the class of drone and bird was interchangeably set as 1 and 0. To ensure class consistency a PS script was executed on the dataset folders.

Furthermore, a simple tool was developed to also fix some conversion inconsistencies within annotation lines. In fact, double spaces and commas were replaced with single space.

Since our dataset was uploaded on open repositories, such as Google Drive, a small number of images got damaged after the process of distributing them into train and test folder. For this reason, before each training process, all the images were scanned for defects with the open-source tool Bad Peggy[5].

[5] https://github.com/coderslagoon/BadPeggy, last access 19.03.2022.

The final step before a training session was the backup of files. Namely, both train and test folder were occasionally backed up. Sometimes, the backup of files was performed before distribution of files into train and test folders.

As per the experimental set up, the default configurations of YOLOv4 model, based on Darknet [8], were utilized. Some Darknet code was modified and compiled to support the correct output format of the Challenge. The Google Colab Pro[6] platform was used for the training and validation of the model. It supports faster GPUs, more memory and longer runtimes as specified in the free version. With Colab one can import the image dataset, train the image classifier and evaluate the model.

Following the iterative process of improving the dataset, our model was trained with its different versions. During the last training around 6000 iterations were made and the whole training process lasted about 8 h. As can be observed from Fig. 4, the training resulted with 68% mAP.

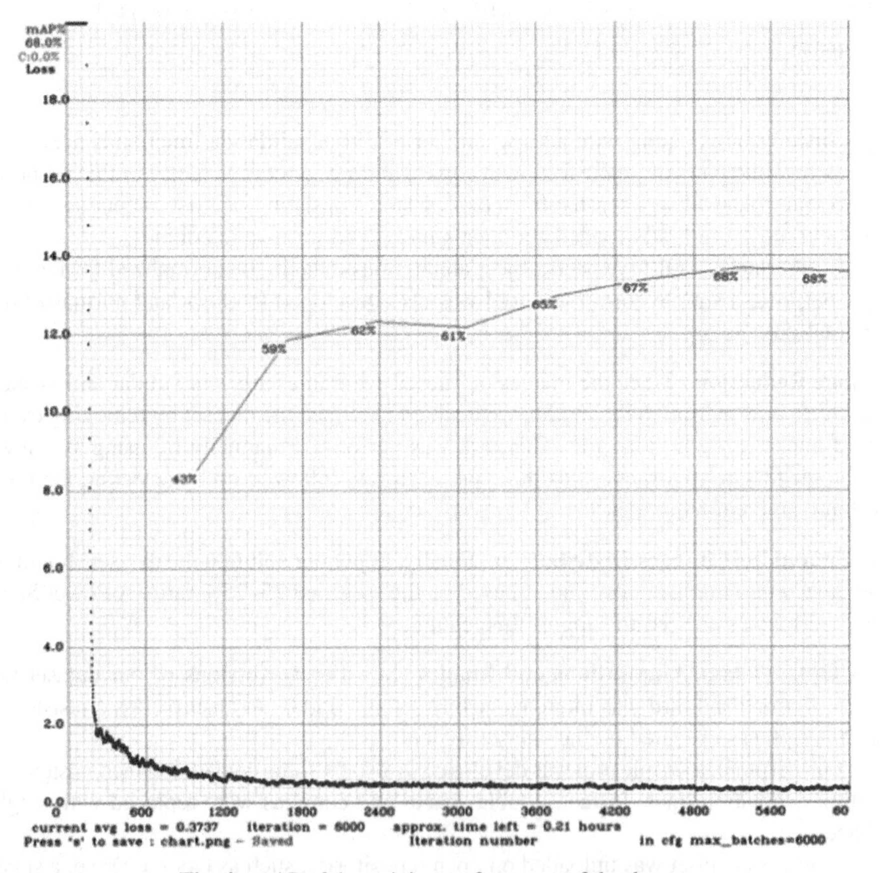

Fig. 4. mAP of the training performance of the dataset

[6] https://colab.research.google.com/, last access 20.03.2022.

4 Experimental Results

As the challenge goal is to detect a drone, experimental results are evaluated only for drone detections and classification by discarding those of birds. In order to describe more correctly the prediction accuracy of our model, the *Precision* metrics are analyzed. Additionally, to better describe the detections of the proposed approach, the *Recall* metrics are utilized.

In order to get a broader perception of the performance of the described approach, a number of test sequences were selected to include the following characteristics: many static objects, complex backgrounds, different target sizes, moving camera and near/far targets. The videos were chosen from the Drone vs. Bird Challenge dataset as they best address these constraints, and also provide ground truth annotations of drone objects. Table 3 lists the selected test sequences with corresponding characteristics, the number of present ground truth objects (#GT) as compared to the number of submitted detections (#Det), resulting recall (#Rec) and precision (#Prec).

Table 3. Description of the test video sequence set with comparison of detection results in terms of number of submitted detections and resulting recall.

Sequence	Characteristics	#GT	#Det	#Rec	#Prec
dji_matrice_210_sky	moving cam; multi-rotor drone; clear sky view; short length;	1318	1470	99.77	92.41
dji_mavick_close_buildings	moving cam; multi-rotor drone; non-sky view; long length;	1501	1034	66.60	100.00
dji_phantom_landing_custom_fixed_takeoff	moving cam; multi-rotor drone; cloudy sky; short length;	2613	2606	92.10	99.63
parrot_disco_zoomin_zoomout	moving cam; fixed-wing drone; clear sky view; short length;	665	252	50.89	81.43

The first test and third sequence (dji_matrice_210_sky and dji_phantom_landing_custom_fixed_takeoff) are not very challenging for the proposed model, because they have a sky view and thus the retrieved results are near ideal. Based on the trained model, which contains a high number of images with clear and cloudy sky view, the high score of recall and precision has turned out as expected. Namely, a static street light has generated a number of false alarms in the first sequence, which has reduced the precision to 92.41%. Unlike this, the performance of the third sequence has resulted with better FPs and thus 99.63% precision, but with greater FNs i.e., recall of 92.10%.

The view of the second sequence (dji_mavick_close_buildings) represents a drone moving on land background. The duration of the sequence is lengthy and the drone appears in every frame, which has resulted with more FNs. Thus, the recall has dropped to 66.60%, which means 1/3 of GTs are missed. However, the precision has remained perfect, because there were no FPs.

A fixed-wing drone is demonstrated in the fourth sequence (par-rot_disco_zoomin_zoomout). Following the clear sky, the resulting precision is perfect. Similar to the second sequence, the recall is again decreased by missing a half of the GTs.

5 Related Work

For getting a better insight about anti-drone YOLO-based approaches, a number of state-of-the-art ones were analyzed. Namely, the following approaches were analyzed:

- Aker et al. [9] describe an end-to-end object detection method to predict the location of the drone in the video frames. The scarce data problem for training the network has been solved by an algorithm for creating an extensive artificial dataset.
- In [11], authors describe an autonomous UAV detection and tracking platform. Namely, a Tiny YOLO detector is integrated into a hunter drone for detecting and chasing another drone.
- Wu et al. [2] propose a video-based detection of drones. To support their approach, they have developed a dataset consisting of 49 videos.
- A combined multi-frame DL detection technique, where the frame coming from the zoomed camera on the turret is overlaid on the wide-angle static camera's frame, is described in [12].
- Lai and Huang [7] have proposed a solution for detecting fixed-wing intruders with YOLOv3.

Each implementation has its own pros and cons. In general, none of the described approaches consider fusing optical data with radar ones. Moreover, even though we currently recognize birds and drones, our dataset next versions will further recognize drones with payload, which is not the case in the approaches. In particular, the solution presented in [9], detects the only drone in the scene and problems occur when the network mixes up a bird with the drone. The rest of the approaches are limited to a single drone class, except [12] who include other classes like: airplane, bird and background. But it does not provide further details about the dataset. The approaches were analyzed on the following different aspects of particular interest.

Network Architecture. The YOLO architecture model is mainly based on Darknet [8], which typically consists of 19 convolutional layers and 5 pooling layers. In general, each approach has applied specific fine-tuning techniques for achieving better performance. For example, the classifier model used in [12] uses 64 × 64 size of the input layer, while vector classification is performed by 2 consecutive fully connected layer with 512 neurons with 0.5 dropout between them [12]. To raise the performance of our approach, we are considering the network modifications in future works.

Dataset. The dataset quantity and quality differ in the approaches, as well as image sizes. In particular, Aker et al. dataset, consisting of 676 534 images with 850 × 480 resolution, combines real drone and bird images with different background videos. Wyder et al. use a synthetically generated dataset[7] of 10,000 images from autonomous drone flying sequences, manually annotated. The same number of images has been generated and used by Lei and Huang for their solution. Our dataset consists of lesser number of images, as we strive to build a high-quality dataset. The proportion of the training versus validation dataset typically ranges between 70–80% and 30%–20%, respectively.

An artificial dataset generation algorithm is described in Aker et al. It describes the process of generation and reduction of the images. However, it does not include the process of image annotation extraction and conversion, as well as image checking for errors.

Annotations. The annotations typically include information about coordinates of the center of the boxes with respect to the grid cell, the width and height in proportion to the whole image, and a confidence score of the detected object within the bounding box. In general, the approaches utilize the existing dataset annotations, or as in our case utilizing several parts from different datasets, and enriching them with new manually labeled and annotated images. Wu et al. have used Kernelized Correlation Filters (KCF) tracker [13] to auto label detected objects. A study about different types of annotation errors examined in a YOLO-based detector is described in [14]. In our approach we have used manual annotation as well as converting to YOLO-based format.

Classes. For better performance results the approaches have mainly considered a single class, i.e., drone, as specified in their model. However, Aker et al. use two classes drones and birds, while Unlu et al. have used four classes in their solution. As previously mentioned, for this paper we have used two classes and will use other ones for differentiating between drone models and carrying or not a payload.

Accuracy. The precision and recall of the approaches are satisfactory, with more than 89% and 85%, respectively. Our approach has resulted with more than 92% and 50% precision and recall, respectively. The autonomous Tiny YOLO-based approach [11] has performed with 77% accuracy in cluttered environments in eight frames per second.

6 Conclusion

Fast and robust detection and recognition is required for the anti-drone domain, because drones have ability to fly with high speed and for a short time can cause huge damage

[7] https://osf.io/jqmk2/, last access 18.03.2021.

to human lives. A lot of research efforts has been dedicated by the image processing community. In particular, the findings of this paper suggest that a methodological approach should be well-defined for the dataset improvement lifecycle. The iterative process includes continuous check, validation and image variety of the dataset. Instead of infusing vast number of images into the dataset, which can cause model confusion, the dataset improvement process should ensure high quality images, which on the other side can lead to a reduced number of false alarms and missed detections.

To date, there is not enough evidence of approaches for detecting killer drones in far distances by combining different data sources. Namely, as can be observed by the results of this paper and based on the described related works, we can conclude that the image processing algorithms do not perform well enough in cases when the background of the view is complex and the distance of the drone is far. For this aim, as per future work of our approach, we propose that the drone detection and recognition should include other technology (i.e., radar RF) and data fusion techniques complemented with optical-based recognition. This will support higher system accuracy and reliability by eliminating the identified obstacles.

Aknowledgments. This work was funded by NATO SPS Programme, approved by Dr. A, Missiroli on 12 June, 2019, ESC(2019)0178, Grant Number SPS.MYP G5633. Total project Grant is Euro 398.000. Total project duration is 36 months, project kick-off date: 25 September 2019. NATO country Project Director Dr. A. Cantelli-Forti, co-director Dr. O. Petrovska, and Dr. I. Kurmashev. We thank Dr. Claudio Palestini, officer at NATO who oversees the project. We thank Prof. Biljana Stojkoska from the Ss. Cyril and Methodius University, for her scientific contribution and tutoring to the authors (Prof. Biljana is not a member of the project and had no access to the instrumental data). We thank author O. Petrovska for funding acquisition.

References

1. White, F..: Data fusion lexicon, San Diego, CA, USA, Code 420 (1991)
2. Wu, M., Xie, W., Shi, X., Shao, P., Shi, Z.: Real-time drone detection using deep learning approach. In: Meng, L., Zhang, Y. (eds.) MLICOM 2018. LNICST, vol. 251, pp. 22–32. Springer, Cham (2018). https://doi.org/10.1007/978-3-030-00557-3_3
3. Coluccia, A., et al.: Drone vs. bird detection: deep learning algorithms and results from a grand challenge. Sensors **21**(8), 2824 (2021)
4. Coluccia, A., Parisi, G., Fascista, A.: Detection and classification of multirotor drones in radar sensor networks: a review. Sensors **20**, 4172 (2020). https://doi.org/10.3390/s20154172
5. Liu, H., Wei, Z., Chen, Y., Pan, J., Lin, L., Ren, Y.: Drone detection based on an audio-assisted camera array. In: IEEE Third International Conference on Multimedia Big Data (BigMM), pp. 402–406. IEEE. (2017)
6. Bochkovskiy, A., Wang, C.Y., Liao, H.Y.M.: YOLOv4: optimal speed and accuracy of object detection. arXiv preprint arXiv:2004.10934 (2020)
7. Lai, Y.C., Huang, Z.Y.: Detection of a moving UAV based on deep learning- based distance estimation. Remote Sens. **12**(18), 3035 (2020)
8. Redmon, J., Farhadi, A.: YOLO9000: better, faster, stronger. In: Proceedings of the IEEE Conference on Computer Vision and Pattern Recognition, pp. 7263–7271 (2017)
9. Aker, C., Kalkan, S.: Using deep networks for drone detection. In: 14th IEEE International Conference on Advanced Video and Signal Based Surveillance (AVSS), pp. 1–6. IEEE (2017)

10. Marez, D., Samuel, B., Lena, N.: UAV detection with a dataset augmented by domain randomization. In: International Society for Optics and Photonics, Geospatial Informatics X, vol. 11398, p. 1139807 (2020)
11. Wyder, P.M., et al.: Autonomous drone hunter operating by deep learning and all- onboard computations in GPS-denied environments. PLoS ONE **14**(11), e0225092 (2019)
12. Unlu, E., Zenou, E., Riviere, N., Dupouy, P.-E.: Deep learning-based strategies for the detection and tracking of drones using several cameras. IPSJ Trans. Comput. Vis. Appl. **11**(1), 1–13 (2019). https://doi.org/10.1186/s41074-019-0059-x
13. Henriques, J.F., Caseiro, R., Martins, P., Batista, J.: High-speed tracking with kernelized correlation filters. IEEE Trans. Pattern Anal. Mach. Intell. **37**(3), 583–596 (2014)
14. Koksal, A., Ince, K.G., Alatan, A.: Effect of annotation errors on drone detection with YOLOv3. In: Proceedings of the IEEE/CVF Conference on Computer Vision and Pattern Recognition Workshops, pp. 1030–1031 (2020)
15. Alqaysi, H., Fedorov, I., Qureshi, F.Z., O'Nils, M.: A temporal boosted yolo-based model for birds detection around wind farms. J. Imaging **7**(11), 227 (2021)

Evaluation of Fully Convolutional One-Stage Object Detection for Drone Detection

Abhijeet Nayak[1,2], Mondher Bouazizi[4,5], Tasweer Ahmad[1,3], Artur Gonçalves[1], Bastien Rigault[1], Raghvendra Jain[5], Yutaka Matsuo[6], and Helmut Prendinger[1](✉)

[1] National Institute of Informatics, Tokyo, Japan
helmut@nii.ac.jp
[2] University of Freiburg, Freiburg im Breisgau, Germany
[3] COMSATS University, Islamabad, Pakistan
[4] Keio University, Yokohama, Japan
[5] Optimays Inc., Tokyo, Japan
[6] The University of Tokyo, Tokyo, Japan

Abstract. In this paper, we present our approach for drone detection which we submitted for the Drone-Vs-Bird Detection Challenge. In our work, we used the Fully Convolutional One-Stage Object Detection (FCOS) approach tuned to detect drones. Throughout our experiments, we opted for a simple data augmentation technique to reduce the amount of False Positives (FPs). Upon observing the results of our early experiments, our technique for data augmentation incorporates adding extra samples to the training sets including the object which generated the most number of FPs, namely other flying objects, leaves and objects with sharp edges. With the newly introduced data to the training set, our results for drone detection on the validation set are as follows: AP scores of 0.16, 0.34 and 0.65 for small-sized, medium-sized and large drones respectively.

Keywords: Object detection · Drone detection · Deep learning · Drone-vs-Bird detection

1 Introduction

With recent technological advancements in the field of Unmanned Aerial vehicles (UAVs), their operational capabilities are ever increasing accompanied by similar decreases in the cost of acquisition. UAVs have been extensively used for surveillance, transportation or content creation. Notwithstanding these applications, there may be many instances where the growing usage of UAVs pose significant security threats to public gatherings, residential buildings, working infrastructure, or classified areas. Additionally, UAVs may also be used for illegal activities such as smuggling, privacy violation, terrorist attacks to name a

The first three authors equally contributed to this work.

P. L. Mazzeo et al. (Eds.): ICIAP 2022 Workshops, LNCS 13374, pp. 434–445, 2022.
https://doi.org/10.1007/978-3-031-13324-4_37

few. Therefore, it becomes imperative to be able to detect flying objects such as drones and UAVs from afar. At the same time, we need a robust mechanism to distinguish between these objects and birds flying in the sky, which may be collected as false positives by the system. The detection capabilities of the system are traded-off with respect to other parameters such as the object distance, object size or detection complexities, for instance.

In order to circumvent the aforementioned challenges, a low-cost solution could be the surveillance and early detection of drones in video footage. This video-based drone detection may not perfectly work for distant drones, but such a low-cost solution is an optimum choice for detecting drones at low-altitude close range targets. Video-based drone detection may come across certain difficulties, considering the following three different scenarios, i) Drone detection in areas with a high likelihood of false positive detections due to flying birds, ii) very low-altitude drones may come across occlusion due to poles and towers in a cityscape, iii) low contrast between the drone and its background may create difficult scenarios. In order to address these challenges, the research community is actively engaged in devising video-based drone detection methods and techniques.

In contrast to the earlier methods for drone detection by means of single shot detectors such as YOLO and SSD, or by the use of region-based networks such as Faster-RCNN, our method mainly investigates the use of Fully Convolutional One-Stage Object Detection (FCOS) [16], which was originally devised as anchor-free single-stage technique for object detection. One of the main advantages of the anchor-free object detector is its reduced computational complexity. This enables faster detection of drones and ensures the use of the system in a real-time environment. Another motivation to incorporate the FCOS method is its remarkable performance for a diverse spectrum of object sizes, which is highly likely in the case of drone detection depending on the distance of the drone from the camera plane.

The main contribution of our paper is that i) we devise a methodology for drone detection using FCOS algorithm and ii) additionally, we investigate various domain adaption techniques (such as transfer learning, data augmentation) which help in robustification of the output of our system. This helps in greatly reducing the number of false positive detections as a result of flying birds or falling leaves, which were earlier detected as drones with a high score.

The remainder of this paper is structured as follows: In Sect. 2, we describe some of the work related to the task of object detection and UAV detection. In Sect. 3, we describe our proposed approach, and give the details of our implementation. In Sect. 4, we present our experimental results, discuss the obtained results and show directions for future work to improve the detection. In the same section, we also address the main issues encountered during the experiments. Finally, in Sect. 5, we conclude our work.

2 Related Work

The competition at hand (Drone-vs-Bird Challenge) has seen multiple interesting submissions in its previous versions [3]. Aker et al. [1] implemented an end-to-end object detection using YOLOv2 [8]. Moreover, the authors used various other data sets to augment the training data. The augmentations included freely available drone data sets, in addition to images of flying birds with varying backgrounds. The inclusion of these data sets improved the diversity of the training data and at the same time, helped provide features that differentiate drone objects from ones that are not. Nalamati et al. [7] proposed a two-stage Faster R-CNN [9] with ResNet-101 [4] as the backbone architecture. This configuration outperformed the Faster R-CNN approach with the Inception-v2 [15] as backbone. Schumann et al. [12] proposed an image processing pipeline for tracking small targets using a wide-view overview camera, and then a mounted zoom camera which focused on the location of interest in order to classify it either as UAV or a distracting class. The authors used Single Shot multi-box Detector (SSD) as the underlying technique for object detection and trained the model on a data set containing actual UAVs and birds as well. [14] introduced a two-stage technique for UAV recognition, where, in the first stage the flying object is detected and in the second stage, the detected object is recognized as a drone, birds or background motion of clouds etc. The authors investigated convolutional neural networks (CNN) in both stages of detection and recognition. [11] used a similar idea as that of [14], where in the first stage background subtraction was exercised to find potential areas of interest, then in the second stage CNN-classifiers were employed to classify the object as either a drone or a different object.

Saqib et al. [10] carried out the task of drone detection by making some adaptations in CNN architectures such as (VGG-16 [13], or Zeiler-Fergus (ZF) [19], etc.). For this work, the authors addressed the data scarcity problem by transfer learning of the model on ImageNet and then fine-tuning. During this research, it was realized that VGG-16 with Region Based Convolutional Neural Networks (R-CNN) out-performs other architectures in consideration. A very unique idea of Super-Resolution (SR) technique for UAV detection was introduced by [6], where in the first stage a deep DCSCN model [18] is applied to zoom into the input image by two-times and then subsequently object detection is applied in the second stage. This method works in an end-to-end manner with joint optimization which resulted in an increased performance. Coluccia et al. [2] document the best methodologies adopted for the 4th edition of the Drone-vs-Bird Detection Challenge, held in 2021.

Most of the recent object detection methods depend upon anchor boxes, however, Tian et al. [16] proposed an idea of anchor-box free Fully Convolutional One-Stage object detector (FCOS). This proposed method is considered to be efficient because it avoids complicated computations for training and hyper-parameter selection due to anchor boxes. FCOS can be implemented by using different backbone architectures, such as ResNet-101-FPN, HRNet-W32-51, ResNeXt-101-FPN etc.

3 Proposed Approach

In this section, we describe the details of our proposal for drone detection, the different challenges encountered and how they have been tackled. As previously stated, in our work, we used FCOS [16], an anchor box free one-stage object detection approach that has proven to be very effective in detecting objects with small size in particular. Nonetheless, being a one-stage object detection method, FCOS allows for a high inference speed, while providing good performance, allowing it to be a good candidate to run on computationally-limited devices. Therefore, we start by summarizing the concept of this approach, how it works and its main key strength points.

3.1 FCOS Description

FCOS is a fully convolutional neural network. In Fig. 1, we show the network architecture as provided in [16]. In brief, the network architecture of FCOS is composed of an input, a backbone, a feature pyramid (proper to FCOS) and a classification + centerness + regression sub-networks. Here, $C3$, $C4$ and $C5$ refer to the feature maps of the backbone network, whereas $F4 \sim F7$ refer to the feature levels used for the final predictions. $H \times W$ correspond to the height and width of the feature maps at each level, and $/s$ refers to the sampling ratio of the feature maps. For instance, for input images of size 800×1024, these numbers correspond to the ones shown in Fig. 1. In the following, we summarize the different sub-networks main key concepts.

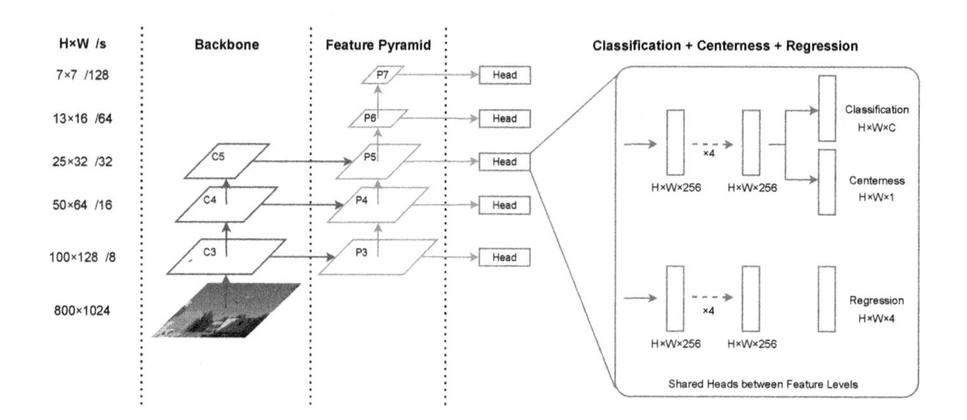

Fig. 1. Architecture of FCOS network.

The Input: This is basically the input to the neural network. As is the case for all other neural network architectures for object detection, the input image is expected to have a certain size $H \times W \times C$, where H and W are the width and height of the image respectively, and C is the number of channels, which, conventionally is set to 3 for RGB.

The Backbone: For it to be compatible with the feature pyramid network (FPN), multi-scale features are to be extracted using an encoder such as ResNet [4], ResNext [17], DenseNet [5] or others. As previously stated, $C1 \sim C5$ correspond to the feature maps extraction at stages $1 \sim 5$.

The Feature Pyramid Network (FPN): The FPN benefits from the scale-invariance of the feature pyramids to detect possible objects at different scales. Generally speaking, early layers have a richer resolution but lower semantic features, whereas later ones are richer semantically, but have lower resolutions. The lateral connections between layers allow the feature fusion between them, making the detection, in particular for smaller objects, much better.

The Detection Heads: FCOS works on a per pixel prediction. A key concept in this context is the detection heads. A prediction is output for 3 heads having a similar architecture: a 2D convolution $\leftarrow$ a group normalization $\leftarrow$ a Rectified Linear Unit (ReLU). These heads' outputs are as follows:

- **The centerness head:** Centerness is a measure of the deviation between the center of an object from the location. In their work, the authors observed that for poorly detected bounding boxes tend to have their centers further away from the ground truth. Thus, they suggested that using this measure, it is possible to improve the performance by learning this centerness scale factor. However, despite its being used or not for training, it is a good measure for evaluation nonetheless. The centerness head outputs the per feature level normalized distance from the center of the object it is in charge of. This translates into the following: the closer the prediction is to a center, the higher the normalized value.
- **The class prediction head:** As its name suggests, this head outputs the per-pixel class probability weighted by the centerness score.
- **The box regression head:** Again, as its name suggests, this head outputs the coordinates of the bounding box, through a regression process.

In summary, FCOS outputs 3 main results: the centerness, the class and the coordinates and dimensions of the bounding box.

3.2 FCOS Parameters

Throughout our work, we used the different values of FCOS parameters shown in Table 1. As can be seen, we used both ResNet-50 [4] and ResNeXt-101 [17]. However, our early experiments show a faster learning with less false positives detected using ResNet-50, leading us to focus on this architecture and run the different folds using it. Later, when we show our experimental results, we will compare both backbones for reference. Other than the base learning rate and the weight decay (as well as the number of classes obviously), we used the default parameters of FCOS.

4 Data Set, Results and Discussion

4.1 Data Set

The data set present for the current version of the Drone-vs-Bird Detection Challenge is composed of 77 short clips of flying drones. The videos include other flying objects such as birds, leaves, etc. In Fig. 2, we show examples of frames captured from different flights containing drones as well as other flying objects. Hereafter, we will use the following words interchangeably: videos, scenarios, and sequences. In Table 2, we present the main statistics of the data set, including the total number of frames, the total number of objects (i.e., drones) as well as their minimum and maximum sizes. In Fig. 3, we show the distribution of sizes of the objects in the training set.

Our early experiments showed that, despite being good at detecting flying objects in general, the approach generates a large number of False Positives (FP), confusing birds and falling leaves, and sharp-edged objects for drones in a considerable number of scenarios. That being the case, we introduced to our training data set a manually selected set of videos publicly available and downloaded from YouTube, containing for the most part enough samples of the objects being confused with the drones. This could help feed the network with more samples of such objects with no annotation to let the network learn that they should belong to the class "background" rather than the class "drone". Examples of images from the downloaded videos used for data augmentation are shown in Fig. 4.

In the remainder of this section, all the results reported are ones with the new data introduced exclusively to the training set.

4.2 Results

Cross-Validation Folds: Throughout our experiments, we split the data set in our hands into 3 sub-sets, and perform a 5-fold cross-validation. Note that,

Table 1. FCOS parameters used for training.

Parameter	Value
Backbone	ResNet-50/ResNeXt-101
Number of workers	4/8
Base learning rate	10^{-2}
Weight decay	10^{-4}
Number of iterations	100,000

Fig. 2. Samples of frames from the data set with the drone objects bounded by boxes in red. (Color figure online)

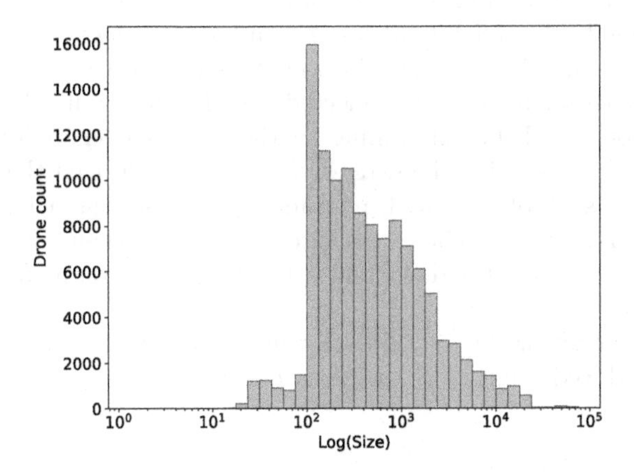

Fig. 3. Count of drone objects as present in the data set with respect to their size (in $\log_{10}$ scale).

during the training of the different folds, the newly added videos are included exclusively in the training set. None of them is used in the validation, as this could wrongly enhance the results.

Fig. 4. Examples of images from the videos used for augmentation: The first row shows images from a video of flying birds. The second row shows images captured showing sharp edges, poles and other objects which were confused with drones during our early experiments. The third row shows examples of moving leaves which also led to missing detection in our early experiments.

In Figs. 5a and 5b, we show the loss and the centerness loss, respectively, per iteration for the different folds for the validation set. Here, Folds 1 to 3 refer to the 3 folds trained using Resnet-50 as a backbone, whereas Fold X refers to the single fold (same training and validation split as Fold 1) which was trained using ResNeXt-101. As we can observe, the results obtained with the ResNet-50 folds are much better than those obtained with ResNeXt-101. What we noticed was a huge number of FPs in the ResNext results. The number of bounding box detections using ResNet-50 were in the vicinity of about 60,000 detections across all probabilities, while the number from ResNext was around 450,000 detections. Both the loss and the centerness loss show very similar behaviors for the 3 folds trained with ResNet-50 as a backbone. However, the model trained in fold 1 shows slightly lower losses, making it the main candidate for our submission.

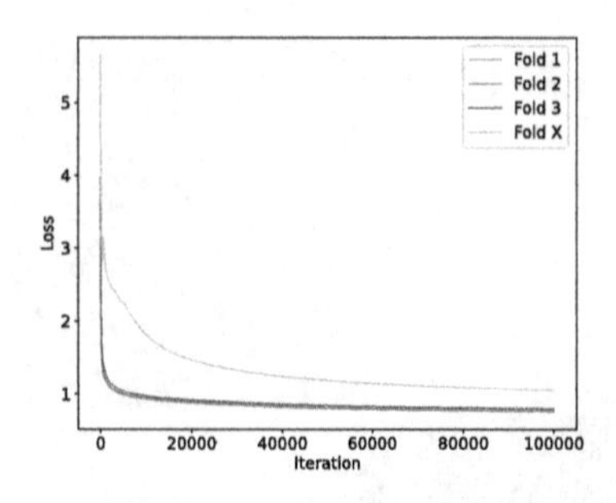

(a) Validation loss for the different folds trained

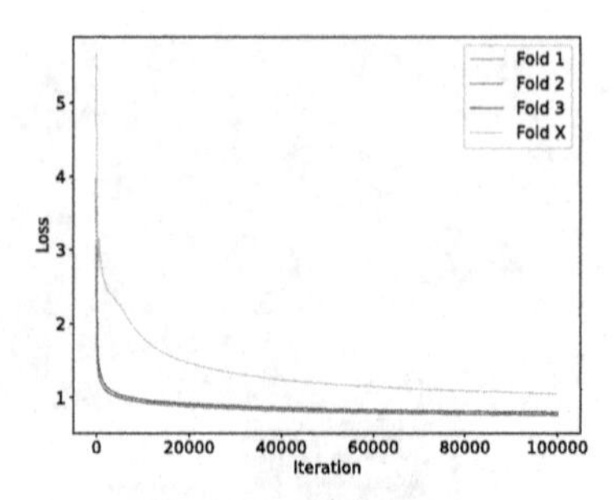

(b) Validation centerness loss for the different folds trained

Fig. 5. Loss and Centerness loss over the training iterations on the different folds for ResNet-50 and ResNeXt-101. Here, Folds 1 – 3 refer to different folds trained using ResNet-50, whereas Fold X refers to the fold trained using ResNeXt-101.

Table 2. Data set specifications.

Total number of images	106,485
Number of images with drones	97,389 (91%)
Number drone objects	119,243
Minimum size	15 pixels
Maximum size	1,394,556 pixels

In Table 3, we give a summary of the detection performance for the different folds at the end of the training, each for its respective validation set. Here, small refers to objects whose size is smaller than 32×32 pixels, medium refers to objects whose size is between 32×32 and 96×96 pixels, and large refers to objects whose size is larger than 96×96 pixels. As can be seen, the performance on Fold 1 and Fold X is very similar. However, the model trained at Fold X generates a huge number of FPs. This led us to exclude this model, as precision is one of the main Key Performance Indicators (KPIs) and was poorly reported with this model.

Table 3. Average precision on small, medium and large objects

Fold	Small objects	Medium objects	Large objects
Fold 1	0.16	0.34	0.65
Fold 2	0.24	0.38	0.48
Fold 3	0.13	0.41	0.54
Fold X	0.21	0.32	0.71

4.3 Discussion

During our experiments, we have observed that the proposed method had trouble identifying drones and generated a large number of FPs as well as a fairly big number of missing identifications (i.e., False Negatives - FNs) in a certain number of flight videos. These include mainly the following cases:

- Low quality videos with very small drones: This is by far the biggest reason of FNs. When the drone object is very small and blends with its background due to the poor quality and the blurriness in the images, our approach tends not to identify the drone.
- Presence of other flying or flying-like objects: Such objects include birds, tree leaves, etc. The presence of such objects led to generating the large number of FPs. Throughout our work, we tried to use data augmentation to give the network more examples of flying objects that are not drones. This technique is limited in the sense where these objects are part of the class "background," which has way more samples than the class "drone." A better option would be to have (a) dedicated class(es) for these objects so that the neural network learns their patterns and manifestations rather than treating them as part of the background. However, this requires acquiring a labeled data set, or labeling the ones we added.
- The poor diversity of the scenarios in the training set: In the data set in hand, a large number of the videos are taken under very similar conditions. This leads to some sort of overfitting, as the networks learns the patterns of drones with reference to their backgrounds in the training samples. Data augmentation by means of adjusting the contrast in the images, for example, could help partially address this issue and improve the performance. However, ultimately, using more diverse videos is a much better option.

Due to time constraints, we were not able to experiment more with the ResNeXt-101 model as well as other ones. In other words, we were not able to exploit the potentials and strong points of each of the backbone architectures, nor did we have the time to evaluate them properly and optimize them for the detection of drones. In a future participation, we would run several rounds of experiments, the first of which aims to identify good candidates. Once good backbone candidates are identified, we would like to experiment with them in more details in later rounds.

5 Conclusions

In this paper, we summarized our proposal for drone detection submitted for the Drone-Vs-Bird Detection Challenge. In our work, we used FCOS which we tuned to detect drones. Upon observing the results of our early experiments, we decided to use a data augmentation technique to reduce the amount of False Positives (FPs). Our technique for data augmentation incorporates adding extra samples to the training sets including the object which generated the most number of

FPs, namely other flying objects, leaves and objects with sharp edges. With the newly introduced data to the training set, our results for drone detection on the validation set were as follows: AP scores of 0.16, 0.34 and 0.65 for small drones, medium-sized drones and large drones respectively.

Acknowledgements. The work was partially supported by The University of Tokyo. The first author conducted this work as part of the NII International Internship Program.

References

1. Aker, C., Kalkan, S.: Using deep networks for drone detection. In: 2017 14th IEEE International Conference on Advanced Video and Signal Based Surveillance (AVSS), pp. 1–6. IEEE (2017)
2. Coluccia, A., et al.: Drone-vs-bird detection challenge at IEEE AVSS2021. In: 2021 17th IEEE International Conference on Advanced Video and Signal Based Surveillance (AVSS), pp. 1–8. IEEE (2021)
3. Coluccia, A., et al.: Drone vs. bird detection: deep learning algorithms and results from a grand challenge. Sensors **21**(8), 2824 (2021)
4. He, K., Zhang, X., Ren, S., Sun, J.: Deep residual learning for image recognition. In: Proceedings of the IEEE Conference on Computer Vision and Pattern Recognition, pp. 770–778 (2016)
5. Huang, G., Liu, Z., Van Der Maaten, L., Weinberger, K.Q.: Densely connected convolutional networks. In: Proceedings of the IEEE Conference on Computer Vision and Pattern Recognition, pp. 4700–4708 (2017)
6. Magoulianitis, V., Ataloglou, D., Dimou, A., Zarpalas, D., Daras, P.: Does deep super-resolution enhance UAV detection? In: 2019 16th IEEE International Conference on Advanced Video and Signal Based Surveillance (AVSS), pp. 1–6. IEEE (2019)
7. Nalamati, M., Kapoor, A., Saqib, M., Sharma, N., Blumenstein, M.: Drone detection in long-range surveillance videos. In: 2019 16th IEEE International Conference on Advanced Video and Signal Based Surveillance (AVSS), pp. 1–6. IEEE (2019)
8. Redmon, J., Farhadi, A.: YOLO9000: better, faster, stronger. In: Proceedings of the IEEE Conference on Computer Vision and Pattern Recognition, pp. 7263–7271 (2017)
9. Ren, S., He, K., Girshick, R., Sun, J.: Faster R-CNN: towards real-time object detection with region proposal networks. In: Advances in Neural Information Processing Systems 28 (2015)
10. Saqib, M., Khan, S.D., Sharma, N., Blumenstein, M.: A study on detecting drones using deep convolutional neural networks. In: 2017 14th IEEE International Conference on Advanced Video and Signal Based Surveillance (AVSS), pp. 1–5. IEEE (2017)
11. Schumann, A., Sommer, L., Klatte, J., Schuchert, T., Beyerer, J.: Deep cross-domain flying object classification for robust UAV detection. In: 2017 14th IEEE International Conference on Advanced Video and Signal Based Surveillance (AVSS), pp. 1–6. IEEE (2017)
12. Schumann, A., Sommer, L., Müller, T., Voth, S.: An image processing pipeline for long range UAV detection. In: Emerging Imaging and Sensing Technologies for Security and Defence III; and Unmanned Sensors, Systems, and Countermeasures, vol. 10799, p. 107990T. International Society for Optics and Photonics (2018)

13. Simonyan, K., Zisserman, A.: Very deep convolutional networks for large-scale image recognition. arXiv preprint arXiv:1409.1556 (2014)
14. Sommer, L., Schumann, A., Müller, T., Schuchert, T., Beyerer, J.: Flying object detection for automatic UAV recognition. In: 2017 14th IEEE International Conference on Advanced Video and Signal Based Surveillance (AVSS), pp. 1–6. IEEE (2017)
15. Szegedy, C., Vanhoucke, V., Ioffe, S., Shlens, J., Wojna, Z.: Rethinking the inception architecture for computer vision. In: Proceedings of the IEEE Conference on Computer Vision and Pattern Recognition, pp. 2818–2826 (2016)
16. Tian, Z., Shen, C., Chen, H., He, T.: FCOS: fully convolutional one-stage object detection. In: Proceedings of the IEEE/CVF International Conference on Computer Vision, pp. 9627–9636 (2019)
17. Xie, S., Girshick, R., Dollár, P., Tu, Z., He, K.: Aggregated residual transformations for deep neural networks. In: Proceedings of the IEEE Conference on Computer Vision and Pattern Recognition, pp. 1492–1500 (2017)
18. Yamanaka, J., Kuwashima, S., Kurita, T.: Fast and accurate image super resolution by deep CNN with skip connection and network in network. In: Liu, D., Xie, S., Li, Y., Zhao, D., El-Alfy, E.S. (eds.) Neural Information Processing, pp. 217–225. Springer, Cham (2017). https://doi.org/10.1007/978-3-319-70096-0_23
19. Zeiler, M.D., Fergus, R.: Visualizing and understanding convolutional networks. In: Fleet, D., Pajdla, T., Schiele, B., Tuytelaars, T. (eds.) ECCV 2014. LNCS, vol. 8689, pp. 818–833. Springer, Cham (2014). https://doi.org/10.1007/978-3-319-10590-1_53

Drone Surveillance Using Detection, Tracking and Classification Techniques

Daitao Xing[1(✉)], Halil Utku Unlu[2], Nikolaos Evangeliou[3], and Anthony Tzes[3]

[1] Department of Computer Science and Engineering, New York University, New York 11201, USA
daitao.xing@nyu.edu

[2] Department of Electrical and Computer Engineering, New York University, New York 11201, USA
utku@nyu.edu

[3] Electrical Engineering, New York University Abu Dhabi and Center for Artificial Intelligence and Robotics, 129188 Abu Dhabi, UAE
{nikolaos.evangeliou,anthony.tzes}@nyu.edu

Abstract. In this work, we explore the process of designing a long-term drone surveillance system by fusing object detection, tracking and classification methods. Given a video stream from an RGB-camera, a detection module based on YOLOV5 is trained for finding drones within its field of view. Although in drone detection, high accuracy and robustness is achieved with the underlying complex architecture, the detection speed is hindered on ultra HD-streams. To solve this problem, we integrate a high efficient object tracker to update target status while avoiding running the detection at each frame. Benefited from lightweight backbone networks with powerful Transformer design, the object tracker achieves real-time speed on standalone CPU devices. Moreover, a drone classification model is applied on the output of the detection and tracking mechanisms to further distinguish drones from other background distractors (birds, balloons). By leveraging inference optimization with TensorRT and ONNX, our system achieves extremely high inference speed on NVIDIA GPUs. A ROS package is designed to integrate the aforementioned components together and provide a flexible, end-to-end drone surveillance tool for real-time applications. Comprehensive experiments on both standard benchmarks and field tests demonstrate the effectiveness and stability of proposed system.

Keywords: Drone detection and classification · Object tracking · Super resolution

1 Introduction

The area of Unmanned aerial vehicles (UAVs) has drawn increasing attention in recent years due to its applications cross diverse fields such as aerial photography [10], mapping and surveying [28], search, rescue and emergency response [1].

P. L. Mazzeo et al. (Eds.): ICIAP 2022 Workshops, LNCS 13374, pp. 446–457, 2022.
https://doi.org/10.1007/978-3-031-13324-4_38

The advent of low-cost small commercial drones led to their deployment into real world, while raising safety, privacy concerns and other types of challenges to the aviation industry [29], border security [8], and critical infrastructures [14]. Therefore, the demand of developing surveillance systems, especially for small drones has risen in the past few years to prevent intentionally or unintentionally misused of drones in urban environments, coastal border, airports and other safety-sensitive areas.

In recent years, there have been a lot of efforts in designing drone surveillance systems [6,26] by adopting effective detection and countermeasure techniques including LiDAR, radio detectors, visual camera and passive acoustic sensors. Among those techniques, visual detection based on Deep Learning methods achieves remarkable progress on both effectiveness and accuracy. With the development of deep learning theory and optimization of hardware, modern object detectors obtain human-level compatible accuracy and operate in real-time speed even on mobile devices. However, drone detection is still a challenging problem due to its small size and fast maneuvers. Other factors caused by illumination change, heavy occlusion and target disappearance from the camera view further hinders drone detection.

To deal with the small object detection problem, recent works [5,19,27] utilize larger and deeper networks with more complex architecture to improve the model discriminative ability. However, constrained by the input size of neural networks, small drones only take less than 100 pixels in HD-frames, providing insufficient information for feature extraction and detection. On the other hand, blurred imaging of small objects from a long distance makes it harder to distinguish drones from other similar distractors like birds and airplanes. The only efficient solution is enlarging their input size to provide more useful information. However, this causes the exponential increment of computational complexity and will use most of the computational power, resulting in the processing delay and detection discontinuity in real applications.

In this work, we build a drone detection module based on YOLOV5. Due to the trade-off between complexity and precision, we choose YOLOV5-m as the base model and restrict the maximum input size to 1280 pixels. To avoid the computational overload caused by the drone detection, the used module only operates in a very low frequency (<1 Hz). Considering the sparsity of drones occurrence in the field of view as well as the flying trajectory continuity, it is not necessary to run the detector on each frame and we use it as an indicator of the first appearance and disappearance of drones in camera view. Once a drone is captured by detection module, a more efficient object tracker running in real-time speed using low-resources will be initialized to update the drone status in the following frames.

Unlike object detection in which the model runs through the whole frame, object tracking, instead, identifies the target object from a local patch, resulting in efficient and accurate schemes. Moreover, modern trackers are optimized for dealing with varying challenging scenarios like fast motion, low-resolution, frequent occlusion, etc. Recent years have witnessed many successful deep learning based object trackers, especially the family of Siamese Network based trackers [2,4,7], which play an important role in the visual tracking community. In this work, we employ a recent Transformer based object tracker, SiamTPN [32], which achieves a desired trade-off between tracking efficiency and accuracy for dealing with varying computational demand. Specifically, the SiamTPN obtains State-Of-The-Art performance while running at real-time speed on both CPU and GPU ends. The outputs from the drone detection modules initializes the tracker module. The trackers after initialization will track the detected objects as they move around frames. Once the target is lost or out of field view, the tracker will be removed from trackers' list. Thanks to the computational efficiency design, the trackers can be easily deployed to track multiple objects in parallel way on single GPU.

However, both detector and trackers may produce false negative predictions. The detector may takes airplanes or birds from a long distance as drones. Meanwhile, the object tracker fails when the object is out of view, occupied or distractors occur. In either case, it demands a robust classifier with strong discriminative ability to determine the final classes for outputs from detector and trackers. Only the objects with a higher confidential score will be kept. To this end, we employ a pre-trained Resnet-50 [12] model and fine tune the final layers with custom classification datasets for drones. To deal with the tiny drones with very low resolution, we deploy a light-weight super resolution method, SRGAN [18], to generate high-resolution patches before feeding them into classifier, which further improve the stability of classification model [24]. In practice, we only apply SRGAN [18] model on small patches with size less than 50×50 pixels.

By integrating the aforementioned components, we propose an efficient, end-to-end drone surveillance system, which can be easily deployed into embedding devices with low computational resources. We further boost the effectiveness by leveraging inference optimization techniques such as TensorRT and ONNX.

2 Related Work

2.1 Object Detection

The deep learning based object detection methods include two branches, like the two-stage methods, including Faster RCNN [25], and single-stage methods using SSD [21], YOLO [15,16] and FCOS [27]. Two-stage methods divide the detection procedure into a coarse classification problem followed by a fine-tuning step, leading to a higher accuracy. Single stage methods, instead, aim to a desired trade-off between efficiency and precision, which is preferred in the systems with limited computational power. To balance the computing resource allocation between object detector and trackers, we employ the single stage YOLOV5 as our detector.

2.2 Object Tracking

The tracking methods can be divided into: a) Discriminative Correlation Filter (DCF) based trackers and b) deep learning-based trackers. DCF based trackers [3,13,23] could run with real-time speed on CPU, but their performance is constrained by the feature representation ability of handcrafted features. In contrast, deep learning based trackers, like the Siamese-based trackers [2,4,7,32] achieve remarkable enhancements in both accuracy and speed by utilizing a high-end GPU device.

2.3 Classification and Super Resolution

Early classification methods like AlexNet [17], and Resnet [12] get higher accuracy in using deeper and wider networks. Among those classifiers, the Resnet family is the most popular framework and is adopted in many computer vision tasks as backbone network. In this work, we use Resnet-50 as our classifier, due to its balance between efficiency and accuracy. We further boost the performance by applying a lightweight super-resolution model, SRGAN, to deal with the small drones with low resolution.

2.4 Inference Optimization

TensorRT is a C++ library that facilities high-performance inference on NVIDIA graphics processing units (GPUs). TensorRT applies graph optimizations, layer fusion, among other optimizations, while also finding the fastest implementation of that model leveraging a diverse collection of highly optimized kernels.

3 Drone Surveillance System

This section presents the drone surveillance system design and the implementation details of each component.

3.1 System Overview

Given the frame f at time t, we first resize the image without crop and maintain the aspect ratio before feeding them into detector $\mathcal{D}$. For images of size 1920×1080 or less, we resize images so that the longer edge equals to 1280 pixels. The object detector returns a new set of recognized drones as $\mathbf{d} = \{\mathbf{d}_1, \mathbf{d}_2, \cdots, \mathbf{d}_n\}$, where $\mathbf{d}_i$ is represented as the concatenation of the bounding box coordinates $\{x, y, w, h\}$ and confidential score $s_{\mathcal{D}}$. For images with higher resolution, we follow the image tilling strategy [30] in which the image is divided into multiple tiles of a fixed size. The tiled images are processed with same detector in a batch manner. The final prediction is the aggregation of outputs from each tile. The detector is set to run at low frequency (<1 Hz) for inference efficiency.

Each tracker $\mathcal{T}$ is responsible for a specific object and returns the updated status as $\mathbf{t}_i = \{x, y, w, h, s_{\mathcal{T}}\}$ where $s_{\mathcal{T}}$ is its confidential score. Together, we have a set of drone candidates $\{\mathbf{d}_1, \mathbf{d}_2, \cdots, \mathbf{d}_n \, \mathbf{t}_1, \mathbf{t}_2, \cdots, \mathbf{t}_m\}$ from the detector and trackers outputs. We crop patches according to those candidates and feed them into the drone classifier $\mathcal{C}$ for further discrimination. A confidential score $s_{\mathcal{C}}$

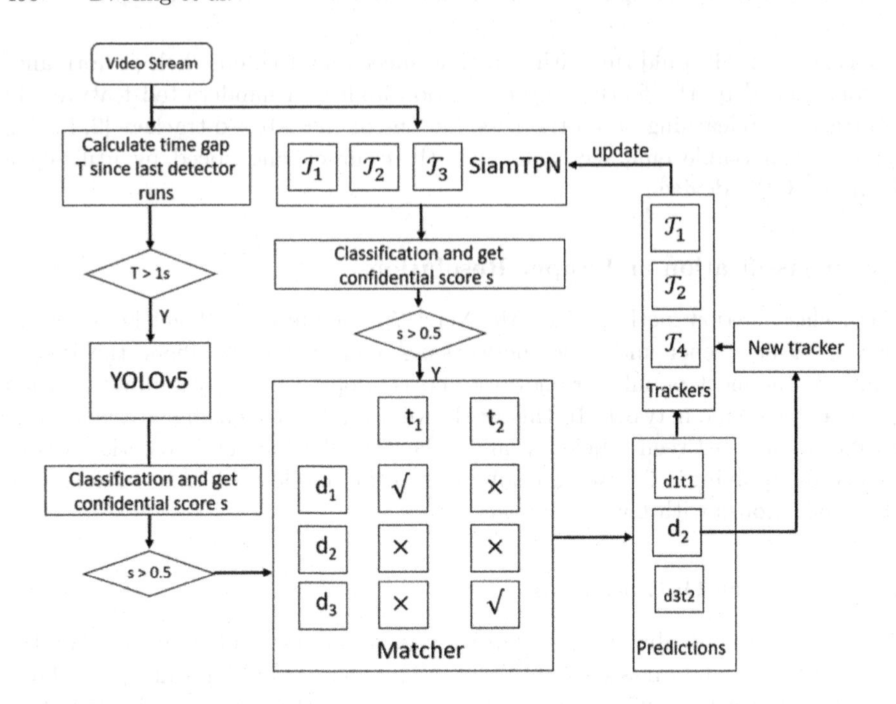

Fig. 1. Drone surveillance system overview. During tracking process, drone 3 is removed during score matching and drone 4 is new drone which will be tracked in the following frames.

is provided for each candidate. Overall, the final confidential score is calculate as $s = s_{\mathcal{C}} \times s_{\mathcal{D},\mathcal{T}}$; only candidates with this score s higher than a threshold will be kept for further processing, while for those candidates whose confidential score falls below the threshold, the corresponding tracker will be removed from tracker list.

A matcher, based on the criterion of maximum Intersection Over Union (IOU), is designed to match the candidates from $\mathbf{d}$ to $\mathbf{t}$. As shown in Fig. 1, the process is similar to non maximum suppression (NMS). Specifically, if a candidate $\mathbf{d}_i$ is matched to $\mathbf{t}_j$, or vice versa, the two instances will be merged and use the one with higher confidential score as final output. If no matches found for instances from $\mathbf{d}$, a new tracker $\mathcal{T}$ will be initialized and added into the tracker list. We should mention that the matcher only works when both detector and trackers are active.

3.2 Detection Module

We select the single-stage object detector YOLOv5 [16] for its efficiency and speed on object detection tasks. Specifically, the COCO [20] pre-trained YOLOv5-m model with input size of 1280 is adopted. In all experiments, the networks were trained using 40 epochs on 4-GPUs with 16 images per batch.

We use the ADAM [22] optimizer with the initial learning rate of 10^{-4}. For the training dataset, we consider the image sequences from Drone vs. Bird Competition [26] and USC drone detection and tracking dataset [31]. We uniformly sampled frames with a fixed rate (5 fps) from each sequence and extracted 32067 images in total for training. We select 4 videos from Drone vs. Bird training dataset for detection and tracking validation purpose.

3.3 Tracking Module

For object tracking, we employ a real-time Siamese Network based deep learning tracker, SiamTPN [32], for its robust performance and real-time speed. As shown in Fig. 2, the SiamTPN utilizes a lightweight backbone and optimized transformer based pyramid network to learn discriminative features from both template and search images. The final prediction is returned after the cross correlation layer. The template image is cropped from image when the detector recognizes a new drone which is not tracked yet. The search image is cropped from the following frames and resized into 256×256. Benefited from the small input size and optimized architecture, the tracker runs at 50 FpS on CPU and over 100 FpS on GPU, where more details can be found in [32]. We compare the performance between the SiamTPN with default trackers provided by OpenCV in Sect. 4. For inference, we directly use the pre-trained model from SiamTPN without further finetuning since the tracker is designed to track any generic objects specified by the template.

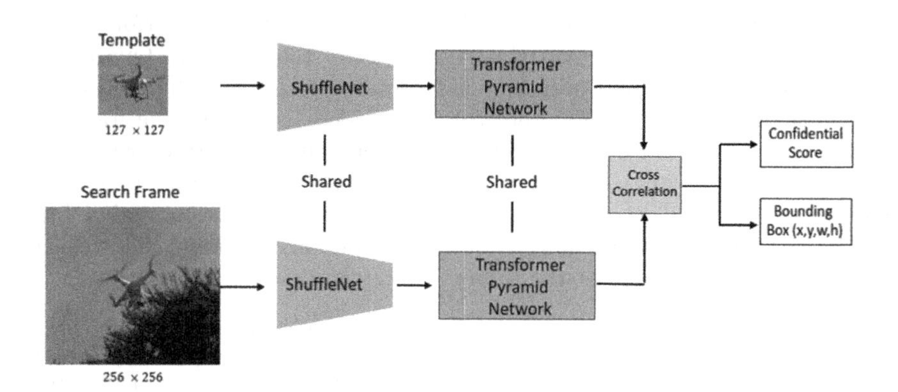

Fig. 2. SiamTPN: Architecture overview

3.4 Classification Module

The drone classification module is fine-tuned on a pre-trained Resnet-50 [12] model. We cropped drones patches and resize them into 224×224, yielding 33733 positive samples. For negative images, we randomly select images from the ImageNet [9] dataset. We found that the negative images are easy to be classified due to the low similarity to drones. Therefore, we run the drone detector on the training dataset from the object detection part and select the false positive predictions as false samples for classification, resulting in a robust performance. During training, we freeze the whole network except the final fully-connected layers for fine-tuning. The network is trained for 50 epochs with 128 images per batch.

4 Experimental Studies

This section first presents the effects of proposed components in the aspects of accuracy and speed. We further apply the system in the field test videos to validate its performance in real-world applications.

4.1 Overall Performance

In order to compare the effect of each proposed component regarding accuracy and speed, we choose 4 videos from the Drone vs. Bird training dataset as validation dataset. We adopt the Average Precision (AP) metric which is extensively used in an object detection task. The prediction outputs are counted as correct when its IOU score with a ground truth bounding box is higher than a threshold. In this study, we test the AP score under varies criterion, including AP under different threshold value (AP, AP_{50}, AP_{75}) and AP for drones with different sizes (AP_S, AP_M, AP_L). All AP scores are calculated with COCO API. Table 1 shows the overall performance on validate set when different components are active. YOLOv5 detector alone shows relative poor performance on video detection, having an AP of 43.4. We notice that the YOLOv5 detector is sensitive to the complex scenarios like object deformation, illumination and object occupation. Since object detector treats videos as independent frames, the predictions shows inconsistency even in adjacent frames. The tracker boosts the performance by 50% by guaranteeing the prediction continuity between frames. The AP_{50} achieves 89 on the validated dataset. The classification module brings relative small performance changes, but it provides an additional check which is useful when the trackers lost the target and return false positive outputs.

Table 1. Overall performance on Validation dataset. Average Precision (AP) scores are calculated with the COCO API. AP_{50}, AP_{75} represent AP with IOU above 0.5, 0.75 respectively. AP_S, AP_M, AP_L represent AP for small, medium, large objects respectively.

	AP	AP_{50}	AP_{75}	AP_S	AP_M	AP_L
YOLOv5	43.4	62.3	52.4	30.5	48.1	64.7
YOLOv5 + SiamTPN	63.1	89.0	77.8	45.9	69.3	91.2
YOLOv5 + SiamTPN + Classification	63.5	89.9	77.9	47.2	69.1	91.5

To further investigate the tracker's performance, we compare the SiamTPN with 3 default trackers from OpenCV, which are: a) CSRT tracker [23], b) KCF tracker [13], and c) MIL tracker [11]. We perform an One-Pass Evaluation (OPE) and measure the precision and success of different tracking algorithms on validating videos. Different from Average Precision, in OPE, the precision is computed by comparing the distance between tracking result and ground truth bounding box in pixels. The success is computed as the IOU scores between tracking result and ground truth bounding box at different threshold levels. Finally, we rank the tracking methods using the Area Under the Curve (AUC). As shown in Fig. 3, the KCF performs poorly on both precision and accuracy. CSRT provides compatible results on precision scores but still have a large gap on success rate compared with SiamTPN. Table 2 shows the speed comparison between those trackers, in which, MIL, KCF and CRST only support CPU while SiamTPN support both CPU and GPU. Overall, SiamTPN achieves best performance in the aspects of speed, accuracy and robustness.

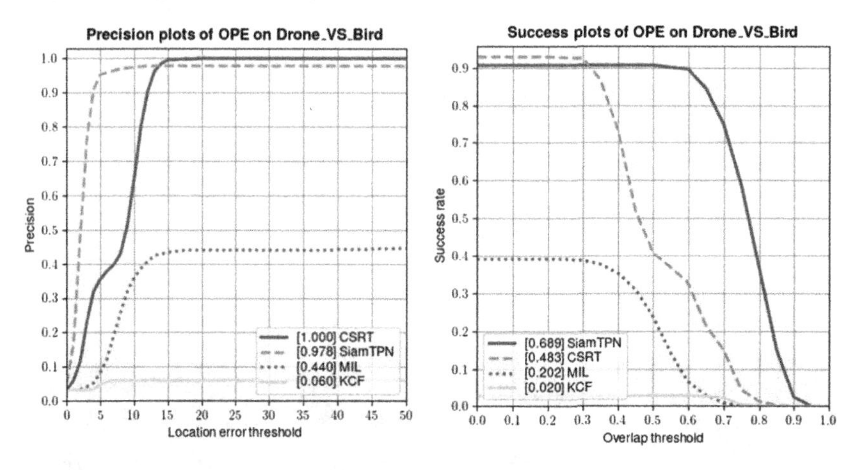

Fig. 3. Tracking performance comparison with OpenCV trackers

Table 2. Speed comparison between trackers

	MIL	KCF	CSRT	SiamTPN	SiamTPN(GPU)
FpS	6	240	45	52	102

4.2 Speed Analysis

In Table 3, we compare the inference speed of each modules and their combination performance. Due to the large input size (1280), YOLOv5m only operates at around 12 FpS on GPU after inference optimization, which is not suitable for applications with real-time requirements. Instead, the optimized SiamTPN achieves 100+ FpS on GPU. Benefiting from small network size and smaller input size, the classification module and super resolution model, SRGAN, require much less computation resources compared with other detection and tracking models. By combining the YOLOv5 and SiamTPN and constraining the detector operation frequency, the inference achieves a real-time speed of 37 FpS. The speed decline comes from the heavy detection module and multiple trackers running in parallel. Nevertheless, the combination of detector and trackers obtains a desired trade-off between accuracy and speed. The classification module and SRGAN introduces a slight computational burden to the system.

Table 3. Speed comparison between individual modules and difference configurations. All models are accelerated with GPU and TensorRT.

#	YOLOv5	SiamTPN	Classification	SRGAN	FpS
1	✓				12
2		✓			102
3			✓		178
4				✓	205
5	✓	✓			37
6	✓	✓	✓		32
7	✓	✓	✓	✓	29

Fig. 4. Visualization of drone surveillance system in field test. The drones are captured by a still camera on the ground (first row) or a camera mounted on a flying drone (second row)

4.3 Field Test Analysis

To validate the reliability of the proposed drone surveillance system in real-world scenarios, we set up several field tests with challenging factors including

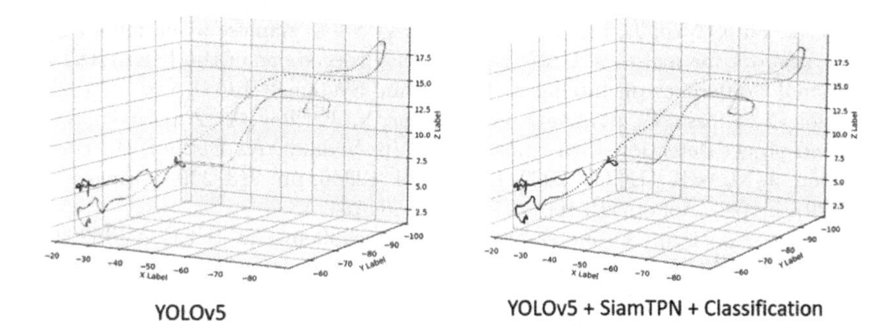

Fig. 5. Drone surveillance performance in real world test.

scale variance, out-of-view, object deformation and partial occlusion. Figure 4 shows the flight status and detection results in difference scenarios. To verify the advantage of proposed system over single detection modules, we compare the drone trajectories coverage percentages based on their prediction results. As shown in Fig. 5, for better visualization, we recorded GPS data and plot the trajectory in 3D as red dots. The dots are labeled as blue only if the drone in this position is correctly recognized. The configuration with YOLOv5, SiamTPN and Classification obtains more consistent predictions than detector alone.

5 Conclusions

In this work, we propose a long-term drone surveillance system which consists of a YOLOv5 based drone detector, real-time object tracker, drone classifier and other auxiliary modules. Those modules are integrated in an efficient way and are optimized with inference acceleration techniques (TensorRT and ONNX) to achieve best performance. Our method ranked second in the 2022 Drone vs. Bird detection challenge. We have also verified our system in real-world test with the preliminary results from both field tests and competition demonstrating the effectiveness of the proposed system.

References

1. Ajith, V., Jolly, K.: Unmanned aerial systems in search and rescue applications with their path planning: a review. In: Journal of Physics: Conference Series, vol. 2115, p. 012020. IOP Publishing (2021)
2. Bertinetto, L., Valmadre, J., Henriques, J.F., Vedaldi, A., Torr, P.H.S.: Fully-convolutional siamese networks for object tracking. In: Hua, G., Jégou, H. (eds.) ECCV 2016. LNCS, vol. 9914, pp. 850–865. Springer, Cham (2016). https://doi.org/10.1007/978-3-319-48881-3_56
3. Bolme, D.S., Beveridge, J.R., Draper, B.A., Lui, Y.M.: Visual object tracking using adaptive correlation filters. In: 2010 IEEE Conference on Computer Vision and Pattern Recognition, pp. 2544–2550. IEEE (2010)

4. Cao, Z., Fu, C., Ye, J., Li, B., Li, Y.: SiamAPN++: siamese attentional aggregation network for real-time UAV tracking. In: Proceedings of the IEEE/RSJ International Conference on Intelligent Robots and Systems (IROS), pp. 1–7 (2021)
5. Carion, N., Massa, F., Synnaeve, G., Usunier, N., Kirillov, A., Zagoruyko, S.: End-to-end object detection with transformers. In: Vedaldi, A., Bischof, H., Brox, T., Frahm, J.-M. (eds.) ECCV 2020. LNCS, vol. 12346, pp. 213–229. Springer, Cham (2020). https://doi.org/10.1007/978-3-030-58452-8_13
6. Coluccia, A., et al.: Drone-vs-bird detection challenge at IEEE AVSS2021. In: 2021 17th IEEE International Conference on Advanced Video and Signal Based Surveillance (AVSS), pp. 1–8. IEEE (2021)
7. Danelljan, M., Bhat, G., Khan, F.S., Felsberg, M.: Atom: accurate tracking by overlap maximization. In: Proceedings of the IEEE Conference on Computer Vision and Pattern Recognition, pp. 4660–4669 (2019)
8. De Cubber, G., et al.: The safeshore system for the detection of threat agents in a maritime border environment. In: IARP Workshop on Risky Interventions and Environmental Surveillance (2017)
9. Deng, J., Dong, W., Socher, R., Li, L., Li, K., Fei-Fei, L.: Imagenet: a large-scale hierarchical image database. In: 2009 IEEE Conference on Computer Vision and Pattern Recognition (2009)
10. Gotovac, D., Gotovac, S., Papić, V.: Mapping aerial images from UAV. In: 2016 International Multidisciplinary Conference on Computer and Energy Science (SpliTech), pp. 1–6 (2016)
11. Grabner, H., Grabner, M., Bischof, H.: Real-time tracking via on-line boosting. In: BMVC, vol. 1, p. 6. Citeseer (2006)
12. He, K., Zhang, X., Ren, S., Sun, J.: Deep residual learning for image recognition. In: Proceedings of the IEEE Conference on Computer Vision and Pattern Recognition, pp. 770–778 (2016)
13. Henriques, J.F., Caseiro, R., Martins, P., Batista, J.: High-speed tracking with kernelized correlation filters. IEEE Trans. Pattern Anal. Mach. Intell. **37**(3), 583–596 (2014)
14. de la Iglesia, D., Mendez, M., Dosil, R., Gonzalez, I.: Drone detection CNN for close-and long-range surveillance in mobile applications. In: Proceedings of the AVSS, Taipei, Taiwan, pp. 18–21 (2019)
15. Jiang, Z., Zhao, L., Li, S., Jia, Y.: Real-time object detection method based on improved yolov4-tiny. arXiv preprint arXiv:2011.04244 (2020)
16. Jocher, G.: ultralytics/yolov5: v3.1 - Bug Fixes and Performance Improvements, October 2020
17. Krizhevsky, A., Sutskever, I., Hinton, G.E.: Imagenet classification with deep convolutional neural networks. In: Advances in Neural Information Processing Systems, vol. 25, pp. 1097–1105 (2012)
18. Ledig, C., et al.: Photo-realistic single image super-resolution using a generative adversarial network. In: Proceedings of the IEEE Conference on Computer Vision and Pattern Recognition, pp. 4681–4690 (2017)
19. Lin, T.Y., Goyal, P., Girshick, R., He, K., Dollár, P.: Focal loss for dense object detection. In: Proceedings of the IEEE International Conference on Computer Vision, pp. 2980–2988 (2017)
20. Lin, T.-Y., et al.: Microsoft COCO: common objects in context. In: Fleet, D., Pajdla, T., Schiele, B., Tuytelaars, T. (eds.) ECCV 2014. LNCS, vol. 8693, pp. 740–755. Springer, Cham (2014). https://doi.org/10.1007/978-3-319-10602-1_48

21. Liu, W., et al.: SSD: single shot MultiBox detector. In: Leibe, B., Matas, J., Sebe, N., Welling, M. (eds.) ECCV 2016. LNCS, vol. 9905, pp. 21–37. Springer, Cham (2016). https://doi.org/10.1007/978-3-319-46448-0_2
22. Loshchilov, I., Hutter, F.: Decoupled weight decay regularization. arXiv preprint arXiv:1711.05101 (2017)
23. Lukezic, A., Vojir, T., Čehovin Zajc, L., Matas, J., Kristan, M.: Discriminative correlation filter with channel and spatial reliability. In: Proceedings of the IEEE Conference on Computer Vision and Pattern Recognition, pp. 6309–6318 (2017)
24. Magoulianitis, V., Ataloglou, D., Dimou, A., Zarpalas, D., Daras, P.: Does deep super-resolution enhance UAV detection? In: 2019 16th IEEE International Conference on Advanced Video and Signal Based Surveillance (AVSS), pp. 1–6. IEEE (2019)
25. Ren, S., He, K., Girshick, R., Sun, J.: Faster R-CNN: towards real-time object detection with region proposal networks. In: Advances in Neural Information Processing Systems, vol. 28 (2015)
26. Svanström, F., Englund, C., Alonso-Fernandez, F.: Real-time drone detection and tracking with visible, thermal and acoustic sensors. In: 2020 25th International Conference on Pattern Recognition (ICPR), pp. 7265–7272. IEEE (2021)
27. Tian, Z., Shen, C., Chen, H., He, T.: FCOS: fully convolutional one-stage object detection. In: Proceedings of the IEEE/CVF International Conference on Computer Vision, pp. 9627–9636 (2019)
28. Tokekar, P., Vander Hook, J., Mulla, D., Isler, V.: Sensor planning for a symbiotic UAV and UGV system for precision agriculture. IEEE Trans. Rob. **32**(6), 1498–1511 (2016)
29. Tsoukalas, A., Xing, D., Evangeliou, N., Giakoumidis, N., Tzes, A.: Deep learning assisted visual tracking of evader-UAV. In: 2021 International Conference on Unmanned Aircraft Systems (ICUAS), pp. 252–257. IEEE (2021)
30. Unel, F., Ozkalayci, B., Cigla, C.: The power of tiling for small object detection. In: CVF Conference on Computer Vision and Pattern Recognition Workshops (CVPRW), Long Beach, CA. IEEE (2019). https://ieeexplore.ieee.org/document/9025422
31. Wang, Y., Chen, Y., Choi, J., Kuo, C.C.J.: Towards visible and thermal drone monitoring with convolutional neural networks. APSIPA Trans. Signal Inf. Process. **8** (2019)
32. Xing, D., Evangeliou, N., Tsoukalas, A., Tzes, A.: Siamese transformer pyramid networks for real-time UAV tracking. In: Proceedings of the IEEE/CVF Winter Conference on Applications of Computer Vision, pp. 2139–2148 (2022)

Medical Imaging Analysis for Covid-19 - MIACOVID 2022

ILC-Unet++ for Covid-19 Infection Segmentation

Fares Bougourzi[1(✉)], Cosimo Distante[1], Fadi Dornaika[2,3], Abdelmalik Taleb-Ahmed[4], and Abdenour Hadid[4]

[1] Institute of Applied Sciences and Intelligent Systems, National Research Council of Italy, 73100 Lecce, Italy
fares.bougourzi@isasi.cnr.it, cosimo.distante@cnr.it

[2] University of the Basque Country UPV/EHU, San Sebastian, Spain
fadi.dornaika@ehu.eus

[3] IKERBASQUE, Basque Foundation for Science, Bilbao, Spain

[4] Univ. Polytechnique Hauts-de-France, Univ. Lille, CNRS, Centrale Lille, UMR 8520 - IEMN, 59313 Valenciennes, France
Abdelmalik.Taleb-Ahmed@uphf.fr

Abstract. Since the appearance of Covid-19 pandemic, in the end of 2019, Medical Imaging has been widely used to analysis this disease. In fact, CT-scans of the Lung can help to diagnosis, detect and quantify Covid-19 infection. In this paper, we address the segmentation of Covid-19 infection from CT-scans. In more details, we propose a CNN-based segmentation architecture named ILC-Unet++. The proposed ILC-Unet++ architecture, which is trained for both Covid-19 Infection and Lung Segmentation. The proposed architecture were tested using three datasets with two scenarios (intra and cross datasets). The experimental results showed that the proposed architecture performs better than three baseline segmentation architectures (Unet, Unet++ and Attention-Unet) and two Covid-19 infection segmentation architectures (SCOATNet and nCoVSegNet).

Keywords: Covid-19 · Segmentation · Deep learning

1 Introduction

Since the appearance of Covid-19 pandemic (at the end of 2019, Wuhan, China), the world has been facing global crisis. The first step in fighting against this disease is to recognize and evaluate the evolution of the infected persons. In fact, the RT-PCR test is considered as the gold standard to recognize the infected persons. However, it has a considerable false-negative rate, especially in early stages of infection [7]. Since, Covid-19 virus mainly affects the respiratory tract, medical imaging are widely used for detecting the infections. These medical imaging modalities include X-ray scans and the CT-scans [3]. Despite that these scanners

Supported by organization x.

P. L. Mazzeo et al. (Eds.): ICIAP 2022 Workshops, LNCS 13374, pp. 461–472, 2022.
https://doi.org/10.1007/978-3-031-13324-4_39

are available in most hospitals even in the less developed countries, there is a need of an expert radiologist to recognize the infection. To solve this issue, many machine learning approaches have been proposed to automatically recognize the infection without the need of an expert radiologist [4,15]. In addition to the ability of the CT-scans to show the infection even in the early stages, it can be used for more purposes. CT-scans were used to segment the infected parts and can be used directly to estimate the infection severity and percentage [5].

Because the infection has high variability in shape, size, and position, segmentation of Covid-19 infection is a very challenging task. Segmentation of Covid-19 infection from CT scans using artificial intelligence has additional challenges. One of the most difficult challenges is the low intensity contrast between the infection and normal tissue, especially in the early days of infection when the infection appears as ground glass opacity (GGO). However, as the infection spreads, it appears as a mixture of GGO and consolidation or consolidation. In this case, it is difficult to distinguish between consolidation and non-lung tissue, especially when the infection adheres to the lung walls. To overcome this challenge, we proposed the ILC-Unet++ segmentation architecture. Our proposed ILC-Unet++ trains the segmentation of infection and lung simultaneously. The main contributions of this work are:

- We propose an ILC-Unet++ segmentation architecture. Our proposed ILC-Unet++ is designed to segment Covid-19 infection and lung simultaneously. The two segmentation tasks infection and lung of ILC-Unet++ share the encoder and intermediate blocks of Unet++. On the other hand, each task has its own decoder.
- To evaluate the performance of our proposed approach, we used both intra and cross-datasets evaluation scenarios of three datasets where we used all slices of CT scans for training and testing.
- To compare the performance of our approach, we used three baseline architectures (Unet, Att-Unet, Unet++) and two state-of-the-art architectures for Covid-19 segmentation (SCOATNet [17] and nCoVSegNet [8]). The experimental results show the superiority of our proposed architecture compared to the basic segmentation architectures as well as the two state-of-the-art architectures in both intra-database and inter-database evaluation scenarios.

2 Related Work

In the recent decade, Deep learning architectures have become dominant in many computer vision tasks [2] including many medical Imaging tasks [4,5,15]. Since the appearance of Covid-19 pandemic, a lot of works have been proposed to segment the Covid-19 infection.

In [18], Xiangyu Zhao et al. proposed dilated dual attention U-Net (D2A U-Net) framework to automatically segment the lung infection in COVID-19 CT slices. To evaluate the performance of their approach, they took just the infected slices from Segmentation dataset nr. 2 [1] and COVID-19-CT-Seg dataset [10], then they used them as training data. As testing data, they used the 100 slices of COVID-19 CT segmentation [1].

Jiannan Liu et al. proposed two-stage cross-domain transfer learning framework [9]. Their framework consists of two main components. First, they proposed nCoVSegNet, which is a deep learning based approach that exploits attention-aware feature fusion and large receptive fields. Second, they trained nCoVSegNet using cross-domain transfer learning strategy, which makes full use of the knowledge from natural images (i.e., ImageNet) and medical images (i.e., LIDC-IDRI) to boost the final training on CT images with COVID-19 infections. They evaluated their approach using MosMedData dataset [11], which used for the last stage transfer learning (40 slices for training and 10 slices testing), then this model tested on [10] dataset.

In [16], Ruxin Wang et al. proposed encoder-decoder CNN-based framework to segment Covid-19 infection from the CT-scans. Their approach is based on aggregating the peer- and cross-level context-aware learning. To capture the complex structure, they used autofocus module to mine and incorporate multiscale contextual information of peer level. Moreover, they proposed panorama module to capture complementary fine details and semantic information. To evaluate the performance of their approach, Ruxin Wang et al. combined different datasets, MosMedData dataset [10,11], COVID-19 CT segmentation [1] and Segmentation dataset nr. 2 [1], where the slice that do not have infection were removed. Then, the obtained dataset was randomly splitted into training and testing splits (it is not clear if patient independent protocol was respected or not).

Deng-Ping Fan et al. proposed Deep Network (Inf-Net) for COVID-19 Lung Infection segmentation [6]. Their Inf-Net approach uses parallel partial decoder to aggregate the high-level features and generate a global map. Moreover, implicit reverse attention and explicit edge-attention are utilized to model the boundaries and enhance the representations. In addition to Inf-Net, Deng-Ping Fan et al. investigated semi-supervised segmentation strategy which exploits random selected propagation framework. The experimental results showed that using the semi-supervised framework can improve the learning ability and achieve better performance.

3 The Proposed Approach

To segment Covid-19 infection, we proposed ILC-Unet++ architecture. Our ILC-Unet++ is designed to segment both infection and lung areas simultaneously. The goal is to guide the training process to look inside the lung regions and distinguish between the infection tissues (especially in the case of consolidation) and the lung walls. Figure 1 shows the difference between our proposed ILC-Unet++ and Unet++ architecture. As shown in Fig. 1(b), ILC-Unet++ has the same encoder and intermediate layer for Infection and Lung Segmentation tasks, while each task has its own decoder. The goal is to learn high-level features for both task using the encoder. Then the decoders exploit these features to segment the infection and the lung regions independently. To make ILC-Unet++

architecture giving more attention to the infection Segmentation than Lung Segmentation, we used compound loss function which gives 0.7 weight for Infection segmentation loss and 0.3 weight for Lung segmentation loss.

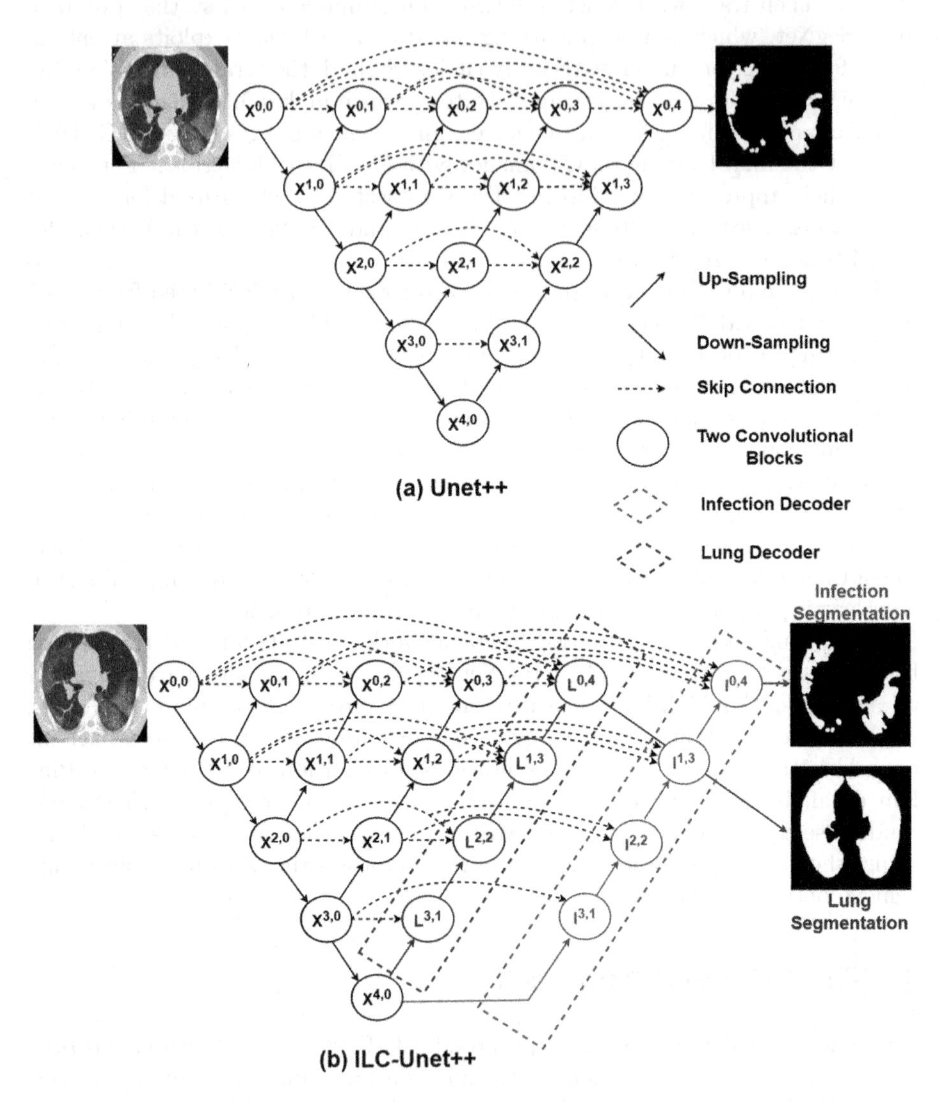

Fig. 1. Comparison of Unet++ (a) and our proposed ILC-Unet++ (b).

In the encoder, we used five Two Convolutional Blocks, extending the number of channels from 3 to 32, 64, 128, 256, and 512, respectively. Similarly, we used five Two Convolutional Blocks that reduce the number of channels in the decoder from 512 to 256, 128, 64, 32, and 1, respectively. The Two Convolutional Blocks consists of two 3 × 3 convolution kernels, each followed by batch normalization and Relu activation function. To compare the performance of the proposed ILC-Unet++ with the baseline architectures, we used the same number of blocks and the number of channels of each block.

4 Performance Evaluation

4.1 Datasets

To evaluate the performance of our proposed approach, we used three publicly available datasets which are COVID-19 CT segmentation [1], Segmentation dataset nr. 2 [1] and COVID-19-CT-Seg dataset [10].

COVID-19 CT segmentation dataset [1] consists of 100 axial CT images (slices) from more than 40 patients with COVID-19 infection. These images were segmented by a radiologist using 3 labels: ground-glass (mask value = 1), consolidation (=2) and pleural effusion (=3). In addition to Covid-19 infection masks, the lung masks were provided.

Segmentation dataset nr. 2 [1] consists of 9 3D CT-scans. Since this dataset was constructed from whole volumes, it contains both positive and negative slices. In total, 829 slices from which 373 slices have been evaluated by a radiologist as positive, then they were segmented. In addition to Covid-19 infection masks, the lung masks were provided.

COVID-19-CT-Seg dataset [10] consists of 20 COVID-19 CT scans. All the cases contain COVID-19 infections. The proportion of infections in the lungs ranges from 0.01% to 59%. This dataset was labelled by different radiologists. Firstly left lung, right lung, and infection were delineated by junior annotators with 1–5 yr experience, then refined by two radiologists with 5–10 yr experience, and finally all the annotations were verified and refined by a senior radiologist with more than 10 yr experience in chest radiology. The whole lung mask includes both normal and pathological regions were labelled. All the annotations were manually performed by ITK-SNAP in a slice-by-slice manner on axial images. In total, there are more than 300 infected slices from more than 1800 slices.

In our experiments, we evaluated both intra-dataset and cross-datasets scenarios. For intra-dataset experiments, we randomly splitted Segmentation dataset nr. 2 [1] and COVID-19-CT-Seg dataset [10] into 70%−30% as training and testing splits, where Patient-Independent is respected in splitting. For the cross-datasets evaluation scenario, we used the trained models on Segmentation dataset nr. 2 (trained on the training data 70%) and test it on COVID-19 CT segmentation [1] and COVID-19-CT-Seg dataset [10], respectively. Similarly, we used the trained models on COVID-19-CT-Seg dataset [10] (trained on the training data 70%) and test it on COVID-19 CT segmentation [1] and COVID-19 CT segmentation dataset [1], respectively.

4.2 Evaluation Metrics

To evaluate the performance of different approaches, we used five evaluation metrics which are: F1-score, Dice-score, Intersection over Union (IoU), sensitivity (Sens), specificity (Spec) and precision (Prec).

F1-score, Intersection over Union (IoU), sensitivity (Sens), specificity (Spec) and precision (Prec) are defined as following.

$$\text{F1-score} = 2 \cdot \frac{TP}{2 \cdot TP + FP + FN} \tag{1}$$

$$IoU = \frac{TP}{(TP + FP + FN)} \tag{2}$$

$$Sens = Recall = \frac{TP}{TP + FN} \tag{3}$$

$$Spec = \frac{TN}{FP + TN} \tag{4}$$

$$Prec = \frac{TP}{TP + FP} \tag{5}$$

where TP is the True Positives, TN is the True Negatives, FP is the False Positives and FN is the False Negative.

The above metrics are micro metrics, i.e.: TP, TN, FP and FN were calculated for all testing images then these metrics were calculated using them. However, Dice-score is the macro version of $F1 - score$. For N testing images it is defined by:

$$\text{Dice-score} = \frac{1}{N}\sum_{i=1}^{N} 2 \cdot \frac{TP_i}{2 \cdot TP_i + FP_i + FN_i} \tag{6}$$

where TP_i, TN_i, FP_i and FN_i are True Positives, True Negatives, False Positives and False Negative for the ith image, respectively. In the experimental results, it is normal to have low dice-score average since in the three datasets there is a large number of uninfected slices, where totally correct prediction of the masks (black mask) will give 0 dice score for these slices which will be included in calculating the average.

4.3 Experimental Setup

For deep learning training and testing, we used the Pytorch [13] library with NVIDIA GPU Device GeForce TITAN RTX 24 GB. The batch size used consists of 6 images, and as a loss function, we used BCE loss function for lung and infection segmentation. We trained the networks for 60 epochs. The initial learning rate is 0.01, which decays by 0.1 after 30 epochs, followed by another decay of 0.1 after 50 epochs. Furthermore, we used active data augmentation techniques which are: rotation using random angle between $-35°$ to $35°$, and random horizontal and vertical flipping.

4.4 Experimental Results on Intra-dataset Scenario

In this section, we evaluate the performance of the proposed ILC-Unet++ and compare it with Unet [14], Att-Unet [12], Unet++ [19], SCOATNet [17] and nCoVSegNet [8]. Tables 1 and 2 show the experimental results using Segmentation dataset nr. 2 [1] and COVID-19-CT-Seg dataset [10], respectively. The results show that the proposed approach achieve better performance than the baseline segmentation architecture, as well as the two proposed architectures for Covid-19 segmentation (SCOATNet [17] and nCoVSegNet [8]). In more details, compared with the original Unet++, the proposed ILC-Unet++ approach improves the micro F1-score results by 12.4 and 14.33 for Segmentation dataset nr. 2 [1] and COVID-19-CT-Seg dataset [10], respectively. On the other hand, we notice that Att-Unet architecture achieved the second best performance on Segmentation dataset nr. 2 [1]. While COVID-19-CT-Seg dataset [10], the proposed architecture in [17] (SCOATNet) achieved the second best performance. Our proposed approach achieved better performance than these architectures by good margin (7.75% and 3.69% for F1-score metric for Segmentation dataset nr. 2 [1] and COVID-19-CT-Seg dataset [10], respectively).

In order to show the effectiveness of the proposed ILC-Unet++, we visualized the segmented masks using Unet [14], Att-Unet [12], Unet++ [19] and ILC-Unet++ as shown in Fig. 2. From the first column, we notice that the proposed approach has the ability to segment the infection in both lungs better than the baseline architectures. However, it is not capable to segment the tiny infections in the upper right lung. From the second column, we can notice that both Unet and Unet++ can wrongly classify the lung boundary tissues as infection, due to low contrast in the boundaries, especially in the consolidation case. Att-Unet also suffers from the same issue, but its mis-segmented boundaries are smaller than the case of using Unet and Unet++. In contrast, our proposed ILC-Unet++ can cope with this issue. From the third example (slice in the bottom of lung scan), we notice that Unet, Att-Unet and Unet++ failed to identify the infection probably because these models are not aware about the lung and non-lung tissues since small part of lung appears in this case. In contrast, our approach is able to distinguish between the lung and non-lung tissues, which lead to accurate Covid-19 infection segmentation. The fourth example shows an example where most of lung regions are infected. In this example, we notice that only our proposed approach has the ability to identify accurately the Covid-19 infection. In contrast, the baseline architectures failed.

4.5 Experimental Results on Cross-Datasets Scenario

In order to compare between different segmentation architectures, it is important to study their generalization ability in cross-datasets scenario. In fact, cross-datasets experiment plays a crucial role to evaluate the effectiveness of each architecture in real scenarios. Tables 3 and 4 show the results of the cross-datasets evaluations by using the best trained model on Segmentation dataset nr. 2 [1] and tested on COVID-19 CT segmentation dataset [1] and COVID-19-CT-Seg

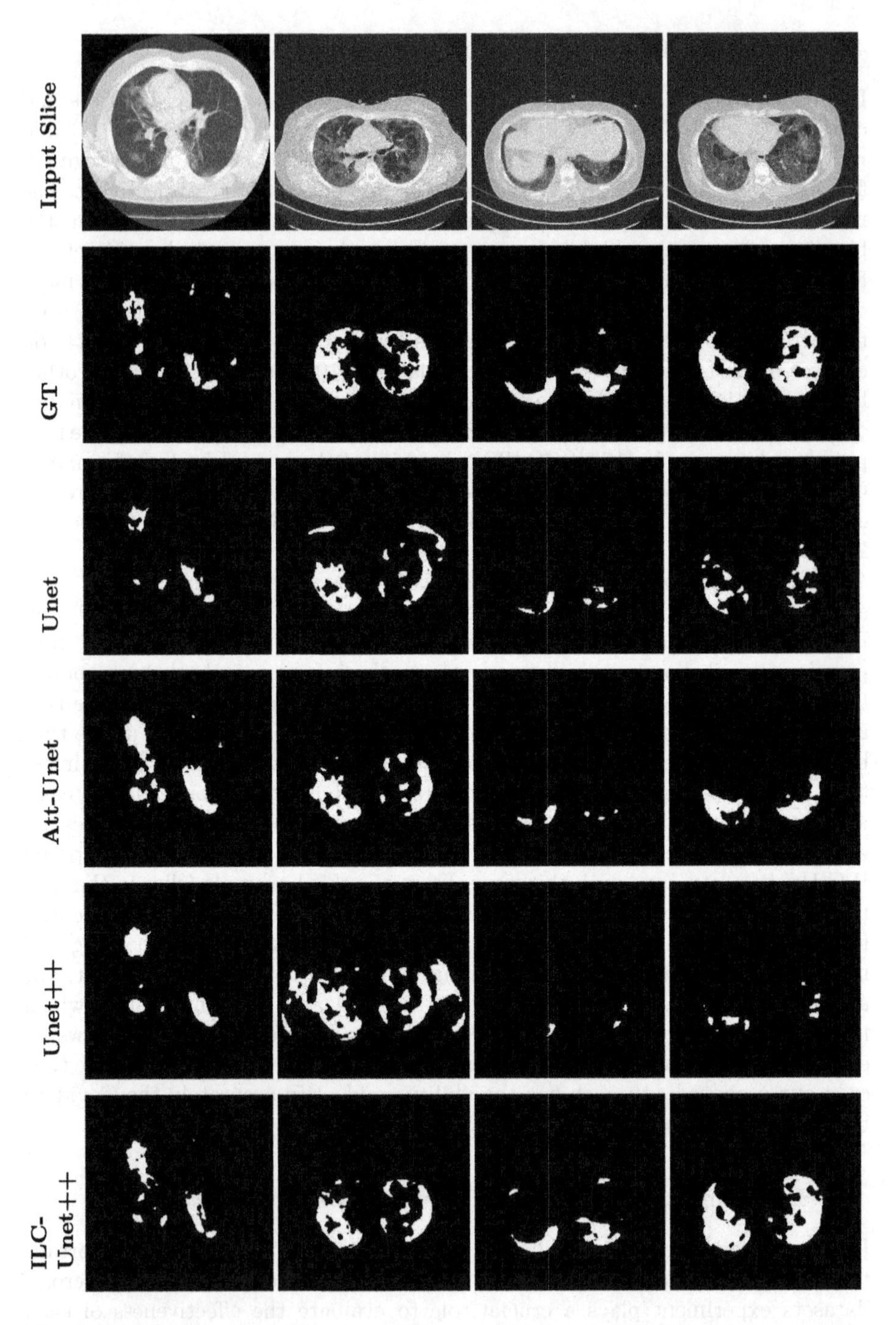

Fig. 2. Visual comparison of a segmentation model trained with different segmentation architectures. The first two column show visualization examples from the validation data of Segmentation dataset nr. 2 [1] and the third and fourth columns show visualization examples from the validation data of COVID-19-CT-Seg dataset [10]

Table 1. Experimental Results of Segmentation dataset nr. 2 [1] for Intra-dataset Scenario.

Model	F1-S	D-S	IoU	Sens	Spec	Prec
SCOATNet [17]	0.660301	0.31233	0.492872	0.622754	0.993232	0.702665
nCoVSegNet [8]	0.663013	0.265537	0.495901	0.617847	0.993685	0.715303
Unet	0.615244	0.303388	0.444298	0.532386	0.994908	0.728646
Att-Unet	0.670075	0.293276	0.503844	0.592339	0.995489	0.771296
UNet++	0.623506	0.333576	0.452967	0.653102	0.986952	0.573475
ILC-Unet++	0.747501	0.381538	0.596807	0.662101	0.997190	0.858194

Table 2. Experimental Results of COVID-19-CT-Seg dataset [10] for Intra-dataset Scenario.

Model	F1-S	D-S	IoU	Sens	Spec	Prec
SCOATNet [17]	0.665088	0.335648	0.498226	0.529936	0.999135	0.892779
nCoVSegNet [8]	0.620931	0.338758	0.450253	0.549374	0.997010	0.713921
Unet	0.618174	0.336321	0.447361	0.485517	0.998842	0.850579
Att-Unet	0.637974	0.354807	0.468401	0.507855	0.998856	0.857737
UNet++	0.558685	0.332053	0.387622	0.409186	0.999244	0.880318
ILC-Unet++	0.701932	0.369856	0.540751	0.627334	0.997825	0.796665

dataset [10], respectively. From the results of these two tables, we notice that our approach achieved the best cross-datasets results in both experiments. Compared with Unet++, our approach improved the performance by 10.5%, and 2.3% respectively. On the other hand, we notice that the second best performance was achieved by different architectures (Att-Unet from Table 3 and Unet++ from Table 4), which means that the other architectures performance changes from set of conditions to another. On the other hand, our proposed approach performs consistently in different conditions.

Table 3. Cross Datasets trained using Segmentation dataset nr. 2 [1] and tested using COVID-19 CT segmentation dataset [1]

Model	F1-S	D-S	IoU	Sens	Spec	Prec
SCOATNet [17]	0.400144	0.354214	0.250113	0.434511	0.945004	0.370816
nCoVSegNet [8]	0.440295	0.339626	0.282294	0.392158	0.970969	0.501905
Unet	0.446756	0.349138	0.287628	0.484711	0.948887	0.414314
Att-Unet	0.545928	0.389001	0.374423	0.496468	0.983022	0.685669
UNet++	0.451955	0.358118	0.291952	0.610437	0.918626	0.358803
ILC-Unet++	0.557752	0.395666	0.386724	0.449845	0.987824	0.733762

Table 4. Cross Datasets results by using the trained model of Segmentation dataset nr. 2 [1] and tested using COVID-19-CT-Seg dataset [10]

Model	F1-S	D-S	IoU	Sens	Spec	Prec
SCOATNet [17]	0.375091	0.275490	0.450838	0.493187	0.998321	0.858192
nCoVSegNet [8]	0.564292	0.279693	0.393041	0.452141	0.997908	0.750435
Unet	0.607886	0.284599	0.436664	0.490115	0.998297	0.800159
Att-Unet	0.515841	0.277080	0.347565	0.406313	0.997649	0.706211
UNet++	0.623017	0.277902	0.463858	0.490003	0.996591	0.706500
ILC-Unet++	0.646257	0.290068	0.477385	0.510575	0.999033	0.880153

Tables 5 and 6 show the cross-datasets results by using the best trained model on COVID-19-CT-Seg dataset [10]and tested on COVID-19 CT segmentation dataset [1] and Segmentation dataset nr. 2 [1] respectively. From the results of these two tables, we notice that our approach achieved the best cross-datasets results in both experiments. Compared with Unet++, our approach improved the performance by 22%, and 8% respectively. On the other hand, we notice that the second best performance achieved by different architectures (Unet from Table 5 and Att-Unet from Table 6), which means that the other architectures performance changes from set of conditions to another.

Table 5. Cross Datasets results by using the trained model of COVID-19-CT-Seg dataset [10] and tested using COVID-19 CT segmentation dataset [1])

Model	F1-S	D-S	IoU	Sens	Spec	Prec
SCOATNet [17]	0.436403	0.430322	0.279102	0.305240	0.993014	0.765222
nCoVSegNet [8]	0.619926	0.543358	0.449198	0.506747	0.541144	0.725555
Unet	0.625757	0.507644	0.455347	0.516686	0.989951	0.793199
Att-Unet	0.512987	0.463873	0.344978	0.383059	0.991765	0.776296
UNet++	0.477812	0.421865	0.313898	0.446254	0.968546	0.514173
ILC-Unet++	0.695481	0.613179	0.533132	0.708385	0.975478	0.683039

Table 6. Cross Datasets trained using COVID-19-CT-Seg dataset [10]) and tested usingSegmentation dataset nr. 2 [1]

Model	F1-S	D-S	IoU	Sens	Spec	Prec
SCOATNet [17]	0.448455	0.188137	0.289037	0.569311	0.991934	0.369925
nCoVSegNet [8]	0.557181	0.217178	0.386175	0.649974	0.994317	0.487573
Unet	0.598438	0.236242	0.426979	0.531154	0.997970	0.685240
Att-Unet	0.601942	0.243752	0.417384	0.633853	0.998127	0.537926
UNet++	0.535670	0.250797	0.365812	0.510533	0.996709	0.563409
ILC-Unet++	0.614716	0.255070	0.443747	0.866751	0.992070	0.476236

5 Conclusion

In this paper, we proposed a Covid-19 infection segmentation approach from CT-scans. Our proposed ILC-Unet++ is based on training both Covid-19 infection and lungs regions segmentation, simultaneously. To prove the efficient of our approach, we evaluated its performance in both within and cross-datasets scenarios. In addition, we compared the performance of our approach with three baseline architectures (Unet, Att-Unet, Unet++) and two state-of-the-art architectures for Covid-19 Segmentation (SCOATNet and nCoVSegNet). The experimental results show the superiority of our proposed architecture compared with the baseline segmentation architectures as well as the two state of the art architectures in both intra and cross-datasets evaluation scenarios.

Acknowledgment. The authors would like to thank Arturo Argentieri from CNR-ISASI Italy for his support on the multi-GPU computing facilities.

References

1. COVID-19 CT-scans segmentation datasets. http://medicalsegmentation.com/covid19/. Accessed 18 Aug 2021
2. Bougourzi, F., Dornaika, F., Taleb-Ahmed, A.: Deep learning based face beauty prediction via dynamic robust losses and ensemble regression. Knowl.-Based Syst. **242**, 108246 (2022). https://www.sciencedirect.com/science/article/pii/S0950705122000740
3. Bougourzi, F., Contino, R., Distante, C., Taleb-Ahmed, A.: CNR-IEMN: a deep learning based approach to recognize COVID-19 from CT-SCAN. In: 2021 IEEE International Conference on Autonomous Systems (IEEE ICAS) (2021)
4. Bougourzi, F., Contino, R., Distante, C., Taleb-Ahmed, A.: Recognition of COVID-19 from CT scans using two-stage deep-learning-based approach: CNR-IEMN. Sensors **21**(17), 5878 (2021). https://doi.org/10.3390/s21175878. https://www.mdpi.com/1424-8220/21/17/5878
5. Bougourzi, F., Distante, C., Ouafi, A., Dornaika, F., Hadid, A., Taleb-Ahmed, A.: Per-COVID-19: a benchmark dataset for COVID-19 percentage estimation from CT-scans. J. Imaging **7**(9), 189 (2021). https://doi.org/10.3390/jimaging7090189. https://www.mdpi.com/2313-433X/7/9/189
6. Fan, D.P., et al.: Inf-net: automatic COVID-19 lung infection segmentation from CT images. IEEE Trans. Med. Imaging **39**(8), 2626–2637 (2020). https://doi.org/10.1109/TMI.2020.2996645
7. Kucirka, L.M., Lauer, S.A., Laeyendecker, O., et al.: Variation in false-negative rate of reverse transcriptase polymerase chain reaction-based SARS-CoV-2 tests by time since exposure. Ann. Internal Med. **173**(4), 262–267 (2020). https://doi.org/10.7326/M20-1495. https://www.acpjournals.org/doi/full/10.7326/M20-1495
8. Liu, J., et al.: COVID-19 lung infection segmentation with a novel two-stage cross-domain transfer learning framework. Med. Image Anal. **74**, 102205 (2021)
9. Liu, J., Dong, B., Wang, S., et al.: COVID-19 lung infection segmentation with a novel two-stage cross-domain transfer learning framework. Med. Image Anal. **74**, 102205 (2021). https://doi.org/10.1016/j.media.2021.102205

10. Ma, J., Wang, Y., An, X., et al.: Toward data efficient learning: a benchmark for COVID 19 CT lung and infection segmentation. Med. Phys. **48**, 1197–1210 (2021). https://doi.org/10.1002/mp.14676. https://ui.adsabs.harvard.edu/abs/2021MedPh.48.1197M
11. Morozov, S., Andreychenko, A., Pavlov, N., et al.: MosMedData: Chest CT Scans with COVID-19 Related Findings Dataset. Radiol. Imaging (2020). https://doi.org/10.1101/2020.05.20.20100362. http://medrxiv.org/lookup/doi/10.1101/2020.05.20.20100362
12. Oktay, O., Schlemper, J., Folgoc, L.L., et al.: Attention U-Net: Learning Where to Look for the Pancreas. arXiv:1804.03999, May 2018. http://arxiv.org/abs/1804.03999
13. Paszke, A., et al.: Pytorch: an imperative style, high-performance deep learning library. In: Advances in Neural Information Processing Systems, pp. 8026–8037 (2019)
14. Ronneberger, O., Fischer, P., Brox, T.: U-Net: convolutional networks for biomedical image segmentation. In: Navab, N., Hornegger, J., Wells, W.M., Frangi, A.F. (eds.) MICCAI 2015. LNCS, vol. 9351, pp. 234–241. Springer, Cham (2015). https://doi.org/10.1007/978-3-319-24574-4_28
15. Vantaggiato, E., Paladini, E., Bougourzi, F., Distante, C., Hadid, A., Taleb-Ahmed, A.: COVID-19 recognition using ensemble-CNNs in two new chest x-ray databases. Sensors **21**(5), 1742 (2021). https://doi.org/10.3390/s21051742. https://www.mdpi.com/1424-8220/21/5/1742
16. Wang, R., Ji, C., Zhang, Y., Li, Y.: Focus, fusion, and rectify: context-aware learning for COVID-19 lung infection segmentation. IEEE Trans. Neural Netw. Learn. Syst. **33**(1), 12–24 (2022). https://doi.org/10.1109/TNNLS.2021.3126305
17. Zhao, S., et al.: SCOAT-Net: a novel network for segmenting COVID-19 lung opacification from CT images. Pattern Recognit. **119**, 108109 (2021)
18. Zhao, X., Zhang, P., Song, F., et al.: D2A U-Net: automatic segmentation of COVID-19 CT slices based on dual attention and hybrid dilated convolution. Comput. Biol. Medi. **135**, 104526 (2021). https://doi.org/10.1016/j.compbiomed.2021.104526. https://www.sciencedirect.com/science/article/pii/S0010482521003206
19. Zhou, Z., Rahman Siddiquee, M.M., Tajbakhsh, N., Liang, J.: UNet++: a nested U-Net architecture for medical image segmentation. In: Stoyanov, D., et al. (eds.) DLMIA/ML-CDS -2018. LNCS, vol. 11045, pp. 3–11. Springer, Cham (2018). https://doi.org/10.1007/978-3-030-00889-5_1

Revitalizing Regression Tasks Through Modern Training Procedures: Applications in Medical Image Analysis for Covid-19 Infection Percentage Estimation

Radu Miron[1,2] and Mihaela Elena Breaban[1(✉)]

[1] Alexandru Ioan Cuza University, Faculty of Computer Science, Iasi, Romania
pmihaela@info.uaic.ro
[2] SenticLab, Iasi, Romania
radu.miron@senticlab.com

Abstract. In order to establish the correct protocol for COVID-19 treatment, estimating the percentage of COVID-19 specific infection within the lung tissue can be an important tool. This article describes the approach we used in order to estimate the COVID-19 infection percentage on lung CT scan slices within the Covid-19-Infection-Percentage-Estimation-Challenge. Our method frames the regression problem as a multi-tasking process and is based on modern training pipelines and architectures that correspond to state of the art models on image classification tasks. It obtained the best score on the validation dataset and ranked third in the testing phase within the competition.

Keywords: Deep regression · Multi-tasking · Covid-19 · Medical image analysis

1 Introduction

The COVID-19 pandemic has become a healthcare crisis around the world since its start in 2019 [1]. Quick discovery of the infected patients is key to positive outcome. Methods like RT-PCR, X-Ray or CT-scans are the to go choice for diagnosis of COVID-19 infection. The last two methods mentioned not only can correctly diagnose a patient with the infection, but they can also give insights into the stage the disease is progressing. The downside of these two methods is the burden an expert radiologist might be put through in order to evaluate a great amount of X-rays or CTs [2]. CT scans have a clear advantage in comparison with X-rays due to their more detailed structure. Signs of early or late stage of infection can be easily detected in CT scans, thus making the decision to follow a certain protocol an easier task for the doctors. Having this into consideration, several AI solutions have been proposed in order to come to the aid of radiologists. The Covid-19 Infection Percentage Estimation competition [3,4]

P. L. Mazzeo et al. (Eds.): ICIAP 2022 Workshops, LNCS 13374, pp. 473–482, 2022.
https://doi.org/10.1007/978-3-031-13324-4_40

establishes a new benchmark that may offer real help into depicting the evolution stage of COVID-19 infection. The organizers have publicly released a dataset which consists of several CT scan slices and the corresponding Covid-19 infection percentage. In the following, we present our solution to the challenge which, on the validation set beats the second place by a large margin and improves with more than 1 the MAE score of the baseline solution provided by the organizers of the competition in [3]. We make the code and models publicly available[1].

2 Related Work

Regression analysis taking as input image data is much less reported in the literature compared to classification, object detection or segmentation tasks, especially when it comes to the medical domain; nevertheless, it can greatly benefit from pre-trained deep models developed to solve the most popular tasks in computer vision. Such methods, that use deep learning (mostly convolutional networks) to build a model able to estimate a numerical response variable given an input image, are generally called deep regression methods.

The latest advancements recorded in deep regression seem to be mostly related to a few datasets that were published as part of some challenges or for benchmarking purposes.

In this regard, Age Estimation or Attractiveness Estimation given as input facial images, attracted a great deal of interest. The authors of [5] improve several results on datasets like ChaLearn (2015/2016) [12], MORPH [13], FGNET [17] or UTKFace [14] for age estimation and SCUT-FBP [15] or CFD [16] for face attractiveness. They jointly learn to maximize the similarity between the target distribution and the generated distribution at training stage and to regress a real number in an end-to-end fashion. The output value for an input x is quantized into a range of possible values instead of just one label. The authors also mention that they pretrain their model on a large corpus of facial images before training on the downstream tasks. The method proposed by these authors is an extension of [6]. For attractiveness estimation no other model was found to report performances on benchmarks. In [7], the authors propose again a multitasking approach, but this time they use extra-training data and infer a posterior distribution for the ages of images given the results of multiple observed events of an annotation process. They use ordinal hyperplane [18] methods which are furthered mapped into posterior distribution using a linear layer with softmax activation. In [8], the authors extend the regression task into binary tasks used for rank prediction, where each task indicates whether the predicted output lies in a certain range or not. For robust results, the authors use for the binary tasks the same weight parameters, but different bias ones. They use weighted cross-entropy to optimize the learning process. In [9] the authors give a two point representation to the age, and consider it as an approximation of adjacent ends of certain bins which split equally the entire domain of ages. Instead of learning directly the age, the model learns the distribution of probability of the

[1] https://github.com/SENTICLABresearch/Covid-19_Percentage_Estimation.

input to be in a certain bin. They also use multi-tasking in an end-to-end fashion, regressing from the learnt distribution the age through a linear layer and a softmax activation function. In [10], the authors use a GAN-like architecture to reconstruct facial images with certain ages. They use the sum of 4 losses in order to finally regress the age from an image. In [11], the authors use an approach similar to Regression via Classification, but instead of projecting the continuous target values into one discrete representation through bins, they do it in multiple ways.

A systematic evaluation and statistical analysis of vanilla deep regression, (i.e. convolutional neural networks with a linear regression top layer) is presented in [30]. The authors use as base architectures VGG-16 [32] and ResNet-50 in the context of three distinct problems: head-pose estimation, facial land- mark detection and human-body pose estimation. They analyze the impact of different network optimizers, batch sizes, batch normalization, dropout and they compare three distinct loss functions: Mean Squared Error, Mean Absolut Error and the Huber loss.

Related to the medical domain, the most popular regression task is estimating bone age from pediatric hand xRays, which was framed as a challenge in 2017, releasing a dataset developed by Stanford University and the University of Colorado that was annotated by multiple expert observers [29]. The best approaches made use of well known pretrained architectures as Inception3 and ResNet-50 along with data augmentation and ensembling.

3 Investigated Approaches

Motivated by the recent great results obtained for image classification which are mainly due to new architectures and new training techniques, we try to revitalize deep regression by resorting to these new state of the art methods in classification.

We have experimented with several neural networks that are aimed at feature extraction: *ResNet* [19], *ResNeXt* [20], *SE-ResNet* [21], *EfficientNet* [22], *SK-ResNeXt* [23] and *ResNeSt* [24]. Jointly, we experimented with several methods for adjusting the final layers of our models.

The first method just adds a linear layer on top of the feature extractor, which outputs a single number, between 0 and 100 (the infection percentage). For this approach we used the smooth *L1* loss, with parameter $\beta = 1$.

The second method adds on top of the feature extractor a linear layer with 101 output cells, followed by a softmax activation layer, thus predicting the probability distribution of each integer percentage. On top of the linear layer we put another layer with 101 input features (the output of the previous step) and 1 output feature (the number we must regress from the input image). We use as loss function the sum of two losses:

$$loss_1 = L1_{smooth}(\sum_{i=0}^{100} i \cdot softmax(f_1)^{(i)}, gt) \tag{1}$$

and

$$loss2 = L1_{smooth}(f_2(0), gt). \tag{2}$$

The first loss is used in order to learn the expectancy of the number we must regress, whereas the second loss is the loss used in the first method. gt stands for ground truth for the current input image. $softmax(f_1)^{(i)}$ represents the ith element of the output of the first added layer on top of the feature extractor after softmax application. $f_2(0)$ represents the output number of the second added layer.

For a third method we add another trick, where instead of approximating the probability expectation, we approximate the probability distribution itself through the KL-divergence loss. If an input image has $p\%$ target, the distribution will be

$$P(output = y) = \begin{cases} 0 \text{ if } y \leq p-3 \text{ or } y \geq p+3 \\ 0.6 \text{ if } y = p \\ 0.15 \text{ if } y = p-1 \text{ or } y = p+1 \\ 0.05 \text{ if } y = p-2 \text{ or } y = p+2 \end{cases} \tag{3}$$

Thus, other than the two losses presented in the previous method, we compute the third loss by being the KL-divergence between the predicted probability distribution and the target one. The final loss will be the sum of the three losses. We also try to improve the power of the feature extractor, in accordance with the idea presented in [5], and we replace its global average pooling with a hybrid pooling mechanism.

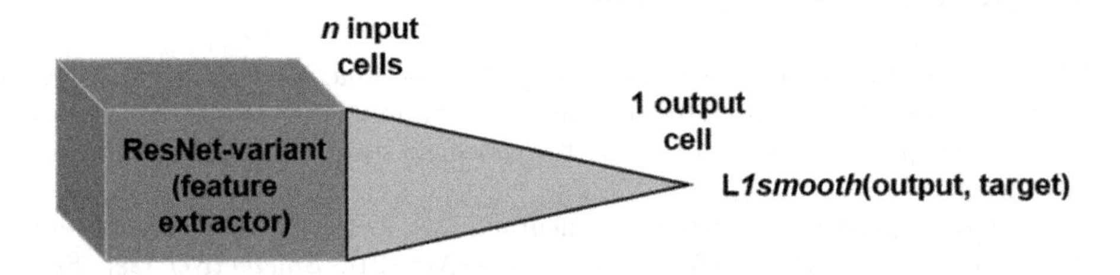

Fig. 1. The first approach, with a simple linear layer with exactly one output cell

3.1 Training Procedure

We believe this step is very important, as we bring modern training techniques used for image classification tasks into regression tasks.

As training procedure we use *SAM + SGD* [25] as optimizer and cosine annealing with warm-up [26] as learning rate scheduler. The initial learning rate is *1e-3*. We train every model for 50 epochs. In order to avoid overfitting, we use Random Augmentation [27] as a strong regularization with the following list of augmentations: rotation between 0 and 30 °C, Color, Contrast, Brightness,

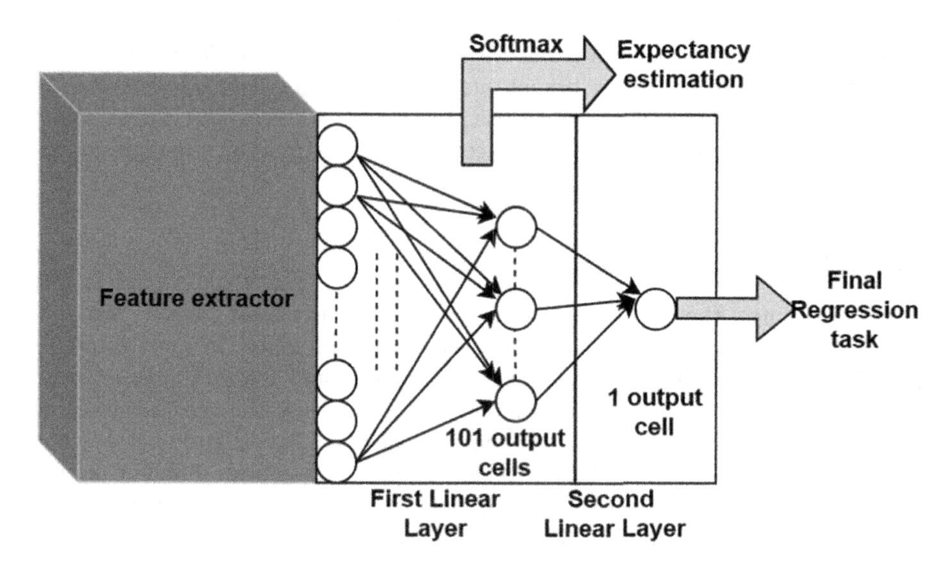

Fig. 2. The second approach, with two linear layers, learning from two tasks

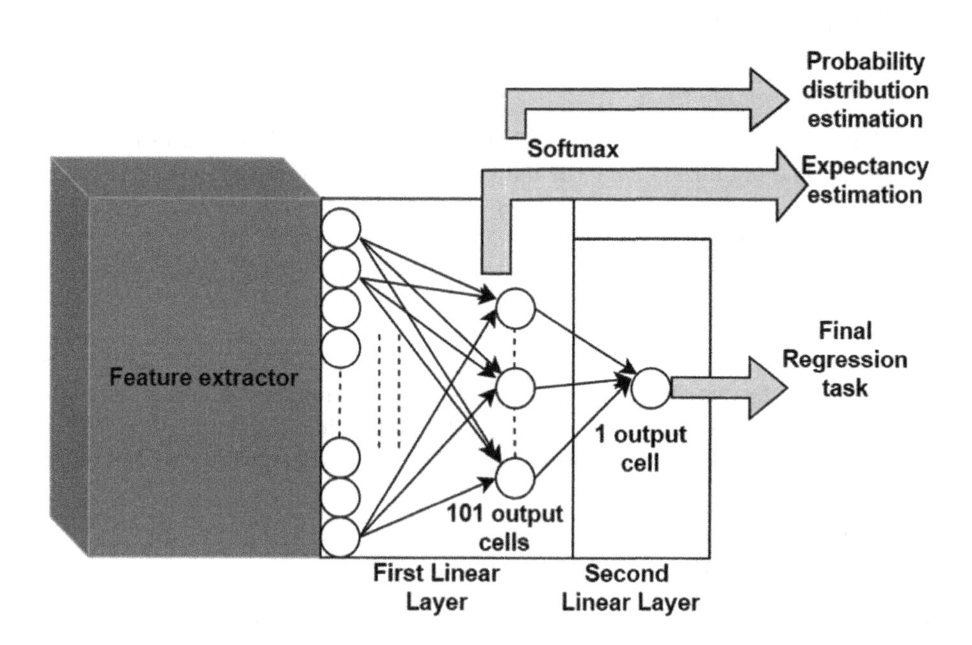

Fig. 3. The third approach, with two linear layers, learning from three tasks

Sharpness, ShearX, ShearY, Cutout, TranslateX, TranslateY. We do not rescale the input and keep the original size of 512×512.

Out of all the feature extractors we tried, we notice that *SK-ResNeXt* and *ResNeSt* (both are based on branch attention [28]) give the best results, no matter what last layer method we use. We stick with *ResNeSt* for the final architecture. The final ensemble also contains a few *ResNeXt* models, as we noticed it boosted the score a little bit more.

3.2 Inference Procedure

During inference, the output is always rounded to the closest integer. We only consider the output of the regression task during inference. We use ensemble models, gathering predictions from models trained on 5 folds we created from the available training set. For each model trained, we use the checkpoints from the last 5 epochs at inference time. We noticed the results are slightly better when we use just two of our folds. When combining predictions from different models into a single prediction, we simply compute the mean of the predictions and round it to the closest integer. The decision to round the final result was made upon small improvements of 0.02 across several submissions on the validation dataset. We also notice that the results get a little bit better if we combine *ResNeSt* models with one *ResNeXt* model.

4 Experimental Analysis

4.1 Dataset

The dataset provided in the competition consists of a selection of slices from 183 CT scans. The organizers split the data in 3 subsets as follows: the training set contains 4355 images, the validation set contains 1301 images and the test set is made of 4449 images.

The organizers of the competition stress the importance of COVID-19% estimation in order to establish the severity of the case. Detailed description on how the dataset was annotated are given in [3].

We split the training set into 5 folds, in a stratified manner taking into consideration the distribution of the labels Normal, Minimal, Moderate, Extent, Severe, and Critical, as described in [3]. This experimental setup, besides its role in model tuning, is intended at building a stable ensemble for the inference phase.

4.2 Ablation Study

Table 1 presents the results on the validation set obtained with the various setups for individual models and training procedures that were described above. The three different setups for addressing regression, which were illustrated in Figs. 1, 2 and 3 and correspond to the single-task method and to the multi-task methods

Table 1. Results on validation dataset, standalone models

Model	Method	MAE
ResNeXt50 (1)	5 folds, 30 epochs, only the regression task	4.601076
ResNeXt50 (2)	2 folds, 30 epochs, only the regression task	4.521138
ResNeSt50 (3)	5 folds, 30 epochs, only the regression task	4.654881
ResNeSt50 (4)	2 folds, 30 epochs, only the regression task	4.516526
ResNeXt50 (5)	2 folds, 50 epochs, 2 tasks, without hybrid pooling	4.554189
ResNeSt50 (6)	2 folds, 50 epochs, 2 tasks, without hybrid pooling	4.343582
ResNeSt50 (7)	2 folds, 50 epochs, 2 tasks, with hybrid pooling	**4.285934**
EfficientNet-b4 (8)	2 folds, 50 epochs, 2 tasks, with hybrid pooling	4.8701
SE-ResNet50 (9)	2 folds, 50 epochs, 2 tasks, with hybrid pooling	4.797079
SK-ResNeXt50 (10)	2 folds, 50 epochs, 2 tasks, with hybrid pooling	4.405842
ResNeSt101 (11)	2 folds, 50 epochs, 2 tasks, with hybrid pooling	4.33359
ResNeSt50 (12)	2 folds, 50 epochs, 3 tasks, without hybrid pooling	4.408916
ResNeSt50 (13)	2 folds, 50 epochs, 3 tasks, with hybrid pooling	4.335127

with, respectively, 2 and 3 objectives, are dedicated three different regions in the table, as demarcated by horizontal lines.

We can notice that *ResNeSt* and *ResNeXt* compete on par when designed with the most simple method of training, that of adding only a linear layer with one output cell (the model in Fig. 1). We can also conclude that using only 2 folds (carefully selected) out of the 5 constructed, improves the results. For the further experiments we only use the two folds that provided the best results from the first 4 experiments. From model (5) we can see that adding extra tasks for *ResNeXt*, does not bring any improvements, whereas for *ResNeSt* (model (6)), the improvements are clear. Using hybrid pooling, instead of global average pooling before the added linear layers, also adds some improvement to the overall result (model 7 vs. model 6). Adding the third task to the training procedure does not seem to bring benefits over the 2-task models, but brings small improvements when used in an ensemble. Hybrid pooling, again, brings benefits (model 13 over 12).

After deciding which were the best models, we started assembling models to improve the overall result. The result obtained on the validation set, which placed us on the first position in the competition in the validation phase, as well as the constituents of the ensemble are reported in Table 2. This ensemble was used in the test phase where it recorded a 4.61 MAE on the test set, being ranked third in the competition.

Self-supervision can be further used to improve the results [31]. We applied pseudo-labeling and extended the original training set with the inclusion of the validation dataset for which we use instead of the ground truth (which is not available) the percentages predicted by the best ensemble model. All models

Table 2. Results on validation dataset, best ensemble

Model	Method	MAE
Models (5, 7, 11, 13)	Each model trained with the configs mentioned in Table 1	**4.171407**

retrained on the extended dataset provided better results compared with their counterpart trained just with the training set, as illustrated in Table 3.

Table 3. Results on validation dataset, best ensemble using self-supervision

Model	Method	MAE
ResNeSt50	2 folds, trained for 50 epochs, 2 tasks, with hybrid pooling	4.189854
ResNeSt50	2 folds, trained for 50 epochs, 3 tasks, with hybrid pooling	4.172175
ResNeSt101	2 folds, trained for 50 epochs, 2 tasks, with hybrid pooling	4.156034
ResNeSt200	2 folds, trained for 50 epochs, 2 tasks, with hybrid pooling	4.448885
ResNeSt101	Same configuration but rounding the percentage at inference	**4.133743**

We noticed during our trials to ensemble the new models that no ensemble is able outperform the best single model trained on the extended training set using pseudo-labeling (self-supervision). Since self-supervision was declared as illegitimate in the competition (even when involving only the validation set) we cannot report results on the test set for these experiments.

5 Conclusions

Deep regression methods, built on existing deep learning models pre-trained for classification tasks in computer vision, may be important tools for assisting medical diagnosis. In this context, re-framing the regression task as multi-task learning proves once again to bring a significant increase in performance.

Acknowledgement. This paper is partially supported by the Competitiveness Operational Programme Romania under project number SMIS 124759 - RaaS-IS (Research as a Service Iasi).

References

1. Gavriatopoulou, M., et al.: Organ-specific manifestations of COVID-19 infection. Clin. Exp. Med. **20**(4), 493–506 (2020). https://doi.org/10.1007/s10238-020-00648-x

2. Kucirka, L.M., Lauer, S.A., Laeyendecker, O., Boon, D., Lessler, J.: Variation in false-negative rate of reverse transcriptase polymerase chain reaction-based SARS-CoV-2 tests by time since exposure. Ann. Intern. Med. **173**, 262–267 (2020)
3. Bougourzi, F., Distante, C., Ouafi, A., Dornaika, F., Hadid, A., Taleb-Ahmed, A.: Per-COVID-19: a benchmark dataset for COVID-19 percentage estimation from CT-scans. J. Imaging. **7**(9), 189 (2021). https://doi.org/10.3390/jimaging7090189
4. Vantaggiato, E., Paladini, E., Bougourzi, F., Distante, C., Hadid, A., Taleb-Ahmed, A.: Covid-19 recognition using ensemble-CNNs in two new chest x-ray databases. Sensors **21**(5), 1748 (2021)
5. Gao, B.-B., Liu, X., Zhou, H.-Y., Wu, J., Geng, X.: Learning Expectation of Label Distribution for Facial Age and Attractiveness Estimation. CoRR,abs/2007.01771 (2020)
6. Gao, B.B., Xing, C., Xie, C.W., Wu, J., Geng, X.: Deep Label Distribution Learning with Label Ambiguity, CoRR, abs/1611.01731 (2016)
7. Zhang, Y., Liu, L., Li, C., Loy, C.C.: Quantifying Facial Age by Posterior of Age Comparisons. CoRR, abs/1708.09687 (2017)
8. Cao, W., Mirjalili, V., Raschka, S.: Consistent Rank Logits for Ordinal Regression with Convolutional Neural Networks. CoRR, abs/1901.07884 (2019)
9. Zhang, C., Liu, S., Xu, X., Zhu, C.: C3AE: Exploring the Limits of Compact Model for Age Estimation. CoRR, abs/1904.05059 (2019)
10. Zhu, H., Zhou, Q., Zhang, J., Wang, J.Z.: Facial Aging and Rejuvenation by Conditional Multi-Adversarial Autoencoder with Ordinal Regression. CoRR, abs/1804.02740 (2018)
11. Berg, A., Oskarsson, M.: Mark O'Connor: Deep Ordinal Regression with Label Diversity. CoRR, abs/2006.15864 (2020)
12. Escalera, S., et al.: ChaLearn looking at people 2015: apparent age and cultural event recognition datasets and results. IEEE Int. Conf. Comput. Vision Workshop (ICCVW) **2015**, 243–251 (2015). https://doi.org/10.1109/ICCVW.2015.40
13. Ricanek, K., Tesafaye, T.: MORPH: a longitudinal image database of normal adult age-progression. In: 7th International Conference on Automatic Face and Gesture Recognition (FGR 2006), pp. 341–345 (2006). https://doi.org/10.1109/FGR.2006.78
14. Zhang, Z., Song, Y., Qi, H.: Age Progression/Regression by Conditional Adversarial Autoencoder. CoRR, abs/1702.08423 (2017)
15. Xie, D., Liang, L., Jin, L., Xu, J., Li, M.: SCUT-FBP: a benchmark dataset for facial beauty perception. In: 2015 IEEE International Conference on Systems, Man, and Cybernetics, pp. 1821–1826 (2015). https://doi.org/10.1109/SMC.2015.319
16. Ma, D.S., Correll, J., Wittenbrink, B.: The Chicago face database: a free stimulus set of faces and norming data. Behav. Res. Methods **47**(4), 1122–1135 (2015). https://doi.org/10.3758/s13428-014-0532-5
17. Fu, Y., Guo, G., Huang, T.S.: Age synthesis and estimation via faces: a survey. IEEE Trans. Pattern Anal. Mach. Intell. **32**(11), 1955–1976 (2010). https://doi.org/10.1109/TPAMI.2010.36
18. Chang, K., Chen, C., Hung, Y.: Ordinal hyperplanes ranker with cost sensitivities for age estimation. CVPR **2011**, 585–592 (2011). https://doi.org/10.1109/CVPR.2011.5995437
19. He, K., Zhang, X., Ren, S., Sun, J.: Deep Residual Learning for Image Recognition. CoRR, abs/1512.03385 (2015)
20. Xie, S., Girshick, R.B., Dollár, P., Tu, Z., He, K.: Aggregated Residual Transformations for Deep Neural Networks, CoRR, abs/1611.05431 (2016)

21. Hu, J., Shen, L., Sun, G.: Squeeze-and-Excitation Networks, CoRR, abs/1709.01507 (2017)
22. M.,Tan and Quoc V. Le: EfficientNet: Rethinking Model Scaling for Convolutional Neural Networks, CoRR, abs/1905.11946 (2019)
23. Li, X., Wang, W., Hu, X., Yang, J.: Selective Kernel Networks. CoRR, abs/1903.06586 (2019)
24. Zhang, H., et al.: ResNeSt: Split-Attention Networks. CoRR, abs/2004.08955 (2020)
25. Foret, P., Kleiner, A., Mobahi, H., Neyshabur, B.: Sharpness-Aware Minimization for Efficiently Improving Generalization. CoRR, abs/2010.01412 (2020)
26. Loshchilov, I., Hutter, F.: SGDR: Stochastic Gradient Descent with Restarts. CoRR, abs/1608.03983 (2016)
27. Cubuk, E.D., Zoph, B., Shlens, J., Le, Q.V.: RandAugment: Practical data augmentation with no separate search. CoRR, abs/1909.13719 (2019)
28. Guo, M.-H., et al.: Attention Mechanisms in Computer Vision: A Survey. CoRR, abs/2111.07624 (2021)
29. Halabi, S.S., Prevedello, L.M., Kalpathy-Cramer, J., et al.: The RSNA pediatric bone age machine learning challenge. Radiology **290**(2), 498–503 (2018)
30. Lathuilière, S., et al.: A comprehensive analysis of deep regression. IEEE Trans. Pattern Anal. Mach. Intell. **42**(9), 2065–2081 (2019)
31. Miron, R., Moisii, C., Dinu, S., Breaban, M.: Evaluating volumetric and slice-based approaches for COVID-19 detection in chest CTs. In: Proceedings of the IEEE/CVF International Conference on Computer Vision, pp. 529–536 (2021)
32. Karen, S., Zisserman, A.: Very deep convolutional networks for large-scale image recognition. In: Computer Vision and Pattern Recognition (cs.CV) (2014)

Res-Dense Net for 3D Covid Chest CT-Scan Classification

Quoc-Huy Trinh(✉), Minh-Van Nguyen, and Thien-Phuc Nguyen-Dinh

Ho Chi Minh University of Science, Ho Chi Minh City, Vietnam
{20120013,20127094}@student.hcmus.edu.vn

Abstract. One of the most contentious areas of research in Medical Image Preprocessing is 3D CT-scan. With the rapid spread of COVID-19, the function of CT-scan in properly and swiftly diagnosing the disease has become critical. It has a positive impact on infection prevention. There are many tasks to diagnose the illness through CT-scan images, include COVID-19. In this paper, we propose a method that using a Stacking Deep Neural Network to detect the Covid 19 through the series of 3D CT-scans images. In our method, we experiment with two backbones are DenseNet 121 and ResNet 101. This method achieves a competitive performance on some evaluation metrics.

1 Introduction

The SARS-COV-2 virus has spread over the world, with a major increase in 2019 and 2020. This global sickness cost nearly all governments in every country across the world a tremendous amount of money in 2020 and the first half of 2021citecovid stat. Thousands of individuals are infected with this disease every day, making it the most hazardous sickness on the planet. Almost all medical photos and records are maintained in a computerized database nowadays. Furthermore, the number of doctors available to diagnose these medical data is restricted, particularly in the case of Covid-19.

With the increasing number of patients and a scarcity of doctors, practically all photographs in the digital database may be used for pre-diagnosis, which helps to speed up diagnosis and improves the doctor's accuracy. This is why, in order to avoid infection, a quick and reliable detection approach is required [13].

COVID-19 has a very severe on the respiratory system of the human. The virus is harbored most commonly with little or no symptoms, but can also lead to rapidly progressive and often fatal pneumonia. With the patients who have COVID-19, the virus can lead negative the patients situation [13].

There are a variety of diagnostic procedures available, including CT scans, chest X-rays, and PCR. They give us with a 3-D perspective of organ creation using CT-scan recording. Due to the lack of overlapping tissues, convenient disease evaluation, and its location, CT scans also provide a more complete overview of the internal structure of the lung parenchyma [4]. As an aspect, it provides a window into pathophysiology that could shed light on several stages of disease detection and evolution.

P. L. Mazzeo et al. (Eds.): ICIAP 2022 Workshops, LNCS 13374, pp. 483–495, 2022.
https://doi.org/10.1007/978-3-031-13324-4_41

Radiologists report COVID-19 patterns of infection with typical features including ground glass opacities in the lung periphery, rounded opacities, enlarged intra-infiltrate vessels, and later more consolidations that are a sign of progressing critical illness [2].

Chest radiography's medical imaging features are typically utilized to detect abnormalities and disorders in tissue such as the brain or lungs. As a result, chest x-ray images or CT-scan images are frequently utilized to detect abnormalities in human tissues.

Patients capture multiple slices during CT-scan recording, but the number of doctors available to diagnose these data is insufficient. These are the reasons why Machine Learning and Deep Learning models must be used in order to facilitate and shorten the diagnostic process. This is why the purpose of the 3D-CT scans pictures classification challenge is to assess several strategies for reliably and efficiently classifying 3D-CT scans images [2].

In this paper, we propose a method that uses Fine-tuning and ensemble Deep Neural Network backbones to classify 3D CT-scan images. With the combination of Deep Neural Network, the model can have higher performance on feature extraction and classification task. In this experiment, we use two backbones, DenseNet 121 and ResNet 101, for evaluating our method on the test dataset with 4355 samples.

We also introduce the Res-Dense net architecture, the experiment, and the assessment mechanism in this section of our proposal. Our strategy earns a competitive score on the training and testing method in this experiment. Furthermore, we suggest various improvements to our methods' performance, and the method can be used to tackle other difficulties in medical imaging, notably in CT-scan images.

2 Related Work

2.1 CT-scan Images

The difference between the CT-Scan images and other medical images, CT-Scan images are created by a series of X-ray images, which are forms of radiation on the electromagnetic spectrum. In addition, as compared to X-ray images, CT scans can provide information on multiple angles of the tissue and show the status of the tissue in each frame of image sequences. Furthermore, the information of the tissue can be clearly depicted through a sequence of photographs taken at the same moment [23].

2.2 Image Classification

Image classification is a task that attempts to classify the image by a specific label. In recent years, the development of computing resources leads to a variety of methods in Image classification. Many deep architectures have been proposed and get competitive results. Moreover, from the classification task, the researcher can use to localize the abnormal on the images, which is very important in the medical diagnosis [21].

2.3 CT-Scan COVID-19 Image Classification

The use of imaging data is illustrated to be a helpful method to diagnose Covid-19. However, computed tomography (CT-Scan) gets a variety of signs and creates difficulty for the doctors. However, with the development of computing and computer vision, there are several methods are proposed to deal with this problem. The flourish of the Deep Learning and Transfer Learning methods are creating a beneficial impact on image classification. By using CNN, Deep Neural Network architectures to classify the Covid-19 CT-scan images, the accuracy of diagnosing achieves competitively. Nowadays, there are many challenges in CT-Scan classification to find the competitive approach and method to apply in a fast and accurate diagnosis to prevent the societal infection [9].

3 Dataset

The dataset we use is from MIA-COVID 19 dataset, which contains the Covid 3D-CT Scan images series from patients that have COVID 19 and patients that do not have COVID 19 [3]. The dataset is split into folders. Each of them is a series of images when doing CT-Scan. All of the images are collected from COVID19-CT-Database. The dataset include the input sequence is a 3-D signal, consisting of a series of chest CT slices, i.e., 2-D images, the number of which is varying, depending on the context of CT scanning. The context is defined in terms of various requirements, such as the accuracy asked by the doctor who ordered the scan, the characteristics of the CT scanner that is used, or the specific subject's features, e.g., weight and age [1].

The COVID19-CT-Database (COV19-CT-DB) consists of chest CT scans that are annotated for the existence of COVID-19. Data collection was conducted in the period from September 1, 2020, to March 31, 2021. Data were aggregated from many hospitals, containing anonymized human lung CT scans with signs of COVID-19 and without signs of COVID-19 [1].

The COV19-CT-DB database consist of about 5000 chest CT scan series, which correspond to a high number of patients (>1000) and subjects (>2000). Annotation of each CT slice has been performed by 4 very experienced (each with over 20 years of experience) medical experts; two radiologists and two pulmonologists. Labels provided by the 4 experts showed a high degree of agreement (around 98%) [2] (Fig. 1).

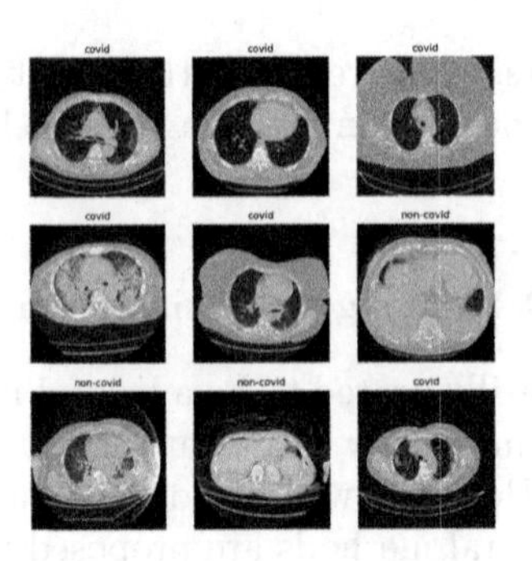

Fig. 1. Sample of dataset

3.1 Evaluation Methods

To evaluate a classification model, we have some basic methods such as Accuracy, Precision, Recall, F1-score, etc. Regarding Precision, this measurement score evaluates the numbers of true positive over the number of false-positive and true-positive while Recall measure the number of positive over the true positive and false negative. However, with the F1-score, which is described as the harmonic mean of the two, the evaluation is similar to the average of Precision and Recall; the measurement score gets sensitive to two inputs having a low value, which helps to make the experiment fair [25].

$$\frac{Precision \times Recall}{Precision + Recall} \tag{1}$$

To evaluate the methods, we use the Macro F1-Score with the following formula [7]:

$$\frac{1}{n} * \sum_{i=0}^{n} F1 - scores_i \tag{2}$$

where:
n: number of classes/labels
i : class/label

4 Method

Despite using CNN with RNN or LSTM to find features of the 3D CT-Scan series, we propose a method that extracts all features of all images in all series. This method can help efficiently reduce the time of training and achieve competitive performance. In the testing phase, we propose to predict all images in the series and calculate the mean score of the series to choose the label of that series.

4.1 Densely Connected Convolutional Network

The demonstration of the recent work has shown that Convolutional Neural Networks can be substantially deeper, more accurate, and more efficient to train if they contain shorter connections between each layer close to the input and those close to the output [19]. DenseNet connects each layer to every other layer in a feed-forward chain. Whereas traditional Convolutional Neural Network architecture with L layers has L connections - one between each layer and its subsequent layer - our network has $\frac{L(L+1)}{2}$ direct connections. For each layer, the feature maps of all preceding layers are used as inputs, and their feature maps are used as inputs into all subsequent layers. DenseNet have several compelling advantages: they alleviate the vanishing gradient problem, strengthens feature propagation, encourages feature reuse, and substantially reduces the number of parameters [6] (Fig. 2).

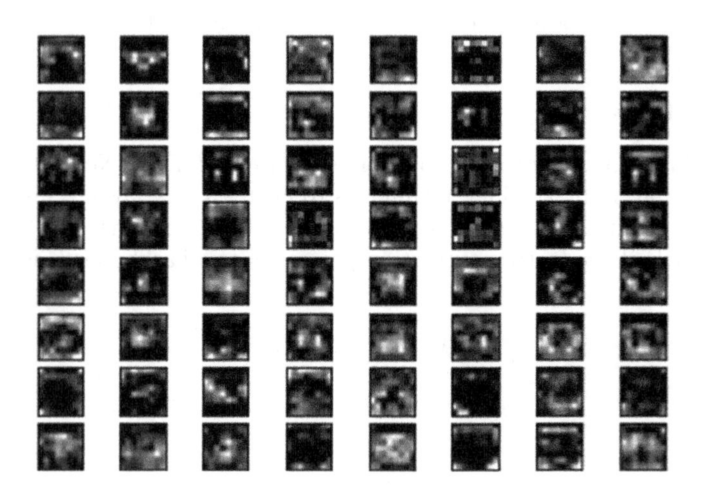

Fig. 2. Feature extraction of DenseNet 121

4.2 Deep Residual Network (ResNet)

Deeper neural networks are difficult to train. Therefore, Deep Residual Network is created to ease the training of networks that are substantially deeper than those used previously [11]. ResNet architecture can explicitly reformulate the layers as learning residual functions regarding the layer inputs, instead of learning unreferenced functions. With this network, it is easier to optimize and can gain accuracy from considerably increased depth. On the ImageNet dataset, the evaluation of residual nets, with a depth of up to 152 layers and eight times deeper than the VGG net, still has lower complexity. An ensemble of these residual nets achieves 3.57 errors on the ImageNet test set. This result won the first place on the ILSVRC 2015 classification task. The architecture also gains a competitive performance on the other dataset such as Cifar, etc. [5] (Fig. 3).

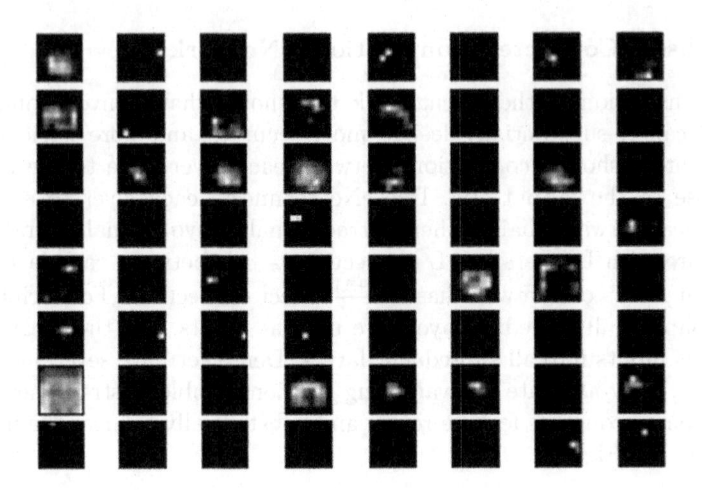

Fig. 3. Feature extraction of ResNet 101

4.3 Res-Dense Net

In our architecture, we propose to use ResNet 101 and DenseNet 121 backbones for the first layers by stacking techniques to create a new layer. This technique can use the merits of these two models to improve the strength and help reduce the drawback of the feature extraction process of two backbones (Fig. 4).

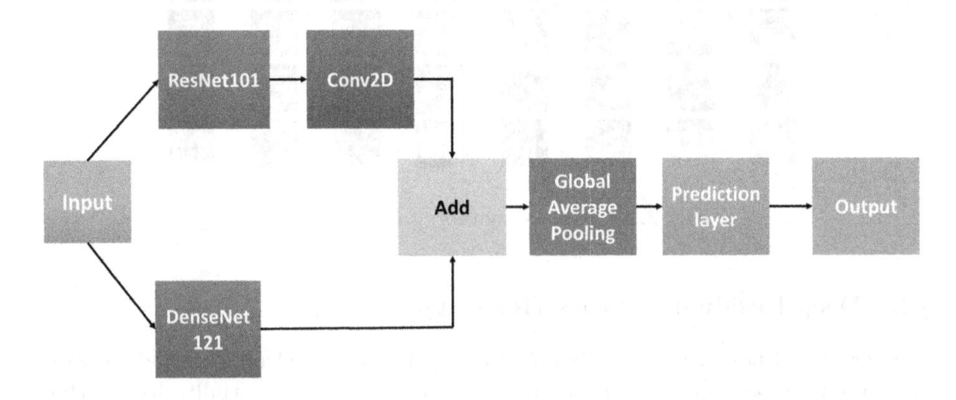

Fig. 4. General network architecture

The input pass in two ways first is the Resnet 101 and a Convolution 2D layer, second, go through the DenseNet 121, then the result is added to the Add layer, then the result feature map then moves to the Global Average Pooling layer and the vector result of this layer is fed to the prediction layer for the output of the model. The figure below will illustrate obviously our work and the model that we design (Fig. 5):

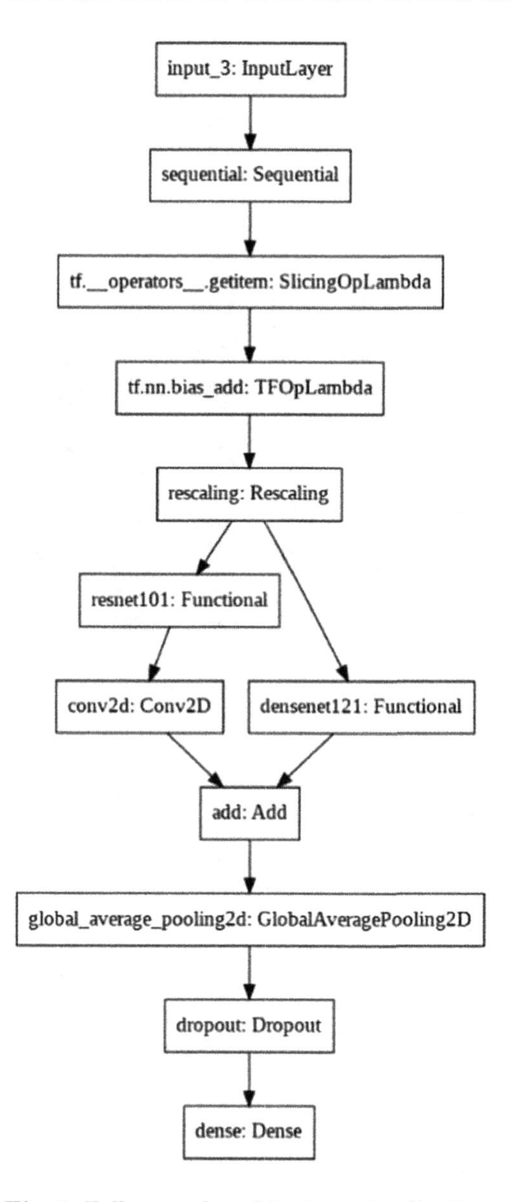

Fig. 5. Full network architecture visualization

One of the merits of our method is when feature maps, which are results of two previous extraction layers, are merged by adding layers to create the regular feature map and coming to the Global Average Pooling layers before coming to the Fully-Connected layer. By this technique, the importance of each area in the images can be defined by projecting weights of the output layer on the Convolution feature map that gain from the previous layer. Two Convolution

Layers are added to make the output shape, and the number of filters of the two outputs from two Deep Neural networks is equal. Moreover, with these layers, we can control the quality of output features that are created by the architecture (Fig. 6).

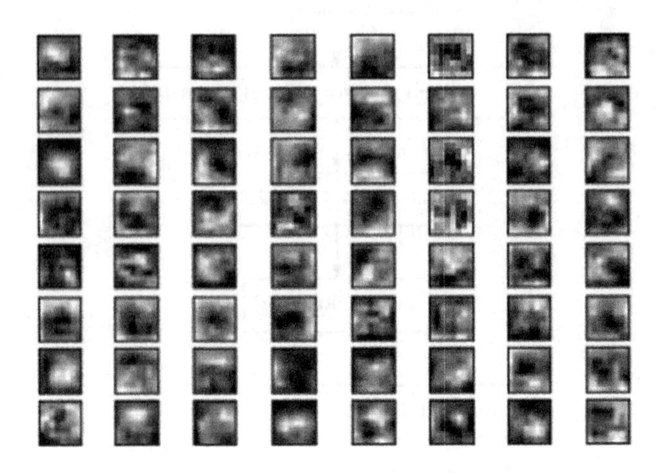

Fig. 6. Feature extraction of Res-Dense net

After that, features pass the Global Average Pooling layers and generate one feature map for each corresponding category of the classification task in the last MLP layer. With two backbones, new feature maps are created by two or more backbones. We propose to use global average pooling to reduce the computation cost.

The following is the Generalized Mean Pooling formula:

$$f^{(g)} = [f_1^{(g)}...f_k^{(g)}...f_K^{(g)}]^T, f_K^{(g)} = (\frac{1}{|\chi_k|}\sum_{x\in\chi_k} x^{P_k})^{\frac{1}{P_k}} \tag{3}$$

In the formula (3), the $f_{(g)}$ is the output of Convolution layers, each feature map from the output is calculated by the average and the result for each will be an element in the vector $f_i^{(g)}$ with $i = [1, k]$. From this formula, the number of nodes for feeding to the multi-layer perceptron reduces.

Instead of adding Flatten layers to create vectors for fully connected layers on top of the feature maps, this layer takes the average of each feature map. Then, the resulting vector is fed directly into the sigmoid activation function. The advantage of Global Average Pooling is there is no parameter to optimize in the global average pooling thus overfitting is avoided at this layer. Furthermore, global average pooling calculates the average out the spatial information, thus it is more robust to spatial translations of the input.

This architecture is inspired by the inception block of GoogleNet, by merging the convolutional layers and using Average pooling layers. This method helps the architecture get deeper but efficiently the computation cost.

4.4 Data Preprocessing

After loading data, we resize all the images to the size (256,256), then we split the dataset into the training set and validation set in the ratio of 0.75:0.25. After resizing and splitting the validation set, we rescale the data pixel down to be in the range $[-1, 1]$. Then we use the application of ResNet to preprocess the input. The Input after preprocess is rescaled to the same input of the ResNet model.

4.5 Data Augmentation

Data Augmentation is vital in the data preparation process. Data Augmentation improves the number of data by adding slightly modified copies of already existing data or newly created synthetic data from existing data to decrease the probability of the Overfitting problem, we use augmentation to generate the data randomly by random flip images and random rotation with an index of 0.2 (Fig. 7).

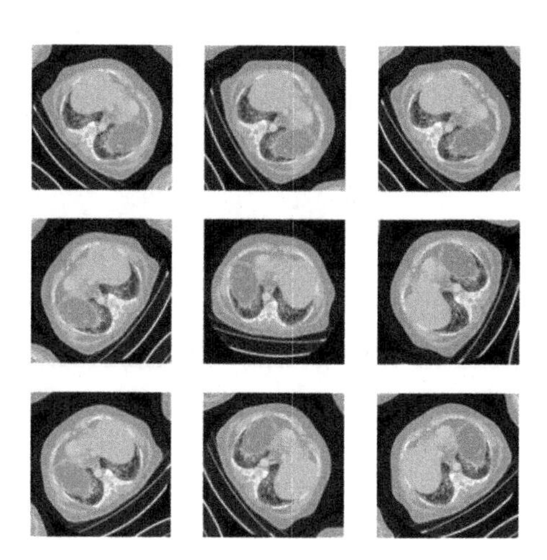

Fig. 7. Data Augmentation result

4.6 Training

Our models are initialized with pre-trained weight from Imagenet. We use a batch size of 32 for training data with an image's size of (256,256). Moreover, we propose to use the RMSprop optimizer with a learning rate is 0.0001 for optimizer and evaluate the training process by accuracy and F1-score. For the loss function, we use Sparse Categorical Cross-entropy. The model is trained with 20 epochs and get the checkpoint that achieves the highest performance (Table 1).

Table 1. The parameter setup for model before training

Parameter	Value
Optimizer	Adam
Learning rate	0.0001
Loss	Binary crossentropy
Metrics	F1-score, accuracy

With a low learning rate, we can ensure that we can find the weight with the competitive results although it costs more time for training.

Firstly, we freeze all the complicated layers of DenseNet and ResNet. Then, we start to train for the first time. Next, we freeze 100 layers before on each backbone. Finally, we start the continuous training process (Fig. 8).

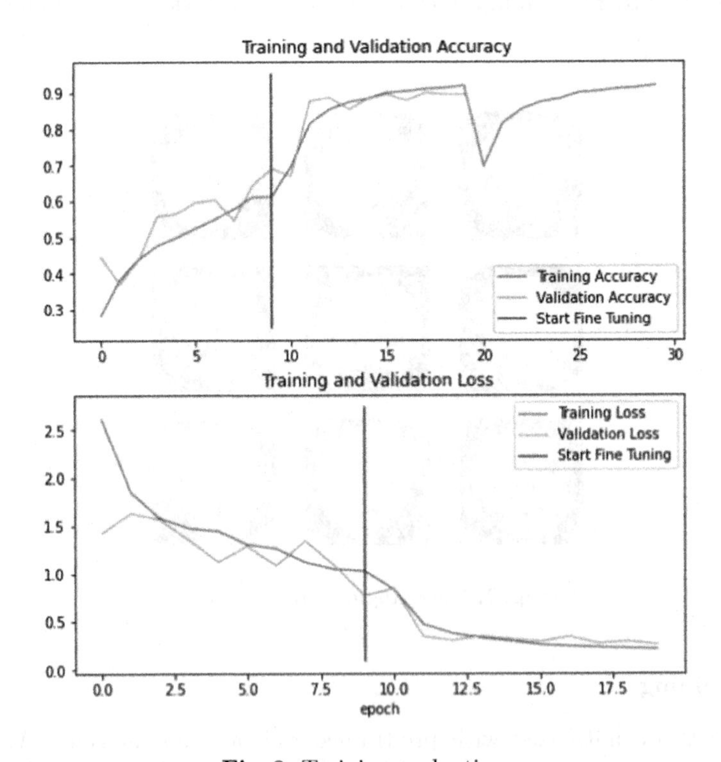

Fig. 8. Training evaluation

After Fine-tuning with 20 epochs, we get the result of a loss of approximately 0.4. The model performs positively on the training dataset. On the evaluation of the Macro F1-score on the test set with 4355 folders of images that contain covid and non-covid, we achieve the competitive score (Table 2):

Table 2. Our result on test set

Score	Value
F1 (COVID)	63.08
F1 (NON-COVID)	93.18
Macro F1	78.13

5 Conclusion

We demonstrated the proposal of using Res-Dense Net with Fine-tuning technique to classify endoscopic images. The result of our research is positive for the F1 score. Moreover, our method can inspire a new approach on classification on 3D images instead of using 3D CNN or Convolution Neural Network with LSTM or RNN. However, there are some drawbacks that we have to do to improve the performance of the model, such as pre-processing data, reduce noise, change the size of the image to train. Furthermore, we can apply ResNet101 V2 or DenseNet 169 backbone, or we can combine with LSTM or RNN modules to have better feature extraction and better performance of the model.

6 Discussion

Our result get 78.13 in Macro F1-score, which gets a higher score than the Traditional 3D CNN and 3D Regnet [17] architecture [16]. However, our model gets the lower result with state of the art architecture [17]. Moreover, because of not increasing the quantity of data, our method also does not get a higher score, we can improve the accuracy of the model by improving data. The architecture that we combine is between ResNet 101 and DenseNet 121. Below is the result comparison between other methods with our method (Table 3).

Table 3. Comparison with other models

Method	Macro F1-score
3D-CNN with Bert [17]	88.22
Res-Dense Net	**78.13**
3D RegNet [16]	71.83s
Shallow Convolution Neural-Network [18]	70.86
Vision transformer [15]	70.5

With the approach of detecting Covid symptoms in each frame of the video, we get a higher result than others with the same approach but with the approach of using the sequence of frames in the video because the architecture uses the power of feature extraction from two backbones are ResNet 101 and DenseNet 121 create the benefit for the prediction process. Nevertheless, our method gets worse in this domain if compare with other methods using Bert or 3D CNN because this method can not find the relation of the frame like approaching sequences like 3D-CNN with Bert method or 3D RegNet method. However, our model has a better score than the 3D RegNet method that uses sequences of frames in the video.

7 Future Work

Although our method gets a competitive score, there are some drawbacks in our methods: the training time gets long with 334 ms/step, we can custom layers in the architecture to accelerate the computing cost. We can get more layers or can ensemble more backbones to achieve higher results. Moreover, we can do a segmentation process on the lung CT-scan to improve the accuracy of training and get a better result for our architecture that is proposed [24].

References

1. Kollias, D., Arsenos, A., Soukissian, L., Kollias, S.: MIA-COV19D: COVID-19 Detection through 3-D Chest CT Image Analysis. ArXiv:2106.07524 (2021)
2. Kollias, D., et al.: Deep transparent prediction through latent representation analysis. ArXiv:2009.07044 (2020)
3. Kollias, D., et al.: Transparent adaptation in deep medical image diagnosis. In: Heintz, F., Milano, M., O'Sullivan, B. (eds.) TAILOR 2020. LNCS (LNAI), vol. 12641, pp. 251–267. Springer, Cham (2021). https://doi.org/10.1007/978-3-030-73959-1_22
4. Kollias, D., Tagaris, A., Stafylopatis, A., Kollias, S., Tagaris, G.: Deep neural architectures for prediction in healthcare. Complex Intell. Syst. **4**(2), 119–131 (2017). https://doi.org/10.1007/s40747-017-0064-6
5. He, K., Zhang, X., Ren, S., Sun, J.: Deep Residual Learning for Image Recognition. CoRR, abs/1512.03385 (2015). http://arxiv.org/abs/1512.03385
6. Huang, G., Liu, Z., Weinberger, K.: Densely Connected Convolutional Networks. CoRR, abs/1608.06993 (2016). http://arxiv.org/abs/1608.06993
7. Opitz, J., Burst, S.: Macro F1 and Macro F1. CoRR, abs/1911.03347 (2019). http://arxiv.org/abs/1911.03347
8. Chollet, F.: Xception: Deep Learning with Depthwise Separable Convolutions. CoRR, abs/1610.02357 (2016). http://arxiv.org/abs/1610.02357
9. Pogorelov, K., et al.: KVASIR: A Multi-Class Image Dataset for Computer Aided Gastrointestinal Disease Detection (2017)
10. Martinez, A.: Classification of COVID-19 in CT Scans using Multi-Source Transfer Learning (2020)
11. Guo, S., Yang, Z.: Multi-Channel-ResNet: an integration framework towards skin lesion analysis. Inform. Med. Unlocked. **12**, 67–74 (2018). https://www.sciencedirect.com/science/article/pii/S2352914818300868
12. Nguyen, N., Tran, D., Nguyen, N., Nguyen, H.: A CNN-LSTM Architecture for Detection of Intracranial Hemorrhage on CT scans (2020)

13. Bonvini, M., Kennedy, E., Ventura, V., Wasserman, L.: Causal Inference in the Time of COVID-19 (2021)
14. Alizadehsani, R., et al.: Risk factors prediction, clinical outcomes and mortality of COVID-19 patients. J. Med. Virol. **93**, 2307–2320 (2020)
15. Gao, X., Qian, Y., Gao, A.: COVID-VIT: Classification of COVID-19 from CT chest images based on vision transformer models (2021)
16. Qi, H., Wang, Y., Liu, X.: 3D RegNet: Deep Learning Model for COVID-19 Diagnosis on Chest CT Image (2021)
17. Tan, W., Liu, J.: A 3D CNN Network with BERT For Automatic COVID-19 Diagnosis From CT-Scan Images (2021)
18. Teli, M.N.: TeliNet: Classifying CT scan images for COVID-19 diagnosis (2021)
19. Huang, G., Liu, Z., Van Der Maaten, L., Weinberger, K.: Densely connected convolutional networks. In: Proceedings of the IEEE Conference on Computer Vision and Pattern Recognition, pp. 4700–4708 (2017)
20. Li, G., Zhang, M., Li, J., Lv, F., Tong, G.: Efficient densely connected convolutional neural networks. Pattern Recogn. **109**, 107610 (2021)
21. Shah, V., Keniya, R., Shridharani, A., Punjabi, M., Shah, J., Mehendale, N.: Diagnosis of COVID-19 using CT scan images and deep learning techniques. Emerg. Radiol. **28**(3), 497–505 (2021). https://doi.org/10.1007/s10140-020-01886-y
22. Ahuja, S., Panigrahi, B.K., Dey, N., Rajinikanth, V., Gandhi, T.K.: Deep transfer learning-based automated detection of COVID-19 from lung CT scan slices. Appl. Intell. **51**(1), 571–585 (2020). https://doi.org/10.1007/s10489-020-01826-w
23. Maghdid, H., Asaad, A., Ghafoor, K., Sadiq, A., Mirjalili, S., Khan, M.: Diagnosing COVID-19 pneumonia from X-ray and CT images using deep learning and transfer learning algorithms. In: Multimodal Image Exploitation And Learning 2021, vol. 11734, pp. 117340E (2021)
24. Miron, R., Moisii, C., Dinu, S., Breaban, M.: COVID Detection in Chest CTs: Improving the Baseline on COV19-CT-DB (2021)
25. Goutte, C., Gaussier, E.: A probabilistic interpretation of precision, recall and F-score, with implication for evaluation. In: Losada, D.E., Fernández-Luna, J.M. (eds.) ECIR 2005. LNCS, vol. 3408, pp. 345–359. Springer, Heidelberg (2005). https://doi.org/10.1007/978-3-540-31865-1_25

Deep Regression by Feature Regularization for COVID-19 Severity Prediction

Davide Tricarico[1,2](✉), Hafiza Ayesha Hoor Chaudhry[1], Attilio Fiandrotti[1], and Marco Grangetto[1]

[1] Università degli Studi di Torino, corso Svizzera 185, Torino, Italy
{davide.tricarico,hafizaayeshahoor.chaudhry,attilio.fiandrotti, marco.grangetto}@unito.it

[2] AITEM Artificial Intelligence Technologies Multipurpose s.r.l., Corso Castelfidardo 36, 10129 Turin, Italy
davide.tricarico@aitemsolutions.com

Abstract. During the COVID-19 worldwide pandemic, CT scan emerged as one of the most precise tool for identification and diagnosis of affected patients. With the increase of available medical imaging, Artificial Intelligence powered methods arisen to aid the detection and classification of COVID-19 cases. In this work, we propose a methodology to automatically inspect CT scan slices assessing the related disease severity. We competed in the ICIAP2021 COVID-19 infection percentage estimation competition, and our method scored in the top-5 at both the Validation phase ranking, with MAE = 4.912%, and Testing phase ranking, with MAE = 5.020%.

Keywords: COVID-19 · Severity prediction · Deep regression

1 Introduction

A new global pandemic Coronavirus Disease (COVID-19) started in 2019 and soon became the center of focus with an unidentified source of start, an exponential growth rate and an insufficient knowledge of the transmission process. The symptoms of this disease vary from mild to severe and in extreme cases may lead to pneumonia, lung failure and ultimately death [11,15]. The epidemic spread of COVID-19 has affected the medical area most and created a shortage of medical supplies all around the world, with varying factors to accessibility of health-care and medical supplies [13]. The most commonly used method to detect COVID-19 is reverse transcription polymerase chain reaction (RT-PCR), where the specimen is taken from lower or upper respiratory tract of the patient. The RT-PCR is cheaper and widely available compared to most of the other methodologies for detecting COVID-19 (like Chest X-rays and CT scans). However, the high rate of false negatives, the lengthy processing time and a low sensitivity rate have made RT-PCR somewhat unreliable [7,21].

P. L. Mazzeo et al. (Eds.): ICIAP 2022 Workshops, LNCS 13374, pp. 496–507, 2022.
https://doi.org/10.1007/978-3-031-13324-4_42

Computed Tomography (CT) on the other hand, gives an in depth knowledge of the patho-physiology, hence giving the possibility of detecting several stages of the disease in light of the affected organs. In common practice CT scan is done after a positive RT-PCR, therefore developing a correlation between both. However a CT scan can also catch early COVID-19 stages in patients who have mild to no symptoms, and those with a negative RT-PCR [1,20]. Due to these advantages CT scan is becoming the ultimate tool in early diagnosis of COVID-19 worldwide.

With an increase in the patients and medical imaging produced, a need for Artificial Intelligence (AI) arises to aid the detection and classification process of COVID-19 patients. CT scans provide a perfect image that could be fed into the Deep Learning (DL) network, generating automated biomarkers to detect and classify COVID-19. The most important factor in deep learning techniques is a diverse, large, and high quality annotated dataset. Another challenge for the medical staff is distinguishing common cold pneumonia from COVID-19 as both affect the lungs, but the COVID-19 pneumonia could be more fatal. Deep learning could not only detect COVID-19 from CT scans but also distinguish it from the common cold pneumonia [12,18]. Using automated deep learning techniques will also take some burden off the manual labor of radiologists and doctors, and help in decreasing the shortage of medical staff.

Using the concept of Multi task learning, feature maps along with an encoder and decoder have been used in detection of the COVID-19 disease, segmentation of the affected areas, and the reconstruction using a large dataset by [2]. Providing promising results, this technique yielded 97% area under the ROC curve. Research has also been done on detecting COVID-19 from CT scans using weakly supervised techniques [9]. The network used in this work was inspired by VGG architecture, consisting of convolution layers, batch normalization and rectified linear units. Despite giving good results on the COVID-19 detection, a big drawback of this work is the high probability of mis-classification of community acquired pneumonia (CAP) as COVID-19.

Apart from CT scans, a lot of research has also been done on other medical imaging including Chest Xrays (C-Xray), Lung Ultrasounds (LUS) and MRI. A deep learning study has been carried out using lung ultrasound videos for the detection of COVID-19 [14]. Besides detection, it also performs localization on each lung ultrasound frame to provide better estimate of the disease. Chest Xray is the other popular imaging technique being used for classification and detection of COVID-19. As it is more accessible and inexpensive than CT scans, many researchers preferred it initially [5,16]. The limitation of Chest Xray is that it has a 69% sensitivity and the best findings from Chest Xrays are retrieved when it is generated after a week of symptom onset [19]. This makes interpreting the early onset Chest Xrays difficult for even the expert radiologist. Other factors as the patient's position (standing or laying down) while going through medical imaging, the presence of tubes in case of severely ill patients, and the type of projection used also plays an important role. In another study, 15% of the Chest Xrays suggested normalcy in the patients who were already affected by

COVID-19 [8]. Proving that CT scans are more reliable and vivid than Chest Xrays and should also be used to follow-up on COVID-19 patients.

In this paper we present a novel technique for the classification and estimation of COVID-19 disease using dataset from a competition proposed by ICIAP 2021 conference. In this challenge, teams are asked to estimate COVID-19 severity score related to transverse plane CT scan slices. For this purpose a set of images and related labels have been provided to the participants. The key contributions of this study are:

- We adopted a contrastive learning method for image regression task, consisting of a training procedure with the goal of learning a feature space where the distance between samples is proportional to their difference in targeted labels;
- We proposed a novel loss function to support the above-mentioned method based on the calculation of distance matrices in the feature and output spaces and the minimization of their relative differences;
- To predict the severity score, we adopted an approach derived from image retrieval and few-shot classification problems, computing the prediction by averaging the score of nearest neighbours in the feature space obtained applying the above-mentioned training procedure and loss function.

2 The Dataset

The dataset used in this project is provided by the ICIAP 2021 conference organizer for the COVID-19 infection percentage estimation competition [4,17], through the Codalab online platform [6]. The dataset includes CT scan slices (simply slices from now on) of patients either infected with COVID-19 or healthy and the problem statement is to estimate the percentage of COVID-19 infection rate. The dataset is divided into three sets, Train, Test and Validation.

The Train set includes 3054 CT scan slices from 132 patients, out of which 128 are diagnosed as infected by COVID-19. The method of diagnosis are CT scan prognosis by expert thoracic radiologists and the positive reverse transcription polymerase chain reaction (RT-PCR). The remaining 4 patients are healthy i.e. not infected by COVID-19. In the Train set, CT scan slices are provided with their respective estimated COVID-19 infection rate, calculated by two expert radiologists. The infection rate is a figure between 0 and 100.

The Validation set includes 1301 slices from 57 patients, among which 55 are COVID-19 positive. This set is blinded, i.e. the COVID-19 ground truth infection rate is not known. The Train and Validation sets have been collected from different patients including both Male and Female patients. The age of patients range from 27 to 70 years old. The data collection was done between June to December 2020, from two hospitals, Hakim Saidane, in Biskra (Algeria), and Ziouch Mohamed, in Tolga (Algeria) [3].

The Test set includes 4449 CT Scan slices from 130 patients. All of the included patients are tested positive for COVID-19 using both diagnosis methods, RT-PCR and thoracic CT scan prognosis. The dimensions of these slices is 630×630.

The test set is also blind, i.e. the ground truth labels are not known. Table 1 summarizes the composition of each set.

Table 1. Summary of the ICIAP 2022 COVID-19 infection percentage estimation competition dataset used in this work.

Dataset	Patients	Positive patients	Total slices
Train	132	128	3054
Val	57	55	1301
Test	130	130	4449

Using the predictions on the Validation set, the top ten teams/participants were invited to test their code on the Test set, after which the final ranking of the challenge would be calculated.

For the evaluation part three metrics are used in this competition, Mean Absolute Error (MAE), Root Mean Square Error (RMSE) and Pearson Correlation coefficient (PC). The priority is given to Mean Absolute Error, RMSE and PC are used as tie breaker between two teams/participants with same MAE. The final ranking of the teams/participants in the competition is calculated from Test set and Validation set results using the following formula:

$$Final\_rank = (0.7 \times Test\_set\_results) + (0.3 \times Validation\_set\_results) \quad (1)$$

3 Proposed Method

The method we propose to predict COVID-19 severity of each CT scan slice is illustrated in Fig. 1 and can be summarized as following. In a nutshell, each slice is projected in a well-behaved lower dimensional *target feature space* such that nearby points in this feature space correspond to slices with similar COVID-19 scores, while distant points correspond to different severity values. Whenever the COVID-19 score shall be predicted for a query slice, it is projected in the target feature space. The query slice projection is compared with the projections of the labelled slices from training set, i.e.e the reference slices. The query slice score is predicted by interpolating the scores of the nearest reference slices in the feature space.

The proposed method is implemented as a 3-stages pipeline as follows:

- Image pre-processing, to normalize the slices pixel intensity and resolution across the different image acquisition settings
- Feature extraction, to project the slice into the above-mentioned well-behaved feature space, and
- Distance based regression, to predict the severity score looking at the query slice neighbors in the feature space from reference set.

In the following, we detail each of the above stages.

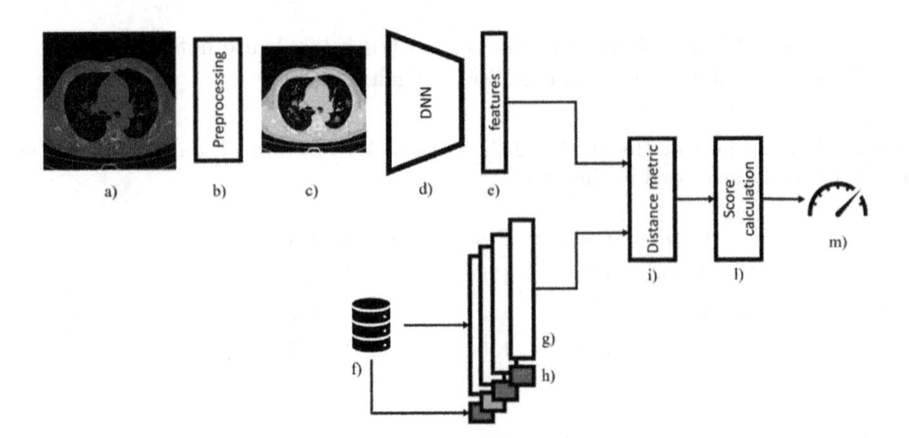

Fig. 1. High level illustration of adopted methodology. **a)** *query*: original CT scan slice; **b)** *image pre-processing*; **c)** pre-processed slice; **d) e)** *feature extraction*; **f) g) h)** *reference set*: database of labelled cases from training set with corresponding projections; **i) l)** *distance based regression*; **m)** final COVID-19 severity prediction

Image Pre-processing. CT scan slices appearance can be strongly affected by the adopted image acquisition process and technology. In particular the contrast, the brightness and the pixel intensity histogram can change when equipment from different vendors are used, or different settings are applied. This variability is evident in Fig. 2, where differences in pixel intensity are present among the samples. It can be also observed, that the image size (Fig. 2(c)) and scale (Fig. 2(f)) can differ significantly between samples.

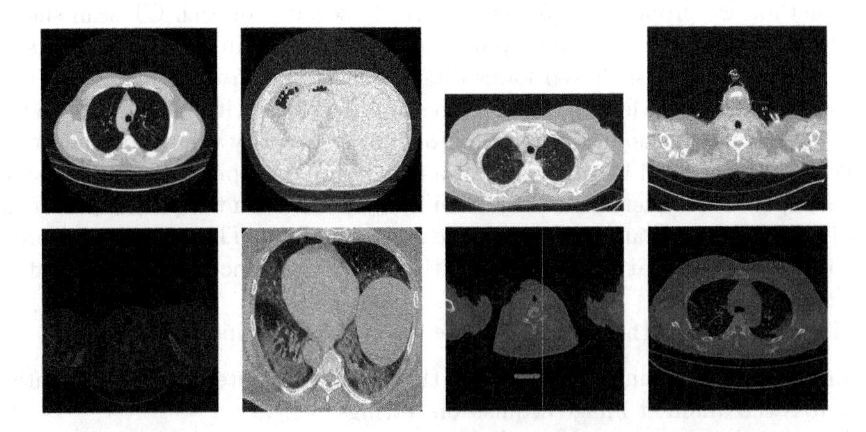

Fig. 2. Samples from dataset provided by ICIAP 2021 Challenge organizers during testing phase. Depending on the production process, slices appearance can strongly differ in term of colour distribution, scale and size. In terms of colour distribution, the slices in the first row are significantly brighter than the last ones. Image size of third slice is different from the others. The sixth slice has a different scale.

Because these sharp differences can lead to poor performance, we put in place a pre-processing strategy composed by the following steps:

- Image resize: first, slices are resized to 512×512 pixels format: our experiments showed that above this resolution no significant gains are found, while the computational complexity remains reasonable;
- Pixel intensity scaling: pixel values are re-scaled in range between 0 to 1, where 1 corresponds to amount equal or higher than 90th percentile of original intensities distribution.

Feature Extraction. The preprocessed slice images are projected into the target feature space using the deep convolutional neural network DenseNet-121 [10]. This architecture is characterized by a sequence of *dense blocks* (concatenation of non linear operators with skip connections to prevent gradient vanishing effect), connected by *transition layers* (concatenation of convolution and pooling layers), finally a *linear layer* is responsible for the class prediction. In our case, the original architecture has been truncated before the output classification layer, resulting in a convolutional feature extractor that yields a 1024-dimensional projection of the 512 × 512 slice image provided in input.

Consider a CT scan slice pre-processed picture x from the set of all possible CT scan slices X, and a deep neural network model $M(\cdot)$, we have:

$$\forall x \in X \rightarrow M(x) = z \in \mathcal{F} \subseteq \mathbb{R}^{1024}, \tag{2}$$

where z is the feature vector corresponding to the CT scan slice x and belongs to the target feature space $\mathcal{F}$. With the notation x_i the i-th sample in X and $z_i = M(x_i)$, consider $y_i \in \mathcal{S} \subseteq \mathbb{R}$, the COVID-19 severity associated with x_i, our goal is to design $M(\cdot)$ such that:

$$\forall i, j, k \in \mathbb{N} : \{||y_i, y_j|| < ||y_i, y_k||\} \rightarrow \{||z_i, z_j|| < ||z_i, z_k||\} \tag{3}$$

where i, j, k indicate three samples in set X and $||a, b||$ is the euclidean distance between points a and b.

To train the model $M(\cdot)$ achieving the sought property in Eq. 3, we minimize a loss function based on the relative distance of the samples in the target feature space $\mathcal{F}$ and *severity scores space* $\mathcal{S}$.

Considering a random subset of n CT scan pictures X^{batch}, related COVID-19 severity scores Y^{batch}, and computed projections in the feature space Z^{batch}, let us define the following distance matrices:

$$D_x \in \mathbb{R}^{n \times n} \rightarrow D_x[i, j] = ||z_i, z_j|| \qquad z_i, z_j \in Z^{batch} \forall i, j \in [1, 2, ...n] \tag{4}$$

$$D_y \in \mathbb{R}^{n \times n} \rightarrow D_y[i, j] = ||y_i, y_j|| \qquad y_i, y_j \in Y^{batch} \forall i, j \in [1, 2, ...n] \tag{5}$$

The *loss function* L_D is defined as the Mean Absolute Error (MAE) between D_x and D_y:

$$L_D = \frac{1}{n^2} \sum_{i=1}^{n} \sum_{j=1}^{n} |D_x[i,j] - D_y[i,j]| \tag{6}$$

Distance Based Regression. To compute the final prediction for COVID-19 severity, our method exploits the distance in the target feature space between the query slice and the reference slices for which the score is known.

Consider a set X^{ref} of n^{ref} references with relative severity score Y^{ref}, such that given a $x_i^{ref} \in X^{ref}$ its disease quantification is y_i^{ref}, $\forall i = [1, 2, ...n_{ref}]$. The methodology computes *feature reference set* Z^{ref} as:

$$Z^{ref} := \{M(x_i^{ref}) \quad \forall i = [1, 2, ...n_{ref}]\} \tag{7}$$

Whenever it is requested to predict the severity score for a query slice x^{query}, its projection in the target feature space is computed as:

$$z^{query} = M(x^{query}) \tag{8}$$

The euclidean distance between the resulting 1024-dimensional vector z^{query} and the set of reference vectors Z^{ref} is then computed.

The elements in feature reference set are sorted from the nearest to the most distant and related elements in Y^{ref} are stored in the sorted list $Y^{ref,sorted}$.

To compute the final prediction $\widehat{y^{query}}$, the proposed algorithm computes:

$$\widehat{y^{query}} = \frac{1}{m} \sum_{i=1}^{m} y_i^{ref,sorted}, \qquad y_i^{ref,sorted} \in Y^{ref,sorted} \quad \forall i = 1, 2, ...n^{ref} \tag{9}$$

where $m \leq n^{ref}$ is a parameter of the algorithm, regulating the number of nearest neighbours considered in the score estimation. To tune it, an iterative process have to be put in place: for increasing values of m, the performance of the method is evaluated, looking for the best trade-off, as we discuss in detail in the experimental section.

4 Results

To verify the performance of the proposed approach, we used the dataset provided by the ICIAP 2021 conference organizer in the context of COVID-19 infection percentage estimation competition [4,17]. We recall that at the moment of the writing of this document, the validation and test set ground truth had not been made available yet, therefore is not possible to perform detailed analysis on the predictions.

We point out that while the training data, we noticed that slice *Image_0736.png* from CT scan 25 looks like a strong outlier: we checked the slices immediately before and after and found values 10 times lower, suggesting a possible mistake in the labelling phase, possibly a typo with decimal separator. To reduce the noise, we chose to exclude this sample from all the experiments below. Training parameters and environment are illustrated in Table 2.

Table 2. Summary of the training setup and relative hyperparameters

Optimization algorithm	Stochastic gradient descent
Learning rate	5e−2
Batch size	12 samples
Epochs	30
Environment	Tensorflow/Python
GPU accelerator	NVIDIA GeForce RTX2080 Ti 12GB

To improve generalization, we adopted online data augmentation strategy, applying the following transformations: *random horizontal flip* and *random crop* of 448×448 pixels patches. Finally, the pictures are resized to the original 512×512 pixels size.

4.1 Preliminary Results on the Training Set

As a preliminary experiment, we find the number of reference neighbors m in Eq. 9 that minimizes the severity score prediction error. We recall that the ground truth for neither the validation or test set is available at the moment of the writing of this document, so we had to rely on the train set solely.

To cope with the lack of ground truth for validation and test, we performed a leave-one-out cross validation (LOOCV). At each iteration, one patient is left out of the train set. Next, the other samples are used to train our neural network responsible for feature extraction minimizing the loss function described in the previous section. The experiments we showed that the loss function converged to about 10^{-2} already after 30 epochs of training. Next, the slices used for training are kept as reference set, while the slices from the left out patient are used as query set. For each query set slice, we compute the Mean Absolute Error (MAE), the Median Absolute Error (MdAE), the Pearson Correlation index (PC) and Root Mean Squared Error (RMSE) comparing the COVID-19 severity score predictions with their ground truth. The same metrics are computed for each slice in the query set. Next, the network is retrained from scratch leaving out another patient and this procedure is repeated for all the 132 patients in the train set. Finally, we find the optimal m that minimizes the MAE across all slices from all left out patients. Our experiments showed best performance for $m = 21$ neighbors, with MAE $= 1.847$, MdAE $= 1.119$, PC $= 0.994$, RMSE $=$ 2.838.

This value for parameter $m = 21$ is used also in the prediction of COVID-19 score for Validation and Test datasets.

To better understand the results, we visualized the distribution of the prediction error and the correlation of predictions with actual values in Fig. 3(a). The prediction error distribution has zero mean and most of its density is in the [−10%, 10%] interval. Figure 3(b) correlates predicted and ground truth COVID-19 severity scores for all the slices in the train set (PC = 0.994).

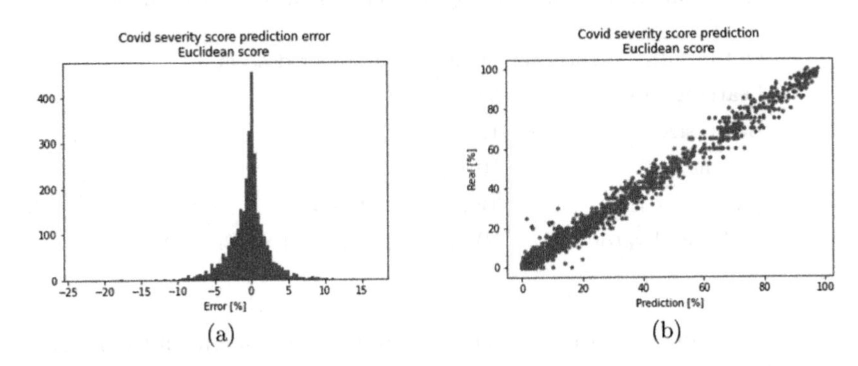

Fig. 3. Right: a) Distribution of prediction error for Train set by using leave-one-out cross-validation. Left: b) Correlation between predicted and actual values for COVID-19 severity score on Train set by using leave-one-out cross-validation

4.2 Validation and Test Set Results

We predicted the COVID-19 severity score for the validation set slices using the network trained on all train set slices and with $m = 21$. All the train set slices have been used as reference set, while the validation set slices are used as query slices. We submitted our predictions to the automatic platform managing the challenge, Codalab [6], that returned us the following results:

- MAE = 4.912
- PC = 0.943
- RMSE = 8.700

In Table 3 is showed the final ranking for the teams with Validation dataset. Our approach resulted in the 5th position.

Then, we repeated the procedure with the test set, and Table 4 shows that our approach resulted in the 5th position.

Table 3. Final competition ranking for Validation dataset. In **bold** the result related to the presented approach

Position	Team	MAE	PC	RMSE
1	SenticLab.UAIC	4.317	0.947	8.359
2	Captain-CSgroup	4.479	0.947	8.406
3	TAC	4.484	0.946	8.547
4	Taiyuan_university_lab713	4.504	0.949	8.097
5	**EIDOSlab_Unito (ours)**	**4.912**	**0.943**	**8.700**
6	IPLab	4.953	0.944	8.604
7	ACVLab	4.993	0.936	9.081

Table 4. Final competition ranking for Test dataset. In **bold** the result related to the presented approach

Position	Team	MAE	PC	RMSE
1	Taiyuan_university_lab713	3.557	0.855	7.510
2	TAC	3.645	0.802	8.571
3	SenticLab.UAIC	4.617	0.763	9.100
4	ACVLab	4.866	0.729	10.275
5	**EIDOSlab_Unito (ours)**	**5.020**	**0.798**	**9.006**
6	Captain-CSgroup	5.168	0.772	8.392
7	IPLab	6.536	0.709	9.976

Table 5. Final competition ranking overall. In **bold** the result related to the presented approach

Position	Team	MAE	PC	RMSE
1	Taiyuan_university_lab713	3.841	0.883	7.920
2	TAC	3.897	0.845	8.554
3	SenticLab.UAIC	4.483	0.819	8.467
4	ACVLab	4.904	0.791	9.439
5	Captain-CSgroup	4.961	0.825	8.402
6	**EIDOSlab_Unito (ours)**	**4.988**	**0.841**	**8.792**
7	IPLab	6.061	0.779	9.016

5 Conclusions

In this paper we proposed a methodology to assess the COVID-19 disease severity score in CT scan slice images based on minimizing a loss function expressing the distance between projected features in a lower dimensional feature space as projected by a convolutional deep neural network. Our approach scored 6th overall in the ICAP2021 COVID-19 infection percentage estimation competition.

The lack of validation and test ground truth prevented us from further exploring our results. For example, the network we trained obtained lowest loss with a learning rate = 0.05, a relatively high value. Also, performance on validation and test are slightly worse than on training with LOOCV, suggesting a possible overfitting situation. As the validation and test set results are made available by the challenge authors, we will compute the relative metrics and take action accordingly (Table 5).

References

1. Ai, T., et al.: Correlation of chest CT and RT-PCR testing for coronavirus disease 2019 (COVID-19) in China: a report of 1014 cases. Radiology **296**(2), E32–E40 (2020)
2. Amyar, A., Modzelewski, R., Li, H., Ruan, S.: Multi-task deep learning based CT imaging analysis for COVID-19 pneumonia: classification and segmentation. Comput. Biol. Med. **126**, 104037 (2020)
3. Bougourzi, F., Distante, C., Abdelkrim, O., Dornaika, F., Hadid, A., Taleb-Ahmed, A.: Per-COVID-19: a benchmark database for COVID-19 percentage prediction from CT-scans (2021)
4. Bougourzi, F., Distante, C., Ouafi, A., Dornaika, F., Hadid, A., Taleb-Ahmed, A.: Per-COVID-19: a benchmark dataset for COVID-19 percentage estimation from CT-scans. J. Imaging **7**(9) (2021). https://doi.org/10.3390/jimaging7090189, https://www.mdpi.com/2313-433X/7/9/189
5. Chen, J.I.Z.: Design of accurate classification of COVID-19 disease in x-ray images using deep learning approach. J. ISMAC **3**(02), 132–148 (2021)
6. CodaLab: Codalab - competition (2021). https://competitions.codalab.org/competitions/35575
7. Fang, Y., et al.: Sensitivity of chest CT for COVID-19: comparison to RT-PCR. Radiology **296**(2), E115–E117 (2020)
8. Hosseiny, M., Kooraki, S., Gholamrezanezhad, A., Reddy, S., Myers, L.: Radiology perspective of coronavirus disease 2019 (COVID-19): lessons from severe acute respiratory syndrome and middle east respiratory syndrome. Am. J. Roentgenol. **214**(5), 1078–1082 (2020)
9. Hu, S., et al.: Weakly supervised deep learning for COVID-19 infection detection and classification from CT images. IEEE Access **8**, 118869–118883 (2020)
10. Huang, G., Liu, Z., Van Der Maaten, L., Weinberger, K.Q.: Densely connected convolutional networks. In: 2017 IEEE Conference on Computer Vision and Pattern Recognition (CVPR), pp. 2261–2269 (2017). https://doi.org/10.1109/CVPR.2017.243
11. Lai, C.C., et al.: Asymptomatic carrier state, acute respiratory disease, and pneumonia due to severe acute respiratory syndrome coronavirus 2 (SARS-CoV-2): facts and myths. J. Microbiol. Immunol. Infect. **53**(3), 404–412 (2020)
12. Li, L., et al.: Using artificial intelligence to detect COVID-19 and community-acquired pneumonia based on pulmonary CT: evaluation of the diagnostic accuracy. Radiology **296**(2), E65–E71 (2020)

13. Ranney, M.L., Griffeth, V., Jha, A.K.: Critical supply shortages-the need for ventilators and personal protective equipment during the COVID-19 pandemic. N. Engl. J. Med. **382**(18), e41 (2020)
14. Roy, S., et al.: Deep learning for classification and localization of COVID-19 markers in point-of-care lung ultrasound. IEEE Trans. Med. Imaging **39**(8), 2676–2687 (2020)
15. Sohrabi, C., et al.: World health organization declares global emergency: a review of the 2019 novel coronavirus (COVID-19). Int. J. Surg. **76**, 71–76 (2020)
16. Tricarico, D., et al.: Convolutional neural network-based automatic analysis of chest radiographs for the detection of COVID-19 pneumonia: a prioritizing tool in the emergency department, phase I study and preliminary "real life"; results. Diagnostics **12**(3) (2022). https://doi.org/10.3390/diagnostics12030570, https://www.mdpi.com/2075-4418/12/3/570
17. Vantaggiato, E., Paladini, E., Bougourzi, F., Distante, C., Hadid, A., Taleb-Ahmed, A.: COVID-19 recognition using ensemble-CNNs in two new chest X-ray databases. Sensors **21**(5), 1742 (2021)
18. Wang, S., et al.: A deep learning algorithm using CT images to screen for corona virus disease (COVID-19). Eur. Radiol., 1–9 (2021)
19. Wong, H.Y.F., et al.: Frequency and distribution of chest radiographic findings in patients positive for COVID-19. Radiology **296**(2), E72–E78 (2020)
20. Xie, X., Zhong, Z., Zhao, W., Zheng, C., Wang, F., Liu, J.: Chest CT for typical coronavirus disease 2019 (COVID-19) pneumonia: relationship to negative RT-PCR testing. Radiology **296**(2), E41–E45 (2020)
21. Yang, Y., et al.: Laboratory diagnosis and monitoring the viral shedding of SARS-CoV-2 infection. Innovation **1**(3), 100061 (2020)

Mixup Data Augmentation for COVID-19 Infection Percentage Estimation

Maria Ausilia Napoli Spatafora(✉), Alessandro Ortis, and Sebastiano Battiato

Department of Mathematics and Computer Science, University of Catania, Catania, Italy
maria.napolispatafora@phd.unict.it, {ortis,battiato}@dmi.unict.it

Abstract. The outbreak of the COVID-19 pandemic considerably increased the workload in hospitals. In this context, the availability of proper diagnostic tools is very important in the fight against this virus. Scientific research is constantly making its contribution in this direction. Actually, there are many scientific initiatives including challenges that require to develop deep algorithms that analyse X-ray or Computer Tomography (CT) images of lungs. One of these concerns a challenge whose topic is the prediction of the percentage of COVID-19 infection in chest CT images. In this paper, we present our contribution to the COVID-19 Infection Percentage Estimation Competition organised in conjunction with the ICIAP 2021 Conference. The proposed method employs algorithms for classification problems such as Inception-v3 and the technique of data augmentation *mixup* on COVID-19 images. Moreover, the *mixup* methodology is applied for the first time in radiological images of lungs affected by COVID-19 infection, with the aim to infer the infection degree with slice-level precision. Our approach achieved promising results despite the specific constrains defined by the rules of the challenge, in which our solution entered in the final ranking.

Keywords: Computer Vision · Inception-v3 · Computer Tomography

1 Introduction

In December 2019 a significant increase of pneumonia cases was reported in Wuhan, Hubei Province, China [9]. These cases are due to an infection with a novel coronavirus. In the following weeks, infections spread across China and other countries around the world [5]. On January 30, 2020, the World Health Organization (WHO) declared the outbreak a Public Health Emergency of International Concern [25]. On February 12, 2020, the WHO baptized the disease caused by the novel coronavirus "coronavirus disease 2019" (COVID-19) [26]. Numerous epidemiological studies have been conducted to model the outbreak and the trend of the pandemic [15,16]. At the same time, a group of international experts, with a range of specialisations, are working to try to contain and

P. L. Mazzeo et al. (Eds.): ICIAP 2022 Workshops, LNCS 13374, pp. 508–519, 2022.
https://doi.org/10.1007/978-3-031-13324-4_43

defeat the pandemic. During these months, the radiological imaging techniques, e.g. X-rays and Computed Tomography (CT), are demonstrating to be the most effective diagnostic tests of this disease including the follow-up assessment and evaluation of disease evolution [3,21] as well as the quick detection of proper interventions especially for asymptomatic cases. Actually, a clinical study with 1014 patients in Wuhan China, has shown that chest CT analysis can achieve 0.97 of sensitivity, 0.25 of specificity, and 0.68 of accuracy for the detection of COVID-19 [1]. Similar observations were also reported in other studies [4,19] suggesting that radiological imaging may help support early screening of COVID-19. Nevertheless, this practice requires a complex effort by radiologists. For this reason, many Computer Vision systems [2,6–8,24] have been developed and proposed to offer a diagnostic tool that can help the decision-makers in the medical and health centers. The goal is a rapid and accurate identification of COVID-19 infection in radiography images (i.e. x-ray or CT imaging) of lungs.

Given the significant scientific and social impact of COVID-19 infection, multiple research initiatives have been proposed worldwide. One of these is the First International Workshop on Medical Imaging Analysis for COVID-19 (MIA COVID) that is associated with 21st International Conference on Image Analysis and Processing (ICIAP 2021). The workshop proposes an associated challenge about the estimation of the percentage of COVID-19 Infection from thoracic CT scans. We as IPLab team[1] of the Department of Mathematics and Computer Science of the University of Catania participated in the challenge. In this paper, we present our results in the MIA COVID challenge, in which we entered in the final top-score ranking (seventh ranked out of 50 participants). The paper illustrates the methodology employed to predict the percentage of COVID-19 infection in CT slices of lungs achieving the best results by the application of the *mixup* data augmentation technique during the training of the neural network. We believe that our method would help radiologists in diagnosing infection related to COVID-19. To encourage research on this topic, we publicly release our codes and models at the following url: https://github.com/ausilianapoli/Percentage-Covid-Estimator.

The paper is structured as follows: Sect. 2 describes related work in relation to our work, Sect. 3 describes all the details of the MIA challenge concerning COVID-19, the methodology and approach used to solve the challenge problem are described in Sect. 4, whereas the experiments are detailed in Sect. 5. Finally, the manuscript concludes with some considerations regarding the results obtained and possible future works in Sect. 6.

2 Related Works

In recent years, there has been an increasing number of algorithmic solutions based on computer vision methods to several tasks and practical applications, among others, significant examples are the applications of computer vision in the field of medical imaging [12,18]. Nowadays, computer vision is used in solving

[1] https://iplab.dmi.unict.it/.

problems involving COVID-19. Research in this area is very fervent, given the pandemic state of the world today. Indeed, the combination of Computer Vision techniques with various imaging modalities can assist to increase the efficiency of COVID-19 detection worldwide [3]. Gudigar et al. [6] have published a survey in which they collected and filtered 184 papers that are the most influential in the field of Computer Vision applied to COVID-19, which deals with classification issues. Generally, in these problems the input data are CT images or X-ray images of patients. For instance, Heidari et al. [7] use chest X-ray images to detect COVID-19 induced pneumonia. They fine-tune a VGG16 [14] based on Convolutional Neural Network (CNN) model pre-trained on ImageNet [20] challenge database yielding 94.5% classification accuracy in three-classes scenario (i.e., pneumonia induced by COVID-19, other pneumonia and normal cases). Moreover, Hellwan et al. [8] finetuned an Imagenet pre-trained DenseNet-201 [31] on a dataset of chest CT images, achieving an accuracy of 97.8% in a two-classes scenario (i.e. COVID-19 vs no COVID-19). Vantaggiato et al. [24] propose an Ensemble-CNNs approach to distinguish chest X-ray images between healthy, COVID-19 and pneumonia resulting in an accuracy of 100%. The CNNs employed are ResNext-50 [28], Inception-v3 [23] and DenseNet-161. Bougourzi et al. [2] introduce a new problem in this research field. It is the estimation of the percentage of COVID-19 infection from CT scans. For this purpose, the authors collected a dataset of 183 CT scans and 3986 slices and the dataset was named Per-COVID-19. The authors employed the Inception-v3 neural network pre-trained on ImageNet to perform transfer learning. However, this work is different from the previous ones because it deals with a problem of regression instead of classification. For this reason, the loss function employed is Huber Loss and performance is measured with Mean Absolute Error (MAE), Pearson Correlation (PC) and Root Mean Square Error (RMSE), obtaining 5.34 MAE, 0.9330 PC and 9.44 RMSE.

We decided to work on the latter problem with a strong emphasis on data augmentation that deals with a suite of techniques that enhance the size and quality of training datasets such that better deep learning models can be developed using them. Traditional methods of data augmentation are proven to be a good practice in many fields [22]. One of these is the color space augmentation which, among other things, leaves plenty of room for creativity as shown in [17,27]. A not conventional type of transformation, e.g. *mixup* [29], regards the combination of two or more original images to generate the new one. The *mixup* authors tested it with state-of-the-art models on some datasets such as ImageNet and CIFAR. These experiments have shown that this data augmentation technique improves the generalisation error. Moreover, the *mixup* technique is employed in chest CT imaging with success. The authors in [30] apply this method on chest CT images to do predictions about pulmonary adenocarcinoma with much improved performance compared to experiments without *mixup*. For these reasons, we choose to employ *mixup* in COVID-19 chest CT images. To the best of our knowledge, this is the first study introducing *mixup* into COVID-19 image computing.

3 Challenge

The MIA COVID 2022 has launched a scientific challenge which task is the estimation of the percentage of COVID-19 infection from thoracic CT scans. The presence of COVID-19 is diagnosed through the reverse transcription polymerase chain reaction. Nowadays, this test is considered as the global standard method for COVID-19 diagnosis [13]. Furthermore, the presence of COVID-19 is confirmed by the CT scan manifestations identified by two experienced thoracic radiologists. In this way, the authors of [2] have built a dataset for the competition's participants (i.e., Per-COVID-19).

The challenge has three sets: Train, Validation, and Test. The Train set contains 132 CT scans, from which 128 CT scans have COVID-19, and the rest 4 CT scans have not any infection type (i.e., healthy). Instead, the Validation set includes 57 CT scans, from which 55 CT scans are affected by COVID-19, and the rest 2 CT scans are healthy. Moreover, no information was provided for the Test set that consists of 4449 CT scans. Finally, the infection labels are available only for the Train set. The Fig. 1 shows some CT scans from the three sets of data.

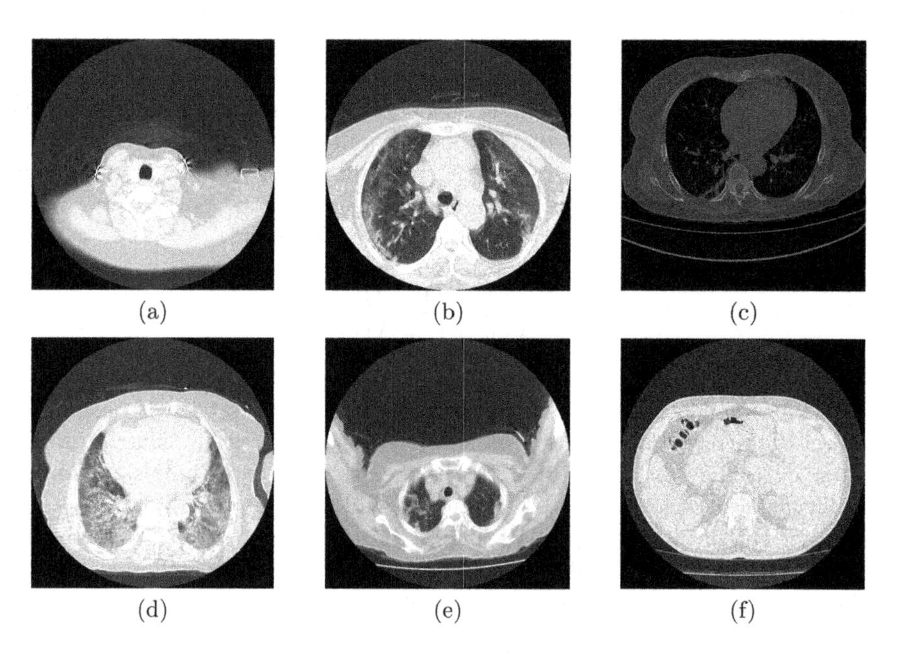

Fig. 1. Sample CT scans included in the three sets of data: **(a)** and **(d)** are train images with respectively 0% and 100% infection percentage; **(b)** and **(e)** are validation images; **(c)** and **(f)** are test images.

The challenge requires that proposed approaches estimate the percentage of COVID-19 infection from each slice using deep learning techniques. Moreover, there are some limitations for the participants and the proposed approaches. Only ImageNet's pre-trained models and lung nodule segmentation models are allowed. Thus, the use of other pre-trained models is not allowed. In addition, it is forbidden to use data other than that provided by the challenge with respect to Train, Validation and Test sets. Finally, the challenge's organisers have established also the evaluation metrics - that are MAE, PC and RMSE - for all approaches. The most important evaluation criterion is the MAE. In the event of two or more competitors achieving the same MAE, the PC and the RMSE are considered as the tie-breaker.

4 Methodology

The COVID-19 pandemic increased considerably the workload in hospitals. This has led to the development of various techniques to make diagnosis faster and more accurate. In this context, computer vision is contributing by developing deep learning algorithms that can predict the rate of infection in chest CT of the patients.

For these reasons, there are an increasing number of research works in this field. Thus, we decided to propose a novel methodology based on a data augmentation technique that firstly is applied on COVID-19 lung images. The technique is *mixup* that creates convex combinations of pairs of samples and their corresponding labels [29]. This means that a *mixup* process is simply averaging out two images and their labels correspondingly as new data. Specifically, *mixup* augmentation is explained with the following equations

$$\begin{aligned} \hat{x} &= \lambda \cdot x_i + (1 - \lambda) \cdot x_j \\ \hat{y} &= \lambda \cdot y_i + (1 - \lambda) \cdot y_j \end{aligned} \tag{1}$$

where $\hat{x}$ is a blend of two images that are x_i and x_j, while $\hat{y}$ is the mixed label of labels y_i and y_j. Instead, λ is a beta distribution generated with a number between 0 and 1 that specifies the weight of the two samples in contributing to the new data. The definition of *mixup* implies that the two samples belong to different classes. Otherwise, this technique would not be applied properly. This is valid for classification problems which are the general and most frequent use case of *mixup*. However, we deal with a regression task and therefore the choice of the two samples is random. Examples of new images generated with *mixup* are shown in Fig. 2.

The *mixup* augmentation creates virtual examples that significantly increase the diversity of data available regardless of the neural architecture employed. Thus, the technique improves the generalisation error of the neural network making it more robust. These are the benefits introduced by the use of this technique that does not compromise the training time of a neural network. Actually, *mixup* inside the training pipeline does not bring computational overhead since it is fast and needs only a few lines of code.

In our approach, we employ the neural network Inception-v3 pre-trained on ImageNet. Inception-v3 is a widely-used neural network for image recognition. Thus, this network is employed in classification problems. Since we deal with a regression task, we replace the last layer of the original Inception-v3 with a fully connected layer with a single output that is the probability of COVID-19 infection. We choose the Huber loss function [10] as criterion. The loss function is defined for N batch size, and $Y = (y_1, y_2, \ldots, y_N)$ are the ground-truth and $\bar{Y} = (\bar{y_1}, \bar{y_2}, \ldots, \bar{y_N})$ are their corresponding estimated percentages. The Huber loss function is defined by

$$L_{Huber} = \frac{1}{N} \sum_{i=1}^{N} l_i \tag{2}$$

where N is the batch size and l_i is defined by

$$l_i = \begin{cases} 0.5\,(x_i - \bar{y_i})^2, \text{ if } |x_i - \bar{y_i}| \leq \delta \\ \delta\,|x_i - \bar{y_i}| - 0.5\,\delta^2, \quad \text{otherwise} \end{cases} \tag{3}$$

where δ is a hyperparameter.

5 Experimental Results

The approach described in the previous Section has been trained and validated by means of the dataset released by the challenge described in Sect. 3. All experiments and setups are reported in detail in the following subsections.

5.1 Neural Network Model Setup

We use the three sets of data provided by the challenge with their respective purposes namely Train set for training our approach, Validation set for validating it and Test set for the final rank. Images have been normalised to the range $[0, 1]$ using the mean and standard deviation of ImageNet challenge database. We employ the *mixup* augmentation as described previously and we generate the beta distribution λ with its parameter $= 0.2$ in Eq. (1). The Fig. 2a is generated from image x_i with 0% infection and image x_j 98% resulting with 38% COVID-19 infection; the Fig. 2b derives from image x_i with 13% infection and image x_j 5% resulting with 10% COVID-19 infection; the last Fig. 2c is mixed from image x_i with 33% infection and image x_j 44% resulting with 37% COVID-19 infection. Moreover, we have used other traditional data augmentation techniques that are:

- gaussian blur with kernel 5×5;
- color jitter;
- random horizontal flipping;
- random vertical flipping;
- random cropping 492×492 followed by a resizing to 512×512;
- rotation with a random degree of $\pm 10°$.

We train our network for 50 epochs with an initial learning rate of 0.0001 with decays by a factor 10 every 10 epochs and batch size equals 20. The weight decay hyperparameter is set to 0.5 to regularize the network. Another hyperparameter varies during the training i.e. δ in the Huber loss function. It decreases from 15 to 1 by 0.5 step every epoch. Finally, we use Stochastic Gradient Descend (SGD) as optimizer.

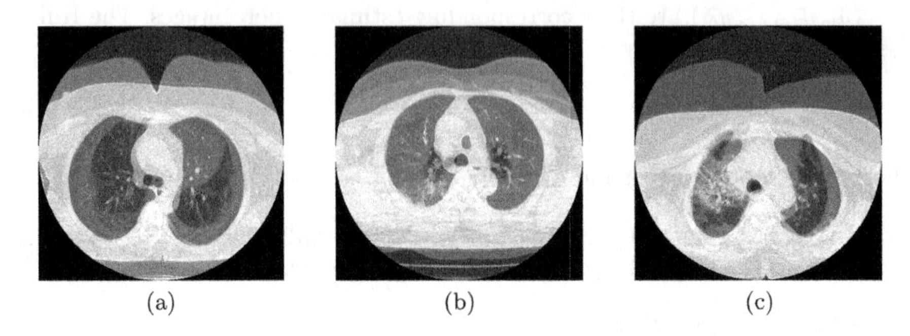

(a) (b) (c)

Fig. 2. Examples of new data generated with mixup augmentation technique from training images

5.2 Results and Discussion

We employ the evaluation metrics suggested by the challenge's organisers i.e. MAE, PC and RMSE. The Table 1 shows the results of our best model for each phase of the challenge. In the final rank our approach is placed seventh. To interpret the network predictions, we also produce saliency maps to visualise the areas of the image most indicative of the infection using CAMERAs [11]. To generate the saliency maps, we feed an image into the fully trained network and choose the final convolutional block of the Inception-v3 network to obtain these maps. The Figs. 3 and 4 show some prediction and their saliency maps respectively for Validation set and Test set. Analysis of the saliency maps shows that the algorithm focuses on the highest density areas, i.e., those with a grey level close to white. This is meaningful because the COVID-19 infection involves an increase in density within the lungs resulting in whitish areas. It is important to note that the lungs are surrounded by other tissue that shows up as a thick whitish area that confuses the algorithm. Actually, it happens that the focus is on these regions where there is no COVID-19 infection.

The experiments show that the proposed method is valid for processing COVID-19 chest CT images. The validation MAE is better than the threshold for admission to the final phase of the challenge (i.e. 5.294). Moreover, in this phase we improved the results achieved by Bougourzi et al. [2] that deal with our same problem but on the entire labeled Per-COVID-19 dataset. Actually, the Train set of the challenge is a subset of the Per-COVID-19 database

Table 1. Results achieved by IPLab team during all challenge's phases.

Phase	MAE	PC	RMSE
Validation results	4.953	0.944	8.604
Testing results	6.536	0.709	9.976
Final ranking ($0.3 * Validation + 0.7 * Testing$)	6.061	0.779	9.016

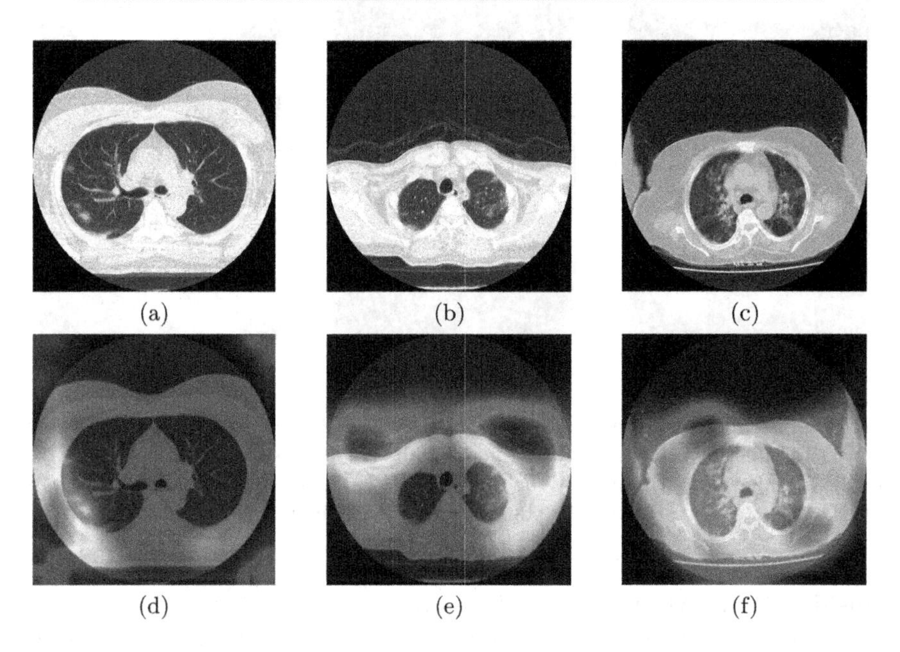

Fig. 3. Some predictions on images from Validation set and their saliency maps: **(a)** has 3% COVID-19 infection estimated by the model while **(d)** is its saliency map; **(b)** has 18% COVID-19 infection estimated by the model while **(e)** is its saliency map; **(c)** has 41% COVID-19 infection estimated by the model while **(e)** is its saliency map.

and labels are provided for only this subset. This limitation together with the impossibility of using external data sources promotes the occurrence of overfitting. Actually, there are signs of overfitting in the performance on the Test set since these are worse than the ones in the Validation set. This is due to the limited amount of data available for the above-mentioned reasons and the diversity of testing images whose details are illustrated in the following. In addition, it was not possible to fine-tune the network after the results about the Test set due to the challenge rules[2]. Despite the limitations imposed and the difficulties exhibited, the proposed method is promising and valid given that the estimated predictions can be plausible as shown in the images reported in this manuscript.

[2] https://github.com/faresbougourzi/Covid-19-Infection-Percentage-Estimation-Challenge.

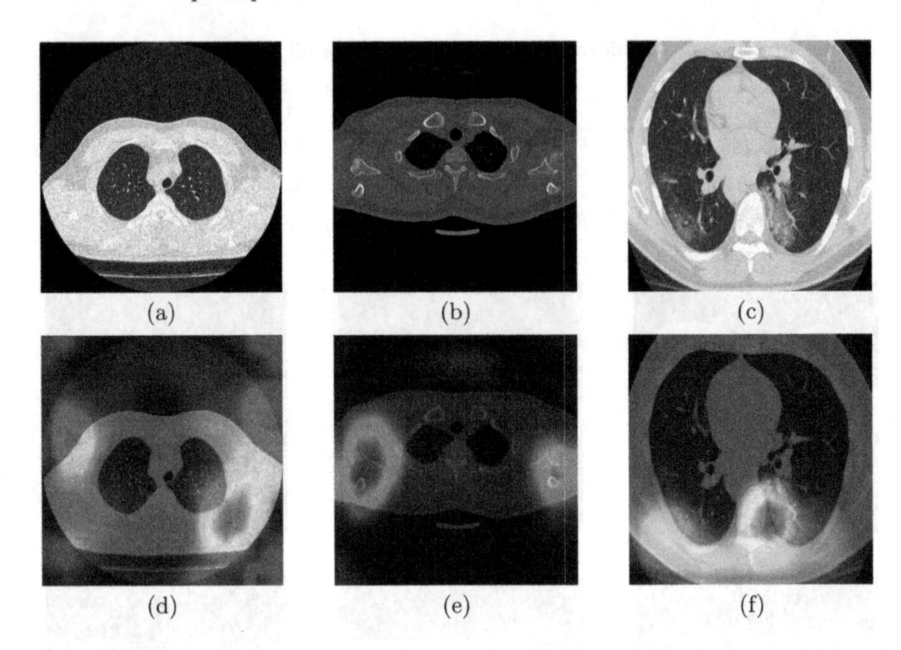

Fig. 4. Some predictions on images from Test set and their saliency maps: **(a)** has 2% COVID-19 infection estimated by the model while **(d)** is its saliency map; **(b)** has 14% COVID-19 infection estimated by the model while **(e)** is its saliency map; **(c)** has 16% COVID-19 infection estimated by the model while **(f)** is its saliency map.

Considerations About Test Set. As previously mentioned, the Test set is different from other datasets. Actually, the Fig. 5 shows that the source of the Testing set is not the same as the other two data sets for the following reasons:

- the size of the images is not always the same (i.e. 512×512) unlike Train set and Validation set as shown in Fig. 5a for which our model predicts 18% COVID-19 infection;
- the colour gamut in Train set and Validation set is the same but not in the Test set for some images as displayed in Fig. 5b for which our approach estimates 7% COVID-19 infection;
- the presence of text in the images (e.g. Fig. 5c for which the model predicts 64% COVID-19 infection), while Train and Validation images have no text.

This negatively affects the performance on the testing set.

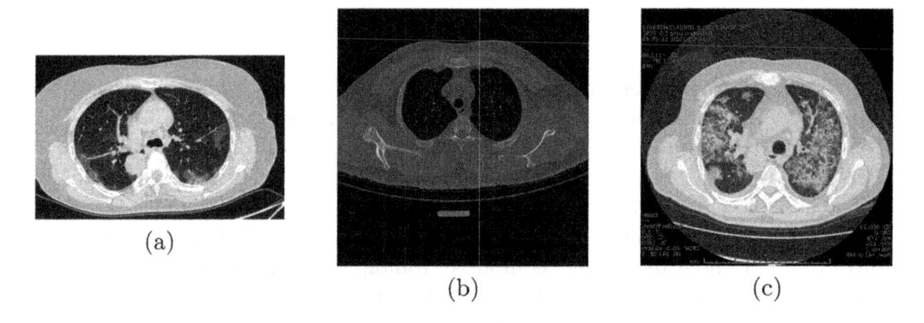

(a) (b) (c)

Fig. 5. Some images of Test set that reveal the different source of the data.

6 Conclusions

The COVID-19 pandemic significantly affected everyone's life, as well as any type of organisation and public service. Hospitals are the most affected target due to the large number of cases of patients admitted for this infection, and indirect consequences to patients affected by other diseases, overload of the healthcare personnel, and the subsequent crisis of any care service. The diagnostic techniques available are good for diagnosing this disease. However, science and research are making a considerable contribution to the development of new diagnostic methods that are faster and more accurate. In this context, computer vision offers tools that could support medical decisions in the presence of radiological images of the lungs. Actually, many research works deal with the COVID-19 detection in chest X-ray or CT resolving a classification task. Moreover, the MIA COVID 2022 workshop has organised a challenge that requires deep learning algorithms for predicting the percentage of COVID-19 infection in chest CT images of patients. This is a regression task to which we have chosen to participate by adopting state-of-the-art classification algorithms. Specifically, we employ the Inception-v3 neural network together with the data augmentation *mixup* technique. The proposed method reaches the final classification by passing the Validation and Testing phases. Our results show that the approach is promising despite the limitations given by the constrains defined by the challenge rules. For these reasons, we plan to overcome the limitations imposed by the challenge such as the prohibition to use external data to those provided by the challenge. As future work, we thought to give as input to our approach the output of a lung segmentation network. Other improvements may be carried out by the exploitation of pre-trained models designed for lungs' micro nodules segmentation, in order to make transfer learning on the available Per-COVID-19 dataset. In addition, we plan further experiments on the *mixup* method to evaluate its advantages and disadvantages in an uncontrolled domain such as misalignment of the chest in the CT images. We would also like to compare *mixup* technique with random noise as the adding noise to the training data could improve generalisation and prevent overfitting.

References

1. Ai, T., et al.: Correlation of chest CT and RT-PCR testing for coronavirus disease 2019 (COVID-19) in China: a report of 1014 cases. Radiology **296**, E32–E40 (2020)
2. Bougourzi, F., Distante, C., Ouafi, A., Dornaika, F., Hadid, A., Taleb-Ahmed, A.: Per-COVID-19: a benchmark dataset for COVID-19 percentage estimation from CT-Scans. J. Imaging **7**(9), 189 (2021). https://doi.org/10.3390/jimaging7090189
3. Cellina, M., Martinenghi, C., Marino, P., Oliva, G.: COVID-19 pneumonia-ultrasound, radiographic, and computed tomography findings: a comprehensive pictorial essay. Emerg. Radiol. **28**, 519–526 (2021)
4. Fang, Y., et al.: Sensitivity of chest CT for COVID-19: comparison to RT-PCR. Radiology **296**(2), E115–E117 (2020)
5. Giovanetti, M., Benvenuto, D., Angeletti, S., Ciccozzi, M.: The first two cases of 2019-nCoV in Italy: where they come from? J. Med. Virol. **92**(5), 518–521 (2020)
6. Gudigar, A., et al.: Role of artificial intelligence in COVID-19 detection. Sensors **21**, 8045 (2021). https://doi.org/10.3390/s21238045
7. Heidari, M., Mirniaharikandehei, S., Khuzani, A., Danala, G., Qiu, Y., Zheng, B.: Improving the performance of CNN to predict the likelihood of COVID-19 using chest X-ray images with preprocessing algorithms. Int. J. Med. Inform. **144**, 104284 (2020). https://doi.org/10.1016/j.ijmedinf.2020.104284
8. Helwan, A., Ma'aitah, M.K.S., Hamdan, H., Ozsahin, D.U., Tuncyurek, O.: Radiologists versus deep convolutional neural networks: a comparative study for diagnosing COVID-19. Comput. Math. Methods Med. **2021**, 5527271 (2021). https://doi.org/10.1155/2021/5527271
9. Huang, C., et al.: Clinical features of patients infected with 2019 novel coronavirus in Wuhan, China. Lancet **395**(10223), 497–506 (2020)
10. Huber, P.J.: Robust estimation of a location parameter. Ann. Math. Stat. **35**, 73–101 (1964)
11. Jalwana, M.A., Akhtar, N., Bennamoun, M., Mian, A.: CAMERAS: enhanced resolution and sanity preserving class activation mapping for image saliency. In: Proceedings of the IEEE/CVF Conference on Computer Vision and Pattern Recognition (2021)
12. Kandel, I., Castelli, M., Popovič, A.: Musculoskeletal images classification for detection of fractures using transfer learning. J. Imaging **6**(11), 127 (2020). https://doi.org/10.3390/jimaging6110127
13. Kucirka, L.M., Lauer, S.A., Laeyendecker, O., Boon, D., Lessler, J.: Variation in false-negative rate of reverse transcriptase polymerase chain reaction-based SARS-CoV-2 tests by time since exposure. Ann. Internal Med. **173**(4), 262–267 (2020). https://doi.org/10.7326/m20-1495
14. Liu, S., Deng, W.: Very deep convolutional neural network based image classification using small training sample size. In: 2015 3rd IAPR Asian Conference on Pattern Recognition (ACPR) (2015). https://doi.org/10.1109/ACPR.2015.7486599
15. Maugeri, A., Barchitta, M., Battiato, S., Agodi, A.: Estimation of unreported novel coronavirus (SARS-CoV-2) infections from reported deaths: a susceptible-exposed-infectious-recovered-dead model. J. Clin. Med. **9**, 1350 (2020). https://doi.org/10.3390/jcm9051350
16. Maugeri, A., Barchitta, M., Battiato, S., Agodi, A.: Modeling the novel coronavirus (SARS-CoV-2) outbreak in Sicily, Italy. Int. J. Environ. Res. Public Health **17**(14), 4964 (2020). https://doi.org/10.3390/ijerph17144964

17. Mikołajczyk, A., Grochowski, M.: Data augmentation for improving deep learning in image classification problem. In: 2018 International Interdisciplinary PhD Workshop (IIPhDW) (2018). https://doi.org/10.1109/IIPHDW.2018.8388338
18. Nayak, D.R., Padhy, N., Mallick, P.K., Bagal, D.K., Kumar, S.: Brain tumour classification using noble deep learning approach with parametric optimization through metaheuristics approaches. Computers **11**, 10 (2022)
19. Ng, M.Y., et al.: Imaging profile of the COVID-19 infection: radiologic findings and literature review. Radiol. Cardiothoracic Imaging **2**(1), e200034 (2020)
20. Russakovsky, O., et al.: ImageNet large scale visual recognition challenge. Int. J. Comput. Vis. **115**(3), 211–252 (2015). https://doi.org/10.1007/s11263-015-0816-y
21. Shi, F., et al.: Review of artificial intelligence techniques in imaging data acquisition, segmentation, and diagnosis for COVID-19. IEEE Rev. Biomed. Eng. **14**, 4–15 (2020)
22. Shorten, C., Khoshgoftaar, T.M.: A survey on image data augmentation for deep learning. J. Big Data **6**(1), 1–48 (2019). https://doi.org/10.1186/s40537-019-0197-0
23. Szegedy, C., Vanhoucke, V., Ioffe, S., Shlens, J., Wojna, Z.: Rethinking the inception architecture for computer vision. In: 2016 IEEE Conference on Computer Vision and Pattern Recognition (CVPR) (2016)
24. Vantaggiato, E., Paladini, E., Bougourzi, F., Distante, C., Hadid, A., Taleb-Ahmed, A.: Covid-19 recognition using ensemble-CNNs in two new chest x-ray databases. Sensors **21**(5), 1742 (2021)
25. World Health Organization: Statement on the second meeting of the International Health Regulations (2005). Emergency Committee regarding the outbreak of novel coronavirus (2019-nCoV). https://www.who.int/news/item/30-01-2020-statement-on-the-second-meeting-of-the-international-health-regulations-(2005)-emergency-committee-regarding-the-outbreak-of-novel-coronavirus-(2019-ncov). Accessed 14 Mar 2022
26. World Health Organization: WHO Director-General's remarks at the media briefing on 2019-nCoV on 11 February 2020. https://www.who.int/director-general/speeches/detail/who-director-general-s-remarks-at-the-media-briefing-on-2019-ncov-on-11-february-2020. Accessed 14 Mar 2022
27. Wu, R., Yan, S., Shan, Y., Dang, Q., Sun, G.: Deep image: scaling up image recognition. arXiv:abs/1501.02876 (2015)
28. Xie, S., Girshick, R., Dollar, P., Tu, Z., He, K.: Aggregated residual transformations for deep neural networks. In: Proceedings of the IEEE Conference on Computer Vision and Pattern Recognition 2017 (2017)
29. Zhang, H., Cissé, M., Dauphin, Y., Lopez-Paz, D.: Mixup: beyond empirical risk minimization. arXiv:abs/1710.09412 (2018)
30. Zhao, W., et al.: Toward automatic prediction of EGFR mutation status in pulmonary adenocarcinoma with 3D deep learning. Cancer Med. **8**(7), 3532–3543 (2019)
31. Zhu, Y., Newsam, S.: DenseNet for dense flow. In: 2017 IEEE International Conference on Image Processing (ICIP) (2017)

Swin Transformer for COVID-19 Infection Percentage Estimation from CT-Scans

Suman Chaudhary(✉), Wanting Yang, and Yan Qiang

Taiyuan University of Technology, Taiyuan, China
Chaudharysuman560@gmail.com, qiangyan@tyut.edu.cn

Abstract. Coronavirus disease 2019 (COVID-19) is an infectious disease that has spread globally, disrupting the health care system and claiming millions of lives worldwide. Because of the high number of Covid-19 infections, it has been challenging for medical professionals to manage this crisis. Estimating the Covid-19 percentage can help medical staff categorize patients by severity and prioritize accordingly. With this approach, the intensive care unit (ICU) can free up resuscitation beds for the critical cases and provide other treatments for less severe cases to efficiently manage the healthcare system during a crisis. In this paper, we present a transformer-based method to estimate covid-19 infection percentage for monitoring the evolution of the patient state from computed tomography scans (CT-scans). We used a particular Transformer architecture called Swin Transformer as a backbone network to extract the feature from the CT slice and pass it through multi-layer perceptron (MLP) to obtain covid-19 infection percentage. We evaluated our approach on the covid-19 infection percentage estimation challenge dataset, annotated by two expert radiologists. The experimental results show that the proposed method achieves promising performance with a mean absolute error (MAE) of 4.5042, Pearson correlation coefficient (PC) of 0.9490, root mean square error (RMSE) of 8.0964 on the given Val set leaderboard and a MAE of 3.5569, PC of 0.8547 and RMSE of 7.5102 on the given Test set Leaderboard. These promising results demonstrate the high potential of Swin Transformer architecture for this image regression task of covid-19 infection percentage estimation from CT-scans. The source code of this project can be found at: https://github.com/suman560/Covid-19-infection-percentage-estimation.

Keywords: COVID-19 · Deep learning · CT-scan · Transformers

1 Introduction

Coronavirus disease 2019 (Covid-19) is a highly contagious respiratory illness caused by severe acute respiratory syndrome coronavirus 2 (SARS-CoV-2). The first case was reported at the end of 2019. However, now it has spread globally, claiming millions of lives worldwide. It has completely disrupted the health care system of many countries due to its highly contagious nature. Early diagnosis of

P. L. Mazzeo et al. (Eds.): ICIAP 2022 Workshops, LNCS 13374, pp. 520–528, 2022.
https://doi.org/10.1007/978-3-031-13324-4_44

covid-19 can play a significant role in saving the lives of patients and preventing the virus from spreading around and causing a health crisis in a particular area.

In order to diagnose COVID-19 in person, various testing methods have been developed. Some of which includes Reverse Transcription Polymerase Chain Reaction (RT-PCR) [10], X-ray scan [19,23,24] and CT scan [2,12]. RT-PCR detects nucleic acid of the virus. Actually, it is considered the standard diagnostic method for covid-19 detection [7,27]. However, the RT-PCR test is very expensive and time-consuming, which prevents it from being the first choice for testing Covid-19 in many parts of the world [10]. X-ray and CT scans can be efficient alternatives for Covid-19 detection [7,19]. However, detecting covid-19 from these scans requires an expert radiologist. Due to the highly contagious nature of covid-19, the number of individuals requiring tests can be high, and the number of expert radiologists might not be sufficient to examine these scans and manage the crisis on time. This is where artificial intelligence (AI) comes into play [18].

AI has become a significant component in many technological and scientific fields, including medical image analysis. Although various computer-based methods have been playing an essential role in analyzing medical images for a long time, the rise of AI methods, especially deep learning-based methods, has significantly advanced the use of AI in the medical field [8,14]. In the past, in order to apply machine learning, it was essential to extract informative features that well represent patterns inherent in data. These meaningful features were mainly extracted by human experts in the respective field. Therefore, designing a computer-based algorithm for medical image analysis was very time-consuming, costly, and challenging. However, in the last few years, deep learning-based algorithms, which require minimum or no feature engineering, have made designing computer vision algorithms much more efficient and faster. The main reason for this sudden rise of these algorithms is the growth of the computational powers of modern computers and large-scale data that are much more readily available than in the past [21].

Most of the proposed approaches for Covid -19 diagnosis from CT focus on two tasks: Covid 19 detection [2,3,11,12,25] and Covid-29 segmentation [7,17,22,27]. However, CT offers many advantages over other approaches. In addition to very accurate Covid-19 detection, it can also be used to quantify infection and monitor disease evolution, which plays an essential role in saving a patient [13]. Moreover, with estimated covid-19 infection percentage from CT scans, the patients can be classified into Normal (0%), Minimal (<10%), Moderate (10–25%), Extent (25–50%), Severe (50–75%), and Critical (>75%) [9]. This categorization can help the intensive care unit (ICU) free up resuscitation beds for critical cases and provide other treatments for less severe cases to manage the healthcare system during a crisis efficiently.

Bougourzi et al. [4] evaluated the performance of three states of the art Deep convolutional neural networks (CNN) in their work for covid-19% estimation from CT-scans. The used CNN architectures are ResNext-50, Densenet161 and Inception-v3. Their results show that the Deep CNN architectures can estimate the COVID-19 infection percentage for monitoring the evolution of the patient state with high accuracy and efficiency. CNN architectures have the ability to capture the local relationship between the input features. However, they cannot capture the global relationship between the input features, preventing them from being highly robust and accurate performers. In 2017 Vaswani et al. [20] proposed another type of deep learning architecture, called transformer architecture, which efficiently captures the global dependencies between input features. It quickly becomes an integral component for natural language processing tasks. However, It consists of a self-attention mechanism whose computational cost grows exponentially with input size. Therefore, it was not easy to implement it for computer vision in a standard way until recently. In 2020 Dosovitsky et al. [6] proposed an efficient way to implement transformer architecture for computer vision. They used a sequence of image patches as input features instead of pixels. This simple technique drastically decreased the size of the input features. They were able to achieve performance as good as the state of the art CNN with minimum computational resources for training. This work, famously known as vision transformer (ViT), opened the path for researchers to find a more efficient way to use a transformer for computer vision. The ViT model had two main limitations. First, it consists of tokens of a fixed scale. However, unlike word tokens, the visual object can have a different scale. Second, its computational complexity grows quadratically with the size of the input, making it unfeasible for tasks involving high-resolution images. In order to overcome these limitations, Liu et al. [15] proposed another transformer-based architecture called the Swin transformer, which is highly flexible at different scales and has linear computational complexity with the input size.

In this paper, we analyzed the performance of Swin transformer architecture for COVID-19% estimation. Results show that Swin transformer, when used as the backbone network for feature extraction, can extract robust features and achieve promising performance in estimating the percentage of Covid-19 infection for monitoring the evolution of the patient state.

2 Methods

2.1 Dataset

In our experiments, we used COVID-19 infection percentage estimation challenge database [4]. It consists of three sets: Train, Val, and Test set. According to the data descriptions provided by challenge organizers, the Train set is obtained from 132 CT scans, from which 128 CT scans has been confirmed to have Covid-19 based on positive reverse transcription-polymerase chain reaction (RT-PCR) and CT scan manifestations identified by two experienced thoracic radiologists. The rest four CT scans do have not any infection type (Healthy). The Val set is obtained from 57 CT scans, from which 55 CT-scans has been confirmed to have Covid-19 based on positive reverse transcription-polymerase chain reaction (RT-PCR) and CT scan manifestations identified by two experienced thoracic radiologists. The rest two CT scans do not have any infection type (Healthy).

There are 3054 CT slices in the Train set, 1301 CT slices in the Val set and 4449 CT slices in the Test set. Some of the CT slices from the Train set with their covid infection percentage is shown below in Fig. 1.

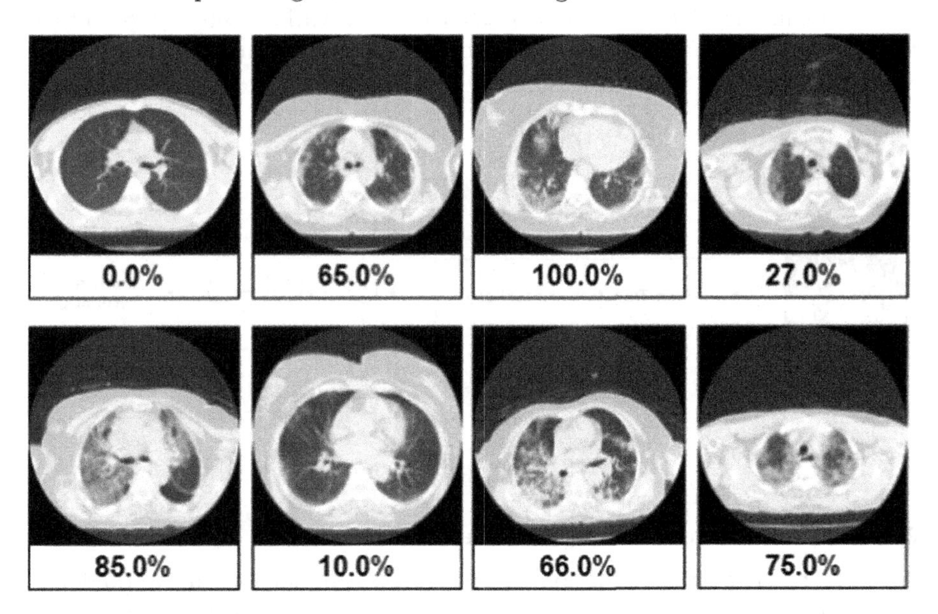

Fig. 1. CT slice with their covid infection percentage

2.2 Swin Transformer

The Swin transformer [15] is a variant of the Vision transformer. It consists of shifted windows that restrict the computation of self-attention to non-overlapping windows while allowing cross-attention. These simple techniques make this model extremely flexible at different scales and have linear computational complexity with respect to the size of the input. As discussed in [15], they proposed four variants of the Swin transformer, namely Swin-T, Swin-S, Swin-B, and Swin-L, based on the channel number of the hidden layer in the first stage (C) and a number of Swin transformer block layers at each stage of the architecture. T, S, B, and L stand for tiny, small, base, and large, respectively. The main distinguishing parameters of these variants are listed below:

1. Swin-T: C = 96, layer numbers = {2, 2, 6, 2}
2. Swin-S: C = 96, layer numbers = {2, 2, 18, 2}
3. Swin-B: C = 128, layer numbers = {2, 2, 18, 2}
4. Swin-L: C = 192, layer numbers = {2, 2, 18, 2}.

2.3 Network Architecture

We used Swin-L as backbone of our final framework. The framework of the implemented algorithms is shown in Fig. 2. Initially, we passed the CT slice into Swin Transformer architecture to extract features. After that, we passed the extracted features through a multi-layer perceptron (MLP) with two hidden layers and one output layer to obtain covid infection percentage. The number of neurons in the first and second hidden layer of MLP is set to 128 and 64, respectively, whereas the output layer consists of only one neuron, as shown in Fig. 2.

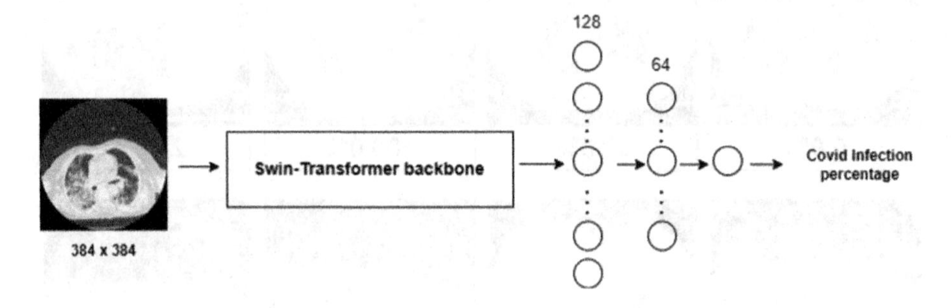

Fig. 2. Proposed framework

2.4 Experimental and Implementation Details

Since the labels for the validation and tests splits are not known, we randomly divided the train set into ten folds and performed training ten times where on each time, nine folds were used for training and one fold for validation. Before passing the CT slice into a model, we down-sampled to size 384×384. We applied two data augmentation HueSaturationValue and RandomBrightnessContrast from albumentation library [5] on the training folds to solve the overfitting problem. Due to the limitation of our GPU memory, the maximum batch size we could set for the algorithm was 8. We used adam optimizer with initial learning rate of 0.0001 to optimize the learning process. Cosine Annealing was used to schedule learning rate after each optimization step. As loss function, we used Mean square error (MSE) to calculate the loss for this image regression problem. Before beginning the training process, the backbone swin transformer was initialized with image net pretrained weight. We used PyTorch [16] deep learning framework in order to perform this experiment. Timm [26] PyTorch library was used to load the swin transformer model, and Tez PyTorch library [1] was used to design the pipeline for the training process. The model's performance was evaluated on the validation set and test set by measuring Mean Absolute Error (MAE), Pearson Correlation Coefficient (PC), Root Mean Square Error (RMSE). The formula for calculating MAE, PC, RMSE is defined below:

$$MAE = \frac{1}{n}\sum_{i=1}^{n} |y_i - \hat{y}_i| \tag{1}$$

$$PC = \frac{\sum_{i=1}^{n}(y_i - \bar{y_i})(\hat{y}_i - \bar{\hat{y}}_i)}{\sqrt{\sum_{i=1}^{n}(y_i - \bar{y_i})^2}\sqrt{\sum_{i=1}^{n}(\hat{y}_i - \bar{\hat{y}}_i)^2}} \tag{2}$$

$$RMSE = \sqrt{\frac{1}{n}\sum_{i=1}^{n}(y_i - \hat{y_i})^2} \quad (3)$$

where $Y = (y_1, y_2, y_3, ..., y_n)$ are the the ground-truth COVID-19 infection percentage and $\hat{Y} = (\hat{y_1}, \hat{y_2}, \hat{y_3}, ..., \hat{y_n})$ are corresponding predicted COVID-19 infection percentage. In Eq. 2, $\bar{y_i}$ and $\bar{\hat{y_i}}$ are the mean value of ground-truth and predicted COVID-19 infection percentages respectively.

3 Results and Discussion

In this work, we evaluated the performance of the swin transformer for the image regression task of covid-19% estimation. Experiments on the covid-19% estimation challenge database show that when swin transformer used as backbone network with image-net pretrained weight can achieve promising performance with MAE of 4.5042, PC of 0.9490, RMSE of 8.0964 on a validation set and MAE of 3.5569, PC of 0.8547 and RMSE of 7.5102 on a test set. Table 1 shows the performance on the validation set by a model trained on different training folds. On each fold, 90% of the data was used for training and 10% for validation. Through multiple experiments, we found that the performance of the Swin Transformer increases with an increase in image size. Another interesting observation is that the ensembling model from 10 folds did not achieve the best result on the validation set, as shown in Table 1. We were able to achieve the best performance on fold 0.

Table 1. Performance of the proposed framework on Val set leaderboard

Fold No.	(MAE)	(PC)	(RMSE)
0	**4.5042**	**0.9490**	**8.0964**
1	4.6950	–	–
2	4.9828	–	–
3	4.6458	–	–
4	4.8468	–	–
5	4.9140	–	–
6	4.5878	–	–
7	4.7681	–	–
8	4.6508	–	–
9	4.7537	–	–
Ensemble	4.5244	–	–

Table 2 shows the slice-level performance comparison of our transformer-based method with the other convolution-based methods on Test set. The performance of convolution-based methods was directly extracted from the benchmark paper published by challenge organizers.

Table 2. Comparison with convolution based methods

Model	(MAE)	(PC)	(RMSE)
ResNeXt-50 with MSE loss [4]	5.61	0.92	10.01
ResNeXt-50 with Huber loss [4]	5.29	0.92	10.10
Densenet-161 with MSE loss [4]	5.48	0.92	9.81
Densenet-161 with Huber loss [4]	5.23	0.93	9.42
Inception-V3 with MSE loss [4]	5.55	0.92	9.87
Inception-V3 with Huber loss [4]	5.10	0.93	9.25
Our proposed model	**3.55**	**0.85**	**7.51**

4 Conclusion

In this paper, we evaluated the performance of the Swin transformer architecture for covid 19 infection percentage estimation for monitoring the evolution of the patient state. We used a swin transformer initialized with image-net pretrained weight to extract the feature from CT slices and passed it through MLP to obtain covid 19 infection percentage. Experiment on covid-19 infection percentage estimation challenge database shows that Swin transformer, when used as a backbone network, can achieve promising performance with a MAE of 4.5042, a PC of 0.9490 and a MSE of 8.0964 on Val set Leaderboard, and a MAE of 3.5569, a PC of 0.8547 and a RMSE of 7.5102 on Test set Leaderboard.

References

1. Abhishek, T.: Tez: a simple PyTorch trainer. https://github.com/abhishekkrthakur/tez
2. Bougourzi, F., Contino, R., Distante, C., Taleb-Ahmed, A.: CNR-IEMN: a deep learning based approach to recognise Covid-19 from CT-scan. In: 2021 IEEE International Conference on Acoustics, Speech and Signal Processing (ICASSP), ICASSP 2021, pp. 8568–8572. IEEE (2021)
3. Bougourzi, F., Contino, R., Distante, C., Taleb-Ahmed, A.: Recognition of Covid-19 from CT scans using two-stage deep-learning-based approach: CNR-IEMN. Sensors **21**(17), 5878 (2021)
4. Bougourzi, F., Distante, C., Ouafi, A., Dornaika, F., Hadid, A., Taleb-Ahmed, A.: Per-Covid-19: a benchmark dataset for Covid-19 percentage estimation from CT-scans. J. Imaging **7**(9), 189 (2021)

5. Buslaev, A., Iglovikov, V.I., Khvedchenya, E., Parinov, A., Druzhinin, M., Kalinin, A.A.: Albumentations: fast and flexible image augmentations. Information **11**(2) (2020). https://doi.org/10.3390/info11020125. https://www.mdpi.com/2078-2489/11/2/125
6. Dosovitskiy, A., et al.: An image is worth 16x16 words: transformers for image recognition at scale. arXiv preprint arXiv:2010.11929 (2020)
7. Fan, D.P., et al.: Inf-Net: automatic Covid-19 lung infection segmentation from CT images. IEEE Trans. Med. Imaging **39**(8), 2626–2637 (2020)
8. Goceri, E., Goceri, N.: Deep learning in medical image analysis: recent advances and future trends (2017)
9. Jalaber, C., Lapotre, T., Morcet-Delattre, T., Ribet, F., Jouneau, S., Lederlin, M.: Chest CT in Covid-19 pneumonia: a review of current knowledge. Diagn. Interv. Imaging **101**(7–8), 431–437 (2020)
10. Kucirka, L.M., Lauer, S.A., Laeyendecker, O., Boon, D., Lessler, J.: Variation in false-negative rate of reverse transcriptase polymerase chain reaction-based SARS-CoV-2 tests by time since exposure. Ann. Intern. Med. **173**(4), 262–267 (2020)
11. Lacerda, P., Barros, B., Albuquerque, C., Conci, A.: Hyperparameter optimization for Covid-19 pneumonia diagnosis based on chest CT. Sensors **21**(6), 2174 (2021)
12. Lassau, N., et al.: Integrating deep learning CT-scan model, biological and clinical variables to predict severity of Covid-19 patients. Nat. Commun. **12**(1), 1–11 (2021)
13. Lei, J., Li, J., Li, X., Qi, X.: CT imaging of the 2019 novel coronavirus (2019-nCoV) pneumonia. Radiology **295**(1), 18 (2020)
14. Litjens, G.: A survey on deep learning in medical image analysis. Med. Image Anal. **42**, 60–88 (2017)
15. Liu, Z., et al.: Swin transformer: hierarchical vision transformer using shifted windows. arXiv preprint arXiv:2103.14030 (2021)
16. Paszke, A., et al.: PyTorch: an imperative style, high-performance deep learning library. Adv. Neural. Inf. Process. Syst. **32**, 8026–8037 (2019)
17. Stefano, A., Comelli, A.: Customized efficient neural network for Covid-19 infected region identification in CT images. J. Imaging **7**(8), 131 (2021)
18. Vaishya, R., Javaid, M., Khan, I.H., Haleem, A.: Artificial intelligence (AI) applications for Covid-19 pandemic. Diab. Metab. Syndr. Clin. Res. Rev. **14**(4), 337–339 (2020)
19. Vantaggiato, E., Paladini, E., Bougourzi, F., Distante, C., Hadid, A., Taleb-Ahmed, A.: Covid-19 recognition using ensemble-CNNs in two new chest X-ray databases. Sensors **21**(5), 1742 (2021)
20. Vaswani, A., et al.: Attention is all you need. In: Advances in Neural Information Processing Systems, pp. 5998–6008 (2017)
21. Voulodimos, A., Doulamis, N., Doulamis, A., Protopapadakis, E.: Deep learning for computer vision: a brief review. Comput. Intell. Neurosci. **2018**, 1–14 (2018)
22. Voulodimos, A., Protopapadakis, E., Katsamenis, I., Doulamis, A., Doulamis, N.: A few-shot U-Net deep learning model for Covid-19 infected area segmentation in CT images. Sensors **21**(6), 2215 (2021)
23. Wang, G., et al.: A deep-learning pipeline for the diagnosis and discrimination of viral, non-viral and Covid-19 pneumonia from chest X-ray images. Nat. Biomed. Eng. **5**(6), 509–521 (2021)
24. Wang, L., Lin, Z.Q., Wong, A.: Covid-Net: a tailored deep convolutional neural network design for detection of Covid-19 cases from chest X-ray images. Sci. Rep. **10**(1), 1–12 (2020)

25. Wang, X., et al.: A weakly-supervised framework for Covid-19 classification and lesion localization from chest CT. IEEE Trans. Med. Imaging **39**(8), 2615–2625 (2020)
26. Wightman, R.: PyTorch image models (2019). https://doi.org/10.5281/zenodo.4414861. https://github.com/rwightman/pytorch-image-models
27. Zhao, X., et al.: D2A U-Net: automatic segmentation of Covid-19 lesions from CT slices with dilated convolution and dual attention mechanism. arXiv preprint arXiv:2102.05210 (2021)

COVID-19 Infection Percentage Prediction via Boosted Hierarchical Vision Transformer

Chih-Chung Hsu(✉), Sheng-Jay Dai, and Shao-Ning Chen

National Cheng Kung University, No. 1, University Rd., East Dist., Tainan City, Taiwan
cchsu@gs.ncku.edu.tw
https://cchsu.info

Abstract. A better backbone network usually benefits the performance of various computer vision applications. This paper aims to introduce an effective solution for infection percentage estimation of COVID-19 for the computed tomography (CT) scans. We first adopt the state-of-the-art backbone, Hierarchical Visual Transformer, as the backbone to extract the effective and semantic feature representation from the CT scans. Then, the non-linear classification and the regression heads are proposed to estimate the infection scores of COVID-19 symptoms of CT scans with the GELU activation function. We claim that multi-tasking learning is beneficial for better feature representation learning for the infection score prediction. Moreover, the maximum-rectangle cropping strategy is also proposed to obtain the region of interest (ROI) to boost the effectiveness of the infection percentage estimation of COVID-19. The experiments demonstrated that the proposed method is effective and efficient.

Keywords: COVID-19 · Computed tomography · Deep learning · Transformer

1 Introduction

In the past decade, deep learning achieved state-of-the-art image recognition tasks compared to conventional machine learning and computer vision techniques. Similarly, deep learning-related schemes were widely adopted in the medical image field. In COVID-19 Computer-Aided Diagnosis (CAD) systems, the recognition issues are the most common and active [1]. However, it is well-known that the different slices in a Computed Tomography (CT) scan have different meanings. Directly predicting the COVID-19 symptoms for all slices in a CT scan might be unreliable since some slices might become meaningless. Therefore, effectively estimating the infection percentage of each slice in a CT scan is the most critical task for diagnosing COVID-19 symptoms.

In general, a U-shaped Network (U-Net) [10] is widely adopted for semantic segmentation of medical images with CAD applications. The main advantage of U-Net is the relatively few parameters in the network, leading to the

P. L. Mazzeo et al. (Eds.): ICIAP 2022 Workshops, LNCS 13374, pp. 529–535, 2022.
https://doi.org/10.1007/978-3-031-13324-4_45

overfitting issue that should be avoided since the number of training medical images is usually less than that of traditional RGB images. In contrast, U-Net might be underfitting with a sufficient number of training images since the network capacity is relatively small. Recently, several different Convolutional Neural Networks (CNNs) have been proposed to improve the effectiveness of feature learning and specific tasks, such as VGG [11], ResNet [5], and EfficientNet [12]. Recently, visual transformer [4] has been proposed to achieve state-of-the-art performance on various tasks. The visual transformer-variants also demonstrated promising results like the Swin-Transformer (Shifted window Transformer) [7] demonstrated excellent performance and was highly efficient on various tasks. In Swin-Transformer, it capably serves as a general-purpose backbone for computer vision tasks. Inspired by CNNs, the feature representation of a hierarchical Visual Transformer is computed with shifted windows. The shifted windowing scheme brings greater efficiency by limiting self-attention computation to non-overlapping local windows while allowing for cross-window connection. Therefore, the computational complexity can be reduced compared to the traditional visual transformer [4]. Therefore, Swin-Transformer is suitable for the Covid-19 infection percentage estimation task.

The CT images of patients with COVID-19 show distinct features such as patchy multi-focal consolidation, ground-glass opacities, interlobular cavities, lobular septum thickening, and a clear indication of fibrotic lesions, peribronchovascular, Pleural effusion, and thoracic lymphadenopathy [9].

Figure 1 shows two CT images which are COVID-19 infected and uninfected patients. The most common finding on chest CT is a "ground-glass opacities" spread throughout the lung. They represent tiny air sacs, or alveoli, that are filled with fluid and become shades of gray on CT scans [6]. As marked by the red arrows in the right CT-scan in Fig. 1. The disease severity was proportionate to lung findings, meaning that more severely ill individuals had more of these opacities in one of both lobes of the lungs in chest CT scans.

The rest of this paper is organized as follows. Section 2, the proposed boosted Swin-Transformer is demonstrated. Section 3 draws the comprehensive experiments for Covid-19 infection prediction. Finally, conclusions are drawn in Sect. 3.

2 Method

2.1 Overview

The framework of the proposed method is depicted in Fig. 2. First, the training samples are cropped by the proposed maximum-rectangle algorithm to obtain the Region Of Interest (ROI). Then, data augmentation is adopted to increase the diversity of the training samples. The regression and classification heads are developed to guide the network to learn the discriminative features by gradient descent algorithm, termed multi-task learning. Here, Swin-Transformer is treated as our backbone network for the infection prediction tasks. Finally, the trained model can be used to predict the infection score of the given CT slice with the regression head.

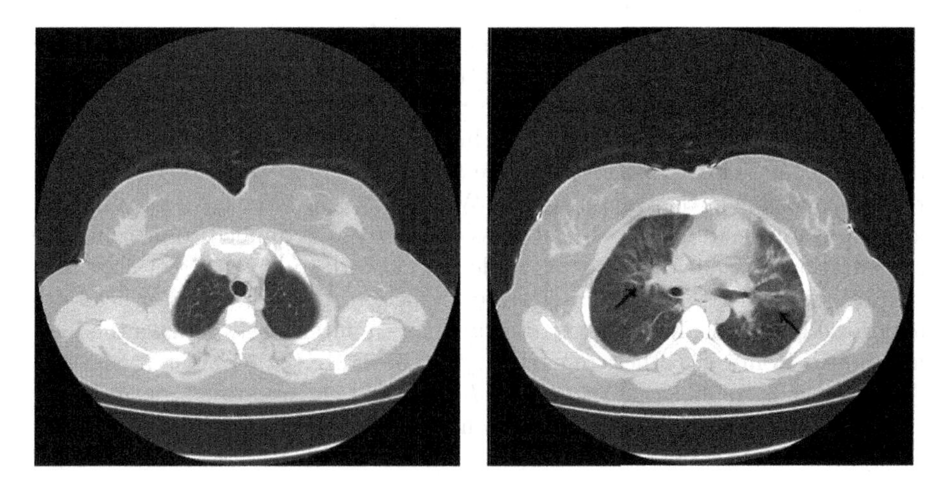

Fig. 1. The difference between positive and negative samples for COVID-19 symptoms.

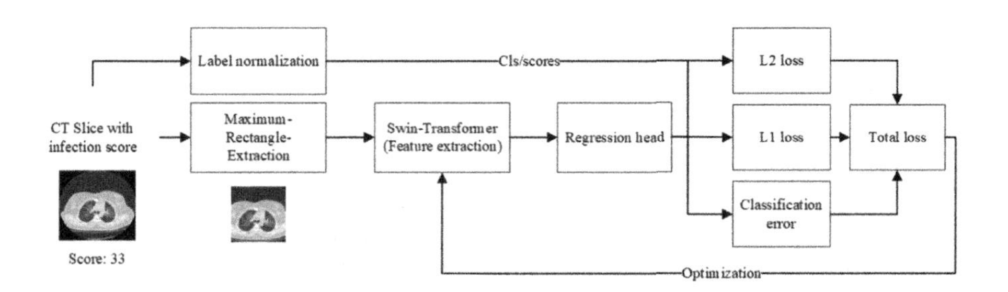

Fig. 2. Overview of the proposed method in the training phase.

2.2 Maximum-Rectangle Extraction

Since we observed that most of the CT-scan will have an ROI where the information only existed within, to retrieve this information without information loss, a maximum-rectangle algorithm is proposed to extract this ROI. Specifically, we take the maximum rectangle out of the circular area to remove the information-less regions (say, black regions/background), as well as keeping the meaningful regions as large as possible. First, we calculate the entropy $I(R)$ of the extracted region R at coordinate (x, y) with width and height (h, w), where we decrease the coordinate and size to obtain the next ROI R_{xyhw}. In this way, we can simply apply the iterative algorithm to obtain the maximum entropy $I(R)$ with the specific coordinate and size. Since this operation can be offline performed on the training and testing samples, it is cost-less in the inference phase.

2.3 Multi-task Learning

Since the number of training samples of the medical images is relatively small compared to the conventional RGB images, it is worthy that more prior information can be adopted to guide the network training for better and more stable purposes. Toward this end, we introduce multi-tasking learning to boost performance. Specifically, we introduce the traditional regression loss and the classification loss to perform better. While regression loss could make the predicted value get closer to the target, the classification loss could be treated as a relaxed condition in the training phase. We quantize the score label into K-level, and the traditional Softmax-based cross-entropy loss is applied to learning the classification task. In this way, we guide the network to have correct classification results rather than the exact score prediction. In the regression head, we adopt the L1 and L2 norm as the loss function simultaneously to have the joint advantages from each other. In summary, we optimize the parameters of Swin-Transformer by minimizing the sum of L_1, L_2, and classification loss computed by Softmax. The L_1 loss, The L_2 loss, and classification loss are computed as

$$L1loss(X_i, y_i) = \frac{1}{n}\sum_{i=1}^{n}|y_i - f(X_i)|, \tag{1}$$

$$L2loss(X_i, y_i) = \frac{1}{n}\sum_{i=1}^{n}(y_i - f(X_i))^2, \tag{2}$$

$$CLSloss(X_i, y_i) = \frac{1}{n}\sum_{i=1}^{n} -log(\frac{e^{f(X_i)}}{\sum_{j=1}^{n} e^{f(X_j)}}), \tag{3}$$

respectively, where f(X_i) is obtained by taking X_i as input CT-scan image, and y_i is the corresponding ground-truth infection percentage.

2.4 Dataset and Criterion

In this section, the datasets used for performance evaluation are provided by organizers [2,3,13]. In this dataset, the training set and validation set contain 3054 and 1301 images, respectively, where the validation set does not provide a label for infection percentages. Therefore, the evaluation results on the validation set can only be retrieved from the competition website. We extract a portion of the training data to avoid overfitting to evaluate performance.

The criterion of the performance evaluation in this study is the Mean Absolute Error (MAE), which is defined as follows:

$$MAE = \frac{1}{n}\sum_{i=1}^{n}|f(X_i) - y_i|, \tag{4}$$

where $f(X_i)$ indicates the estimation of infection percentage and its corresponding ground truth is y_i.

3 Experimental Result

In this study, the initial learning rate is $1e-6$ with Cosine Annealing decay scheduling in the training phase. The total number of epochs of the training phase is 260, which is determined empirically. The weight decay is 0.005, and the optimizer is AdamW [8] with the momentum 0.9. The number of Linear Embedding and Swin-Transformer-Block in each stage (i.e., each block in Fig. 3) for conventional Swin-Transformer [7] are 2, 2, 6, 2, respectively. Since the number of the training images in dataset [3] is relatively larger than that of the traditional medical images, we increase the number of the blocks in the 3rd unit from 6 to 20 to increase the network capacity. Additionally, we also increase the number of the attention head in Swin-Transformer to further improve the global attention ability.

Table 1. Performance comparison between the proposed Swin-Transformer and other methods for the validation set.

Method	MAE
ResNet-50 [5]	17.6033
VGG19 [11]	7.4596
Method in [3]	5.2943
Our method	**4.9926**

Table 1 presents the comparison between the proposed Swin-Transformer with the multi-task learning and the other conventional CNNs, including ResNet-50 and VGG19 without multi-task learning. It is clear that the Swin-Transformer achieves state-of-the-art performance on the validation set, implying that Swin-Transformer well captures the hierarchical Transformer whose representation is computed with shifted windows.

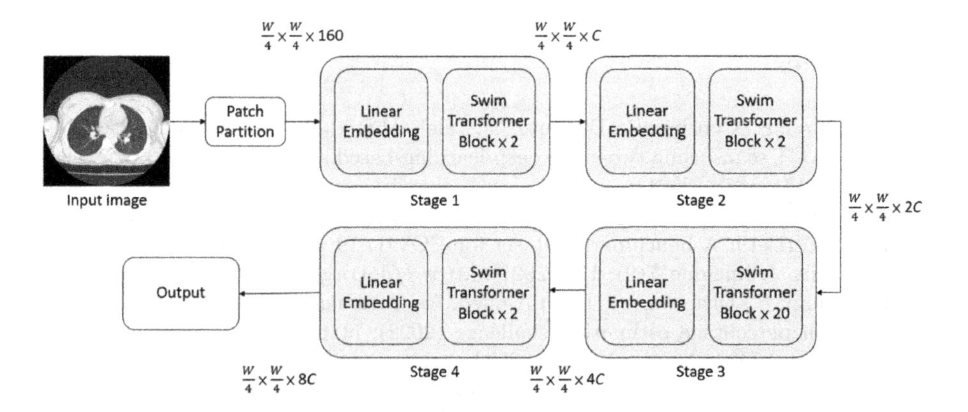

Fig. 3. The proposed fixed Swin-transformer.

3.1 Ablation Study

In this section, we draw the advantages of the proposed multi-task learning and the maximum-rectangle algorithm in Table 2. It is remarkable that each part of the proposed method is significantly contributing to the performance of the infection score prediction for COVID-19 CT scans.

Table 2. Ablation study of the proposed method.

Method	Multi-task learning	Maximum-rectangle	MAE
Swin-transformer			5.2533
Swin-transformer	v		5.1194
Swin-transformer		v	5.0932
Full model	v	v	**4.9926**

4 Conclusion

In this paper, the boosted Swin-Transformer with both proposed multi-task learning and the maximum-rectangle algorithm has been proposed to predict the infection score of COVID-19 CT scans effectively. First, the promising result is the multi-task learning incorporated with the classification and regression heads. Second, a maximum-rectangle algorithm is also proposed in this paper to retrieve the complete information from CT slices to guide our Swin-Transformer's feature learning better. A more significant number of the blocks in the third stage of Swin-Transformer also verified that the performance could be further improved when the number of the training samples is relatively sufficient. The experimental results prove the efficiency of using the proposed fixed Swin Transformer to improve estimating Covid-19 infection percentage estimation from the CT scans compared with other approaches.

References

1. Bougourzi, F., Contino, R., Distante, C., Taleb-Ahmed, A.: Recognition of COVID-19 from CT scans using two-stage deep-learning-based approach: CNR-IEMN. Sensors **21**(17), 5878 (2021)
2. Bougourzi, F., Distante, C., Ouafi, A., Dornaika, F., Hadid, A., Taleb-Ahmed, A.: Per-COVID-19: A benchmark dataset for COVID-19 percentage estimation from CT-scans. J. Imaging **7**(9), 189 (2021). https://doi.org/10.3390/jimaging7090189
3. Bougourzi, F., Distante, C., Taleb-Ahmed, A., Dornaika, F., Hadid, A.: COVID-19 infection percentage estimation challenge (2022). https://sites.google.com/view/covid19iciap2022. Accessed 20 Apr 2022
4. Dosovitskiy, A., et al.: An image is worth 16 × 16 words: Transformers for image recognition at scale. arXiv preprint arXiv:2010.11929 (2020)

5. He, K., Zhang, X., Ren, S., Sun, J.: Deep residual learning for image recognition. In: Proceedings of the IEEE Conference on Computer Vision and Pattern Recognition, pp. 770–778 (2016)
6. Kundu, R., Singh, P.K., Ferrara, M., Ahmadian, A., Sarkar, R.: ET-Net: an ensemble of transfer learning models for prediction of COVID-19 infection through chest CT-scan images. Multimed. Tools App. **81**(1), 31–50 (2022)
7. Liu, Z., et al.: Swin transformer: Hierarchical vision transformer using shifted windows. arXiv preprint arXiv:2103.14030 (2021)
8. Loshchilov, I., Hutter, F.: Fixing weight decay regularization in Adam (2018)
9. Perumal, V., Narayanan, V., Rajasekar, S.J.S.: Prediction of COVID-19 with computed tomography images using hybrid learning techniques. Dis. Mark. **2021**, 552279 (2021)
10. Ronneberger, O., Fischer, P., Brox, T.: U-Net: convolutional networks for biomedical image segmentation. In: Navab, N., Hornegger, J., Wells, W.M., Frangi, A.F. (eds.) MICCAI 2015. LNCS, vol. 9351, pp. 234–241. Springer, Cham (2015). https://doi.org/10.1007/978-3-319-24574-4_28
11. Simonyan, K., Zisserman, A.: Very deep convolutional networks for large-scale image recognition. arXiv preprint arXiv:1409.1556 (2014)
12. Tan, M., Le, Q.: EfficientNet: rethinking model scaling for convolutional neural networks. In: International Conference on Machine Learning, pp. 6105–6114. PMLR (2019)
13. Vantaggiato, E., Paladini, E., Bougourzi, F., Distante, C., Hadid, A., Taleb-Ahmed, A.: COVID-19 recognition using ensemble-CNNs in two new chest X-ray databases. Sensors **21**(5), 1742 (2021)

Novel Benchmarks and Approaches for Real-World Continual Learning - CL4REAL

Catastrophic Forgetting in Continual Concept Bottleneck Models

Emanuele Marconato[1,3(✉)], Gianpaolo Bontempo[2,3(✉)], Stefano Teso[1(✉)], Elisa Ficarra[2(✉)], Simone Calderara[2(✉)], and Andrea Passerini[1(✉)]

[1] University of Trento, Trento, Italy
{emanuele.marconato,stefano.teso,andrea.passerini}@unitn.it
[2] University of Modena and Reggio Emilia, Modena, Italy
{gianpaolo.bontempo,elisa.ficarra,simone.calderara}@unimore.it
[3] University of Pisa, Pisa, Italy

Abstract. Almost all Deep Learning models are dramatically affected by Catastrophic Forgetting when learning over continual streams of data. To mitigate this problem, several strategies for Continual Learning have been proposed, even though the extent of the forgetting is still unclear. In this paper, we analyze Concept Bottleneck (CB) models in the Continual Learning setting and we investigate the effect of high-level features supervision on Catastrophic Forgetting at the representation layer. Consequently, we introduce two different metrics to evaluate the loss of information on the learned concepts as new experiences are encountered. We also show that the obtained Saliency maps remain more stable with the attributes supervision. The code is available at https://github.com/Bontempogianpaolo1/continualExplain

Keywords: Continual Learning · Explainable Artificial Intelligence · Catastrophic Forgetting · Concept Bottleneck models

1 Introduction

In recent years, a significant number of important successes have been reached in computer vision for the classification task. Many of them are grounded on the i.i.d. hypothesis and to achieve good generalization results, huge quantities of data are required. However, in a more realistic case, data are received in small amounts over continuous streams or they cannot be revisited once trained on them, e.g. for privacy reasons. Several studies have shown that standard models are not able to generalize when they learn incrementally as they tend to forget past information. This phenomenon is referred to as Catastrophic forgetting [8](CF) and resolving this issue would be a milestone towards more human-like AIs.

Over years, this problem has been addressed in Continual Learning (CL) research [6,11,13,20]. However, despite some marginal results, state of the art

E. Marconato and G. Bontempo—Equal contribution.

P. L. Mazzeo et al. (Eds.): ICIAP 2022 Workshops, LNCS 13374, pp. 539–547, 2022.
https://doi.org/10.1007/978-3-031-13324-4_46

CL models still can not achieve as good performances as iid approaches. Even more, they struggle when compared to trivial approaches in CL, such as GDUMB [16].

All the state-of-the-art CL models are based on advanced Deep Learning models, thus their missing interpretation complicates more the understanding of what is forgotten [22]. In order to solve the interpretability issue, the field of eXplainable Artificial Intelligence (XAI) is promoting the research for explanation methods on how a Machine Learning model returns a certain prediction [9]. These methods are divided into two categories: the so-called "transparent" methods, which treats models that are interpretable by design, and "post-hoc" methods [2]. A class of XAI models, see e.g. [3] and references therein, integrates explainability in the learning pipeline and forces the model to extrapolate interpretable high-level features. After that, the decision process is based solely on them with simple rules [17]. In particular, several works showed that achieving better interpretability does not imply drops in downstream performance [1,5,12].

To the best of our knowledge, no work has yet addressed the learning of intermediate concepts in Continual Learning nor how human-understandable concepts are affected by CF. In this article, we mind the gap between existing literature on XAI based on Deep Interpretable models and Continual Learning by integrating Concept Bottleneck models into Continual Learning. To summarize, our contribution is two-fold:

1. We introduce Concept Bottleneck models in Class-Incremental learning and we design two different metrics for evaluating CF of concepts Sect. 3;
2. Considering only the fine-tune strategy, we analyze how CF affects the model both with and without supervision on attributes.

This study paves the way for a more grounded use of Deep Interpretable models for Continual Learning.

2 Related Works

Deep Interpretable models are a class of algorithms introduced to make more explainable usual Deep Learning models [17]. Based on [3], they are a "grayfication" of black boxes where the classifier becomes "white" (interpretable), while the concept extractor remains a black-box. They usually consist of two stacked parts: the first one is a concept extractor while the second is a classification module. The inherent interpretability of these models is guaranteed once simple rules are applied on the extracted concepts to predict the target class, e.g. by combining high-level features with linear weights. Depending on the type of supervision on the concept layer, different models have been proposed [1,5,12].

Continual Learning explores the relaxation of the i.i.d. hypothesis to sequences of data seen in several experiences: in each of these, the available dataset reduces only to the observed stream. While many possible CL scenarios exist, CL tier research has specialized in three categories [20]. The *Task-Incremental Learning* considers different experiences containing data points that do not share previously seen classes. In this scenario, the experience identity is always available and, therefore, multi-head approaches are possible given the experience label. On the other hand, in the *Class-Incremental Learning* the task identity is omitted: models in this scenario are based on the single-head approach. *Domain-Incremental Learning* differs from other scenarios as the shift occurs in the observed domain, where, in principle, previously seen classes can still be encountered. In this case, also, the experience identity is not available.

At the present moment, few works have explored the inception of XAI methods in CL [7,10]. Among them, the authors in [7] proposed a replay method which saves the most relevant samples and correspondent saliency maps for each experience.

3 Methodology

In the following, we consider the class prediction on a dataset of images, each one provided by a fixed number of human concepts. Let $\mathcal{D} = \{(\mathbf{x}^n, y^n), n = 1, ..., N\}$ be the observed labelled samples for that we want to infer a predictive model, and let $\mathcal{S}$ be the additional set provided by the human-annotated concepts.

In general, a Concept-Learning model [3] consists in an embedding function $\mathcal{E} : \mathbb{R}^D \rightarrow \mathbb{R}^d$ mapping the input $\mathbf{x} \in \mathbb{R}^D$ to its encoding $\mathbf{z} \in \mathbb{R}^d$ with a classifier on top $\mathcal{F} : \mathbb{R}^d \rightarrow \mathcal{K}$, where $\mathcal{K} = \{1, ..., K\}$ contain all provided K classes. In our case, the encoding outputs the percentage of each concept in the instance $\mathbf{x}$:

$$\mathbf{z} = \text{sigmoid}(\mathcal{E}(x))\,.$$

We consider a linear classifier $\mathcal{F}_W : [0,1]^d \rightarrow \mathbb{R}^K$, where W denotes the weights. The probability over the predicted class is given by

$$p(y|\mathbf{x}) = \text{softmax}(\mathcal{F}(\mathbf{z}))$$

We study the Class-Incremental scenario: for each experience t, the task dataset $\mathcal{D}^{(t)}$ contains only a restrained subset of classes $\mathcal{Y}^{(t)}$, such that $\mathcal{Y}^{(i)} \cap \mathcal{Y}^{(j)} = \emptyset,\ \forall i \neq j$.

3.1 Concept Bottleneck Models

When referring to Concept Bottleneck models [12], the information related to visual concepts is added to the dataset $\mathcal{D} \equiv \{\mathcal{X}, \mathcal{S}, \mathcal{Y}\}$. In our case, the additional information is made of binary values, each one referring to the presence or absence of a discriminative attribute for the classification task.

Following [12], we force a semantic structure in the encoding space $\mathbb{R}^d$ by constraining each dimension to match all d supervised concepts. The overall loss is defined as:

$$\begin{aligned}
\mathcal{L}(\mathbf{x}, \mathbf{c}, y) &= \mathcal{L}_{CE}(\mathbf{x}, y) + \mathcal{L}_{enc}(\mathbf{x}, \mathbf{c}) \\
&= -\log p(y|\mathbf{x}) - \frac{1}{N_C}\sum_{j=1}^{N_C}[c_j \log z_j + (1 - c_j)\log(1 - z_j)]
\end{aligned} \tag{1}$$

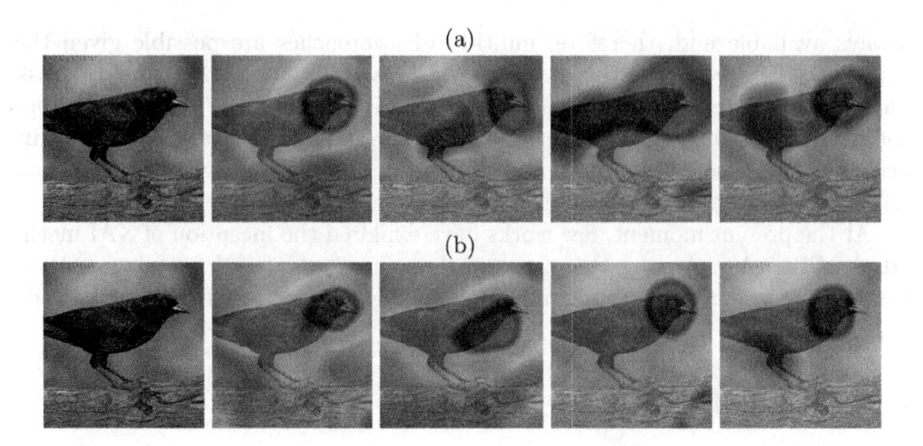

Fig. 1. Saliency maps of a bird image belonging to the first experience classes. a) shows the evolution over the tasks without the supervision on the attributes, while b) with full supervision on concepts.

where $\mathcal{L}_{CE}$ is the categorical loss for classification and $\mathcal{L}_{enc}$ is the Binary Cross Entropy for the encoding $\mathbf{z} = \mathcal{E}(\mathbf{x})$ over the expected $\mathbf{c}$. In this case, the activation $\mathbf{z} \in [0, 1]^d$ can be interpreted as a collection of d disjoint percentages of each concept presence.

3.2 Continual Learning

In Class-Incremental Learning, we have access to only one dataset $\mathcal{D}^{(t)} = \{\mathcal{X}^{(t)}, \mathcal{S}^{(t)}, \mathcal{Y}^{(t)}\}$ for each experience t. The Continual strategy we considered consists in the optimal train of the CB model on the data points of each experience: no regularization nor replay strategies are adopted. This procedure is usually addressed as the fine-tuning strategy [6] and it constitutes the baseline over which new CL strategies are constructed. Let $\theta \equiv (\theta_{\mathcal{E}}, \theta_{\mathcal{F}} \equiv W)$ be the parameters of the Concept Bottleneck model, after each experience we obtain:

$$\theta_t^* = \arg\min_{\theta} \mathcal{L}_\theta(\mathcal{D}^{(t)}) \tag{2}$$

4 Concept Shift Detection

In this section, we introduce the metrics to quantify the Catastrophic Forgetting effect on the learned concepts. Unlike those already present in the literature, see e.g. [6,13,20], the ones we have considered take into account the concept distribution and can be evaluated w.r.t the ground-truth thanks to the availability of extra annotation.

4.1 Concepts Accuracy

The average accuracy over seen class-specific attributes is the first quantity that can be evaluated. For each concept j, we define the average accuracy on all experiences based on the concept-extractor model $\mathcal{E}_\tau$ at task τ as:

$$\mathcal{A}_j(\tau) = \frac{1}{\tau}\sum_{t=1}^{\tau} \mathbb{E}_{\mathbf{x},\mathbf{c}\sim\mathcal{D}_t}\left[\mathbb{1}\{(\mathcal{E}_\tau(\mathbf{x}))_j - \mathbf{c}_j\}\right] \tag{3}$$

where $\mathcal{D}_t$ denotes the observed dataset at the experience t. Taking the average over all concepts in Eq. (3), we obtain the mean accuracy for the fine-tuned model at task τ:

$$\langle\mathcal{A}(\tau)\rangle = \frac{1}{N_C}\sum_{j=1}^{N_C} \mathcal{A}_j(\tau) \tag{4}$$

The same can be done by restricting the calculation to each observed experience $e \in \{1, ..., \tau\}$, obtaining:

$$\mathcal{A}_j^{(e)}(\tau) = \mathbb{E}_{\mathbf{x},\mathbf{c}\sim\mathcal{D}_e}\left[\mathbb{1}\{(\mathcal{E}_\tau(\mathbf{x}))_j - \mathbf{c}_j\}\right], \quad \langle\mathcal{A}^{(e)}(\tau)\rangle = \frac{1}{N_C}\sum_{j=1}^{N_C} \mathcal{A}_j^{(e)}(\tau)\,. \tag{5}$$

Thus, Eq. (5) reduces to Eqs. (3) and (4) when taking the time average of $\mathcal{A}^{(e)}$ over e.

4.2 Concepts Divergence

As an additional quantifier for the concept drift in CL, we introduce the *concept divergence* built on the concepts posterior distribution at a given experience t. In order to acquire information on the class encoding, we estimate the frequency on each concept j for every class l within past and present experiences:

$$p_j^{(l)}(t) = \frac{1}{|C_l|}\sum_{\mathbf{x}\in C_l} \mathcal{E}_t(\mathbf{x})_j$$

where $C_l = \{\mathbf{x}, \mathbf{c}, y \in \mathcal{D} | y = l\}$. For the sake of simplicity, we approximate the concept distribution for the class l at time t with the disjoint probability distribution:

$$P^{(l)}(\mathbf{c}; t) = \prod_{j=1}^{N_c} p_j^{(l)}(t) \tag{6}$$

Then, to quantify the distance between two different temporal concept distributions we take the Kullback-Lieber divergence:

$$D_l(t, t') = \sum_{\{\mathbf{c}\}} P^{(l)}(\mathbf{c}, t) \log \frac{P^{(l)}(\mathbf{c}, t)}{P^{(l)}(\mathbf{c}, t')} \tag{7}$$

where the sum is done over all possible realizations of the concept vector $\mathbf{c}$. By means of Eq. (6) we obtain:

$$D_l(t, t') = \sum_{j=1}^{N_C} \left[p_j^{(l)}(t) \log \frac{p_j^{(l)}(t)}{p_j^{(l)}(t')} + (1 - p_j^{(l)}(t)) \log \frac{1 - p_j^{(l)}(t)}{1 - p_j^{(l)}(t')} \right] \tag{8}$$

To obtain information on the average KL divergence it is sufficient to equate on the observed classes at the reference time t_0. In this way, the average KL divergence becomes:

$$D(t_0, t) = \frac{1}{|\mathcal{Y}_{t_0}|} \sum_{l \in \mathcal{Y}_{t_0}} D_l(t_0, t) \tag{9}$$

5 Experiments

5.1 Implementation Details

Data: The entire set of experiments has been performed on the CUB-200 dataset [21], which contains images of 200 different birds species with almost 30 instances per class. Following the procedure in CB [12], a subset of the most significant attributes per class is extracted in order to reduce their sparsity, passing from 312 annotated attributes to 112 per image. Also, the same set of concepts is assigned to each class images by a majority voting.
Model: The model architecture is composed of:

- a pretrained Inception-v3 model [19] as the backbone for processing the low-level features;
- a Multi Layer Perceptron with ReLU activations in the middle layers and Sigmoid activation in the last one for mapping the low-level features to concepts;
- a final linear layer with Softmax activation for mapping concepts to labels.

Finally, in order to reproduce continual learning experiments, we based the implementation on Pytorch library [15], Mammoth [4] and Avalanche [14] frameworks.

5.2 Results

We performed experiments with two versions of the dataset: one without the presence of attributes, whereas the other with full supervision. They are the two extreme cases, whether, for a realistic setting, only a fraction of concepts would

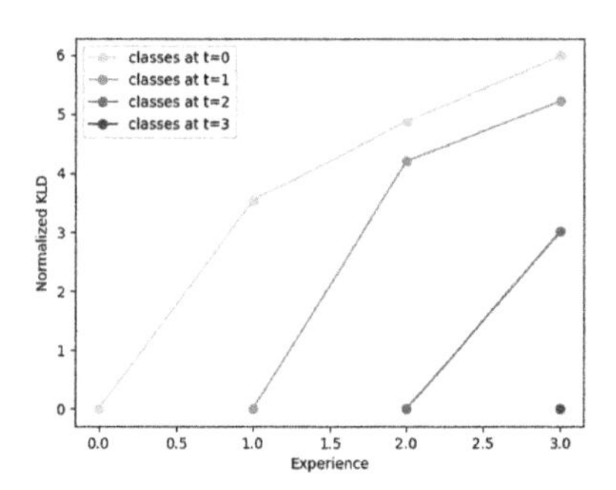

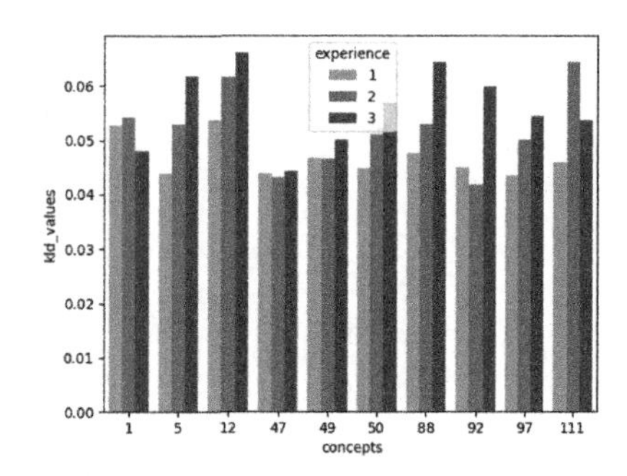

Fig. 2. Results without supervision on the concepts: (left) the average accuracy on concepts as new experiences are encountered, (right) the average concept shift over the classes learned at experience zero.

be provided to some images. This complicates more the scenario under study and it is left for future work.

For the concept-unsupervised case, we measured the concepts divergence over the experiences and plot the worst 10 attributes at each experience. In Fig. 2, we report the result obtained with the concept shift measure (8) on a single run over a total of 4 experiences. The left figure shows that the learned concepts representations at each reference experience are almost immediately forgotten at successive tasks, as expected. Each curve refers to the average concept shift on the classes encountered at every experience $t \in \{0, 1, 2, 3\}$. Instead, the right figure displays the 10 attributes most affected by CF effect after the training at experience zero: almost all of them get slightly worse as successive tasks are encountered. The values are calculated on the instances encountered in the first task.

In the concept-supervised scenario, we only measured the accuracy on the task-specific attributes w.r.t. their ground truth values and reported the results in Fig. 3. Surprisingly, it is possible to see that CF does not worsen too much the accuracy on the learned concepts as new experiences are met.

As a qualitative investigation, we applied GradCAM [18] on the concept layer to look at the Saliency map obtained from the prediction of the ground-truth class and reported them in Fig. 1. In the concept-supervised case, as the tasks pass, it can be noticed a smoother transition of the activation map than in the unsupervised one. This indicates a small improvement on the learned representation when concepts annotations are available.

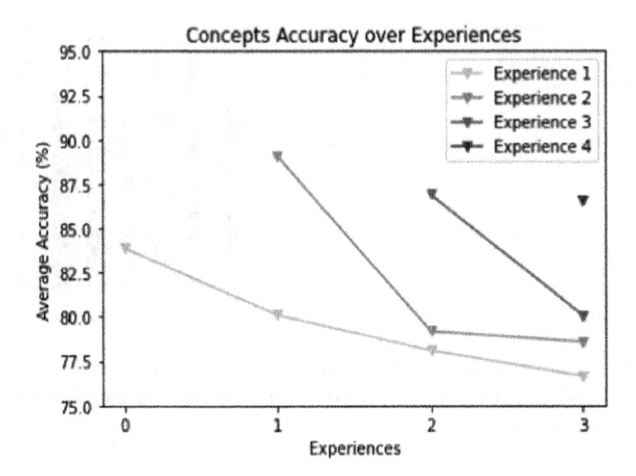

Fig. 3. Average accuracy on attributes. Each line refer to the average on the data encountered at the starting experience.

6 Conclusion

In the following work, we proposed a further investigation of CF in the Class-Incremental scenario focusing at the concept-level representations. Adopting the Concept Bottleneck architecture, we showed two different methods to evaluate the Catastrophic Forgetting at that level, with either the presence or the absence of attributes. As expected, the concept shift measure of Eq. (8) increases in the unsupervised concept setting, thus indicating a severe drift from the initial representation. On the other hand, when supervision on concepts is available, the accuracy obtained is not dramatically affected by forgetting: this is due to the presence of shared concepts over different experiences. It must be noticed that in the CL scenario we considered, the model is dramatically affected by the CF effect and, as is, its performances are the lower bound for any non-trivial CL strategy.

In future work, these metrics can be included in the learning process in order to mitigate CF and to improve the representation over experiences. Moreover, some existing CL strategies may show lesser forgetting as attributes are included: we expect this kind of supervision to play an important role for learning a consistent representation over time.

References

1. Alvarez-Melis, D., Jaakkola, T.S.: Towards robust interpretability with self-explaining neural networks (2018)
2. Francesco, B., Fosca, G., Riccardo, G., Francesca, N., Dino, P., Salvatore, R.: Benchmarking and survey of explanation methods for black box models (2021)
3. Bontempelli, A., Giunchiglia, F., Passerini, A., Teso, S.: Toward a unified framework for debugging gray-box models (2021)

4. Boschini, M., Bonicelli, L., Buzzega, P., Porrello, A., Calderara, S.: Class-incremental continual learning into the eXtended DER-verse. arXiv preprint arXiv:2201.00766 (2022)
5. Chen, C., Li, O., Tao, C., Barnett, A.J., Su, J., Rudin, C.: This looks like that: deep learning for interpretable image recognition. Curran Associates Inc., Red Hook, NY, USA (2019)
6. De Lange, M., et al.: A continual learning survey: defying forgetting in classification tasks. IEEE Trans. Pattern Anal. Mach. Intell. **44**, 3366–3385 (2022)
7. Ebrahimi, S., et al.: Remembering for the right reasons: explanations reduce catastrophic forgetting. In: International Conference on Learning Representations (2021)
8. French, R.M.: Catastrophic forgetting in connectionist networks. Trends Cogn. Sci. **3**(4), 128–135 (1999)
9. Guidotti, R., Monreale, A., Ruggieri, S., Turini, F., Giannotti, F., Pedreschi, D.: A survey of methods for explaining black box models. ACM Comput. Surv. **51**(5), 1–42 (2019)
10. He, Y.: Adaptive explainable continual learning framework for regression problems with focus on power forecasts (2021)
11. Kirkpatrick, J., et al.: Overcoming catastrophic forgetting in neural networks. Proc. Nat. Acad. Sci. **114**(13), 3521–3526 (2017)
12. Pang, W.K., et al.: Concept bottleneck models (2020)
13. Lesort, T., Lomonaco, V., Stoian, A., Maltoni, D., Filliat, D., Díaz-Rodríguez, N.: Continual learning for robotics: definition, framework, learning strategies, opportunities and challenges. Inf. Fusion **58**, 52–68 (2020)
14. Lomonaco, V.: Avalanche: an end-to-end library for continual learning. In: Proceedings of IEEE Conference on Computer Vision and Pattern Recognition, 2nd Continual Learning in Computer Vision Workshop (2021)
15. Paszke, A., et al.: PyTorch: an imperative style, high-performance deep learning library. In: Wallach, H., Larochelle, H., Beygelzimer, A., d' Alché-Buc, F., Fox, E., Garnett, R. (eds.) Advances in Neural Information Processing Systems, vol. 32, pp. 8024–8035. Curran Associates Inc. (2019)
16. Prabhu, A., Torr, P.H.S., Dokania, P.K.: GDumb: a simple approach that questions our progress in continual learning. In: Vedaldi, A., Bischof, H., Brox, T., Frahm, J.-M. (eds.) ECCV 2020. LNCS, vol. 12347, pp. 524–540. Springer, Cham (2020). https://doi.org/10.1007/978-3-030-58536-5_31
17. Rudin, C.: Stop explaining black box machine learning models for high stakes decisions and use interpretable models instead. Nat. Mach. Intell. **1**(5), 206–215 (2019)
18. Selvaraju, R.R., Cogswell, M., Das, A., Vedantam, R., Parikh, D., Batra, D.: Grad-CAM: visual explanations from deep networks via gradient-based localization. In: Proceedings of the IEEE International Conference on Computer Vision, pp. 618–626 (2017)
19. Szegedy, C., Vanhoucke, V., Ioffe, S., Shlens, J., Wojna, Z.: Rethinking the inception architecture for computer vision. In: Proceedings of the IEEE Conference on Computer Vision and Pattern Recognition, pp. 2818–2826 (2016)
20. van de Ven, G.M., Tolias, A.S.: Three scenarios for continual learning. arXiv arXiv:1904.07734 (2019)
21. Welinder, P., et al.: Caltech-UCSD birds 200. Technical report CNS-TR-2010-001, California Institute of Technology (2010)
22. Zhang, C., Bengio, S., Recht, B., Vinyals, O., Hardt, M.: Understanding deep learning requires rethinking generalization (2016)

Practical Recommendations for Replay-Based Continual Learning Methods

Gabriele Merlin[1(✉)], Vincenzo Lomonaco[1], Andrea Cossu[1,2], Antonio Carta[1], and Davide Bacciu[1]

[1] Department of Computer Science, University of Pisa, Pisa, Italy
[2] Scuola Normale Superiore, Pisa, Italy

Abstract. Continual Learning requires the model to learn from a stream of dynamic, non-stationary data without forgetting previous knowledge. Several approaches have been developed in the literature to tackle the Continual Learning challenge. Among them, Replay approaches have empirically proved to be the most effective ones [16]. Replay operates by saving some samples in memory which are then used to rehearse knowledge during training in subsequent tasks. However, an extensive comparison and deeper understanding of different replay implementation subtleties is still missing in the literature. The aim of this work is to compare and analyze existing replay-based strategies and provide practical recommendations on developing efficient, effective and generally applicable replay-based strategies. In particular, we investigate the role of the memory size value, different weighting policies and discuss about the impact of data augmentation, which allows reaching better performance with lower memory sizes.

Keywords: Continual learning · Replay-based approaches · Catastrophic forgetting

1 Introduction

Traditional machine learning models learn from independent and identically distributed samples. In many real-world environments, however, such properties on training data cannot be satisfied. As an example, consider a robot learning a sequence of different tasks. For artificial neural networks, learning a new task causes a deterioration of performance on the previous one. This phenomenon is known as Catastrophic Forgetting [18]. Continual learning [19] is a branch of machine learning which focuses on learning from a sequence of tasks while at the same time preventing catastrophic forgetting. Although many approaches have been developed with different degrees of success, preventing catastrophic forgetting is still a difficult task. Moreover it is difficult to compare these approaches since there is not a standard evaluation protocol [8].

P. L. Mazzeo et al. (Eds.): ICIAP 2022 Workshops, LNCS 13374, pp. 548–559, 2022.
https://doi.org/10.1007/978-3-031-13324-4_47

The aim of this work is to deepen our understanding of replay-based strategies [2,21,23,25], a specific category of continual learning strategies, and provide *practical recommendations* to achieve a better efficiency-efficacy trade-off in their implementation. Replay strategies avoid forgetting by training the model on both current samples and some samples of the past tasks. In this paper, we extensively compare replay-based strategies on different benchmarks and settings to better characterize the role played by their main components in the mitigation of forgetting. We explore three main research directions. The first (Sect. 5) concerns the role of memory size. We extensively test the most popular replay strategies varying this parameter, finding out that the memory size value depends not only on the size of the dataset but also on the difficulty of the tasks and the number of classes involved in the learning process. The second direction (Sect. 6) is related to the balancing of the memory buffer. In the literature the replay buffer is usually balanced to have an equal amount of samples of each past task or class. We propose many weighting policies to distribute samples, unbalanced by task. We discover recent memories are more useful with respect to others, confirming the observation on the human brain [3]. Finally, we test the role of data augmentation [31] in a continual learning scenario (Sect. 7). We find out that performance increases by augmenting the memory, particularly with a low memory budget.

2 Related Works

The problem of learning from a sequence of tasks was posed since the origin of artificial intelligence [28,29]. However, only in 1989 Closkey [18] dealt with catastrophic forgetting directly. In 1995 a new method was proposed to prevent it named Replay [25]. This simple method consists of storing in a buffer some samples and presenting them during consecutive tasks. During the last few years we have witnesses a significant interest in this area and many strategies have been developed. Replay-base approaches have proved to be effective [1,2,5,24] and they differ mainly by the selection algorithm. Buzzega et al. in [5], proved the effectiveness of the standard Replay strategy [25] using a set of "tricks", even without changing the selection algorithm. Moreover Replay-based approaches are biologically-plausible: previous experiences rehearsal is believed to be important for stabilizing new memories [30].

Despite this prolific research paper production, none of these works compares and investigates replay-based strategies extensively.

3 Design Choices

Replay-based approaches rely on a simple yet effective mechanism: replay some previous samples to avoid catastrophic forgetting. However this apparently simple mechanism hides many possible modifications. In this section we describe three possible choices and variations concerning continual learning and replay-based strategies.

3.1 Replay Buffer

Replay buffer is the principal component of a replay-based strategy.

The *buffer structure* defines how samples are distributed. Samples can be balanced in the buffer by task or class. In this case the amount of samples belonging to the same task/class is the same. This structure rely on the assumption that the task label is known.

Selection and Discarding procedures are the principal components of a replay-based strategy. The standard Replay strategy [25] assumes that every sample is important for learning, thus select and discard randomly samples, taking into consideration the buffer structure. More advanced approches are possible and demonstrate to be effective with more realistic beanchmarks. ASER [26] uses data shapley values [9] to score samples and keep only the most informative ones. Selecting examples in GSS [2] consists of maximizing the diversity of samples in the replay buffer as suggested in [20] but using the gradient values. ICarl [23] instead uses an hearding strategy to select and discard samples.

3.2 Memory Size

The size of the memory buffer is a common parameter among all the replay-based strategies. Despite the importance of this parameter only few papers test extensively its impact using different continual learning strategies.

In a realistic application this parameter depends on the hardware resources or time constraints for training. When applying a continual learning strategy in a new setting it is essential to know the amount of samples sufficient to have good performances. For this purpose we investigated the influence of this parameter using different strategies and benchmarks (Sect. 5). The aim is to provide some practical recommendation useful to apply a continual learning strategy in new domains.

3.3 Weighting Policies

In literature, selection policies do not takes into account the importance of each task. However learning could be difficult for some tasks and it could require more replay of samples.

In a realistic scenario, using a random buffer, we don't have a-priori knowledge of information such as the nature of the current task, the represented classes, the number of samples or the difficulty of the current task. In this setting we have only the possibility to balance the amount of samples belonging to each previous task. For this reason we experiment with some weighting policies to verify the effects of recent and old memories in the learning process (Sect. 6).

This experiment is motivated by some recent findings on the human episodic memory [3], suggesting that episodic encoding occurs preferentially at the end of events.

3.4 Augmentation

Data augmentation is a helpful machine learning technique to help improving the generalization capabilities of a deep network [31]. In a continual learning scenario, using data augmentation, we can store original samples in the buffer and then augment them at training time to have more variety and hopefully increase accuracy. In this way, intuitively we can have a smaller buffer size. Data augmentation in continual learning is explored by Buzzega et al. in [5], in this case, *crops* and *horizontal-flips* are applied in the input stream and in the replay buffer. This augmentation leads to an increment in the test accuracy using the Replay strategy. In a realistic scenario, the training set augmentation is not always possible: the training time increases with an augmented dataset. We investigated the augmentation technique in Sect. 7. The aim of this experiment is to verify whether the augmentation of the training set (and which in particular) is indeed needed to achieve better performance and which augmentation strategy is most impactful.

4 Experimental Setup

The goal of this section is to describe benchmarks, models and replay-based strategies used in the experiments. For the experimental part we used Avalanche [15] the reference continual learning framework based on PyTorch. The goal of this library is to provide a shared and collaborative open-source codebase for fast prototyping, training and reproducible evaluation of continual learning algorithms.

4.1 Benchmarks and Models

Continual learning algorithm are evaluated by *benchmarks*: they specify how the stream of data is created by defining the originating dataset(s), the amount of samples, the criteria to split the data in different *tasks* or *experiences* [6] and so on. In literature, different benchmarks are used to evaluate results.

We select benchmarks belonging to the *New Classes* scenario i.e. data samples contained in the training set at time-step i are related to a new dependent variable Y to be learned from the model. We select three three of them for our experiments: *Split-MNIST* [27], *Split-CIFAR-10* [32] and *Split-TinyImagenet* [17]. These benchmark are derived respectively from MNIST [7], CIFAR-10 [12] and TinyImagenet [13] datasets. We also include *CORe50-NC* [14] in our experiments, a benchmark specifically designed for continual learning. This benchmark is divided in 9 tasks, the first task contains 10 classes, the remaining 8 classes. In our experiments, we set the number of tasks of each benchmark to 5, except for CORe50-NC, with a random order of classes.

Concerning the neural network models, for Split-MNIST we use a Multi-Layer Perceptron with 3 layers and 300 ReLU units at each layer. For Split-Cifar10, Split-TinyImagenet and CORe50-NC we exploit the ResNet-18 model pretrained on Imagenet [10] instead.

4.2 Strategies

We selected four strategies among the most popular and promising rehearsal approaches.

Replay. We select Replay [22,25] because it is powerful, simple and easily adjustable. It is also a simple way to prevent catastrophic forgetting, and it performs better with respect to more complicated strategies [5]. In our experiment we use random sampling and we randomly choose the samples to discard, to maintain simplicity.

GDumb. *Greedy Sampler and Dumb Learner* (GDumb) [21] is a simple approach that is surprisingly effective. The model is able to classify all the labels since a given moment t using only samples stored in the memory. Whenever it encounters a new task, the sampler just creates a new bucket for that task and starts removing samples from the one with the maximum number of samples. Samples are removed randomly. Compared to others, with the same memory size, this strategy is more efficient, in terms of execution time and resources. In particular setting this simple strategy can outperforms other approaches. However, it is not a valid continual learning strategy, since for each new task the model does not adapt, it must be re-trained from scratch.

ICarl. *Incremental Classifier and Representation Learning* (ICarl) [23] is a hybrid approach between rehearsal and regularization. The model parameters are updated by minimizing both a classification loss and a distillation loss. The replay memory is managed by a herding strategy: a sample is added if it causes the average feature vector over all exemplars to best approximate the average feature vector over all training examples. The order of its elements matters, with exemplars earlier in the list being more important. Reducing the exemplar set means discarding the less important samples. We selected ICarl because it is an effective hybrid strategy, in particular with low memory budget.

GSS. *Gradient based Sample Selection* [2] is a replay-based strategy. The selection of the memory buffer population is seen as a constraint selection problem. The goal is to optimize the loss on the current examples without increasing the losses on the previously learned ones. Selecting examples consists of maximizing the diversity of samples in the replay buffer using the gradient. The first way to select samples is based on integer quadratic programming, the second solution consists of a faster greedy-alternative and it is sufficient to achieve good performances. Scores for each sample is based on the maximal cosine similarity with a fixed number of others random samples in the buffer.

Avalanche[15] includes many Continual Learning strategies. It has been necessary to validate the strategies used in the experiments. We made sure to reproduce results of the original paper with the new Avalanche implementation.

5 Memory Size Experiment

This experiment is designed to understand the impact of memory size for every selected strategy and have an insight on the amount of samples sufficient to have good performances in a classification task as we propose in Sect. 3.2. In our work, we analyze a vast set of results and try to generalize those across different benchmarks and strategies.

5.1 Grid Search and Final Models

We select the models through a grid search and we choose a fixed order of classes. The selected parameters for the grid search are chosen following the parameters used in other works [2,4,5,11,17,21]. The memory buffer is balanced by task i.e. the memory contains an equal number of samples belonging to each task. For each benchmark we use 10% of the training set as validation set and a batch size of 32 examples. We use 4 epochs for Split-MNIST, 50 epochs for Split-CIFAR10, 100 epochs for Split-Tiny-Imagenet and CORe50-NC. GSS takes up to 10x higher execution time with respect to other strategies. As a result it was necessary to simplify the grid search for Split-MNIST benchmark and we did not test it using other benchmarks.

We have averaged the results of final models over 3 runs changing in each of them the classes order in a random manner. We plot the accuracy values in Figs. 1, 2, 3, 4. For each curve we calculate the elbow point, depicted with a black square. In this case, it indicates the optimal trade-off between accuracy and memory size. These values give us an idea of the memory sizes useful to have good performance.

5.2 Discussion

Our results show that the Replay strategy is a powerful and simple mechanism that most of the time is able to achieve good performance. Instead, ICarl has a particular behaviour: it performs well with lower memory size. This is due to the herding strategy as confirmed in other works [4,23]. In the following sections we analyze more in detail these results.

Split-MNIST. Replay strategy achieved the best performance with respect to the others. However, GDumb is able to reach good performance with high memory size and a considerably lower training time. ICarl is valid and effective using a smaller memory size. Concerning GSS, the performance are worse than others strategies, but the parameters used for grid search are fewer.

Split-CIFAR-10. Interestingly, in Split-CIFAR-10 the Replay strategy is effective only for high memory sizes. Instead, ICarl is much more effective with low memory sizes, it reaches with only 200 samples in memory the same accuracy of Replay strategy with 800 samples in memory. GDumb is not effective in this more challenging benchmark.

Split-Tiny-Imagenet. The performance of various strategies are poor. Instead, ICarl gains accuracy as memory size increases. This behaviour is different with respect to Split-MNIST and Split-CIFAR-10. This is due to the difference in their tasks, since, contrarily to Cifar and MNIST, Tiny-Imagenet has 200 classes.

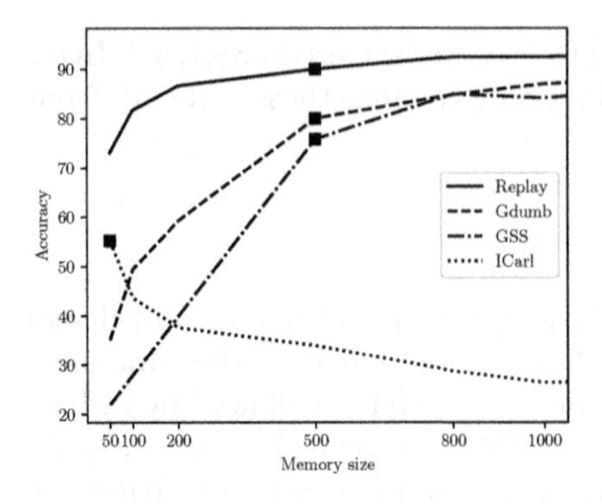

Fig. 1. Split-MNIST memory-accuracy curve

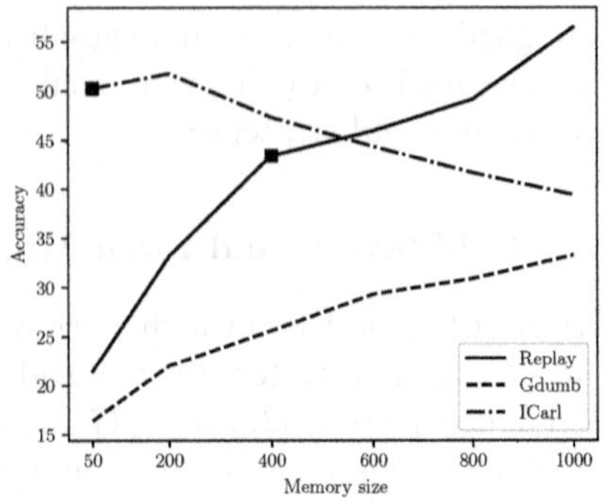

Fig. 2. Split-CIFAR-10 memory-accuracy curve

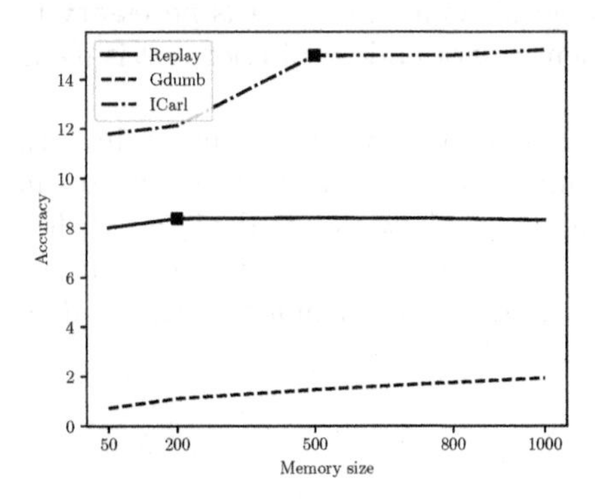

Fig. 3. Split-TinyImagenet memory-accuracy curve

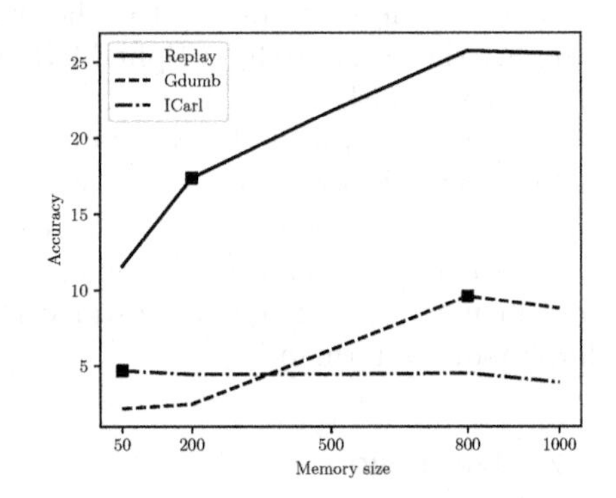

Fig. 4. CORe50-NC memory-accuracy curve

In fact, if a benchmark includes more classes than another, a greater memory size should be granted.

CORe50-NC. In this benchmark, Replay strategy is the most effective with both low and high memory sizes. GDumb and ICarl are ineffective with this benchmark. ICarl slowly increases its accuracy as the memory size increases up to 800. In this case, we can observe a trend inversion in the accuracy values.

This experiment give an insight on the memory size value needed to have good performance. Results show that in most of the case 1% of the training set is sufficient to achieve reasonable results.

6 Examples Weighting

The goal of this experiment is to verify the effects of recent and old memories in the learning process. Recent or old memories can have a different impact on the learning process as we declared in Sect. 3.3. We propose and investigate 7 alternatives to the balanced policy over tasks.

6.1 Weighting Policies

We propose different weighting policies i.e. methods to distribute samples, for the replay strategy, unbalanced by task. We report them in Table 1 with their abbreviations, instead in Fig. 5 we depict some of them. Except for *Balanced* policy, all the policies are parametrized by a $factor$ parameter that regulates the relevance of a task with respect to the others. For example, in *Increasing* policy the number of memory samples for each task is $factor - time$ greater than the number of memory samples for the previous task. If the amount of samples of first task is x, there will be $factor * x$ samples for the second, $factor^2 * x$ for the third and so on. A particular case is the *Middle* policy that works assigning greater weights to middle distance tasks. Once a new task arrives, some previous tasks may need more weight than before, as a result the medium policy does not exploit the full buffer due to those re-calibration. Contrarily, the *Middle+replications* replicates some random samples to fill the buffer. *MiddleHigh* policy gives more weight to middle and low distance samples. In this case the amount of samples of low distance task is the same as the one of middle distance task. The weight of other task is e regulated by the *factor* parameter. Using the same priciples we prosose *MiddleLow* and *MiddleLow+replications*.

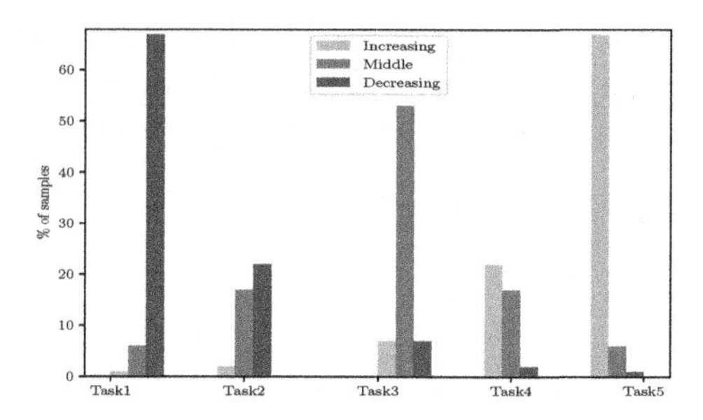

Fig. 5. Weighting Policies

6.2 Grid Search and Final Models

We exploit the same grid search parameter and architecture adopted in the first experiment described in Sect. 5.1 for the *Split-MNIST* [27] benchmark with 5 experiences. We average the final results over 6 runs using 3 as *factor* parameter. In Table 1 we report the accuracy and standard deviation of these policies varying the memory size.

6.3 Discussion

The results highlight that the balanced policy is the best among all. Besides this, the results are interesting. Let us analyze results starting with the most simple weighting policies: Decreasing, Increasing and Middle. For most memory

Table 1. Accuracy and std of the final models, averaged over 6 runs. Bal.=Balanced; Dec.=Decreasing; Inc.=Increasing; Mid.=Middle; Mid.+=Middle+replications; Mid.Hig.=MiddleHigh; Mid.Low=MiddleLow; Mid.Low+=MiddleLow+replications

	Bal.	Dec.	Inc.	Mid.	Mid.+	Mid.Hig.	Mid.Low	Mid.Low+
50	*74.54±3.64*	60.36±4.03	62.72±2.43	**71.63±4.14**	66.02±3.07	71.05±5.55	70.59±5.51	66.51±5.22
100	*82.35±1.24*	72.80±3.71	73.56±3.22	77.48±3.32	77.55±1.77	**81.01±1.55**	77.41±2.31	77.92±2.37
200	*85.59±0.69*	79.37±4.83	77.24±2.82	83.18±1.30	82.93±1.52	84.07±2.30	**85.14±1.25**	83.72±3.01
500	*90.69±0.42*	84.08±5.04	83.89±3.62	89.09±0.72	**89.57±0.68**	89.02±1.07	88.94±0.68	89.48±1.18
800	*91.83±0.43*	87.19±3.27	86.89±3.07	91.48±0.55	90.78±0.55	91.33±0.79	90.64±0.75	**91.70±1.05**
1k	*92.94±0.36*	87.18±2.04	88.60±1.73	91.36±0.6	92.52±0.78	**92.70±0.43**	92.48±0.45	91.92±0.94
2k	*94.69±0.31*	90.60±1.71	91.28±1.32	94.15±0.17	93.58±0.67	**94.58±0.23**	94.50±0.71	93.66±1.08
4k	*95.56±0.21*	92.43±1.83	93.01±0.79	95.03±0.57	94.97±0.32	95.35±0.19	95.35±0.67	**95.38±0.24**
5k	*95.76±0.22*	93.54±1.49	94.14±0.82	95.25±0.31	94.71±0.65	**95.71±0.26**	95.39±0.34	95.34±0.30

size values, the Increasing policy achieves better results than the Decreasing, but lower with respect to the Middle policy. From this observation we can infer that the most valuable samples are those from low and middle distances from the current task. We continue our analysis with the other policies. We observe that the best policy among all is MiddleHigh, confirming our previous statement. The performance of this policy is similar to those of the Balanced strategy. Concerning the policies with replications, results are not better with respect to the same policy without replications. This might depend on the fact that we replicate data without further transformations decreasing the diversity of the data.

7 Augmentation

The aim of this experiment is to investigate on the augmentation technique in a continual learning scenario as we propose in Sect. 3.4. Inspired by Buzzega et al. [5] we test different augmentation strategy applied only in the memory samples. The goal is to verify if the augmentation of the training set is indeed needed or if augmenting the buffer memory is sufficient to achieve better performance.

7.1 Settings and Results

The experiments have been performed with the Split-CIFAR-10 benchmark. Model, epochs, and batch size are the same described in Sect. 4. In this experiment, we fix the learning rate to 0.01 and the momentum to 0. We average the results over 4 runs using different memory size and varying the type of augmentation: Vertical-Flip, Horizontal-Flip, Resize-Crop and Rotation. Results are reported in Table 2.

Table 2. Experiment 3. Accuracy and std of final models averaged over 4 runs

	Memory size					
	20	50	250	500	750	1000
Original	19.80±0.54	22.04±0.77	38.74±3.92	44.97±4.08	53.14±1.36	57.42±3.39
Vertical	20.48±1.13	23.12±1.29	37.86±2.22	**45.85±2.17**	**53.33±1.57**	54.28±1.06
Horizontal	20.02±0.33	**23.73±1.31**	35.09±4.73	44.56±3.30	50.11±2.82	55.46±0.57
Crop	19.59±0.91	23.07±0.83	36.63±5.82	45.43±4.29	51.71±2.03	56.92±1.49
Rotation	**20.50±0.57**	23.07±0.83	36.63±5.82	45.43± 4.29	51.71±2.03	56.92±1.49

7.2 Discussion

In this case, the data augmentation shows just a slight increment in accuracy. We suppose that this is due to the training set's lack of augmentation and the transformation used. However, our experimental results reflect the findings reported in [5]: data augmentation is effective with low memory size. With 20 and 50 of memory size, accuracy is significantly higher.

8 Conclusion and Future Works

This work aims to deepen replay-based strategies, providing some insights and practical recommendations on specific implementation issues. We have validated many replay-based strategies already implemented in Avalanche. We investigated multiple aspect of continual learning strategies by means of three experiments.

Concerning the memory size experiment we extensively investigated the behavior of each strategy and benchmark varying the memory size. For each benchmark and for each strategy we found the amount of samples sufficient to have reasonably good results and we provided a general guideline to set this parameter in unseen benchmarks. We understand the role of memory samples of different tasks, testing different weighting policies. The variation of the standard balanced policy has proved to be useful to understand the impact of samples belonging to different tasks. We found out that Middle and Low distance tasks are more important than others. This paves the way to other experiments regarding this aspect, as well as to the development of new strategies exploiting this discovery.

We explored the usage of data augmentation in continual learning. We confirmed the results presented in [5], remarking the importance of augmenting not only the memory buffer, but also the training set. However, augmenting only the memory buffer helps to improve the accuracy, in particular with lower memory size. More experiments concerning these strategies could be performed. Concerning the weighting experiment in Sect. 6, it could be interesting to test the weighting policies with more challenging benchmarks. In our experiment we fixed the *factor* parameter but it could be interesting to test other values. Another possible modification is changing the type of augmentation, since we simply replicate some samples. Regarding the augmentation experiment, it could be interesting to observe the same phenomenon on more challenging benchmarks. Concerning the type of augmentation, we test only simple augmentation techniques. It could be interesting to test other neural-based technique.

Acknowledgements. This work has been partially supported by the H2020 TEACHING project (GA 871385).

References

1. Aljundi, R., et al.: Online continual learning with maximal interfered retrieval. In: Advances in Neural Information Processing Systems, vol. 32 (2019)
2. Aljundi, R., Lin, M., Goujaud, B., Bengio, Y.: Gradient based sample selection for online continual learning. In: Advances in Neural Information Processing Systems, vol. 32 (2019)
3. Ben-Yakov, A., Henson, R.N.: The hippocampal film editor: sensitivity and specificity to event boundaries in continuous experience. J. Neurosci. **38**(47), 10057–10068 (2018)
4. Buzzega, P., Boschini, M., Porrello, A., Abati, D., Calderara, S.: Dark experience for general continual learning: a strong, simple baseline. Adv. Neural. Inf. Process. Syst. **33**, 15920–15930 (2020)
5. Buzzega, P., Boschini, M., Porrello, A., Calderara, S.: Rethinking experience replay: a bag of tricks for continual learning. In: 2020 25th International Conference on Pattern Recognition (ICPR), pp. 2180–2187. IEEE (2021)
6. Carta, A., Cossu, A., Lomonaco, V., Bacciu, D.: Ex-model: Continual learning from a stream of trained models (2021). arXiv preprint arXiv:2112.06511
7. Deng, L.: The mnist database of handwritten digit images for machine learning research. IEEE Signal Process. Mag. **29**(6), 141–142 (2012)
8. Díaz-Rodríguez, N., Lomonaco, V., Filliat, D., Maltoni, D.: Don't forget, there is more than forgetting: new metrics for continual learning (2018). arXiv preprint arXiv:1810.13166
9. Ghorbani, A., Zou, J.: Data shapley: Equitable valuation of data for machine learning. In: International Conference on Machine Learning, pp. 2242–2251. PMLR (2019)
10. He, K., Zhang, X., Ren, S., Sun, J.: Deep residual learning for image recognition. In: Proceedings of the IEEE Conference on Computer Vision and Pattern Recognition, pp. 770–778 (2016)
11. Hsu, Y.C., Liu, Y.C., Ramasamy, A., Kira, Z.: Re-evaluating continual learning scenarios: A categorization and case for strong baselines (2018). arXiv preprint arXiv:1810.12488
12. Krizhevsky, A., Hinton, G., et al.: Learning multiple layers of features from tiny images (2009)
13. Le, Y., Yang, X.: Tiny imagenet visual recognition challenge. CS 231N **7**(7), 3 (2015)
14. Lomonaco, V., Maltoni, D.: Core50: a new dataset and benchmark for continuous object recognition. In: Conference on Robot Learning, pp. 17–26. PMLR (2017)
15. Lomonaco, V., et al.: Avalanche: an end-to-end library for continual learning. In: Proceedings of the IEEE/CVF Conference on Computer Vision and Pattern Recognition, pp. 3600–3610 (2021)
16. Lomonaco, V., et al.: Cvpr 2020 continual learning in computer vision competition: approaches, results, current challenges and future directions. Artif. Intell. **303**, 103635 (2022)

17. Mai, Z., Li, R., Jeong, J., Quispe, D., Kim, H., Sanner, S.: Online continual learning in image classification: an empirical survey. Neurocomputing **469**, 28–51 (2022)
18. McCloskey, M., Cohen, N.J.: Catastrophic interference in connectionist networks: the sequential learning problem. In: Psychology of learning and motivation, vol. 24, pp. 109–165. Elsevier (1989)
19. Parisi, G.I., Kemker, R., Part, J.L., Kanan, C., Wermter, S.: Continual lifelong learning with neural networks: a review. Neural Netw. **113**, 54–71 (2019)
20. Pomponi, J., Scardapane, S., Lomonaco, V., Uncini, A.: Efficient continual learning in neural networks with embedding regularization. Neurocomputing **397**, 139–148 (2020)
21. Prabhu, A., Torr, P.H.S., Dokania, P.K.: GDumb: a simple approach that questions our progress in continual learning. In: Vedaldi, A., Bischof, H., Brox, T., Frahm, J.-M. (eds.) ECCV 2020. LNCS, vol. 12347, pp. 524–540. Springer, Cham (2020). https://doi.org/10.1007/978-3-030-58536-5_31
22. Ratcliff, R.: Connectionist models of recognition memory: constraints imposed by learning and forgetting functions. Psychol. Rev. **97**(2), 285 (1990)
23. Rebuffi, S.A., Kolesnikov, A., Sperl, G., Lampert, C.H.: icarl: incremental classifier and representation learning. In: Proceedings of the IEEE conference on Computer Vision and Pattern Recognition, pp. 2001–2010 (2017)
24. Riemer, M., et al.: Learning to learn without forgetting by maximizing transfer and minimizing interference (2018). arXiv preprint arXiv:1810.11910
25. Robins, A.: Catastrophic forgetting, rehearsal and pseudorehearsal. Connect. Sci. **7**(2), 123–146 (1995)
26. Shim, D., Mai, Z., Jeong, J., Sanner, S., Kim, H., Jang, J.: Online class-incremental continual learning with adversarial shapley value. In: Proceedings of the AAAI Conference on Artificial Intelligence. vol. 35, pp. 9630–9638 (2021)
27. Shin, H., Lee, J.K., Kim, J., Kim, J.: Continual learning with deep generative replay. In: Advances in Neural Information Processing Systems, vol. 30 (2017)
28. Turing, A.M.: Computing machinery and intelligence. In: Epstein, R., Roberts, G., Beber, G. (eds.) Parsing the Turing Test, pp. 23–65. Springer, Dordrecht (2009). https://doi.org/10.1007/978-1-4020-6710-5_3
29. Weng, J., et al.: Autonomous mental development by robots and animals. Science **291**(5504), 599–600 (2001)
30. Wilson, M.A., McNaughton, B.L.: Reactivation of hippocampal ensemble memories during sleep. Science **265**(5172), 676–679 (1994)
31. Wong, S.C., Gatt, A., Stamatescu, V., McDonnell, M.D.: Understanding data augmentation for classification: when to warp? In: 2016 International Conference on Digital Image Computing: Techniques and Applications (DICTA), pp. 1–6. IEEE (2016)
32. Zenke, F., Poole, B., Ganguli, S.: Continual learning through synaptic intelligence. In: International Conference on Machine Learning, pp. 3987–3995. PMLR (2017)

Author Index

Printed in the United States
by Baker & Taylor Publisher Services